5th Edition

CRIMINOLOGICAL
THEORY

type?

¯SITY OF

For our children and grandchildren
Catherine and Robert
Jordan
Charlie and Mike
Stephen, Christopher, Taylor, and Justin

5th Edition

CRIMINOLOGICAL
CONTEXT AND CONSEQUENCES THEORY

J. Robert Lilly
Northern Kentucky University

Francis T. Cullen
University of Cincinnati

Richard A. Ball
Pennsylvania State University—Fayette

Los Angeles | London | New Delhi
Singapore | Washington DC

For information:

SAGE Publications, Inc.
2455 Teller Road
Thousand Oaks, California 91320
E-mail: order@sagepub.com

SAGE Publications Ltd.
1 Oliver's Yard
55 City Road
London EC1Y 1SP
United Kingdom

SAGE Publications India Pvt. Ltd.
B 1/I 1 Mohan Cooperative Industrial Area
Mathura Road, New Delhi 110 044
India

SAGE Publications Asia-Pacific Pte. Ltd.
33 Pekin Street #02-01
Far East Square
Singapore 048763

Printed in the United States of America

Library of Congress Cataloging-in-Publication Data

Lilly, J. Robert.
Criminological theory : context and consequences/J. Robert Lilly, Francis T. Cullen, Richard A. Ball.—5th ed.
    p. cm.
Includes bibliographical references and index.
ISBN 978-1-4129-8145-3 (pbk.)
    1. Criminology. 2. Crime—United States. 3. Criminal behavior—United States. I. Cullen, Francis T.
II. Ball, Richard A., 1936- III. Title.

HV6018.L55 2011
364.973—dc22   2010033655

This book is printed on acid-free paper.

11 12 13 14 10 9 8 7 6 5 4 3

Acquisitions Editor:	Jerry Westby
Editorial Assistant:	Nichole O'Grady
Production Editor:	Laureen Gleason
Copy Editor:	Kristin Bergstad
Typesetter:	C&M Digitals (P) Ltd.
Proofreader:	Gretchen Treadwell
Indexer:	Michael Ferreira
Cover Designer:	Glenn Vogel
Marketing Manager:	Erica DeLuca

Contents

Preface

The idea for this book was birthed during the mid-1970s when the United States and criminology on both sides of the Atlantic were experiencing immense changes. Between that time and the appearance of the first edition of the book in 1989, much of our individual energies were devoted to establishing and maintaining our careers and to our changing family responsibilities. At times, it seemed as though the circumstances needed to sustain the type of collective effort required for *Criminological Theory* were so elusive as to prevent the book from ever being written. Yet, the idea of a book that went beyond explaining criminological theory—one that used a sociology of knowledge perspective to explain the origins, developments, and consequences of criminological theory—remained very much alive. We were certain that few works like it in criminology had been written before. Then and now, we were committed to demonstrating that ideas about the causes of crime have consequences.

Criminological Theory has now been an ongoing project for two decades. During this time, the book has nearly doubled in size—a fact that reflects both the increasing richness of theorizing about crime and our efforts to add substantive value as we authored each new edition. Thus, the second edition in 1995 included empirical updates, substantial rewriting, and a new chapter devoted to fresh directions in critical thinking about crime. The emphasis on a sociology of knowledge perspective remained the same. The third edition, which appeared in 2002, attempted to capture novel theoretical developments that had occurred within both mainstream and critical theoretical paradigms. The fourth edition, published in 2006, expanded the book from 9 to 14 chapters and identified new theoretical trends in the United States and in Europe.

As we crafted this fifth edition, we remained excited to have the opportunity to chronicle the major advances within criminological theory, ranging from biosocial to cultural criminology. As with each previous revision, we updated materials and sought to make the book more informative, interesting, and accessible. Here are the most important changes that we have included in the fifth edition:

- A new chapter that reviews theories of white-collar crime, from the writings of Edwin Sutherland to more contemporary developments.

- A substantially revised chapter on biosocial criminology that includes the latest research in this growing area.

- Seven new tables that summarize theoretical developments and that can serve as useful study guides.

- A new discussion in Chapter 1, to help guide readers through the book, that gives an overview of the main changes in American society and their relationship to theoretical developments.

- More biographical information, drawn from new sources, on theorists so as to show how context influenced their theorizing.

- Discussions of new theoretical developments, ranging from Hirschi's control theory and behavioral economics to critical and feminist perspectives.

- New sources that assess the empirical status of the major theories.

- Updates of crime control policies and their connection to criminological theory.

Because criminology is an evolving field of study, we are convinced that the contents of the shifting contexts of the social world from which criminology comes will continue to influence its theoretical explanations for crime and the policy responses to it. It is our hope, however, that criminology never will be a mere reflection of the world around it.

There are far too many people to whom we owe debts for the success of *Criminological Theory* to be properly thanked here. For this reason, we mention only two. First, the late James A. Inciardi, who gave us the opportunity to write for Sage Publications, deserves our gratitude for his faith in our efforts and patience when it seemed as though the first edition never would see the light of day. Second, Jerry Westby, the current Sage editor, has shown unwavering confidence in our project across multiple editions, always providing just the right dollop of enthusiasm and wise advice to enable us to bring our work to fruition.

Finally, we want to express our appreciation to the many criminologists—and their students—who have embraced our efforts to tell the story of the development of criminological theory. Without your continued support, *Criminological Theory* would not be in its fifth edition. It has been a privilege to share our ideas with you.

J. Robert Lilly

Francis T. Cullen

Richard A. Ball

Acknowledgments

S AGE Publications gratefully acknowledges the contributions of the following individuals:

Timothy Austin
Indiana University of Pennsylvania

Mike Costelloe
Northern Arizona University

Mathieu Deflem
University of South Carolina

C. Nana Derby
Virginia State University

M. George Eichenberg
Tarleton State University

Andrea Lange
Washington College

Emmanuel Onyeozili
University of Maryland Eastern Shore

Martin D. Schwartz
Ohio University

Amy Thistlethwaite
Northern Kentucky University

Scott Vollum
James Madison University

Jennifer Wareham
Wayne State University

1

The Context and Consequences of Theory

The Thinker
by Auguste Rodin
1840–1917
French artist and sculptor

Crime is a complex phenomenon, and it is a demanding, if intriguing, challenge to explain its many sides. Many commentators—some public officials come to mind—often suggest that using good common sense is enough to explain why citizens shoot or rob one another and, in turn, to inform us as to what to do about such lawlessness. Our experience—and, we trust, this book as well—teaches that the search for answers to the crime problem is not so easy. It requires that we reconsider our biases, learn from the insights and mistakes of our predecessors who have risked theorizing about the causes of crime, and consider clearly the implications of what we propose.

But the task—or, as we see it, the adventure—of explaining crime is an important undertaking. To be sure, crime commentary frequently succumbs to the temptation to exaggerate and sensationalize, to suggest that crimes that are exceptionally lurid and injurious compose the bulk of America's lawlessness, or perhaps to suggest that most citizens spend their lives huddled behind barricaded doors and paralyzed by the fear that local thugs will victimize them. There is, of course, an element of truth to these observations, and that is why they have an intuitive appeal. Yet most Americans, particularly those living in more affluent communities, do not have their lives ripped apart by brutal assaults or tragic murders. And although many citizens lock their doors at night, install burglar alarms, and perhaps buy weapons for protection, they typically say that they feel safe in and close to their homes (Cullen, Clark, & Wozniak, 1985; Scheingold, 1984).

But these cautionary remarks do not detract from the reality that crime is a serious matter that, we believe, deserves study and understanding. Most Americans escape the type of victimization that takes their lives or destroys their peace of mind, but too many others do not share this good fortune. Thus, media reports of Americans killing Americans are sufficiently ubiquitous that many of us have become so desensitized to the violence in our communities that we give these accounts scarcely more attention than the scores from the day's sporting events. And it is likely that most of us have friends, or friends of friends, who have been seriously assaulted or perhaps even murdered.

Statistical data paint an equally bleak picture. Each year, the Federal Bureau of Investigation (FBI) publishes the *Uniform Crime Reports* in which it lists the numbers of various crimes that have become known (mostly through reports by citizens) to the nation's police departments. According to these statistics, since the year 2000, an average of more than 16,300 U.S. residents were murdered annually. Although there has been a recent decline in crime, each year there still are about 1.4 million Americans robbed, raped, or seriously assaulted and nearly 10 million whose houses are burglarized or whose property is damaged or stolen (Blumstein & Wallman, 2000; Federal Bureau of Investigation [FBI], 2010).

It is disturbing that these statistics capture only part of the nation's crime problem. Many citizens, perhaps one in every two or three who are victimized, do not report crimes against them to the police; thus, these acts do not appear in the *Uniform Crime Reports*. For example, the National Crime Victimization Survey, a study in which citizens are asked whether they have been victimized, estimates that residents over 12 years of age experienced approximately 21.3 million crimes in 2008, more than one fifth of which were violent victimizations (Rand, 2009).

Furthermore, these FBI statistics do not include drug-related offenses, which are commonplace. They also measure mainly serious street crimes. Yet we know that minor crimes—petty thefts, simple assaults, and so on—are even more widespread. "Self-report" surveys, in which the respondents (typically juveniles) are asked to report how many offenses they have committed, consistently indicate that the vast majority of people have engaged in some degree of illegality. But more important, there are other realms of criminality—not only quite prevalent but also quite serious—that traditionally have not come to the attention of police because they are not committed on the streets. Domestic violence—child abuse, spousal assault, and so on (i.e., the violence that occurs "behind closed doors")—is one of these areas (Straus, Gelles, & Steinmetz, 1980), as are sexual assaults that occur on dates and against people who know one another (Fisher, Daigle, & Cullen, 2010). Another such area is white-collar crime, that is, the crimes committed by professional people in the course of their occupations (Sutherland, 1949). As recent revelations suggest (recall the massive frauds at Enron and WorldCom), corruption in the business and political communities takes place regularly and has disquieting consequences (Cullen, Maakestad, & Cavender, 1987; Simon & Eitzen, 1986).

More statistics and observations could be added here, but this would only belabor the point that crime is a prominent feature of our society. Indeed, lawlessness— particularly lethal violence—in the United States rivals or surpasses that in other industrialized Western nations (Currie, 1985, 2009; Lynch, 2002; Messner & Rosenfeld, 2001; Zimring & Hawkins, 1997). Making cross-cultural comparisons is difficult; for example, nations differ in what they consider to be illegal and in their methods of col-lecting crime data. Even so, Currie's (1985) review of available statistical information revealed that, as of the late 1970s, "about ten American men died by criminal violence for every Japanese, Austrian, West German, or Swedish man; about fifteen American men died for every Swiss or Englishman; and over twenty [American men died] for every Dane" (p. 25). Similar differences remain today (Currie, 1998b; Rosenfeld, 2009). According to Currie (2009), "in most other affluent industrial societies, the deliberate killing of one person by another is an extremely rare event. . . . Their neighborhoods are not torn by drive-by shootings or by the routine sound of police helicopters in the night. There are no candles at shrines for homicide victims" (p. 3). Furthermore, crime is not evenly distributed within the United States. As Blumstein (2000) noted, in 1996 only "ten cities (New York, Chicago, Los Angeles, Detroit, Philadelphia, Washington, New Orleans, Baltimore, Houston, [and] Dallas, in order of decreasing numbers of homicides) accounted for fully one quarter of all the nation's homicides" (p. 36). Striking differences in criminality also are found within communities.

But why is crime so prevalent in the United States? Why is it so prevalent in some of our communities but not others? Why do some people break the law, whereas others are law abiding? Why do the affluent, and not just the disadvantaged, commit illegal acts? How can these various phenomena be explained?

Over the years, theorists have endeavored to address one or more of these questions. In this book, we attempt to give an account of their thinking about crime—to exam-ine its context, its content, and its consequences. Before embarking on this story of criminological theorizing, however, it is necessary to discuss the framework that will inform our analysis.

Theory in Social Context

Most Americans have little difficulty in identifying the circumstances they believe cause people to engage in wayward conduct. When surveyors ask citizens about the causes of crime, only a small percentage of the respondents say that they "have no opinion." The remainder of those polled usually remark that crime is caused by factors such as unem-ployment, bad family life, and lenient courts (Flanagan, 1987; see also Roberts & Stalans, 2000; Unnever, Cochran, Cullen, & Applegate, 2010).

Most people, then, have developed views on why crime occurs; that is, they have their "theories" of criminal behavior. But where do such views, or such theories, come from? One possibility is that citizens have taken the time to read extensively on crime, have sifted through existing research studies, and have arrived at informed assess-ments of why laws are disregarded. But only exceptional citizens develop their views on crime—or on any other social issue—in this way. Apart from criminologists who study crime for a living, most people have neither the time nor the inclination to inves-tigate the crime problem carefully.

This observation might not seem particularly insightful, but it is important in illu-minating that most people's opinions about crime are drawn less from sustained thought and more from the implicit understandings that they have come to embrace during their lives. Attitudes about crime, as well as about other social issues, can come from a variety of sources—parents, church sermons, how crime is depicted on television, whether one has had family members or friends who have turned to crime, whether one has experimented with criminal activity oneself or perhaps been victimized, and so on. In short, social experiences shape the ways in which people come to think about crime.

This conclusion allows us to offer three additional points. First, members of the general public are not the only ones whose crime theories are influenced by their life experiences. Academic criminologists and government officials who formulate crime policy have a professional obligation to set aside their personal biases, read the existing research, and endorse the theory that the evidence most supports. To an extent, criminologists and policy makers let the data direct their thinking, but it is equally clear that they do not do so fully. Like the general public, they too live in society and are shaped by it. Before ever entering academia or public service, their personal experiences have provided them with certain assumptions about human nature and about the ways in which the world operates; thus, some will see themselves as liberals and others as conservatives. After studying crime, they often will revise some of their views. Nonetheless, few ever convert to a totally different way of thinking about crime; how they explain crime remains conditioned, if only in part, by their experiences.

Second, if social experiences influence attitudes about criminality, then as society changes—as people come to have different experiences—views about crime will change as well. We illustrate this point throughout this book, but a few brief examples might help to clarify matters for immediate purposes.

It will not surprise many readers to learn that Americans' views on crime have changed markedly since the settlers first landed on the nation's shores. Indeed, at different times in U.S. history, Americans have attributed the origins of crime to spiritual demons and the inherent sinfulness of humans, to the defective biological constitution of inferior people in our midst, to the denial of equal opportunity, and to the ability of the coldly rational to calculate that crime pays. As we will see, each of these theories of crime, and others as well, became popular only when a particular set of circumstances coalesced to provide people with the experiences that made such reasoning seem logical or believable.

Thus, for colonists living in a confining and highly religious society, it "made sense" for them to attribute crime to the power of demons to control the will of those who fell prey to the temptations of sin. For those of the late 1800s who witnessed the influx of foreigners of all sorts and learned from the social Darwinists that natural selection determined where each individual fell in the social hierarchy, it made sense that people became poor *and* criminal because they were of inferior stock. For those of the 1960s who were informed that systematic barriers had prevented minorities from sharing in the American dream, it made sense that people became criminal *because they were poor*—because they were denied equal opportunity. During more recent times, as society has taken a turn in a conservative direction and it has become fashionable to blame social ills on a permissive society, it has made sense to more and more Americans that people commit crimes because they know that they risk only a "slap on the wrist" if they are caught.

In short, social context plays a critical role in nourishing certain ways of theorizing about crime. If the prevailing social context changes and people begin to experience life differently, then there will be a corresponding shift in the way in which they see their world and the people in it. Previous theories of crime will lose their appeal, and other perspectives will increasingly make sense to larger numbers of people. Note that all of this can take place—and, indeed, usually does take place—without systematic analysis of whether the old theory actually was wrong or whether the new theory represents an improvement.

But does any of this relate to you, the reader? Our third point in this section is that your (and our) thinking about crime undoubtedly has been conditioned by your social experiences. When most of us look to the past, we wonder with a certain smugness how our predecessors could have held such strange and silly views about crime or other things. In making this type of remark, however, we not only fail to appreciate how their thoughts and actions were constrained by the world in which they lived but also implicitly assume that our thoughts and actions are unconstrained by our world. Our arrogance causes us to accept our interpretations—our theories—as "obviously" correct. We forget that future generations will have the luxury of looking at us and assessing where *we* have been strange and silly.

This discussion suggests the wisdom of pausing to contemplate the basis of your beliefs. How have your social experiences shaped the way in which you explain crime? Asking and seeking answers to this question, we believe, opens the possibility of lifting the blinders that past experiences often strap firmly around one's eyes. It creates, in short, the exciting opportunity to think differently about crime.

Theory and Policy: Ideas Have Consequences

Theory often is dismissed as mere empty ruminations—fun, perhaps, but not something for which practical men and women have time. But this is a shortsighted view, for as Thomas Szasz (1987) cautioned, *ideas have consequences* (see also Weaver, 1948). Theory matters.

When it comes to making criminal justice policy, there is ample evidence of this maxim (Sherman & Hawkins, 1981). Lawlessness is a costly problem; people lose their property and sometimes their lives. The search for the sources of crime, then, is not done within a vacuum. Even if a theorist wishes only to ruminate about the causes of theft or violence, others will be ready to use these insights to direct efforts to do something about the crime problem. Understanding why crime occurs, then, is a prelude to developing strategies to control the behavior. Stephen Pfohl (1985) captured nicely the inherent relationship between theory and policy:

> Theoretical perspectives provide us with an image of what something is and how we might best act toward it. They name something this type of thing and not that. They provide us with the sense of being in a world of relatively fixed forms and content. Theoretical perspectives transform a mass of raw sensory data into understanding, explanations, and recipes for appropriate action. (pp. 9–10)

This discussion also leads to the realization that different theories suggest different ways of reducing crime. Depending on what is proposed as the cause of illegal behavior, certain criminal justice policies and practices will seem reasonable; others will seem irrational and perhaps dangerously irresponsible. Thus, if offenders are viewed as genetically deranged and untrainable—much like wild animals—then caging them would seem to be the only option available. But if offenders are thought to be mentally ill, then the solution to the problem would be to treat them with psychotherapy. Or if one believes that people are moved to crime by the strains of economic deprivation, then providing job training and access to employment opportunities would seem to hold the promise of diminishing their waywardness.

This is not to assert that the relationship between theory and policy is uncomplicated. Sometimes theories emerge, and then the demand to change policy occurs. Sometimes policies are implemented, and then attempts are made to justify the policies by popularizing theories supportive of these reforms. Often, the process is interactive, with the theory and policy legitimating one another. In any case, the important point is that support for criminal justice policies eventually will collapse if the theory on which they are based no longer makes sense.

An important observation follows from this discussion: As theories of crime change, so do criminal justice policies. At the turn of the 20th century, many Americans believed that criminals were "atavistic reversions" to less civilized evolutionary forms or, at the least, feebleminded. The call to sterilize offenders so that they could not pass criminogenic genes on to their offspring was widely accepted as prudent social action. Within two decades, however, citizens were more convinced that the causes of crime lay not within offenders themselves but rather in the pathology of their environments. The time was ripe to hear suggestions that efforts be made to "save" slum youths by setting up neighborhood delinquency prevention programs or, when necessary, by removing juveniles to reformatories, where they could obtain the supervision and treatment that they desperately needed. In more recent decades, numerous politicians have jumped onto the bandwagon, claiming that crime is caused by the permissiveness that has crept into the nation's families, schools, and correctional system. Not surprisingly, they have urged that efforts be made to "get tough" with offenders— to teach them that crime does not pay by sending them to prison for lengthier stays and in record numbers.

But we must remember not to decontextualize criminological theory. The very changes in theory that undergird changes in policy are themselves a product of transformations in society. As noted earlier, explanations of crime are linked intimately to social context—to the experiences people have that make a given theory seem silly or sensible. Thus, it is only when shifts in societal opinion occur that theoretical models gain or lose credence and, in turn, gain or lose the ability to justify a range of criminal justice policies.

We also hope that you will find the discussion in this book of some personal relevance. We have suggested that thought be given to how your own context may have shaped your thinking. Now we suggest that similar thought be given to how your thinking may have shaped what you have thought should be done about crime. The challenge we are offering is for you to reconsider the basis and consistency of your views on crime and its control—to reconsider which theory you should embrace and the consequences that this idea should have. We hope that this book will aid you as you embark on this adventure.

Context, Theory, and Policy: Plan of the Book

"Perhaps the clearest lesson to be learned from historical research on crime and deviance," Timothy Flanagan (1987) reminded us, "is that the approach to crime control that characterizes any given era in history is inexorably linked to contemporaneous notions about crime causation" (p. 232). This remark is instructive because it captures the central theme of this book—the interconnection among social context, criminological theory, and criminal justice policy making. As we progress through subsequent chapters, this theme forms the framework for our analysis. We discuss not only the content of theoretical perspectives but also their contexts and consequences.

But the scope of the enterprise should be clarified. Our purpose here is to provide a primer in criminological theory—a basic introduction to the social history of attempts, largely by academic scholars, to explain crime. In endeavoring to furnish an accessible and relatively brief guide to such theorizing, we have been forced to leave out historical detail and to omit discussions of the many theoretical variations that each perspective on crime typically has fostered. As a result, this book should be viewed as a first step to understanding the long search for the answer to the riddle of crime. We hope that our account encourages you to take further steps in the time ahead.

Our story of criminological theory commences, as most stories do, at the beginning, with the founding of criminology and early efforts, to use Rennie's (1978) words, "to search for the criminal man." Our story has 15 chapters and traces the development of criminological theory up to the present time. These chapters are thus arranged largely in chronological order. Because some theories arose at approximately the same time, the chapters should not be seen as following one another in a rigid, lockstep fashion. Further, inside each chapter, the ideas within a theoretical tradition are often traced from past to present—from the originators of the school of thought to its current advocates. Still, the book is designed to allow readers to take an excursion across time and historical context to see how thinking about crime has evolved.

Table 1.1 provides a handy guide that tells how *Criminological Theory: Context and Consequences* is arranged. This guide, much like a roadmap, is intended to be clear and simple. As readers travel through our volume, they may wish to consult Table 1.1 as a way of knowing where they are. When it comes to theory, the field of criminology has an embarrassment of riches—diverse theories competing to explain crime. The very complexity of human conduct and society perhaps requires numerous theoretical perspectives, with each capturing a part of reality ignored by competing approaches. Regardless, readers have the challenge of keeping all the theories straight in their minds as the story of criminological theory unfolds in the pages ahead. Table 1.1 should help in this important task.

Table 1.1   Criminological Theory in Context

Social Context	Criminological Theory	Chapters in This Book
Enlightenment—mid-1700s to late 1700s	Classical school	2
Rise of social Darwinism, science, and medicine—mid-1800s into 1900s	Early positivist school—*biological positivism*	2
Mass immigration, the Great Depression, and post–World War II stability—1900 to the early 1960s	Chicago school, anomie-strain, control—*mainstream criminology*	3, 4, 5, and 6
Social turmoil—1965 to late 1970s	Labeling, conflict, Marxist, feminist, white-collar—*critical criminology*	7, 8, 9, 10, and 15
Conservative era—1980 to the early 1990s, and beyond	Deterrence, rational choice, broken windows, moral poverty, routine activity, environmental—*rejecting mainstream and critical criminology*	12 and 13
	Peacemaking, left realism, new European, cultural, convict—*rejecting conservative theory and policy*	9 and 10
The new century—2000 to today	Biosocial, life-course/developmental—*becoming a criminal*	14 and 15

Before commencing with our criminological storytelling, let us preview in some detail what the chapters will cover. Table 1.1 presents the outline of how different social contexts are related to the emergence of different theories. The chapters in which these theories are contained also are listed.

INVENTING CRIMINOLOGY: MAINSTREAM THEORIES

Chapter 2 reviews the two theoretical perspectives that are generally considered to be the foundation of modern criminology. The *classical school* arose in the

Enlightenment era. It emphasized the rejection of spiritual or religious explanations of crime in favor of the view that offenders use their reason—the assessment of costs and benefits—in deciding whether a potential criminal act pays and should be pursued. The classical school argued that the criminal law could be reformed so that it would be fair (everyone treated equally) and just punitive enough to dissuade people from breaking the law (the crime would not be profitable). This approach is the forerunner of more contemporary theories of rational choice and deterrence.

Chapter 2 is mainly devoted, however, to the *positivist school*, which emphasized the scientific study of criminals. Led by Cesare Lombroso, positivism flourished in Italy in the late 1800s and into the 1900s. These ideas also were popular in the United States, where a similar tradition arose. These scholars assumed that there was something different about those who offended that distinguished them from those who did not offend. In medicine, we ask what makes someone sick; similarly, they thought we should ask what makes someone criminal. As in medicine, they felt that the key to unlocking this puzzle was to study offenders scientifically—to probe their bodies and their brains for evidence of individual differences. Influenced by Darwinism and medicine, they largely concluded that the criminally wayward possessed biological traits that determined their behavior. Crime was not due to a sinful soul or chosen freely but rather was predetermined by a person's constitutional makeup.

Starting in the 1930s, however, American criminology embarked on an alternative path. The positivist school's advocacy of using science to study crime continued to be embraced. But scholars increasingly suggested that the answers to crime were to be found not within people but rather in the social circumstances in which people must live. The United States was making its transition to a modern, industrial, urban nation. As waves of immigrants came to our shores and settled in our cities, scholars wondered whether their subsequent experiences might prove criminogenic. The Chicago school of criminology rose to prominence by pioneering the study of urban areas and crime (see Chapter 3).

When scholars peered into impoverished inner-city neighborhoods—buffeted by the misery inflicted by the Great Depression—they saw the breakdown of personal and social controls, the rise of criminal traditions, and barriers to the American dream for success that all were taught to pursue. Scholars of this generation thus developed three core ways of explaining crime: *control theory*, which explored how crime occurs when controls weakened; *differential association theory*, which explored how crime occurs when individuals learned cultural definitions supportive of illegal conduct; and *anomie-strain theory*, which explored how crime occurs when people endure the strain of being thwarted in their efforts to achieve success. The first two of these theories had their origins in the Chicago school of criminology; the third had its origins in the writings of Robert K. Merton. These perspectives are reviewed in Chapters 3, 4, 5, and 6.

Taken together, these three theories are sometimes called *mainstream criminology*. For more than 80 years, they have occupied the center of American criminology. In the aftermath of World War II, they were particularly dominant. During this period, the youth population began to expand and youth culture rose in prominence—developments that triggered concerns about juvenile delinquency. These perspectives were used to explain why some youngsters committed crime and others did not and why gangs were found in some neighborhoods and not others. Often, control, differential association, and anomie-strain theories were tested against one another in self-report studies conducted with high school students (see, e.g., Hirschi, 1969). Even today, these early works and their contemporary extensions remain at the core of the discipline (e.g., self-control theory, social learning theory, general strain theory).

The centrality and enduring influence of control, differential association, and anomie-strain theory is one reason why these perspectives are said to comprise mainstream criminology. But the term "mainstream" is used in another sense as well. Developed in a period when the United States was becoming a dominant world power and flourishing in the relative stability of post–World War II America, these perspectives remained in the *political mainstream*: They did not fundamentally challenge the organization of the social order. To be sure, these three theories identified problems in American society and were used to suggest policies that might address them. But for the most part, they stopped short of criticizing the United States as being rotten at its core—of being a society in which inequalities in power, rooted in a crass capitalism, created crimes of the poor that were harshly punished and crimes of the rich that were ignored. In short, control, differential association, and anomie-strain theories were mainstream because they tended to favor reform of the status quo in America rather than its radical transformation.

SOCIAL TURMOIL AND THE RISE OF CRITICAL THEORIES

Starting in the mid-1960s, however, scholars increasingly sought to identify how conflict and power were inextricably involved in the production of crime and in the inequities found in the criminal justice system. They were influenced by the changing context of American society. During the 1960s and into the 1970s, the United States experienced contentious movements to achieve civil rights and women's rights. Americans witnessed riots in the street, major political figures assassinated, widespread protests over the Vietnam War culminating with students shot down at Kent State University, and political corruption highlighted most poignantly by the Watergate scandal. These events sensitized a generation of criminologists to social and criminal injustices that compromised the American dream's promise of equality for all and that led to the abuse of state power. Given this jaundiced view of American society, the new brand of theorizing that they developed was called *critical criminology*.

Although not yet fully developed, the seeds of critical criminology can be traced in part to *labeling theory*, which is discussed in Chapter 7. Scholars in this perspective offered the bold argument that the main cause of stable involvement in crime is not society per se but rather the very attempts that are made to reduce crime by stigmatizing offenders and processing them through the criminal justice system. The roots of critical criminology are discussed more deeply in Chapter 8, which reviews theorists called *conflict* or *radical* scholars. These theorists illuminated how power shapes what is considered to be a crime and who is subjected to arrest and imprisonment. They went so far as to suggest that the embrace of capitalism is what induces high rates of lawlessness among both the rich and the poor.

Chapter 10 explores another line of inquiry encouraged by critical criminology: the development of *feminist theory*. This perspective has led to the "gendering of criminology" in North America and Britain. In light of the changing social context surrounding gender, we trace how understandings of female criminality shifted from theories highlighting the individual defects of women to explanations illuminating how gender roles shape men's and women's illegal conduct. An attempt is made to capture the rich diversity of feminist thinking as we examine how scholars have linked crime to such factors as patriarchy, masculinities, and the intersection of race, class, and gender.

Finally, in Chapter 11, *theories of white-collar crime* are examined. Although not all of these perspectives are critical in content, the very inquiry into this topic was spurred by critical criminology's concern with inequality and injustice. Thus, theories of white-collar crime illuminate and explain the crimes of the powerful. They are built on the very premise that, although the poor might monopolize prisons, they do not monopolize crime. In fact, scholars have shown the immense cost of white-collar criminality—especially that committed by corporations—and have explored why this injurious conduct occurs.

CRIMINOLOGICAL THEORY IN THE CONSERVATIVE ERA

Although many criminological theories emerged in response to the social context of the 1960s and 1970s, especially the concern with prevailing inequities in money and power, America turned to the political right during the Reagan and Bush years of the 1980s and beyond. During this time, new criminologies emerged claiming that crime was due not to the faults in society but rather to the faults of individuals. To at least some degree, these explanations may be seen as attempts to revitalize—dressed in new language and with more sophisticated evidence—the models of crime that were popular a century ago. These theories vary in their scientific merit, but they are consistent in suggesting that the answer to crime rests largely in harsher sanctions—especially the expanded use of imprisonment—against offenders. In this sense, these theories are best considered *conservative* explanations of crime. They are reviewed in Chapter 12.

Other theories of this era were not conservative in content or temperament. They did not depict offenders as wicked super-predators who required imprisonment or as crass calculators who required harsh deterrence. However, they were also skeptical both of critical criminology for its utopian and impractical policies (they doubted a socialist revolution was on the horizon) and of mainstream criminology for its exclusive focus on offenders (rather than on the opportunities needed for a crime to take place). They claimed that we needed an approach that understood the elements of crime and how to manipulate them to prevent such acts from occurring. For them, practical thinking that led to effective crime prevention was the only way to stem the conservative call for mass imprisonment.

Thus, Chapter 13 investigates *routine activity theory* or *environmental criminology*, which argues that crime is best understood as an "event" that involves not only a motivated offender but also the "opportunity" to break the law (the presence of a suitable target to victimize and the absence of guardianship to prevent the victimization). Although mainstream criminological theory historically has focused on what motivates people to commit crime, it has not systematically assessed how variations in the opportunity to offend affect the amount and distribution of criminality in American society. Furthermore, this perspective maintains that crime will best be diminished not by efforts to change offenders but rather by making the social and physical environment less hospitable to offending (e.g., installing a burglar alarm in a house; hiring a security guard in a bank; placing a camera to watch a parking lot). This is often called *situational crime prevention* because the focus is on reducing opportunities for crime within a particular situation.

Chapter 13 also explores perspectives that investigate the thinking and decision making of offenders, including *rational choice theory* (an approach that is compatible with the opportunity paradigm and calls for situational crime prevention) and *perceptual deterrence theory*. Given that these perspectives see crime as a choice shaped by objective or perceived costs and benefits, they have elements compatible with conservative theory. However, depending on how they are set forth, they do not necessarily justify harsh criminal justice penalties.

During the 1980s, most criminologists—both in the United States and abroad—opposed conservative criminology and its preference for mass imprisonment as the key weapon in the war on crime (Currie, 1985). They were dismayed as the daily count of Americans behind bars grew from around 200,000 in the early 1970s to about 2.4 million today (Newburn, 2007). Although not as dramatic, similar trends occurred in some European nations. Over this 40-year period, the rejection of conservative criminology was voiced perhaps most loudly and consistently by critical criminologists. Critical views have their roots in the 1960s and 1970s, but they were nourished by the need to deconstruct conservative crime ideology and to unmask the harm caused by "get tough" policies.

These perspectives are presented in Chapter 9, which builds on the discussions of conflict theory in Chapter 8. The focus here is on *new directions in critical theory*. These contributions, which include the insights of British and other European scholars, enrich our understanding of crime by challenging traditional interpretations of social reality and especially the efficacy and justice of repressive state policies favored by conservatives. This section thus examines postmodern thought, the early development and the extension of Britain's *new criminology* into what is now known as *left realism,* the new European criminology, cultural criminology, and convict criminology.

CRIMINOLOGICAL THEORY IN THE 21ST CENTURY

Contemporary criminological theory is a mixture of old and new ways of thinking. Powerful theoretical traditions may age but they tend not to die. Once they emerge, these paradigms may fluctuate in the allegiance they inspire, but they often remain integral to the criminological enterprise. Further, their core ideas are at times elaborated into more sophisticated and empirically defensible perspectives (e.g., Sutherland's differential association theory transformed into Akers's social learning theory). When this occurs, seemingly dormant perspectives can be revitalized and generate renewed attention (e.g., strain theory revitalized by Agnew and by Messner and Rosenfeld). Still, there are ways of thinking that emerge that are innovative, rival older ways of thinking, and offer the possibility of renovating how criminological theory and research are undertaken.

In this regard, the final two chapters of the book explore theoretical models that are shaping thinking about crime during the 21st century in important ways. To a degree, these theories lack a clear ideological or political slant, and in this sense they might be considered *new mainstream* criminological perspectives. They reflect a social context in which grand solutions to crime and other social problems are being relinquished in favor of more middle-range or practical efforts to improve the crime problem. These preferences are also reflected in the growing popularity of environmental criminology and its focus on situational crime prevention, a perspective mentioned above.

Thus, Chapter 14 discusses the resurgence of biological theorizing or, as it is more often called today, the *biosocial perspective*. Although still controversial to a degree, the prevalence of research on brains, genetics, and other biological factors is bringing biological thinking back toward the center of criminology. There is now a renewed search for the "criminal man"—that is, a search for the biological traits that differentiate offenders from nonoffenders. This research often is nuanced and involves explorations of how biological factors interact with social factors to shape behavior. Its policy implications are potentially complex, since they might justify efforts to incapacitate or cure those whose criminality is rooted in their bodies.

Finally, Chapter 15 discusses a paradigm that is increasingly dominating American criminology: *life-course* or *developmental criminology*. This approach focuses its attention on how the roots of crime can be traced to childhood. This perspective also argues that the key to understanding crime is in studying how people develop into offenders and how they escape from their lives of crime. These theories are potentially important in suggesting a progressive policy agenda because they show the complex factors that place youngsters at risk for crime and call for policies aimed not at punishment but at early intervention.

Conclusion

With this prelude shared, it is now time to embark on an exploration of criminological theory. We are, in a way, the guides in this intellectual tour across criminology. The core challenge is to reveal the diverse attempts scholars have made to explain the mystery of why crime occurs. We will show that scholars have probed how the causes of criminal conduct might reside in our bodies, minds, and social relationships. And we will illuminate where scholars agree and where they disagree. At the journey's end, we trust that we will have provided an enriched knowledge of crime's origins.

Again, the subtitle of *Criminological Theory* was carefully chosen: *Context and Consequences*. As we hope to convey, theory construction is a human enterprise. It reflects not only the detached scientific appraisal of ideas and evidence but also a scholar's unique biography situated within a unique historical period. Accordingly, understanding the evolution of criminological theory requires us to consider the social context in which ideas are formulated, published, and accepted as viable. Further, theories matter; ideas have consequences. Every effort to control crime is pregnant with an underlying theory. Our explanations of crime thus provide the impetus and justification for the crime control policies that we pursue.

2

The Search for the "Criminal Man"

Cesare Lombroso

1836–1909

University of Pavia and University of Turin, Italy

Often called "the father of modern criminology"

Before we examine the content of this chapter, it is important to remember a few of the cautionary comments offered in Chapter 1. By keeping these ideas in mind, we more than likely will be successful in accomplishing the goal of introducing you to the context and consequences of criminological theory.

We want you to remember that the search for explanations of criminal behavior is not easy because we constantly must guard against our biases, mistaken perceptions, and prejudices. Unless we maintain our intellectual guard against these problems, our learning will be severely limited. This will become obvious as we study the following chapters and learn that many theories of crime that have experienced popularity with the public and professional criminologists also have been criticized for having serious blind spots. Unfortunately, the blind spots often have contributed to the creation and implementation of official policies that have produced results as undesirable as crime itself. Although it is impossible to develop perfect policies, we must keep in mind the fact that theories do influence the policies and practices found in criminal justice systems.

It is important to remember that the explanations of crime, whether they are created by the public or by professional criminologists, are influenced by the social context from which they come. This means that the social context will consist of perceptions and interpretations of the past as well as the present. It might also mean that the explanations of crime include some thoughts about what crime and society will look like in the not too distant future. This is illustrated by what Bennett said in 1987 about what crime might look like during the next 20 to 50 years. Now, nearly a quarter century since her predictions, we can assess what she wrote but first a brief examination of the social context that influenced her writing of *Crimewarps* will emphasize what we mean by the importance of social context. As you will see, the context includes general sociological factors such as time and place. It also includes the author's career experiences and opportunities.

By the time Bennett began to work on *Crimewarps,* she had completed a doctorate in sociology and was an accomplished scholar, researcher, teacher, and journalist with more than 20 years of work on the topic of crime. In addition, she was an associate of the Center for Policy Research and of the Center for Investigative Reporting. She also had worked as a network correspondent for NBC News, and she had been a talk show host for PBS television. In other words, we can say that Bennett was experienced and, therefore, prepared to study major trends. In fact, her book was an outgrowth of having been asked by the Insurance Information Institute to be a consultant and media spokesperson on the topic of "the state of crime in the future" (Bennett, 1987, p. vii). Bennett's experiences and the consulting work for the Insurance Information Institute occurred within a shifting social context that she connected to crime by using the term *crimewarps.* She used this term to refer to "the bends in today's trends that will affect the way we live tomorrow" (p. xiii). Essentially, her thesis was that much of what we have come to regard as basic demographic features of our society's population and crime trends are changing dramatically. She referred to these crimewarps as representing a "set of major social transformations" (p. viii). Altogether, she identified six "warps." For example, she labeled one warp "the new criminal." This refers to the fact that today's "traditional" criminal is a poor, undereducated, young male. By relying on demographic information and dramatic news accounts of current crimes, Bennett argued that traditional criminals will be displaced by older, more upscale offenders. These offenders will include, among other trends, more women involved in white-collar crime and domestic violence. In addition, she argued that teenagers will commit fewer crimes, and senior citizens "will enter the crime scene as geriatric delinquents" (p. xiv).

A re-examination of Bennett's predictions demonstrates how difficult it is to make long-term crime trend predictions. At best, her view of the future of crime was a mixed bag. Today, what she called a "traditional criminal" has not been replaced by older, upscale offenders. There is also no evidence that women are more involved with white-collar crime than in 1987. The Federal Bureau of Investigation does report that across the United States, about 6% of all known bank robbers have been women in recent years. The change is a slight uptick from 5% recorded in 2002 (Morse, 2010). Nor is it at all clear that teenagers are committing less crime today than in 1987 or that senior citizens have become "geriatric delinquents." Whereas some indicators of some forms of domestic violence have declined, such as the number of wives who murdered their husbands between 1976 and 2000, other patterns of domestic violence have remained unchanged. Domestic homicide for African Americans of both genders remains well above the White rate (Bureau of Justice Statistics, 2007; Rennison, 2003).

In 2009 and early 2010, certain crime rates were surprisingly low in view of the seriousness of the banking crisis and financial recession. Although violent crime was expected to increase during this time, it thus far has not. Major crimes—violent or not—went down between "7 percent and 22 percent over the same period last year [2008]" (Dewan, 2009). The number of homicides in Chicago dropped 12%, while in Charlotte, North Carolina, the total fell 38%. The same cannot be said for other cities, including Philadelphia. Whereas it is too soon to tell if the decline will last throughout the current recession with a double-digit unemployment rate, it is clear that no single lens—such as Bennett's use of demographics—is adequate for explaining crime trends. Crime trends are very hard to predict accurately.

The importance of this example does not lie in the accuracy of Bennett's claims, her career experiences, or the fact that she was invited by the Insurance Information Institute to work on the future of crime at a time when our society was experiencing dramatic demographic transformations. Rather, the lesson to keep in mind is that all of these factors coalesced in such a manner as to allow Bennett to write a book on crime that makes sense because it is timely in view of what we know about society today and what we think it might look like in the future. Remember that writers, like ideas, are captives of the time and place in which they live. For this reason alone, it is impossible to understand criminological theory outside of its social context (Rennie, 1978). It remains to be seen how Bennett's book will be evaluated in the coming decades, and it remains to be seen whether the "criminal man" of the future will fit Bennett's predictions.

One more cautionary comment needs to be made. Just as the social context of the late 1980s made a book like Bennett's possible, the context of previous historical eras made different kinds of theories about crime possible. Early theories of crime tended to locate the cause of crime not in demographic shifts (as did Bennett) but rather within individuals—in their souls (spiritualism/demonology), their wills (classical school), or their bodily constitutions (positivist school). We examine each of these theories in this chapter. We are now ready to begin our search for the criminal man. We start with the earliest explanation—spiritualism.

Spiritualism

As an explanation of criminal behavior, spiritualism provides a sharp contrast to the scholarly explanations used today. Unlike today's theories, spiritualism stressed the conflict between absolute good and absolute evil (Tannenbaum, 1938). People who committed crimes were thought to be possessed by evil spirits, often referred to as sinful demons.

Although the genesis of this perspective is lost in antiquity, there is ample archaeological, anthropological, and historical evidence that this explanation has been around for many centuries. We know, for example, that primitive people explained natural disasters such as floods and famines as punishments by spirits for wrongdoings. This type of view also was used by the ancient Egyptians, Greeks, and Romans. Much later, during the Middle Ages in Europe, spiritualistic explanations had become well organized and connected to the political and social structure of feudalism. One important reason for this particular development is that, originally, crime was a private matter between the victim (or the family of the victim) and the offender(s). Unfortunately, this means of responding to offenses had a tendency to create long blood feuds that could destroy entire families. There also was the problem of justice: A guilty offender with a strong family might never be punished.

To avoid some of these problems, other methods were constructed for dealing with those accused of committing crimes. Trial by battle, for example, permitted the victim or some member of his or her family to fight the offender or some member of the offender's family (Vold & Bernard, 1986). It was believed that victory would go to the innocent if he or she believed in and trusted God. Unfortunately, this arrangement permitted great warriors to continue engaging in criminal behavior, buttressed by the belief that they always would be found "innocent." Trial by ordeal determined guilt or innocence by subjecting the accused to life-threatening and/or painful situations. For example, people might have huge stones piled on them. It was believed that if they were innocent, then God would keep them from being crushed to death; if they were guilty, then a painful death would occur. People also were tied up and thrown into rivers or ponds. It was believed that if they were innocent, then God would allow them to float; if they were guilty, then they would drown (Vold & Bernard, 1986).

Compurgation represented another means of determining innocence or guilt based on spiritualism. Unlike trial by battle or ordeal that involved physical pain and/or the threat of death, compurgation allowed the accused to have reputable people swear an oath that he or she was innocent. The logic was based on the belief that no one would lie under oath for fear of God's punishment (Vold & Bernard, 1986).

The same fear of God's punishment formed the explanation of crime and deviance for what Erikson (1966) called the "wayward Puritans"—citizens in the early American Massachusetts Bay colony. And later when our penitentiaries were constructed, they were thought of as places for "penitents who were sorry for their sins" (Vold & Bernard, 1986, p. 8). During the last three decades we have had many groups and individuals who believe that crimes and other wrongs can be explained by the devil. For example, in 1987, when "prime-time preacher" Jim Bakker of the famed PTL Club (Praise The Lord and People That Love) confessed to an adulterous one-night stand with a former church secretary, some of his followers said that it was the result of the devil's work ("God and Money," 1987). And on one occasion when Internal Revenue Service auditors revealed that several millions of dollars were unaccounted for by the PTL organization, Bakker's then wife Tammy Faye Bakker said that the devil must have gotten into the computer ("T.V. Evangelist Resigns," 1987). Others who have been caught for criminal acts have turned to God for cures for their behavior. Charles Colson of Watergate fame, for example, took the Christian message to prisoners as a solution to their problems.

More recently, three evangelical Christians from the United States traveled to Uganda and taught that there was a dark and hidden gay agenda that posed a threat to Bible-based values and the traditional African family (Gettleman, 2010). High-profile televangelist Pat Robertson and host of the *The 700 Club* announced just days after the

January 2010 earthquake in Haiti that it occurred because the Haitians had sworn a pact with the devil to get rid of those who had colonized and enslaved them. "But ever since they have been cursed by one thing after another" (Robertson, 2010, p .1; see also Miller, 2010, p. 14).

It is important to remember that even though people might criticize the argument that "the devil made me do it" as quaint or odd, it nevertheless makes sense for some people who try to understand and explain crime. In fact, until recently it was argued that the public's interest in this type of explanation grew at the same rate as our population growth. In the late 1980s, according to the National Council of Churches, church membership expanded gradually at approximately the same rate as the nation's population. At that time, nearly 70% of the nation's people reported that they were involved with churches, a figure that had remained steady during recent years ("U.S. Churches," 1987, p. A13). By early 2009, this trend had changed, perhaps in conjunction with the decline in the "cultural wars" over such things as "family values" and the re-assertion of science-driven policies over faith-based strategies (Rich, 2009).

According to an early 2009 American Religious Identification (ARI) Survey (see Grossman, 2009), almost all religious denominations had lost ground since the first ARI research in 1990. "The percentage of people who call themselves in some way Christian has dropped more than 11% in a generation" (Grossman, 2009, p. 1). Whereas the percentage of people who in 1990 had selected "no religion" was an estimated 8.2% of the population, this figure had jumped to 15% in 2008. As Grossman (2009) observed, this "category now outranks every other major U.S. religious group except for Catholics and Baptists" (p. 1). The major problem with spiritualistic explanations, nonetheless, is that they cannot be tested scientifically. Because the cause of crime, according to this theory, is otherworldly, it cannot be verified empirically. It is primarily for this reason that modern theories of crime and social order rely on explanations that are based on the physical world. These theories are called natural explanations.

Naturalistic theories and spiritualistic explanations have in common their origin in the ancient world. Despite this common origin, the two perspectives are very different. Thus, by focusing on the physical world of facts, naturalistic theories seek explanations that are more specific and detailed than spiritualistic theories. This approach to thinking and explanation was very much part of the Greeks, who early in their search for knowledge philosophically divided the world into a dualistic reality of mind and matter. This form of thinking still is prevalent in the Western world, as evidenced by reasoning that restricts explanations of human behavior to either passion or reason.

An early example of a naturalistic explanation is found in "Hippocrates' (460 B.C.) dictum that the brain is the organ of the mind" (Vold, 1958, p. 7). Additional evidence of efforts to explain phenomena by naturalistic reasoning was present approximately 350 years later in the first century B.C. in Roman thought, which attempted to explain the idea of "progress" with little reliance on demons or spirits (p. 7). The existence of this reasoning, however, does not mean that demonic and spiritualistic explanations had begun to wane by the time of the Roman Empire. In fact, these explanations reigned high well into the Middle Ages.

On the other hand, naturalistic explanations persisted in spite of the spiritualistic perspective's dominance, and by the 16th and 17th centuries several scholars were studying and explaining humans in terms known to them (Vold, 1958). Their efforts are identified collectively as the classical school of criminology. Later in this chapter, we consider a second influential naturalistic theory, the positivist school of criminology.

The Classical School: Criminal as Calculator

The most important feature of the classical school of thought is its emphasis on the individual criminal as a person who is capable of calculating what he or she wants to do. This idea was supported by a philosophy that held that humans had free will and that behavior was guided by hedonism. In other words, individuals were guided by a pain-and-pleasure principle by which they calculated the risks and rewards involved in their actions. Accordingly, punishment should be suited to the offense, not to the social or physical characteristics of the criminal.

If this sounds familiar, then you should not be surprised. One of the basic tenets of the United States' legal heritage is that people should be given equal treatment before the law. People should not be punished or rewarded just because they happen to have the right names or be from powerful families. Equality is one of the powerful ideas that was endorsed widely by the 18th- and 19th-century Enlightenment writers who influenced our Founding Fathers. Of particular interest here is the scholar most often identified as the leader of the classical school of criminology, Cesare Bonesana Marchese de Beccaria (1738–1794), an Italian mathematician and economist. It is Beccaria who pulled together many of the most powerful 18th-century ideas of democratic liberalism and connected them to issues of criminal justice.

Although it is true that Beccaria was born into an aristocratic family and had the benefit of a solid education in the liberal arts, there is little (if any) evidence in his background that would have predicted that his one small book on penal reform, *On Crimes and Punishments,* eventually would be acknowledged to have had "more practical effect than any other treatise ever written in the long campaign against barbarism in criminal law and procedure" (Paolucci, 1764/1963, p. ix; see also Beccaria, 1764/1963). Indeed, one biographical overview of Beccaria indicates that his education failed to produce a modicum of enthusiasm for scholarship except for some attraction to mathematics (Monachesi, 1973). This interest soon passed, however, and what seems to have emerged is a discontented young man with strong arguments against much of the status quo, including his father's objections to his marriage in 1761 (Paolucci, 1764/1963, p. xii). To understand Beccaria's great contribution, we must examine the social context of his life.

Unlike the United States' concern for protecting its citizens through equal protection, due process, and trial by their peers, the criminal justice system of Beccaria's Europe, especially the ancient regime in France, was "planned to ruin citizens" (Radzinowicz, 1966, p. 1), a characterization that applied to the police, criminal procedures, and punishment. The police of Paris, for example, were "the most ruthless and efficient police marching in the world" (p. 2). They were allowed by the French monarchy to deal not only with criminal matters but also with the morals and political opinions of French citizens. They relied heavily on spies, extensive covert letter opening, and the state-approved capacity not only to arrest people without warrants but also to pass judgment and hold people in custody indefinitely on unspecified charges.

Once arrested, the accused had few legal protections. He or she was cut off from legal assistance, subjected to torture, and hidden from family and friends. Witnesses against the accused testified in secret. Once guilt was determined, punishments were severe, "ranging from burning alive or breaking on the wheel to the galleys and many forms of mutilation, whipping, branding, and the pillory" (Radzinowicz, 1966, p. 3). Death by execution in early 18th-century London took place every 6 weeks, with 5 to 15 condemned hanged on each occasion (Lofland, 1973, p. 35).

Beccaria became familiar with these conditions through the association and friendship of Alessandro Verri, who held the office of protector of prisoners in Milan, Italy, where Beccaria lived. Outraged by these conditions and recently having become familiar with the writings of scholars such as Montesquieu, Helvetius, Voltaire, Bacon, Rousseau, Diderot, and Hume, Beccaria was encouraged by a small group of intellectuals to take up his pen on behalf of humanity. He was not eager to write, however, because he did not enjoy writing and because he worried about political reprisals for expressing his views. He so feared persecution from the monarchy for his views that he chose to publish his book anonymously (Monachesi, 1973). It took 11 months to write, but once published, the volume excited all of Europe as if a nerve had been exposed. By 1767, when the book was first translated into English, it already had been through several French and Italian editions. But what did it say? What caused all of the excitement?

Beccaria's tightly reasoned argument can be summarized in relatively simple terms (Radzinowicz, 1966; Vold, 1958). First, to escape war and chaos, individuals gave up some of their liberty and established a contractual society. This established the sovereignty of a nation and the ability of the nation to create criminal law and punish offenders. Second, because criminal laws placed restrictions on individual freedoms, they should be restricted in scope. They should not be employed to enforce moral virtue. To prohibit human behavior unnecessarily was to increase rather than decrease crime. Third, the presumption of innocence should be the guiding principle in the administration of justice, and at all stages of the justice process the rights of all parties involved should be protected. Fourth, the complete criminal law code should be written and should define all offenses and punishments in advance. This would allow the public to judge whether and how their liberties were being preserved. Fifth, punishment should be based on retributive reasoning because the guilty had attacked another individual's rights. Sixth, the severity of the punishment should be limited and should not go beyond what is necessary for crime prevention and deterrence. Seventh, criminal punishment should correspond with the seriousness of the crime; the punishment should fit the crime, not the criminal. For example, fines would be appropriate for simple thefts, whereas the harsher sanctions of corporal punishment and labor would be acceptable for violent crimes. Eighth, punishment must be a certainty and should be inflicted quickly. Ninth, punishment should not be administered to set an example and should not be concerned with reforming the offender. Tenth, the offender should be viewed as an independent and reasonable person who weighed the consequences of the crime. The offender should be assumed to have the same power of resistance as nonoffenders. Eleventh, for Beccaria, the aim of every good system of legislation was the prevention of crime. He reasoned that it was better to prevent crimes than to punish those who commit them.

Beccaria was not, however, the only scholar of his time to consider these issues. Jeremy Bentham (1748–1832), an English jurist and philosopher, also argued that punishment should be a deterrent, and he too explained behavior as a result of free will and "hedonistic calculus" (Bentham, 1948). John Howard (1726–1790), also English and a contemporary of Beccaria and Bentham, studied prisons and advocated prison reform (Howard, 1792/1973). His work often is credited with having influenced the passage of England's Penitentiary Act of 1779, which addressed prison reform.

The influence of these writers, however, went far beyond the passage of specific laws. Their ideas inspired revolutions and the creation of entirely new legal codes. The French Revolution of 1789 and its famous Code of 1791 and the U.S. Constitution each was influenced by the classical school. But by the 1820s, crime still was flourishing, and the argument that bad laws made bad people was being questioned seriously (Rothman, 1971). Also, the argument that all criminal behavior could be explained by hedonism was weakening as the importance of aggravating and mitigating circumstances increased. Nor did the new laws provide for the separate treatment of children. Nevertheless, the classical school did make significant and lasting contributions. The calls for laws to be impartial and specific and for punishment to be for crimes instead of criminals, as well as the belief that all citizens should be treated fairly and equally, now have become accepted ideas. But what caused crime remained a troubling question unanswered by the Enlightenment's "rather uncomplicated view of the rational man" (Sykes, 1978, p. 11), a view based primarily on armchair thinking. The result was a new search for the "criminal man," with emphasis given to action being determined instead of being the result of free will. The advocates of this new way of thinking created what came to be known as the positivist school of criminology.

The Positivist School: Criminal as Determined

The most significant difference between the classical school and the positivist school is the latter's search for empirical facts to confirm the idea that crime was determined by multiple factors. This is a clear shift away from the reasoning of Beccaria and Bentham, who thought that crime resulted from the free will and hedonism of the individual criminal. As will become apparent, the 19th century's first positivists wanted scientific proof that crime was caused by features within the individual. They primarily emphasized the mind and the body of the criminal, to some extent neglecting social factors external to the individual. (These factors later became the focus of sociological explanations of crime. This type of explanation is discussed in more detail later.)

But the search for causes of crime, in fact, did not begin with the 19th-century positivists. Early examples in literature, for example, connected the body through the ideas of beauty and ugliness to good and evil behavior. Shakespeare's *Tempest* portrayed a deformed servant's morality as offensive as his appearance, and Homer's *Iliad* depicted a despised defamer as one of the ugliest of the Greeks. This form of thinking still is present today, as evidenced by contemporary female beauty contests in which contestants vie with one another not only by publicly displaying their scantily clothed bodies but also through rendering artistic performances of some sort. A beautiful woman is expected to do good things. Although an example of sexist thinking, the connection between physical features and behavior is less than scientific.

THE BIRTH OF THE POSITIVIST SCHOOL: LOMBROSO'S THEORY OF THE CRIMINAL MAN

The modern search for multiple-factor explanations of crime usually is attributed to Cesare Lombroso (1835–1909), an Italian who often is called the "father of modern criminology" (Wolfgang, 1973, p. 232). A clue to his work and the social context of his life is gleaned from his self-description as "a slave to facts," a comment that could not have been made by writers of the classical school. During the century that separated Beccaria's graduation from the University of Pavia (in 1758) and Lombroso's graduation from the same institution (in 1858) with a degree in medicine, secular, rational-scientific thinking and experimentation had become increasingly more acceptable ways of analyzing reality.

An essential clue to understanding Lombroso's work is to recognize that, during the last half of the 19th century, the answer to the age-old question, "What sort of creatures are humans?" had begun to depart from theological answers to answers provided by the objective sciences, particularly biology. It was here that humans' origins as creatures were connected to the rest of the animal world through evolution (Vold, 1958). No other 19th-century name is associated with this connection more often than Charles Darwin (1809–1882), the English naturalist who argued that humans evolved from animals. His major works—*Origin of Species* (Darwin, 1859/1981), *The Descent of Man* (Darwin, 1871), and *The Expressions of the Emotions in Man and Animals* (Darwin, 1872)—all predated Lombroso's. For Lombroso, the objective search for explaining human behavior meant disagreement with free will philosophy. He became interested in psychiatry "sustained by close study of the anatomy and physiology of the brain" (Wolfgang, 1973, p. 234).

Lombroso's interest in biological explanations of criminal behavior developed between 1859 and 1863 when he was serving as an army physician on various military posts. During this time, he developed the idea that diseases, especially cretinism and pellagra, contributed to mental and physical deficiencies "which may result in violence and homicide" (Wolfgang, 1973, p. 236). He also used his position as a military physician to systematically measure approximately 3,000 soldiers so as to document the physical differences among inhabitants from various regions of Italy. From this study, Lombroso made "observations on tattooing, particularly the more obscene designs which he felt distinguished infractious soldiers" (p. 235). Later, Lombroso used the practice of tattooing as a distinguishing characteristic of criminals. He started to publish his research on the idea that biology—especially brain pathologies—could explain criminal behavior in a series of papers that first started to appear in 1861. By 1876 when he published his findings in *On Criminal Man* (Lombroso, 1876), his work contained not only a biological focus but an evolutionary one as well. This book went through several Italian editions and foreign language translations.

The central tenet of Lombroso's early books on crime is that criminals represent a peculiar physical type distinctively different from that of noncriminals. In general terms, he claimed that criminals represent a form of degeneracy that was manifested in physical characteristics reflective of earlier forms of evolution. He described criminals as atavistic, throwbacks to an earlier form of evolutionary life. For example, he thought that ears of unusual size, sloping foreheads, excessively long arms, receding chins, and twisted noses were indicative of physical characteristics found among criminals.

Lombroso classified criminals into four major categories: (1) *born criminals* or people with atavistic characteristics; (2) *insane criminals* including idiots, imbeciles, and paranoiacs as well as epileptics and alcoholics; (3) *occasional criminals* or *criminaloids,* whose crimes are explained primarily by opportunity, although they too have innate traits that predispose them to criminality; and (4) *criminals of passion,* who commit crimes because of anger, love, or honor and are characterized by being propelled to crime by an "irresistible force" (Wolfgang, 1973, pp. 252–253).

To Lombroso's credit, he modified his theory throughout five editions of *On Criminal Man,* with each new edition giving attention to more and more environmental explanations including climate, rainfall, sex, marriage customs, laws, the structure of government, church organization, and the effects of other factors. But he never completely gave up the idea of the existence of a born criminal type, and although he most often is thought of as the person who connected biological explanations to criminal behavior, he was not the first person to do so.

This idea was developed, for example, during the 1760s by the Swiss scholar Johann Kaspar Lavater (1741–1801), who claimed that there was a relationship between facial features and behavior. Later, Franz Joseph Gall (1758–1828), an eminent European anatomist, expanded the idea and argued that the shape of an individual's head could explain his or her personal characteristics. This explanation was called *phrenology,* and by the 1820s it was stimulating much interest in the United States. One book on the subject, for example, went through nine editions between 1837 and 1840 (Vold, 1958). It is instructive to note that as phrenology increased in popularity, its explanatory powers were expanded; more and more different forms of human conduct, it was claimed, could be explained by the shape of the head. The importance of phrenology for us is that it indicates the popularity of biological explanations of behavior nearly 50 years before Lombroso received his degree in medicine in 1858 and long before he ever was called the father of modern criminology. In fact, Ellis (1913) identified nearly two dozen European scholars who had pointed to the relationship between criminals' physical and mental characteristics and their behavior before Lombroso. These included Henry Mayhew and John Binny, investigative reporters from London; Scottish prison physician J. Bruce Thomson; and Henry Maudsley, a fashionable London psychiatrist (Rafter, 2008). So, why should Lombroso be studied and remembered?

Although his biological explanation of crime is considered simple and naïve today, Lombroso made significant contributions that continue to have an impact on criminology. Most noteworthy here is the attention that he gave to a multiple-factor explanation of crime that included not only heredity but also social, cultural, and economic variables. The multiple-factor explanation is common in today's study of crime. Lombroso also is credited with pushing the study of crime away from abstract metaphysical, legal, and juristic explanations as the basis of penology "to a scientific study of the criminal and the conditions under which he commits crime" (Wolfgang, 1973, p. 286). We also are indebted to Lombroso for the lessons he taught regarding methods of research. He demonstrated the importance of examining clinical and historical records, and he emphasized that no detail should be overlooked when searching for explanations of criminal behavior.

He also left a legacy of ideas and penological plans for how new late 19th-century nations could deal with the disruptive populations—criminals, the insane, and other deviants—that were associated with industrialization, immigration, urbanization, and war. These contributions were influenced in large measure by the fact that whereas northern and southern Italy had been geographically united in 1870, the country was "nonetheless characterized by political and radical disunity, a country more in name than in fact, with subjects who identified with their hometowns rather than with a new national government" (Rafter, 2008, p. 85). With his home country long fragmented by religion and politics, Lombroso anticipated the problems that would ensue, and he was deeply interested in creating clear and rational plans to deal with the disorderly. He also had another, perhaps his most significant, contribution as a criminologist. According to Rafter, he was the only criminologist of his time who can qualify as a "paradigm shifter" (2008, p. 85). He took the topic of the causes of crime away from sin and placed it in the realm of science, where it remains today.

LOMBROSO'S LEGACY: THE ITALIAN CRIMINOLOGICAL TRADITION

Enrico Ferri. Lombroso's legacy of positivism was continued and expanded by the brilliant career and life of a fellow Italian, Enrico Ferri (1856–1929). Born into the family of a poor salt-and-tobacco shopkeeper, Ferri came to be one of the most influential figures in the history of criminology (Sellin, 1973). In the words of Thorsten Sellin, one of the world's most eminent criminologists, who as a young man in November 1925 heard Ferri lecture at 70 years of age: "Ferri the man is as fascinating as Ferri the scholar" (Sellin, 1973, p. 362). Ferri was a scholar with brilliant ideas and strong passions who believed that life without an ideal, whatever it might be, was not worth living. By 16 years of age, Ferri was developing his lifelong commitment to the "scientific orientation," having come under the influence of a great teacher who himself was of strong convictions. Ferri gave up the clerical robe for the study of philosophy.

Ferri was just 21 years old when he published his first major work, *The Theory of Imputability and the Denial of Free Will* (Vold, 1958, pp. 32–33). It was an attack on free will arguments and contained a theoretical perspective that was to characterize much of Ferri's later work on criminality as well as his political activism. Unlike Lombroso, who gave more attention to biological factors than to social ones, Ferri gave more emphasis to the interrelatedness of social, economic, and political factors that contribute to crime (Vold, 1958). He argued, for example, that criminality could be explained by studying the interactive effects among physical factors (e.g., race, geography, temperature), individual factors (e.g., age, gender, psychological variables), and social factors (e.g., population, religion, culture). He also argued that crime could be controlled by social changes, many of which were directed toward the benefit of the working class. He advocated subsidized housing, birth control, freedom of marriage, divorce, and public recreation facilities, each reflective of his socialist belief that the state is responsible for creating better living and working conditions. It is not surprising that Ferri also was a political activist.

He was elected to public office after a much publicized lawsuit in which he successfully defended a group of peasants accused of "incitement to civil war" after a dispute with wealthy landowners (Sellin, 1973, p. 377). He was reelected 11 times by the Socialist party and stayed in office until 1924. Throughout his career, Ferri attempted to integrate his positivist approach to crime with political changes. For example, he tried unsuccessfully to have the new Italian Penal Code of 1889 reflect a positivist philosophy instead of classical reasoning. And after Mussolini came to power during the early 1920s, Ferri was invited to write a new penal code for Italy (Vold, 1958). It reflected his positivist and socialist orientation, but it too was rejected for being too much of a departure from classical legal reasoning. After nearly 50 years as a socialist liberal, Ferri changed his philosophy and endorsed fascism as a practical approach to reform. According to Sellin (1973), fascism appealed to Ferri because it offered a reaffirmation of the state's authority over excessive individualism, which he had criticized often.

Although it is puzzling to try to understand Ferri's shift from socialism to fascism, some insight is provided by considering that he was living during a time of great social change and that Ferri, as a person with humble origins, wanted the changes to produce a better society. He believed that, to accomplish this reform, individuals must be legally responsible for their actions instead of being only morally responsible to God. This approach to responsibility represented a radical departure from tradition because it was offered within a theoretical framework that called for "scientific experts" not only to explain crime but also to write laws and administer punishment. In essence, it was a call for the state to act "scientifically" in matters of social policy.

Ferri's call for legal responsibility was offered when Italy was experiencing much unrest caused mostly by industrialization during the late 1800s and later by the social disorder stemming from World War I. Evidence of Ferri's response to the changing conditions in Italy is found in the fact that in the first four editions of *Sociologia Criminale,* he listed only five classes of criminals: (1) the born or instinctive criminal whom Lombroso had identified as the atavist, (2) the insane criminal who was clinically identified as mentally ill, (3) the passion criminal who committed crime as a result of either prolonged and chronic mental problems or an emotional state, (4) the occasional criminal who was more the product of family and social conditions than of abnormal personal physical or mental problems, and (5) the habitual criminal who acquired the habit from the social environment. For the fifth edition of *Sociologia Criminale*, Ferri (1929–1930) added a new explanation of crime—the involuntary criminal. Ferri explained this phenomenon as "becoming more and more numerous in our mechanical age in the vertiginous speed of modern life" (Sellin, 1973, p. 370).

We also must understand that Ferri's interest in fascism did not occur in a social void absent of public support. It took only 5 years from the time when Mussolini formed the Fasci del Combattimento in 1919 until the first elections in Italy under the fascists in 1924, when Mussolini received 65% of the vote. When Mussolini asked Ferri to rewrite the Italian code, Ferri's response was consistent with the mood of the times.

Raffaele Garofalo. After Lombroso and Ferri, Raffaele Garofalo (1852–1934) was the last major contributor to the positivist or Italian school of criminology. Unlike Lombroso's emphasis on criminals as abnormal types with distinguishable anatomic, psychological, and social features, and unlike Ferri's emphasis on socialist reforms and social defenses against crime, Garofalo is remembered for his pursuit of practical solutions to concrete problems located in the legal institutions of his day and for his doctrine of "natural crimes."

In many ways, Garofalo's work represents the currents of interests in late 19th-century Europe more clearly than does either Lombroso's or Ferri's. This is the result of three different but interconnected phenomena. First, Garofalo was born only 6 years before Lombroso received his degree in medicine. Thus, by the time Garofalo was an adult, enough time had passed since the publication of Lombroso's and Ferri's major works to permit some degree of reactive evaluation. Second, Garofalo was both an academician and a practicing lawyer, prosecutor, and magistrate who faced the practical problems of the criminal justice system on a daily basis. Accordingly, he was in an excellent position to be familiar with the great attention that Lombroso's and Ferri's work had received in both academic and penal circles and with the practical policy implications of their writings. Third, at the time when Garofalo (1885) published the first edition of *Criminology* at 33 years of age, the social Darwinian era was at the peak of its existence, with numerous suggestions from biology, psychology, and the social sciences on how society could guarantee the survival of the fittest through criminal law and penal practice (Hawkins, 1931).

Garofalo's theoretical arguments on the nature of crime and on the nature of criminals were consistent with social Darwinism. For example, he argued that because society is a "natural body," crimes are offenses "against the law of nature." Therefore, criminal action was a crime against nature. Accordingly, the "rules of nature" were the rules of right conduct revealed to humans through their reasoning. It is obvious here that Garofalo's thinking also included some influence from the classical school and its emphasis on reasoning. For Garofalo, the proper rules of conduct came from thinking about what such rules should allow or prohibit. He nevertheless identified acts that no society could refuse to recognize as criminal and to repress by punishment—natural crimes. These offenses, according to Garofalo, violated two basic human sentiments found among people of all ages: the sentiments of probity and pity (Allen, 1973, p. 321; Vold, 1958, p. 37). *Pity* is the sentiment of revulsion against the voluntary infliction of suffering on others. *Probity* refers to the respect for the property rights of others.

The social Darwinian influence on Garofalo's thinking also is apparent in his explanation of where the sentiments of probity and pity could be found. They were basic moral sensibilities that appear more or less in "advanced form in all civilized societies" (Allen, 1973, p. 321), meaning that some societies had not evolved to the point of advanced moral reasoning. Similarly, the Darwinian influence is present in Garofalo's argument that some members of society might have a higher than average sense of morality because they are "superior members of the group" (p. 321).

Garofalo's notion of the characteristics of the criminal also revealed a Darwinian influence, but less so than when he addressed the issue of punishment and penal policies. In developing this portion of his theoretical arguments, he first reconsidered the Lombrosian idea of crime being associated with certain anatomical and physical characteristics and concluded that although the idea had merit, it had not been proved. Sometimes physical abnormalities were present, and sometimes they were not. He argued instead for the idea that true criminals lacked properly developed altruistic sentiments (Allen, 1973). In other words, true criminals had psychic or moral anomalies that could be transmitted through heredity. But the transmission of moral deficiencies through heredity was a matter of degree. This conclusion led Garofalo to identify four criminal classes, each one distinct from the others because of deficiencies in the basic sentiments of pity and probity.

Murderers were totally lacking in both pity and probity, and they would kill or steal when given the opportunity. Lesser criminals, Garofalo acknowledged, were more difficult to identify. He divided this category based on whether criminals lacked sentiments of either pity or of probity. Violent criminals lacked pity, which could be influenced very much by environmental factors such as alcohol and the fact that criminality was endemic to the population. Thieves, on the other hand, suffered from a lack of probity, a condition that "may be more the product of social factors than the criminals in other classes" (Allen, 1973, p. 323). His last category contained cynics or sexual criminals, some of whom would be classified among the violent criminals because they lacked pity. Other lascivious criminals required a separate category because their actions stemmed from a "low level of moral energy" rather than from a lack of pity (p. 329).

Nowhere is Garofalo's reliance on Darwinian reasoning more obvious than when he considered appropriate measures for the social defense against crime. Here he again used the analogy of society as a natural body that must either adapt to the environment or be eliminated. He reasoned that because true criminals' actions reveal an inability to live by the basic human sentiments necessary for society to survive, they should be eliminated. Their deaths would contribute to the survival of society (Barnes, 1930). For lesser criminals, he proposed that elimination take the form of life imprisonment or overseas transportation (Allen, 1973).

It is clear that deterrence and rehabilitation were secondary considerations for Garofalo. But he favored "enforced reparation" and indeterminate sentences, indicating that Garofalo's social defenses against crime were modeled to some extent on the psychic characteristics of the offender. In this regard, his position on punishment is more in line with the free will reasoning of the classical scholars than Garofalo might admit.

One conclusion about Garofalo is clear, however, and that is his position on the impor- tance of society over the individual. To him, the individual represents but a cell of the social body that could be exterminated without much (if any) great loss to society (Allen, 1973). By giving society or the group supremacy over the individual, Garofalo and Ferri were willing to sacrifice individual rights to the opinions of "scientific experts," whose decisions might not include the opinions of those they were evaluating and judging or the opinions of the public. Not surprisingly, their work was accepted by Mussolini's regime in Italy because it lent the mantle of scientific credence to the ideas of racial purity, national strength, and authoritarian leadership (Vold, 1958).

The work of the Italian positivists also suffered from serious methodological research problems. For example, their work was not statistically sophisticated. As a result, their conclusions about real or significant differences between criminals and noncriminals were, in fact, highly speculative.

This problem was addressed by Goring's (1913) study of 3,000 English convicts and a control group of nonconvict males. Unlike Lombroso, Ferri, and Garofalo, Goring employed an expert statistician to make computations about the physical dif- ferences between criminals and noncriminals. After 8 years of research on 96 differ- ent physical features, Goring concluded that there were no significant differences between criminals and noncriminals except for stature and body weight. Criminals were found to be slightly smaller. Goring interpreted this finding as confirmation of his hypothesis that criminals were biologically inferior, but he did not find a physical criminal type.

THE CONTINUING SEARCH FOR THE INDIVIDUAL ROOTS OF CRIME

Body Types and Crime. The search for a constitutionally determined criminal man did not stop with Goring's (1913) conclusions. Kretschmer (1925) took up the theme as the result of his study of 260 insane people in Swabia, a southwestern German town. He was impressed with the fact that his subjects had definite types of body builds that he thought were associated with certain types of psychic dispositions. First published in German in 1922 and translated into English in 1925, Kretschmer's study identified four body types: asthenic, athletic, pyknic, and some mixed unclassifiable types. He found *asthenics* to be lean and narrowly built, with a deficiency of thickness in their overall bodies. These men were so flat-chested and skinny that their ribs could be counted easily. The *athletic* build had broad shoulders, excellent musculature, a deep chest, a flat stomach, and powerful legs. These men were the 1920s' counterpart of the modern "hunks" of media fame. The *pyknics* were of medium build with a propensity to be rotund, sort of soft appearing with rounded shoulders, broad faces, and short stubby hands. Kretschmer argued that the asthenic and athletic builds were associated with schizophrenic personalities, whereas the pyknics were manic-depressives.

Four years after the English translation appeared in the United States, Mohr and Gundlach (1929–1930) published a report based on 254 native-born White male inmates in the state penitentiary at Joliet, Illinois. They found that pyknics were more likely than asthenics or athletics to have been convicted of fraud, violence, or sex offenses. Asthenics and athletics, on the other hand, were more likely to have been convicted of burglary, robbery, or larceny. But Mohr and Gundlach were unable to demonstrate any connection among body build, crime, and psychic disposition.

Ten years later, the search for physical types that caused crime was taken up by Earnest A. Hooton, a Harvard University anthropologist. He began with an extensive critique of Goring's research methods and proceeded to a detailed analysis of the measurements of more than 17,000 criminals and noncriminals from eight different states (Hooton, 1939). In his three-volume study, Hooton (1939) argued that "criminals are inferior to civilians in nearly all of their bodily measurements" (Vol. 1, p. 329). He also reported that low foreheads indicated inferiority and that "a depressed physical and social environment determines Negro and Negroid delinquency to a much greater extent than it does in the case of Whites" (p. 388).

These and similar conclusions generated severe criticism of Hooton's work, especially the racist overtones and his failure to recognize that the prisoners he studied did not represent criminal offenders who had not been caught or offenders who had been guilty but not convicted. His control group also was criticized for not being representative of any known population of people. This group consisted of Nashville firefighters and members of the militia, each of whom could be expected to have passed rigorous physical examinations that would distinguish them from average males. He also included in his control group beachgoers, mental patients, and college students. He offered no explanation as to why these disparate categories of people represented normal physical types. Hooton also was criticized for treating some small differences in measurement as greatly significant and for ignoring other differences that were found.

It is important to notice that despite the stinging criticism received by Hooton and by others who were searching for biological explanations, the search continued and expanded into the 1940s and 1950s. The work by William H. Sheldon, for example, shifted attention away from adults to delinquent male youths. Sheldon (1949) studied 200 males between 15 and 21 years of age in an effort to link physiques to temperament, intelligence, and delinquency. By relying on intense physical and psychological examinations, Sheldon produced an Index to Delinquency that was used to give a quick and easy profile of each male's problems. A total score of 10 meant that a boy's case was severe enough to require total institutionalization, a score of 7 meant that the case was borderline, and a score of 6 was interpreted as favoring adjustment and independent living outside of an institution.

Sheldon (1949) classified the boys' physiques by measuring the degree to which they possessed a combination of three different body components: endomorphy, mesomorphy, and ectomorphy. Each could dominate a physique. Endomorphs tended to be soft, fat people, mesomorphs had muscular and athletic builds, and ectomorphs had skinny, flat, and fragile physiques. Sheldon also rated each of the 200 youths' physiques by assigning a score of 1 to 7 for each component. For example, the average physique score for the 200 males was 3.5–4.6–2.7, representing a rather husky male (p. 727). Overall, Sheldon concluded that because youths came from parents who were delinquent in very much the same way that the boys were delinquent, the factors that produce delinquency were inherited.

William Sheldon's findings were given considerable support by Sheldon Glueck and Eleanor Glueck's comparative study of male delinquents and nondelinquents (Glueck & Glueck, 1950). As a group, the delinquents were found to have narrower faces, wider chests, larger and broader waists, and bigger forearms and upper arms than the nondelinquents. An examination of the overall ratings of the boys indicated that approximately 60% of the delinquents and 31% of the nondelinquents were predominantly mesomorphic. The authors included this finding in their list of outstanding factors associated with male delinquency. As with each of the previous scholars who attempted to explain criminal behavior by relying primarily on biological factors, their findings neglected the importance of sociological phenomena. It is unclear, for example, whether the Gluecks' mesomorphs were delinquents because of their builds and dispositions or because their physiques and dispositions are conceived socially as being associated with delinquency. This, in turn, could create expectations about illegal activities that males might feel pressured to perform.

Efforts to connect body shape and behavior were not limited to crime alone. In 1995, it was revealed that between the 1940s and the 1960s, Hooton and Sheldon had been involved with a eugenics experiment that took "posture photos" of freshmen as they entered some of the nation's most prestigious Ivy League schools, including Harvard University, Yale University, and Wellesley College. They were pursuing the now discredited idea that body shape and intelligence are somehow connected. Some of "America's Establishment" photographed for the experiment included New York Senator Hillary Rodham Clinton, former president George H. W. Bush, New York Governor George Pataki, University of Oklahoma President David Boren, and television journalist Diane Sawyer. Although some of the photographs had been destroyed since the experiment was ended during the late 1960s, in 1995 the Smithsonian museum in Washington, D.C. still had as many as 20,000 photographs of men and 7,000 photographs of women. Since then, the photographs have been destroyed ("Naked Truth Returns," 1995; "Naked Truth Revealed," 1995; "Smithsonian Destroys," 1995).

Psychogenic Causes of Crime. At this point, we turn our attention to another form of positivism, one that places no emphasis on types of physiques as causes of crime. Here the search for the causes of crime is directed to the mind. These theories often are referred to as the "psychogenic school" because they seek to explain crime by focusing attention on the personality and how it was produced. In this way, the analysis is "dynamic" rather than "constitutional," as was the emphasis from biological positivism. This school of thought developed along two distinct lines: one stressing psychoanalysis and the other stressing personality traits.

We begin with Sigmund Freud (1859–1939) and the psychoanalytical approach. Freud, a physician, did not directly address the question of what caused criminal behavior. He was interested in explaining all behavior, including crime. He reasoned that if an explanation could be found for normal behavior, then surely it also could explain crime.

At the core of the theories of Freud and his colleagues is the argument that all behavior is motivated and purposive. But not all desires and behavior are socially acceptable, so they must be repressed into the unconsciousness of the mind for the sake of morality and social order. The result is that tensions exist between the unconscious id, which is a great reservoir of aggressive biological and psychological urges, and the conscious ego, which controls and molds the individual. The superego, according to Freud (1920, 1927, 1930), is the force of self-criticism that reflects the basic behavioral requirements of a particular culture. Therefore, crime is a symbolic expression of inner tensions that each person has but fails to control. It is an "acted out" expression of having learned self-control improperly.

Franz Alexander, a psychoanalyst, and William Healy, a physician, both applied Freud's principles in their study of criminal behavior (Alexander & Healy, 1935). For example, they explained one male criminal's behavior as the result of four unconscious needs: (1) overcompensation for a sense of inferiority, (2) the attempt to relieve a sense of guilt, (3) spite reactions toward his mother, and (4) gratification of his dependent tendencies by living a carefree existence in prison.

Freud's colleague, August Aichhorn, wrote that many children continued to act infantile because they failed to develop an ego and superego that would permit them to conform to the expectations of childhood, adolescence, and adulthood. Aichhorn (1936) contended that such children continued to operate on the "pleasure principle," having failed to adapt to the "reality principle" of adulthood. Kate Friedlander, a student of Freud and Aichhorn, also focused on the behavior of children and argued that some children develop antisocial behavior or faulty character that makes them prone to delinquency (Friedlander, 1949). Redl and Wineman (1951) advanced a similar argument stating that some children develop a delinquent ego. The result is a hostile attitude toward adults and aggression toward authority because the children have not developed a good ego and superego.

The search for personality traits, the second tradition of investigation that attempted to locate the cause of crime within the mind, was started by attempting to explain mental faculties biologically. Feeblemindedness, insanity, stupidity, and dullwittedness were thought to be inherited. This view was part of the efforts during the late 19th century to explain crime constitutionally. It became a popular explanation in the United States after *The Jukes* was published (Dugdale, 1877). *The Jukes* described a family as being involved in crime because its members suffered from "degeneracy and innate depravity."

Interest in explaining family-based mental deficiencies by heredity continued through the end of the 19th century and well into the first quarter of the 20th century. Goddard, for example, published *The Kallikak Family* in 1912, and a follow-up study on the Jukes by Estrabrook (1916) appeared 4 years later. Unfortunately, these studies were very general and avoided advanced and comparative statistics.

More exacting studies of the mind came from European research on the measurement of intelligence. French psychologist Alfred Binet (1857–1911), for example, first pursued intelligence testing in laboratory settings and later applied his findings in an effort to solve the problem of retardation in Paris's schools. Aided by his assistant, Theodore Simon, Binet revised his IQ tests in 1905, 1908, and 1911, and when the scale appeared in the United States, it was revised once again. Common to each revision was the idea that an individual should have a mental age that could be identified with an intelligence quotient or IQ score.

Goddard (1914, 1921) usually is credited as the first person to test the IQs of prison inmates. He concluded that most inmates were feebleminded and that the percentage of feeblemindedness ranged from 29% to 89%. His research was plagued, however, by the difficult issue of determining what score(s) should be used to define feeblemindedness.

The legitimacy and practical value of IQ testing were given great support when the U.S. Army Psychological Corps decided to use this method to determine who was fit for military service in World War I (Goddard, 1927). The result was that, at one point, nearly a third of the draft army was thought to be feebleminded. This conclusion was modified later, but faith in IQ testing as a means of explaining crime continued well past World War I and into the 1920s and 1930s. Eventually, the increasing use of sophisticated research methods began to produce results indicating that when inmates were compared to the general population, only slight differences in feeblemindedness were found. Today, very little (if any) research tries to explain crime as the result of feeblemindedness, although recent scholars have argued that low IQ is a central cause of criminal behavior (Herrnstein & Murray, 1994).

The Consequence of Theory: Policy Implications

THE POSITIVIST SCHOOL AND THE CONTROL OF THE BIOLOGICAL CRIMINAL

The most obvious orientation displayed by positivists of the mid-19th century through the first quarter of the 20th century was their placement of the causes of crime primarily within individual offenders. This is not at all surprising once we recognize that the early positivists—especially Lombroso, Ferri, and Garofalo in Italy; Freud, Aichhorn, and Kretschmer in Austria and Germany; and Alexander and Healy in the United States—were all educated in medicine, in law, or in both fields. Each of these disciplines places great emphasis on individuals as the explanation of behavior. It is this flavor of emphasis that sheds much light on the policy consequences of their explanations of crime. But the disciplinary perspectives alone are inadequate for the purposes of fully understanding these consequences because each discipline was influenced by the general temper of the times. For the positivists considered in this chapter, the temper of the times was Darwinism strongly flavored by Victorianism.

The magnitude of the impact of the Darwinian argument is difficult to describe in a few paragraphs or pages and, indeed, even in a score of books. Nevertheless, an effort must be attempted before we proceed. In the simplest of terms, Darwin's evolutionary thesis represents one of the most profound theories of all times. It not only offered revolutionary new knowledge for the sciences but also helped to shatter many philosophies and practices in other areas. It commanded so much attention and prestige that the entire literate community felt "obliged to bring his world outlook into harmony with their findings" (Hofstadter, 1955b, p. 3). According to Hofstadter (1955b), Darwin's impact is comparable in its magnitude to the work of Nicolaus Copernicus

(1473–1543), the European astronomer; Isaac Newton (1642–1727), the English mathematician and physicist; and Freud, the Austrian psychoanalyst. In effect, all of the Western world had to come to grips with Darwin's evolutionary scheme.

Although there was much discussion and controversy about the social meaning of Darwin's theory of the "struggle for survival" and the "survival of the fittest," social Darwinists generally agreed that the theory's policy implications were politically conservative. It was argued that any policies that advocated government-sponsored social change would, if executed, actually be an interference with nature. The best approach was minimal involvement. "Let nature take its course" became a frequent refrain uttered by social Darwinists. It carried the clear message that accelerated social change was undesirable. Policies designed to accomplish "equal treatment," for example, were opposed strongly. Social welfare programs, it was argued, would perpetuate the survival of people who were negligent, shiftless, silly, or immoral while, at the same time, retarding individual and national economic development. Hard work, saving, and moral constraint were called on as the solutions to individual and collective social and economic good fortune.

It is important to note an irony regarding the social Darwinists. Although many of the nation's leaders during the late 19th and early 20th centuries were opposed to programs of social change because they feared that these programs would threaten the nation's survival, they often were the same people who were changing our economy radically and plundering our natural resources through speculation, innovation, and daring. They also were the same leaders who were introducing "new economic forms, new types of organization, [and] new techniques" (Hofstadter, 1955b, p. 9). Changes that benefited their interests were acceptable; changes for the less fortunate were not. The stage was set for "scientifically justified" forms of control that would contain or eliminate crime. This control came in the form of the genetics movement with the blessings of the Victorian concern for morality and purity (Pivar, 1973). Unfortunately the stage also was set for scientifically justified policies that, in fact, resulted in selective abuse and neglect.

When social Darwinism was used to formulate crime control policies, major themes appeared. On the one hand, the "born criminal" legacy from Lombroso and his students, and especially Garofalo's policy of "elimination" for certain criminal offenders, produced a penal philosophy that stressed incapacitation. Clearly, the emphasis was on removing criminals from the community to prevent them from committing any additional biologically determined harm. Therefore, it was inappropriate to attempt to reform or rehabilitate criminal offenders. Warehousing convicted offenders was considered a sufficient socially responsible response to the problem of what to do with lawbreakers.

On the other hand, the second means of controlling crime allowed for a type of rehabilitation. It was based on medical reasoning that viewed individuals as biological objects that needed treatment and allowed some of the most repressive state policies in the history of American penology. The worst of these policies were justified by the study of genetics and what has been called the eugenics movement (Beckwith, 1985).

As a science during the early 1900s, the study of eugenics claimed that inheritance could explain the presence of simple and complex human behavioral characteristics. Thus, it reinforced the ideas of biological determinism and contributed greatly to the argument that many of the social problems of the late 19th century, such as the conflicts over wages and working conditions, could be traced to the genetic inferiority of foreigners who were working in the United States. The 1886 Haymarket bombings and riots in Chicago, for example, were thought to be caused by "inferior foreigners." This theme was advanced by industrialists and by newspapers such as the *New York Times,* which described labor demonstrations as "always composed of foreign scum, beer-smelling Germans, ignorant Bohemians, uncouth Poles, and wild-eyed Russians" (Beckwith, 1985, p. 317).

With the support of leading industrialists such as the Carnegies and leading education institutions such as Harvard, centers for the study of eugenics soon were established, and efforts were undertaken to study the nation's "stock." Accordingly, many states passed laws designed to permit the application of the eugenicists' arguments. Between 1911 and 1930, for example, more than 30 states established laws requiring sterilization for behavioral traits thought to be determined genetically (Beckwith, 1985, p. 318). The laws targeted behavior such as criminality, alcoholism, sodomy, bestiality, feeblemindedness, and the tendency to commit rape. The result was the sterilization of at least 64,000 people. Many of the same states also passed laws permitting psychosurgeries including the now infamous frontal lobotomy. The total number of these types of operations is unknown.

The laws passed under the influence of the eugenics movement were not restricted to sterilization and psychosurgery, although these were the two most brutal and repressive measures sponsored. The miscegenation laws endorsed in 34 states made it illegal for African Americans and Whites to marry each other; some states also forbade marriage between Whites and Asians (Provine, 1973). These laws also called vigorously for the passage of immigration laws to be based on a quota system calculated on the proportion of people living in the United States from a specific country in 1890. And in 1924, the U.S. Congress, after hearing leading eugenicists testify that our stock was being weakened by the influx of people from southern and eastern European countries, passed the Immigration Restriction Act of 1924. It was directed explicitly to a population of people who were thought to be biologically inferior. As such, it was a racist law and a forerunner of the 1930s and 1940s eugenic policies of Nazi Germany.

Despite the clear implications of eugenic policies for repressing a targeted population, the practice of sterilization and psychosurgery continued in the United States until well into the 1970s. Between 1927 and 1972 in Virginia alone, more than 8,000 individuals were sterilized because they were identified as feebleminded (Katz & Abel, 1984, p. 232), and more than 20,000 people were sterilized in California. But as Katz and Abel (1984) pointed out, the real reason for sterilization was not feeblemindedness but rather class: "The one characteristic that did have to be demonstrated was indigency. . . . Commitment papers had to certify that the individual was both without means of support and without any acquaintance who would give bond" (p. 233). To allow the poor to propagate, it was claimed, would increase the numbers of feebleminded offspring, the poor, alcoholics, criminals, and prostitutes.

THE POSITIVIST SCHOOL AND CRIMINAL JUSTICE REFORM

In pointing out the worst effects of biologically oriented theories, we are not unmindful that the positivist school also helped to usher in an approach to policy that was reformative rather than punitive in impulse. To be sure, the conclusion that offenders are characterized by unchangeable bodily or psychological characteristics leads to the conclusion that offenders should be either eliminated, caged indefinitely (incapacitated), or altered physically through intrusive measures. And "crime prevention," as we have seen, becomes a matter of not allowing the "defectives" to multiply. But if one assumes that the causes determining crime are *changeable*—for example, unemployment or the emotional turmoil from family conflict—then the policy implications are much more optimistic. The challenge becomes one of diagnosing the forces that moved a person to break the law and then developing a strategy—perhaps job training or family counseling—to help the person overcome these criminogenic factors. In short, the challenge is to rehabilitate offenders so that they might rejoin society as "normal" citizens.

Lombroso was aware and supportive of this logic. In addition to his efforts to apply scientific principles to explaining the causes of crime, he also dreamed of "revamping the criminal justice system so that it would incorporate his criminology and react to offenders according to their degree of innate dangerousness" (Rafter, 2008, p. 83). For instance, he argued for "probation, juvenile reformatories, and other intermediate punishments that would keep offenders who were not particularly dangerous out of ordinary prisons" (Rafter, 2008, p. 83).

Notably, as America pushed into the 20th century, the appeal of biologically oriented theories eventually began to diminish. In their place, more optimistic positivist theories emerged that drew their images of offenders from psychology and especially from sociology. These newer approaches argued that the troubles of criminals could be rectified through counseling or by fixing the social environments in which they lived.

This new way of thinking about crime had a large effect on policy. Indeed, during the first two decades of the 20th century, this thinking helped to shape a campaign that renovated the criminal justice system. Reformers, called Progressives, argued that the system should be arranged not to punish offenders but rather to rehabilitate them. Across the nation, state after state established a separate juvenile court to "save" children from lives of crime (Platt, 1969). Similarly, efforts were undertaken to make release from prison based on the extent to which a person had been rehabilitated, not on the nature of the crime (as the classical school would mandate). Accordingly, states passed laws that made sentencing more indeterminate (e.g., offenders were given sentences that might range from 1 to 5 years) and created parole boards to decide which offenders had been "cured" and should be returned to the community. Probation, a practice through which offenders were to be both supervised and helped by officers of the court, also was implemented widely. More generally, criminal justice officials were given great discretion to effect the individualized treatment of offenders (Rothman, 1980).

We will return to these themes in Chapter 3 and later in the book. Even so, we need to emphasize two points. First, controversy still exists today on whether the policies instituted in the name of rehabilitation made criminal justice systems more humane or more repressive. Some criminologists believe that the discretion given to criminal justice officials allowed offenders to be abused (e.g., Rothman, 1980), whereas other criminologists insist that rehabilitation has helped to humanize a system that is puni-tive by nature (e.g., Cullen & Gilbert, 1982). Second, although the ideas of the early positivist theorists declined in popularity—but certainly did not disappear, as the Virginia sterilization example shows—we now are seeing a renewed interest in the idea that the origins of crime lie in unchangeable characteristics of individuals. As Chapter 11 will discuss, the 1980s brought a revitalization of the view that criminals are wicked by nature, a view that has had questionable, if not disquieting, policy implications.

While not (yet?) a part of the argument that criminals are wicked by nature, DNA as a powerful crime-solving and preventing tool was born during the 1980s (Wambaugh, 1989). Since then, it has created heated ethical and legal policy debates that again demonstrate the need to be cautious when implementing innovative biology-based policies. The most recent international debate surrounding the use of DNA originated in 2001 in England and Wales where it was then a matter of policy to gather and store fingerprints and the genetic footprint on all criminal suspects, including those who were found innocent. At that time, Britain's DNA Database contained profiles on more than 4.6 million people, some 860,000 of whom did not have criminal records. In late 2008, the European Court of Human Rights ruled unanimously that Britain's policy was a violation of the human right to privacy (Lyall, 2008, p. A19; see also Donovan & Klahm, 2009; Lynch, Cole, McNally, & Jordan 2008). Some people think that the United States should have a national DNA database that houses DNA on everyone—including the innocent and those guilty of crimes (Seringhaus, 2010). DNA is more fully dis-cussed in Chapter 14.

Conclusion

It should be clear that to understand the policies that are created to address the prob-lems of crime, we must realize that they are influenced greatly by the theories that guide them. We also must remember to examine the social context in which theories and policies are constructed. This lesson is important because it points out that theo-ries are neither value free nor free from the times in which they are advocated. Theories also have an impact on people. Sometimes the scientifically justified policies are allowed to do great harm because the social context in which they are advocated and implemented is ignored.

Closely related to this point is the claim made by some critical criminologists (see Chapter 9) that, although the work of the classical and positivist schools has a history of advancing the humanitarian ideas developed by the contributors to the Enlightenment, this is actually just one interpretation, albeit widely accepted. According to critical criminologists, one problem with this historical assessment is that it overlooks that the Enlightenment had a dark underbelly. "In the long-run," observes Lynch (2000), "Enlightenment perspectives can be seen as mechanisms that justify and legitimize capitalist social relationships, and it is no accident of history that

Enlightenment scholarship arose along with capitalist systems of production" (p. 148; see also Gaukroger, 2006). What this means is that while much of Enlightenment writing was about human freedoms, individualism, and human dignity, some of it can be interpreted as being based instead on the assumption that the "nature of man" was deterministic, not volitional. There are important implications that flow from this view of the Enlightenment and the early development of criminology.

For one thing, a deterministic view of humankind meant that scientific knowledge could be employed to control workers as well as criminals in order for society to be stable and for capitalism to be efficient and productive. According to some critical criminologists, the nascent "sciences of man"—especially Lombroso's work—were not humanitarian but were in fact tools to facilitate exploitation and oppression. To tell the story of the history of the Enlightenment and early criminology without this contextualization gives support to another objection supported by some critical criminologists.

The issue here—and the lesson to be learned—is that to rely too heavily on telling the story of criminology chronologically can give the appearance of it being a value-free objective science when in fact it is not. Rethinking the work of Lombroso and other positivists is illustrative of this point. On the surface, the positivists' contributions had the appearance of reflecting rigorous science with its emphasis on cataloging and measuring. In fact, more than this was likely involved. As Morrison (2004) observes, the positivists' approach was more akin to artwork that relied on "a cultural practice of presenting and representing *which implied* that the abnormal and the dangerous could be recognized and mapped in physical space and evolutionary time" (p. 68, emphasis added). What Lombroso and others actually did, it is argued, was to *attribute* or *project* their own perceptions about the meaning of physical characteristics, including tattoos, onto criminals. So, instead of actually identifying criminals, the "scientific method" that Lombroso and his followers used generated classed-based projections that described the physical features of members of the dangerous classes that threatened capitalism and bourgeois sensibilities. Criminology is thus part of the "apparatus of the 'science of oppression' established by Enlightenment philosophy" (Lynch, 2000, p. 147). In this way, critical scholars argue, the scientific method with its emphasis on measurement—coupled with the use of the camera to objectify native people and advance imperialism—produced pictures of the criminal as the "other" (Morrision, 2006; see also Gaukroger, 2006).

We are keenly aware that despite its many advantages in advancing criminological knowledge, the scientific method's misuse potential to identify, objectify, and oppress remains part of modern social life. The contemporary use of science to not define sexual and racial minorities as "the other" is illustrative (see Henry & Tator, 2002; Mohr, 2008–2009; Tucker, 1994). One of most important tasks for criminologists is to expose, critique, and condemn such usage wherever it is found.

Clifford R. Shaw
1895–1957
Chicago Area Project
Author of Social Disorganization Theory

3

Rejecting Individualism

The Chicago School

Although exceptions exist, most early theories located the sources of crime within the individual. These theories differed markedly on where precisely the source of waywardness lay. Was it in the soul? The mind? The body's very biological makeup? Even so, these theories shared the assumption that little insight on crime's origins could be gained by studying the social environment or context external to individuals. In one form or another, these early theories blamed individual offenders—not society—for the crime problem.

But as the United States entered the 20th century, a competing and powerful vision of crime emerged—a vision suggesting that crime, like other behavior, was a social product. The earlier theories did not vanish immediately or completely; indeed, in important ways, they continue to inform current-day thinking. But they did suffer a stiff intellectual challenge that greatly thinned the ranks of their supporters. This major theoretical shift, one that rejected individualist explanations of crime in favor of social explanations, might have been expected. Society was undergoing significant changes, and people's experiences were changing as well. The time was ripe for a new understanding of why some citizens break the law.

By the end of the 1930s, two major criminological traditions had been articulated that sought, in David Matza's (1969) words, to "relocate pathology; it was moved from the personal to the social plane" (p. 47). The first of these traditions, the Chicago school of criminology, argued that one aspect of American society, the city, contained potent criminogenic forces. The other tradition, Robert K. Merton's (1938) strain theory, contended that the pathology lay not in one ecological location (e.g., the city) but rather in the broader cultural and structural arrangements that constitute America's social fabric. Although they differed in how they believed that society created lawbreakers, these theories agreed that the key to unlocking the mystery of crime was in understanding its social roots. Taken together, they offered a strong counterpoint to explanations that blamed individuals for their criminality.

The effects of these two schools of thought have been long-lasting. Even today, more than seven decades after their initial formulation, the Chicago school and strain theories continue to be of interest to criminologists and to shape correctional policies. They deserve careful consideration. Accordingly, in Chapter 4 to follow, we consider the origins and enduring influence on criminological theory of strain theory. First, however, we explore here in Chapter 3 how a group of scholars located in the Chicago area sought to understand the concentration of crime in certain neighborhoods. As we will see, their investigations would result in a major school of criminology and lay the groundwork for important contemporary theories of crime.

The Chicago School of Criminology: Theory in Context

What made it seem reasonable—why did it make sense—to blame the city for the nation's crime problem? Why would such a vision become popular in the 1920s and 1930s and, moreover, find special attention in Chicago?

The answers to these questions can be found in part in the enormous changes that transformed the face of the United States and made the city—and not the "little house on the prairie"—the nation's focal point. During the latter half of the 1800s, cities grew at a rapid pace and became, as Palen (1981) observed, "a controlling factor in national life" (p. 63). Between 1790 and 1890, for example, the urban population grew 139-fold; by 1900, 50 cities existed with populations in excess of 100,000 (p. 63).

But Chicago's growth was particularly remarkable. When the city incorporated in 1833, it had 4,100 residents; by 1890, its population had risen to 1 million; and by 1910, the count surpassed 2 million (Palen, 1981, p. 63). But such rapid expansion had a bleaker side. Many of those settling in Chicago (and in other urban areas) carried little with them; there were waves of immigrants, displaced farmworkers, and African Americans fleeing the rural South. For most newcomers, the city—originally a source of much hope—brought little economic relief. They faced a harsh reality—pitiful wages; working 12-hour days, 6 days a week, in factories that jeopardized their health and safety; living in tenements that "slumlords built jaw-to-jaw . . . on every available space" (p. 64). Writing on the meatpacking industry in Chicago, Upton Sinclair (1906) gave this environment a disquieting label: "the jungle."

Like other citizens, criminologists during the 1920s and 1930s witnessed—indeed, lived through and experienced—these changes that created bulging populations and teaming slum areas. It was only a short leap for them to believe that growing up in the city, particularly in the slums, made a difference in people's lives. In this context, crime could not be seen simply as an individual pathology; it made more sense when viewed as a social problem.

This conclusion, moreover, was reinforced by a broad liberal reform movement that arose early during the 1900s—the Progressive movement. Although they believed in the essential goodness of America and so rejected calls for radical change, Progressives were critical of the human costs wrought by America's unbridled industrial growth. They were troubled particularly by the plight of the urban poor, a mushrooming population of the system's casualties who had few prospects of stable or rewarding lives. They worried, as Rothman (1980) wrote, that the "promise" of the American system "did not extend evenly to all segments of the society; it did not penetrate the ghetto or the slum. Thus, an understanding of the etiology of crime demanded a very close scrutiny of the conditions in these special enclaves" (p. 51).

Criminologists in the Chicago school would echo this conclusion. The Progressives rejected the social Darwinists' logic that the poor, and the criminals among them, were biologically inferior and had fallen to society's bottom rung because they were of lesser stock. The Progressives preferred a more optimistic interpretation: The poor were pushed by their environment—not born—into lives of crime. Accordingly, hope existed that changing the context that nurtures offenders would reverse the slums' negative effects and transform these individuals into law-abiding citizens. In particular, the goal was to save the poor, particularly their children, by providing social services—schools, clinics, recreational facilities, settlement houses, foster homes, and reformatories (if necessary)—that would lessen the pains of poverty and teach the benefits of middle-class culture (Platt, 1969; Rothman, 1980).

But the moral imperative was to act on this belief—and the Progressives did, creating what came to be known as the "age of reform" (Hofstadter, 1955a, 1963). The linchpin of their agenda was the assumption that the government could be trusted to create and administer agencies that would effect needed social reform. The Progressives campaigned to have the state guide the nation toward the common good by controlling the greed of industry and by providing the assistance that the poor needed to reach the middle class. In the area of criminal justice, their efforts led to the creation of policies and practices that were intended to allow the state to treat the individual needs and problems of offenders—the juvenile court, community supervision through probation and parole, and indeterminate sentences (Rothman, 1980).

Thus, during the first decades of the 1900s, the city became a dominant feature of American life, and a pervasive movement arose warning that the social fabric of urban slums bred crime. Still, the question remains as to why Chicago became a hotbed of criminological research. As suggested, part of the answer can be found in this city's status as an emerging economic and population center. But the other piece of the puzzle lies in the existence, at the University of Chicago, of the nation's oldest sociology program, established in 1892 (Bulmer, 1984).

By the 1920s, "surrounded . . . with ever-present reminders of the massive changes that were occurring within American society," the department's faculty and students had embarked on efforts to systematically study all aspects of the urban laboratory that lay before them (Pfohl, 1985, p. 143). Robert E. Park, a newspaper reporter-turned-sociologist, was particularly influential in shaping the direction of this work. He commented, "I expect I have actually covered more ground tramping about in cities in different parts of the world than any other living man" (cited in Madge, 1962, p. 89). These journeys led Park to two important insights.

First, Park concluded that the city's development and organization, like any ecological system, were not random or idiosyncratic but rather patterned and, therefore, could be understood in terms of basic social processes such as invasion, conflict, accommodation, and assimilation. Second, he observed that the nature of these social processes and their impact on human behavior, such as crime, could be ascertained only through careful study of city life. Accordingly, he urged students and colleagues to venture into Chicago and to observe firsthand its neighborhoods and diverse conglomeration of peoples (Madge, 1962). Several scholars, most notably Clifford R. Shaw and Henry D. McKay, embraced Park's agenda and explored how urban life fundamentally shaped the nature of criminal activity. In so doing, they laid the foundation for the Chicago school of criminology.

Shaw and McKay's Theory of Juvenile Delinquency

Shaw and McKay were not faculty members at the University of Chicago; rather, they were employed as researchers for a state-supported child guidance clinic. Even so, they enjoyed close relationships with the sociology department—they had been students there but did not finish their doctorates—and were influenced profoundly by its theorizing (Snodgrass, 1976). In particular, they were persuaded that a model of the city formulated by Ernest Burgess, Park's colleague and collaborator, provided a framework for understanding the social roots of crime. Indeed, it was Burgess's model that led them to the conclusion that neighborhood organization was instrumental in preventing or permitting delinquent careers (Gibbons, 1979; Pfohl, 1985). We will review this general model of urban growth and then consider how it guided Shaw and McKay's approach to studying delinquency in Chicago.

BURGESS'S CONCENTRIC ZONE THEORY

As cities expand in size, how do they grow? One answer is that the growth is haphazard—not according to any set pattern. But Burgess, like Park, rejected this view in favor of the hypothesis that urban development is patterned socially. He contended that cities "grow radially in a series of concentric zones or rings" (Palen, 1981, p. 107).

As Figure 3.1 shows, Burgess (1967/1925) delineated five zones. Competition determined how people were distributed spatially among these zones. Thus, commercial enterprises were situated in the "loop" or central business district, a location that afforded access to valuable transportation resources (e.g., railroads, waterways). By

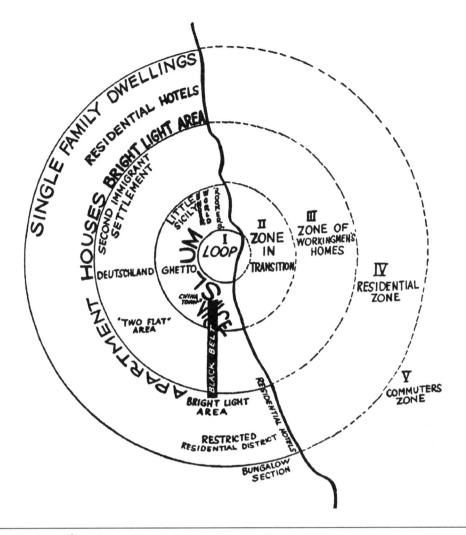

Figure 3.1 Urban Areas

contrast, most high-priced residential areas were in the outer zones, away from the bustle of the downtown, away from the pollution of factories, and away from the residences of the poor.

But the zone in transition was a particular cause for concern and study. This zone contained rows of deteriorating tenements, often built in the shadow of aging factories. The push outward of the business district, moreover, led to the constant displacement of residents. As the least desirable living area, the zone had to weather the influx of waves of immigrants and other migrants who were too poor to reside elsewhere.

Burgess observed that these social patterns were not without consequences. They weakened the family and communal ties that bound people together and resulted in *social disorganization*. Burgess and the other Chicago sociologists believed that this disorganization was the source of a range of social pathologies, including crime.

DISORGANIZATION AND DELINQUENCY

Burgess's model was parsimonious and persuasive, but did it really offer a fruitful approach to the study of crime? Would it stand up to empirical testing?

Shaw and McKay took it on themselves to answer these questions. As a first step, they sought to determine whether crime rates would conform to the predictions suggested by Burgess's model—highest rates in the zone in transition, with this rate declining progressively as one moved outward to the more affluent communities. Through painstaking research, they used juvenile court statistics to map the spatial distribution of delinquency throughout Chicago.

Shaw and McKay's data analysis confirmed the hypothesis that delinquency flourished in the zone in transition and was inversely related to the zone's affluence and corresponding distance from the central business district. By studying several decades of Chicago's court records, they also were able to show that crime was highest in slum neighborhoods regardless of which racial or ethnic group resided there. Further, they were able to show that as groups moved to other zones, their rates decreased commensurately. This observation led to the inescapable conclusion that it was *the nature of the neighborhood—not the nature of the individuals within the neighborhood*—that regulated involvement in crime.

But what social process could account for this persistent spatial distribution of delinquency? Borrowing heavily from Burgess and the other Chicago sociologists, Shaw and McKay emphasized the importance of neighborhood organization in preventing or permitting juvenile waywardness. In more affluent communities, families fulfilled youths' needs and parents carefully supervised their offspring. But in the zone in transition, families and other conventional institutions (e.g., schools, churches, voluntary associations) were strained, if not broken apart, by rapid and concentrated urban growth, people moving in and out (transiency), the mixture of different ethnic and racial groups (heterogeneity), and poverty; *social disorganization prevailed*. As a consequence, juveniles received neither the support nor the supervision required for wholesome development. Left to their own devices, slum youths were freed from the type of social controls operative in more affluent areas; no guiding force existed to stop them from seeking excitement and friends—perhaps the wrong kind of friends—in the streets of the city.

This view of delinquency causation, it should be noted, likely resonated with the personal experiences of Shaw and McKay. As Snodgrass (1976) observes, they were:

> two farm boys who . . . were both born and brought up in rural mid-western areas of the United States, both received Christian upbringings, and both attended small, denominational country colleges. Shaw was from an Indiana crossroads that barely constituted a town, and McKay was from the vast prairie regions of South Dakota. (p. 2)

As a result, they were raised in communities where people were alike (homogeneity), shared the same Christian values, lived most of their lives (stability), knew one another well if not intimately, and made sure that youngsters were good kids lest their parents be told of this misconduct. In short, their childhood communities were organized and marked by social control. It thus "made sense" that the absence of the routines, social intimacy, and virtues of small town life in the slums of Chicago—the *disorganization* that prevailed—would be implicated in the causation of delinquency (Snodgrass, 1976).

Importantly, Shaw and McKay's focus on how *weakening controls* make possible a delinquent career allowed them to anticipate a criminological school that eventually would become known as *control* or *social bond* theory (see Chapters 5 and 6). As Kornhauser (1978) observed, however, they believed that another social circumstance also helped to make slum neighborhoods especially criminogenic. We turn next to this aspect of their thinking.

TRANSMISSION OF CRIMINAL VALUES

Shaw and McKay did not confine their research to the epidemiology of delinquency. Following Park's admonition, they too "tramped" about Chicago. As we will see, they were activists who were involved in efforts to prevent delinquency. They also attempted to learn more about why youths become deviant by interviewing delinquents and compiling their autobiographies in a format called life histories. These efforts led to the publication of titles such as *The Jack-Roller: A Delinquent Boy's Own Story* (Shaw, 1930), *The Natural History of a Delinquent Career* (Shaw, 1931), and *Brothers in Crime* (Shaw, 1938; see also Shaw & McKay, 1972).

These life histories contained an important revelation: Juveniles often were drawn into crime through their association with older siblings or gang members. This observation led Shaw and McKay (1972) to the more general conclusion that disorganized neighborhoods helped to produce and sustain "criminal traditions," which competed with conventional values and could be "transmitted down through successive generations of boys, much the same way that language and other social forms are transmitted" (p. 174). Thus, slum youths grew up in neighborhoods characterized by "the existence of a coherent system of values supporting delinquent acts" (p. 173) and could learn these values readily in their daily interactions with older juveniles. By contrast, youths in organized neighborhoods, where the dominance of conventional institutions precluded the development of criminal traditions, remained insulated from deviant values and peers. Accordingly, for them, delinquent careers were an unlikely option.

THE EMPIRICAL STATUS OF SOCIAL DISORGANIZATION THEORY

As noted previously, Shaw and McKay were able to collect data showing that crime was distributed across neighborhoods in a pattern consistent with social disorganization theory. Still, this theory now is decades old, and criminologists today have more sophisticated multivariate techniques to assess whether the factors identified by Shaw and McKay are able to explain why some geographical areas have higher rates of offending than others. In contemporary times, does Shaw and McKay's work have anything meaningful to tell us about communities and crime?

Pratt and Cullen (2005; see also Pratt, 2001) completed a comprehensive meta-analytic review of the existing research on social disorganization theory. As Pratt and Cullen noted, a difficulty in assessing this theory is that most research has examined the structural causes of social disorganization—poverty, racial and ethnic heterogeneity, residential mobility, urbanism/structural density, family disruption, and so on—but not social disorganization itself directly. With this qualification, Pratt and Cullen's analysis reveals that the variables specified by Shaw and McKay generally are related to crime rates in the predicted direction.

But perhaps the strongest support for Shaw and McKay's theory comes from the now classic study of Sampson and Groves (1989), which measured not only structural variables but also social disorganization. Sampson and Groves tested the theory using data drawn from the 1982 British Crime Survey that covered 238 localities in England and Wales and included more than 10,000 respondents. Their empirical model included measures of low socioeconomic status, heterogeneity, mobility, family disruption, and urbanism. It also included three measures of whether a locality was socially organized or disorganized: the strength of local friendship networks, residents' participation in community organizations, and the extent to which the neighborhood had unsupervised teenage peer groups. Consistent with Shaw and McKay's theory, Sampson and Groves found that structural factors increased social disorganization and that, in turn, disorganized areas had higher levels of crime than organized areas. Notably, their analysis was replicated using a later version of the British Crime Survey (Lowenkamp, Cullen, & Pratt, 2003; but see also Veysey & Messner, 1999). Taken together, these findings lend support to Shaw and McKay's social disorganization theory.

SUMMARY

Shaw and McKay believed that juvenile delinquency could be understood only by considering the social context in which youths lived—a context that itself was a product of major societal transformations wrought by rapid urbanization, unbridled industrialization, and massive population shifts. Youths with the misfortune of residing in the socially disorganized zone in transition were especially vulnerable to the temptations of crime. As conventional institutions disintegrated around them, they were given little supervision and were free to roam the streets, where they likely would become the next generation of carriers for the neighborhood's criminal tradition. In short, when growing up in a disorganized area, it is this combination of (1) a breakdown of control and (2) exposure to a criminal culture that lures individual youngsters into crime and, across all juveniles, that creates high rates of delinquency.

Later in this chapter, we will see how this vision of crime led Shaw and McKay to assert that delinquency prevention programs must be directed at reforming communities, not simply reforming individuals. First, however, we consider how their work laid the groundwork for Edwin H. Sutherland's classic theory of *differential association*.

Sutherland's Theory of Differential Association

In 1906, Sutherland departed his native Nebraska and traveled to the University of Chicago, where he enrolled in several courses in the divinity school. He was also persuaded to register for Charles R. Henderson's course, "Social Treatment of Crime." Henderson took an interest in his new student—an interest that proved mutual. It was not long before Sutherland decided to enter the sociology program. He went on to devote the remainder of his career to exploring the social roots of criminal behavior (Geis & Goff, 1983, 1986; Schuessler, 1973).

After receiving his doctorate in 1913, Sutherland held a series of academic positions at Midwestern institutions, including the University of Illinois and the University of Minnesota. In 1930, he was offered and accepted a professorship at the University of Chicago. His stay in Chicago proved short-lived. Apparently disenchanted with his position—he cited "certain distractions" as the reason for his departure—Sutherland left 5 years later to join the sociology department at Indiana University, a post he held until his death in 1950. Even so, he maintained contact with his friends in Chicago including McKay (Geis & Goff, 1983, p. xxviii; Schuessler, 1973, pp. xi–xii).

Although Sutherland spent most of his career away from the city and its university, the Chicago brand of sociology intimately shaped his thinking about crime. Indeed, as we will see next, much of his theorizing represented an attempt to extend and formalize the insights found in the writings of Shaw and McKay as well as other Chicago school scholars (see, e.g., Thrasher, 1927/1963).

DIFFERENTIAL SOCIAL ORGANIZATION

Like most Chicago criminologists, Sutherland (1939) rejected individualist explanations of crime. "The neo-Lombrosian theory that crime is an expression of psychopathology," he claimed, "is no more justified than was the Lombrosian theory that criminals constitute a distinct physical type" (p. 116). Instead, he was convinced that social organization—the context in which individuals are embedded—regulates criminal involvement.

Shaw and McKay had used the term *social disorganization* to describe neighborhoods in which controls had weakened and criminal traditions rivaled conventional institutions. At the suggestion of Albert Cohen, however, Sutherland substituted for social disorganization the concept of *differential social organization*, a term that he believed was less value laden and captured the nature of criminal areas more accurately. Thus, Sutherland (1942/1973) contended that social groups are arranged differently; some are organized in support of criminal activity, whereas others are organized against such behavior. In turn, he followed Shaw and McKay's logic in proposing that lawlessness would be more prevalent in those areas where criminal organization had taken hold and where people's values and actions were shaped on a daily basis.

DIFFERENTIAL ASSOCIATION

Although Sutherland incorporated into his thinking the thesis that community or group organization regulates rates of crime, he built more systematically on Shaw and McKay's observation that delinquent values are transmitted from one generation to the next. For Sutherland, to say that the preference for crime is "culturally transmitted" was, in effect, to say that criminal behavior is *learned* through social interactions.

To describe this learning process, Sutherland coined the concept of *differential association*. Much like other Chicago school scholars, he noted that, especially in the inner-city areas, there was culture conflict. Two *different* cultures—one criminal, one conventional—vied for the allegiance of the residents; the key was which culture, which set of definitions, an individual most closely associated with. Thus, Sutherland contended that any person would inevitably come into contact with "definitions favorable to violation of law" and with "definitions unfavorable to violation of law." The *ratio* of these definitions or views of crime—whether criminal or conventional influences are stronger in a person's life—determines whether the person embraces crime as an acceptable way of life.

Sutherland held that the concepts of differential association and differential social organization were compatible and allowed for a complete explanation of criminal activity. As a social-psychological theory, differential association explained why any given individual was drawn into crime. As a structural theory, differential social organization explained why rates of crime were higher in certain sectors of American society: Where groups are organized for crime (e.g., in slums), definitions favoring legal violations flourish; therefore, more individuals are likely to learn—to differentially associate with—criminal values.

Sutherland's theory of differential association went through various stages of development, but by 1947 he was able to articulate in final form a set of nine propositions. These propositions compose one of the most influential statements in criminological history on the causes of crime:

1. Criminal behavior is learned.

2. Criminal behavior is learned in interaction with other persons in a process of communication.

3. The principal part of the learning of criminal behavior occurs within intimate personal groups.

4. When criminal behavior is learned, the learning includes (a) techniques of committing the crime, which sometimes are very complicated, sometimes are very simple; [and] (b) the specific direction of motives, drives, rationalizations, and attitudes.

5. The specific direction of motives and drives is learned from definitions of legal codes as favorable and unfavorable.

6. A person becomes delinquent because of an excess of definitions favorable to violation of law over definitions unfavorable to violation of law. This is the principle of differential association.

7. Differential associations may vary in frequency, duration, priority, and intensity.

8. The process of learning criminal behavior by association with criminal and anti-criminal patterns involves all the mechanisms that are involved in any other learning.

9. While criminal behavior is an expression of general needs and values, it is not explained by those general needs and values since noncriminal behavior is an expression of the same needs and values. (Sutherland & Cressey, 1970, pp. 75–76)

THEORETICAL APPLICATIONS

Taken together, these propositions convey an image of offenders that departs radically from the idea that criminals are pathological creatures driven to waywardness by demons, feeble minds, deep-seated psychopathology, and/or faulty constitutions. Instead, Sutherland was suggesting that the distinction between lawbreakers and law-abiding people lies not in their personal fiber but rather in the content of what they have learned. Those with the good fortune of growing up in conventional neighborhoods will learn to play baseball and attend church services; those with the misfortune of growing up in slums will learn to rob drunks and roam the streets looking to do mischief.

But could the theory of differential association account for all forms of crime? Sutherland believed that he had formulated a general explanation that could be applied to very divergent types of illegal activity. Unlike Shaw and McKay, he did not confine his investigations to the delinquency of slum youths. For example, he compiled his famous life history of Chic Conwell, a "professional thief" (Sutherland, 1937). This study showed convincingly that differential association with thieves was the critical factor in determining whether a person could become a pickpocket, a shoplifter of high-priced items, or a confidence man or woman. Such contact was essential because it provided aspiring professional thieves with the tutelage, values, and colleagues needed to learn and perform sophisticated criminal roles.

But more provocative was Sutherland's (1949) claim that differential association could account for the offenses "committed by a person of respectability and high social status in the course of his [or her] occupation" (p. 9)—illegal acts for which Sutherland (1940) coined the term "white-collar crimes." His investigations revealed that lawlessness is widespread in the worlds of business, politics, and the professions. As Sutherland (1949/1983) put it, "Persons in the upper socioeconomic class engage in much criminal behavior" (p. 7). Indeed, his own research into the illegal acts by large American corporations revealed that they violated legal standards frequently and that most could be termed "habitual criminals" (Sutherland, 1949).

This empirical reality, Sutherland (1940) observed, presented special problems for most theories of his day, which assumed that "criminal behavior in general is due either to poverty or to the psychopathic and sociopathic conditions associated with poverty." After all, most "white-collar criminals . . . are not in poverty, were not reared in slums or badly deteriorated families, and are not feebleminded or psychopathic" (pp. 9–10). By contrast, the principle of differential association can explain the criminality of the affluent.

Thus, in many occupations, illegal practices are widely accepted as a way of doing business. White-collar workers, Sutherland (1940) noted, might "start their careers in good neighborhoods and good homes [and then] graduate from colleges with some idealism." At that point, however, they enter "particular business situations in which criminality is practically a folkway and are inducted into that system of behavior just as into any other folkway" (p. 11). Similar to slum youths and offenders who become professional thieves, their association with definitions favorable to violation of law eventually shapes their orientations and transforms them from white-collar workers into white-collar criminals. In effect, a criminal tradition has been transmitted. We revisit Sutherland's insights on these topics in Chapter 11, where we consider theories of white-collar crime.

The Chicago School's Criminological Legacy

The Chicago school has not escaped the critical eye of subsequent scholars. One limitation that critics often note, for example, is that the Chicago criminologists emphasized the causal importance of the transmission of a "criminal culture" but offered much less detail on the precise origins of this culture. Similarly, although deploring the negative consequences of urban growth, such as crime and delinquency, the Chicago theorists tended to see the spatial distribution of groups in the city as a "natural" social process. This perspective diverted their attention from the role that power and class domination can play in creating and perpetuating slums and the enormous economic inequality that pervades urban areas.

Scholars also have questioned whether the Chicago school can adequately account for all forms of crime. The theory seems best able to explain involvement in stable criminal roles and in group-based delinquency but is less persuasive in providing insights on the cause of "crimes of passion" or other impulsive offenses by people who have had little contact with deviant values. Sutherland's theory of differential association, moreover, has received special criticism. The formulation is plausible and perhaps correct, but can it be tested scientifically? Would it ever be possible to accurately measure whether, over the course of a lifetime, a person's association with criminal definitions outweighed his or her association with conventional definitions (Empey, 1982; Pfohl, 1985; Vold & Bernard, 1986)?

Despite these limitations, few scholars would dispute that the Chicago school has had a profound influence on criminology. At the broadest level, the Chicago criminologists leveled a powerful challenge—backed by a wealth of statistics—to explanations that saw crime as evidence of individual pathology. They captured the truth that where people grow up and with whom they associate cannot be overlooked in the search for the origins of crime.

The Chicago school also laid the groundwork for the development of two perspectives that remain vital to this day. On the one hand, as indicated previously, Shaw and McKay's premise that weakening social controls permit delinquency to take place was an early version of what since has become known as *control* or *social bond* theory (see Chapters 5 and 6). On the other hand, the Chicago criminologists' thesis that criminal behavior occurs as a consequence of cultural transmission or differential association gave rise to *cultural deviance* theory, a perspective that assumes that people become criminal by learning deviant values in the course of social interactions (Empey, 1982).

COLLECTIVE EFFICACY

Robert Sampson has perhaps done the most to revitalize Shaw and McKay's view that the degree of informal control exercised by residents will affect the extent of a community's crime problem (see, e.g., Sampson, 1986a; Sampson & Groves, 1989). While a graduate student at the University at Albany in the late 1970s, he enrolled in a seminar with Travis Hirschi. As will be seen in Chapter 6, Hirschi was an advocate of control theory, arguing at this time that weakened bonds to parents, schools, and other aspects of conventional society permitted delinquency. This view was a rejection of that side of the Chicago school that linked crime to the transmission or learning of cultural values (Sampson, 2011).

Sampson was persuaded by his professor's logic. However, he was not interested in why one individual rather than another engaged in crime, which was Hirschi's concern. This is sometimes called a micro-level question, because the focus is on individual differences in criminal conduct. Rather, Sampson was more fascinated by the core macro-level question that had preoccupied scholars like Shaw and McKay: why one neighborhood or area (called a macro-level unit) rather than another had higher crime rates.

Importantly, in Hirschi's seminar, Sampson was introduced to a brilliant, newly published book by Ruth Kornhauser (1978). In *Social Sources of Delinquency*, Kornhauser shared Hirschi's critique of cultural theories of crime—likely one reason why Hirschi assigned the book in his class! But unlike Hirschi, she mainly focused on the macro-level control side of social disorganization theory as an alternative explanation. Kornhauser's distinctive focus led Sampson to the works of Shaw and McKay and of other scholars within the Chicago school. Sampson (2011) thus made the career choice to focus on macro-level control theory. As Sampson notes, he divides his academic life in this way:

> before I read Kornhauser and after I read Kornhauser. . . . I can still feel the intellectual jolt. The occasion was a graduate seminar with Travis Hirschi on theories of deviance. It was in that seminar that a light bulb turned on. The switch was our close reading of Ruth Kornhauser's *Social Sources of Delinquency*. . . .She went on in devastating fashion to critique criminological theories in what she called cultural deviance and strain

traditions. . . . More than Kornhauser's critique inspired me, however. She also articulated a positive theoretical framework that drew on the classic Chicago School. . . .As I saw it, her argument was closely aligned with a form of Hirschi's control theory pitched at the macro level. (2011, pp. 64, 67–68)

As seen above, Sampson contributed a classic test of Shaw and McKay's work (Sampson & Groves, 1989). Following this study, he continued to explore these issues. In 1997, with Stephen Raudenbush and Felton Earls, he set forth an exciting reconceptualization of the social disorganization framework called *collective efficacy theory* (see also Sampson, 2006). This perspective continues to shape criminological thinking and research today.

Sampson et al. (1997) observed that neighborhoods vary in their ability to "activate informal social control." Informal social control involves residents' behaving proactively—not passively—when they see wayward behavior, such as by calling police authorities, coming to the rescue of someone in trouble, and telling unruly teenagers to quiet down and behave. The likelihood that residents will take such steps, however, is contingent on whether there is "mutual trust and solidarity among neighbors" (p. 919). As a result, in neighborhoods where such cohesiveness prevails, residents can depend on one another to enforce rules of civility and good behavior. Such places have "collective efficacy, defined as social cohesion among neighbors combined with their willingness to intervene on behalf of the common good" (p. 918).

With his colleagues, Sampson (2011) carefully selected the words "collective" and "efficacy" to make up the construct of *collective efficacy*. These terms were meant to capture culture and social action. Thus, collective was meant to suggest that the residents in an area had a *shared expectation* for control—that is, they could count on neighbors to agree that certain situations—for example, a group of boisterous teens—were inappropriate and deserving of a reaction. In turn, efficacy was meant to suggest that the residents could count on their neighbors to exert *human agency* and actually to do something to solve the problem (e.g., go out on the street and tell the youths to quiet down; demand that the police close down a drug market). Put another way, neighbors with collective efficacy define problems similarly and are capable of jointly taking action to solve them.

Sampson et al. (1997) also argued that collective efficacy is not evenly distributed across neighborhoods. Rather, in communities marked by a concentration of immigrants, residential instability, and the grinding economic deprivation of "concentrated disadvantage," collective efficacy is weak (see also Sampson, Morenoff, & Earls, 1999). Sampson et al. (1997) predicted that these communities will not have the social capital to assert informal social controls and to keep the streets safe.

Importantly, Sampson et al. (1997) provided data to back up these theoretical claims. In 1995, their research team interviewed 8,762 people who lived in 343 Chicago neighborhoods. Controlling for the personal characteristics of the respondents (sometimes called "composition effects"), the authors found that collective efficacy was a "robust" predictor of levels of violence across neighborhoods. Their analysis also revealed that collective efficacy "mediated" much of the relationship between crime and the neighborhood characteristics of residential stability and concentrated disadvantage, a finding that "is consistent with a major theme in neighborhood theories of social orga-nization" (p. 923).

The theory of collective efficacy, therefore, appears to be a promising explanation of why urban neighborhoods differ in their levels of criminal behavior (see also Pratt & Cullen, 2005; Sampson, 2006). A remaining issue, however, is whether the concept of collective efficacy offers a truly novel concept or whether it is really just the "opposite side" of social disorganization. A distinctive feature of the concept appears to be its focus not merely on the degree of neighborhood organization but also on the willing-ness of residents to *activate* social control. In other words, the concept of "efficacy" implies not merely a state of being socially organized but rather a state of being ready for *social action*. The future of this theory is likely to hinge on whether Sampson et al. continue to clarify the concept of collective efficacy and demarcate its components.

CULTURAL DEVIANCE THEORY

Theoretical Variations. Cultural deviance theory has evolved along a number of paths, but we can identify three particularly influential versions. First, some criminologists have asserted that lower-class culture as a whole—not subcultures within lower-class areas—is responsible for generating much criminality in urban areas. Walter Miller offered perhaps the clearest and most controversial of these theories. According to Miller (1958), urban gang delinquency is not a product of intergenerational poverty per se but rather is a product of a distinct lower-class culture whose "focal concerns" encourage deviance rather than conformity. If the focal concerns of middle-class cul-ture are achievement, delayed gratification, and hard work, then the lower-class coun-terparts are trouble, smartness, toughness, fate, and autonomy. As a result, middle-class youths are oriented toward good grades, college, and career; lower-class youths are oriented toward physical prowess, freedom from any authority, and excite-ment on the streets. Miller contended that, not surprisingly, youths who adhere to such *"cultural practices which comprise essential elements of the total life pattern of lower class culture automatically [violate] legal norms"* (p. 18, emphasis in original).

Second, by contrast, a number of criminologists have explored how delinquent *sub-cultures* arise in particular sectors of society (urban lower-class areas). These subcul-tures are, in effect, relatively coherent sets of antisocial norms, values, and expectations that, when transmitted or learned, motivate criminal behavior (Cloward & Ohlin, 1960; Cohen, 1955). We will return to this line of analysis later in Chapter 4.

Third, other researchers have developed a similar theme in arguing for the exis-tence of *subcultures of violence*. Wolfgang and Ferracuti (1982), for example, noted that in areas where such a subculture has taken hold (e.g., urban slums), people acquire "favorable attitudes toward . . . the use of violence" through a "process of dif-ferential learning, association, or identification." As a result, "the use of violence . . . is not necessarily viewed as illicit conduct, and the users do not have to deal with feel-ings of guilt about their aggression" (p. 314). In this regard, there is considerable the-oretical and empirical debate over whether high rates of violent crime and homicide in certain sectors of society, such as the inner cities and the South, can be attributed to a geographically based subculture of violence (Cao, Adams, & Jensen, 1997; Hawley & Messner, 1989).

Code of the Street. An important application of the subculture of violence concept can be found in Elijah Anderson's acclaimed book, *Code of the Street.* This work is based on Anderson's (1999) 4 years of ethnographic research in Philadelphia in which he took up the problem of "why it is that so many inner-city young people are inclined to commit aggression and violence toward one another" (p. 9). In essence, he argued that the answer to this problem lies in the violent "code" that prevails in the inner city and that governs the choices that adolescents make in their daily lives.

According to Anderson (1999), minority youths in the inner city are culturally isolated—cut off from conventional society—and face daunting economic barriers. Most families struggle to be "decent" and to try to impart the values of hard work and civility to their children. Youths in other households, which Anderson calls "street families," are less fortunate. They are born into families that are disrupted and dysfunctional. Most often, these families are headed by single mothers who at times might have drug problems. The children are neglected and, when disciplined, typically receive harsh, physical, and erratic punishment. With little prospect of participating meaningfully in mainstream society, they grow up alienated and embittered. Weakly bonded to conventional institutions, angry, and ineffectively parented, the youths of these families turn to the streets. There, they spend their days and, frequently, their nights.

With so little in life available to them, these youths' major project is to "campaign for respect." They seek to display their status on the streets through flamboyant dress, through a very masculine demeanor, and (most important) by developing reputations for "nerve"—as youths who are "not to be messed with" and who are "bad." But self-respect based on reputation is precarious because it is open to challenge and can be taken from a youth in a zero-sum contest. If one teen "disrespects" another, then a failure to respond constitutes a failure to show "nerve" and a loss of status.

In such situations, a "code of the street" shapes how the "disrespected" party should react. Anderson (1999) defined this code as

> a set of informal rules governing interpersonal public behavior, particularly violence. The rules prescribe both proper comportment and the proper way to respond if challenged. They regulate the use of violence and so supply a rationale allowing those inclined to aggression to precipitate violent encounters in an approved way. (p. 33)

In other sectors of society, inadvertent or slight affronts might be overlooked or might evoke verbal responses showing unhappiness. But the code of the street brooks no such affronts. As Anderson (1999) observed, "In street culture, respect is viewed as almost an external entity, one that is hard-won but easily lost—and so must constantly be guarded" (p. 33). Thus, the code demands that virtually any form of disrespect should be met with the immediate threat or application of physical violence, lest respect be forfeited. It also mandates that physical violence should be met with violence. In such contexts, violent encounters are an ongoing possibility, and when they occur, they are at risk of escalating into lethal exchanges.

Anderson (1999) cautioned that the code affects not only kids from street families but also those from decent families. The "decent" youths cannot be sheltered by their families forever; they eventually must venture into public spaces where youths from all families mix. If they disobey the code and fail to show the willingness to use violence (i.e., if they achieve no respect), then they will be easy prey for other youths to victimize—to insult, take their possessions, and assault. Such decent kids thus must learn and obey the code to survive, showing enough willingness to use violence to deter the constant threat of victimization. The cost of embracing these street subcultural values, however, is that they risk being drawn into violent confrontations that are inconsistent with their decent way of life. Therefore, youths from all families—street and decent kids alike—are encapsulated by the code of the street and suffer its consequences.

Anderson (1999) argued that the code of the street, although firmly entrenched, is not intractable. Its strength and sway over inner-city youths are rooted in the structural conditions that expose these youngsters to hurtful deprivations and strip them of any meaningful way of gaining respect through conventional avenues. This situation is exacerbated by a lack of trust in the police and by the accompanying sense that problems, including criminal victimization, must be dealt with alone. In the end, the code of the street is a "cultural adaptation" to the conditions prevailing in destitute urban communities. If there is a ray of hope in this portrait, it is that different conditions might evoke different cultural adaptations. For Anderson, the most salient step is giving youths hope and a meaningful stake in conformity. He concluded, "Only by reestablishing a viable mainstream economy in the inner city, particularly one that provides access to jobs for young inner-city men and women, can we encourage a positive sense of the future" (p. 325).

Although widely read, Anderson's work has been subjected to relatively few empirical tests (Swartz, 2010). This dearth of research likely occurs because measuring the cultural landscape of inner-city areas would require extensive surveys of residents on the codes they do and do not embrace. Even so, beginning studies in this area are supportive of Anderson's contention that that the code of the street has structural sources and contributes to involvement in violent delinquency (Brezina, Agnew, Cullen, & Wright, 2004; Stewart & Simons, 2006). Alternatively, in a study of 720 African American youths from 259 neighborhoods, Stewart, Schreck, and Simons (2006) found no support for Anderson's thesis that adopting the code, presumably to display a tough posture that masks vulnerability, increases safety. They discovered that embracing "the street code exacerbates the risk of victimization beyond what would be the case from living in a dangerous and disorganized neighborhood" (p. 427).

Despite the daunting methodological challenges, Anderson's depiction of inner-city culture is innovative and offers rich research opportunities. Given the acclaim it has received, his work should be replicated using both quantitative and qualitative methods.

AKERS'S SOCIAL LEARNING THEORY

Extending Sutherland: Differential Social Reinforcement. Ronald Akers set forth the most influential contemporary extension of Sutherland's differential association perspective with his *social learning theory* (see, e.g., Akers, 1977, 1998, 2000; Akers, Krohn, Lanza-Kaduce, & Radosevich, 1979; Burgess & Akers, 1966). As noted previously, the Chicago theorists emphasized that criminal values are learned through associations. Even so, these theorists had little to say about precisely how this acquisition of antisocial definitions occurs. In his social learning theory, Akers addressed this issue and attempted to specify the *mechanisms* and *processes* through which criminal learning takes place.

Akers provided little systematic analysis of the structural origins of criminal values and learning except to observe that social location differentially exposes individuals to learning environments conducive to illegal conduct (Sampson, 1999). Nonetheless, Akers made a major contribution in illuminating *how* people learn to become offenders.

His interest in this issue began in 1965 when he accepted his first faculty appointment at the University of Washington in Seattle. In graduate school at the University of Kentucky, Akers had undertaken a dissertation in the sociology of law under the guidance of Richard Quinney, a well-known conflict theorist (see Chapter 8). But upon arriving at the University of Washington, he encountered another new professor, Robert Burgess, who was interested in applying the principles of behavioral psychology to sociological topics. Given Akers's background in criminology, they soon agreed that one possible target for Burgess's project might be Sutherland's differential association theory, which saw learning as the basis of criminal behavior.

> Burgess and I agreed that Sutherland's theory was the most fruitful place to explore the ramifications of behavioral psychology for sociology.... Sutherland's was self-consciously a "learning" theory and operant conditioning was self-consciously a "learning" theory. They both were theories of behavioral acquisition, maintenance, performance, and change. Sutherland provided the social interaction context of the learning and Skinner provided the mechanisms of learning. (Akers, 2011, p. 357)

Akers's collaboration with Burgess resulted in a reformulation of Sutherland's theory titled "A Differential-Reinforcement Theory of Criminal Behavior" (Burgess & Akers, 1966). This essay received considerable attention but eventually receded in significance. Its enduring importance, however, was in enticing Akers into the study of crime, with a special focus on extending Sutherland's work by illuminating the mechanisms through which criminal learning occurs. Eventually, this led Akers to formulate his social learning theory.

In this perspective—and similar to Sutherland—Akers noted the importance of differential association in shaping the "definitions" that can prompt wayward behavior. Akers moved beyond Sutherland in specifying the dimensions of these definitions. Some definitions are "general," such as religious values on right and wrong, and some definitions are "specific" and, therefore, pertain to whether crime is permissible in certain situations. Some definitions are "negative" and some are "positive" toward criminal behaviors; still others are "neutralizing" in the sense that they encourage offending by "justifying or excusing it" (Akers, 2000, p. 77).

Sutherland's theory implies that definitions, once internalized, continue to regulate people's decisions. Akers, however, elaborated this model. First, he noted that, in addition to definitions, people can become involved in crime through imitation—that is, by modeling criminal conduct. Second, and most significant, Akers contended that definitions and imitation are most instrumental in determining initial forays into crime. At this juncture, another theoretical issue arises: Why do people continue to commit illegal acts and become stabilized in a criminal way of life? Borrowing from operant psychology, he proposed that social reinforcements—rewards and punishments—determine whether any behavior is repeated. The continued involvement in crime, therefore, depends on exposure to social reinforcements that reward this activity. The stronger and more persistent these reinforcements (i.e., the more positive the consequences), the greater the likelihood that criminal behavior will persist. Akers called this "differential social reinforcement."

The Empirical Status of Social Learning Theory. Akers's social learning theory has been subjected to extensive empirical testing, mostly in studies where measures of social learning are used to account for self-reported delinquency. Overall, the research is supportive of the perspective, including studies in which social learning theory was tested against competing explanations of crime such as social bond theory (Akers, 1998, 2000; Akers & Jensen, 2003, 2006; Akers & Sellers, 2004; see also Kubrin, Stucky, & Krohn, 2009). A recent meta-analysis of 133 studies revealed consistent support for the theory (Pratt et al., 2010). A meta-analysis of predictors of criminal recidivism also showed that, in line with social learning theory, antisocial values and peer associations are strong predictors of reoffending (Andrews & Bonta, 1998). Furthermore, the theory has been shown to account for variation in crime among felony offenders of both genders (Alarid, Burton, & Cullen, 2000). Finally, evaluations of correctional rehabilitation programs conclude—again consistent with social learning theory—that programs that target and change antisocial values and peers (typically "cognitive-behavioral" interventions) are effective in lowering recidivism (Cullen & Gendreau, 2000; Cullen, Wright, Gendreau, & Andrews, 2003; see also Andrews, 1980).

In the existing research, the strongest predictor of criminal involvement typically is differential association as measured by the "number of delinquent friends" reported by a survey respondent. Critics of social learning theory assert that the close association between delinquent friends and crime is spurious. They argue that, rather than delinquent friends causing wayward behavior, this really is a case of "birds of a feather flocking together"—of delinquent kids hanging around with one another because they share the common trait of being delinquent. Research indicates that such self-selection into peer groups does occur. But studies also suggest that even with self-selection, the continued association with other antisocial peers can amplify delinquent involvement (see, e.g., Warr, 2002; Wright & Cullen, 2000; see also Akers, 1998, 2000). As Akers (1999) reminded us, whereas birds of a feather may flock together, it also is the case that "if you lie down with dogs, you get up with fleas" (p. 480).

The Consequences of Theory: Policy Implications

CHANGE THE INDIVIDUAL

As we will see shortly, the logic of Shaw and McKay's social disorganization theory led to the conclusion that the most effective way to reduce crime was to *reorganize communities*. However, the Chicago school's emphasis on cultural learning suggests that crime can be countered by treatment programs that attempt to *reverse offenders' criminal learning* (Andrews & Bonta, 2003, 2010). This emphasis on altering an offender's social learning is particularly consistent with Akers's theory (Akers & Sellers, 2004, pp. 101–108). Note that although there is a focus on the individual offender, the focus is not on altering some inherent or underlying pathology but on changing the values and ways of thinking that the offender has acquired in prior social interactions with parents, siblings, peers, and other actors in society.

In this regard, intervention based on differential association and social learning theory often attempts to remove offenders from settings and people that encourage crime and to locate them in settings where they will receive prosocial reinforcement. This might involve, for example, placing youths in a program that uses positive peer counseling or in a residential facility that uses a "token economy" in which conformist behavior earns juveniles points that allow them to purchase privileges (e.g., home visits, ice cream, late curfew). Furthermore, there is now growing evidence that cognitive-behavioral programs are among the most effective treatment interventions in reducing a range of waywardness, including crime (Lipsey, Chapman, & Landenberger, 2001; MacKenzie, 2006; Wilson, Bouffard, & MacKenzie, 2005; see also Spiegler & Guevremont, 1998). These programs assume that "cognitions" (what Sutherland might call "definitions") lead to behavior. The key is to change those cognitions, such as antisocial values, that are criminogenic. Finally, there is evidence from the family literature that harsh and erratic child-rearing techniques can lead parents to reinforce aggression and other problematic responses and to ignore prosocial conduct. Programs that teach parents better management skills that involve the reinforcement of "good" rather than "bad" behavior have proven to reduce antisocial behavior (Farrington & Welsh, 2007; Reid, Patterson, & Snyder, 2002).

CHANGE THE COMMUNITY

As we have noted, the early Chicago criminologists rejected prevailing individualist biological and psychiatric explanations in favor of elucidating crime's social roots. Consistent with this theoretical perspective, they offered the "first systematic challenge to the dominance of psychology and psychiatry in public and private programs for the prevention and treatment of juvenile delinquency" (Schlossman, Zellman, & Shavelson, 1984, p. 2). The solution to youthful waywardness, they contended, was not to eradicate the pathologies that lie within individuals but rather to eradicate the pathologies that lie within the very fabric of disorganized communities.

Beginning during the early 1930s, Shaw thus embarked on efforts to put his theory into practice, establishing one of the most famous interventions in the history of

American criminology: *the Chicago Area Project* (CAP). Shaw's strategy was for CAP to serve as a catalyst for the creation of neighborhood committees in Chicago's disorganized slum areas. Committee leaders and the project's staff would be recruited not from the ranks of professional social workers but rather from the local community. The intention was to allow local residents the autonomy to organize against crime. Shaw believed that unless the program developed from the "bottom up," it would neither win the community's support nor have realistic prospects for successful implementation (Kobrin, 1959; Schlossman et al., 1984).

CAP took several approaches to delinquency prevention. First, a strong emphasis was placed on the creation of recreational programs that would attract youths into a prosocial environment. Second, efforts were made to have residents take pride in their community by improving the neighborhood's physical appearance. Third, CAP staff would attempt to mediate on behalf of juveniles in trouble. This might involve having discussions with school officials on how they might reduce a youth's truancy or appealing to court officials to divert a youth into a CAP program. Fourth, CAP used staff indigenous to the area so as to provide "curbside counseling." In informal conversations, as opposed to formal treatment sessions, these streetwise workers would attempt to persuade youths that education and a conventional lifestyle were in their best interest. As Schlossman et al. (1984) observed, "They served as both model and translator of conventional social values with which youths . . . had had little previous contact" (p. 15).

Was Shaw's project effective? Unfortunately, the lack of a careful evaluation using a randomized control group precludes a definitive answer. Even so, in 1984, Schlossman et al. provided a "fifty-year assessment of the Chicago Area Project." They concluded that the different kinds of evidence they amassed, "while hardly foolproof, justify a strong hypothesis that CAP has long been effective in reducing rates of reported juvenile delinquency" (p. 46; see also Kobrin, 1959, p. 28). CAP reminds us that "despite never-ending hard times and political powerlessness, some lower-class, minority neighborhoods still retain a remarkable capacity for pride, civility, and the exercise of a modicum of self-governance" (Schlossman et al., 1984, p. 47).

Notably, CAP and Shaw's legacy have endured into the 21st century. The Chicago Area Project continues to exist today, emphasizing advocacy for youths, the provision of direct services, and community organizing to improve the quality of life in disadvantaged urban neighborhoods (Chicago Area Project, 2010). It coordinates more than 40 projects and "affiliates" (i.e., community organizations that embrace the CAP mission). Thus, "projects and affiliates are mandated to positively impact areas in the Chicago vicinity with high rates of juvenile delinquency or other symptoms of social disorganization" (Chicago Area Project, 2010, p. 2).

Conclusion

The Chicago school of criminology had a defining influence on the development of American criminology. The school's theorists elevated to prominence sociological over individual-trait explanations of crime. With rapidly changing Chicago as their labora-tory, they situated offenders within the broad context of the transformation of the United States from an agrarian to an urban, industrial society. In their view, individu-als were born into neighborhoods that differed in their organization. Youngsters raised in outlying city areas were encapsulated by strong institutions that exerted control over their behavior and exposed them predominantly to conventional beliefs. For most, a life in crime—whether in jack-rolling or in professional thievery—was inconceivable. But those born into inner-city areas faced the special challenges of social disorganiza-tion. Weakened social institutions, including the family, were unable to control their conduct or to prevent criminal traditions, rackets, and groups (e.g., gangs) from tak-ing root in the neighborhood. Youths thus experienced two competing ways of life—the conventional and the criminal. They could—and many did—"differentially associate" and learn to prefer crime over conformity.

The Chicago school also made important methodological advances. Quantitatively, they showed the value of mapping crime by geographic area. Anticipating by decades what would later be called "hot spots" of crime, they showed that criminal acts were not randomly distributed but highly concentrated. For them, place mattered. What was it about places with a lot of crime that differentiated them from places without much crime? However, they also valued qualitative methods. The members of the Chicago school were not armchair criminologists but rather walked inner-city streets and interviewed offenders about their personal histories. These revelations allowed their statistics to come to life. Each spot on their maps was not simply a data point but a delinquent with his or her own story. They did not lose touch with the humanity of those they studied, which is perhaps one reason why they sought solutions to crime in social reform rather than in prison construction.

4

Crime in American Society

Anomie and Strain Theories

Robert K. Merton
1910–2003
Columbia University
Author of Anomie-Strain Theory

I n a 1940 article published in the *American Anthropologist,* Robert K. Merton joined with M. F. Ashley-Montagu to offer a stinging rebuttal to Earnest Hooton's (1939) biological theory of crime. As might be recalled from Chapter 2, where we previously reviewed his criminology, Hooton proposed that comparisons between prisoners and noncriminals (e.g., Nashville firefighters) showed that offenders possessed a distinct set of bodily features. Taken together, these characteristics were held to be evidence of their supposed biological inferiority, which in turn was the source of their criminality. Hooton was not reticent about the policy implications of his conclusions, proposing that crime prevention required that society rid itself of the "organically inferior." "It follows that the elimination of crime," observed Hooton, "can be effected only by the extirpation of the physically, mentally, and morally unfit, or by their complete segregation in a socially aseptic environment" (quoted in Merton & Ashley-Montagu, 1940, p. 391). In another context, Hooton referred to the need for "biological housecleaning" (Bruinius, 2006, p. 239).

Fearing that Hooton's writings would "occupy as conspicuous a place in the history of criminology as the works of his predecessors in the field, Lombroso and Goring" (p. 384), Merton and Ashley-Montagu used biting wit and the hard reanalysis of the study's data to reveal the unsteady foundation on which Hooton's theorizing rested. As

it turned out, the appeal of Hooton's criminology was short-lived as American criminology moved away from a vision of offenders as biologically inferior. In fact, Merton
and Ashley-Montagu's critical article was a prominent reason why Harvard University
Press subsequently declined to publish more of Hooton's writings on the topic (Ashley-
Montagu, 1984). Still, at the time, it was clear that Merton and Ashley-Montagu felt that
the troubling view that offenders were not only different but also trapped in *inferior
bodies*—bodies that made them candidates for *extirpation*—was sufficiently popular
at this time in history to demand a systematic rejoinder.

Merton and Ashley-Montagu were at their best, perhaps, when they revisited
Hooton's data that provided bodily measurements such as the breadth and depth of the
chest, weight, head length and circumference, and nose height. To assess inferiority, they
categorized these bodily features in terms of how closely they matched the characteristics of apes—our supposedly inferior ancestors on the evolutionary tree. Much to their
amusement, we suspect, they showed that in many ways certain types of offenders had
fewer ape-like bodily features and thus arguably were more organically advanced than
civilians! More to the point, although not dismissing the possibility that biology played
a role in offending, Merton and Ashley-Montagu (1940) suggested that the sources of
criminal behavior were decidedly cultural and social. This entertaining comparison of
angels to criminals illuminates their sociocultural theoretical preference:

> Actually, what we wish to do here is to suggest that the differences between the angels and
> the criminals are only skin deep; that the criminal may not have sprouted wings as the
> angels have done, not because it was not in them to do so, but because their wings were
> clipped before they were ready to try them. (Merton & Ashley-Montagu, 1940, p. 385)

But what was it precisely that "clipped the wings" of individuals and thus transformed
them, regardless of bodily constitution, into criminals? Merton might have embraced the
most prominent sociological theory of the day—that of the Chicago school. He was
familiar with their writings—familiar enough, in fact, to ask Edwin Sutherland to comment on a prepublication version of the essay he authored with Ashley-Montagu
(Merton, 1984). Merton, however, would not seek to embrace or to elaborate the insights
of the Chicago school. It is not so much that he rejected the theorizing of Shaw and
McKay and of Sutherland but that he thought that other factors—conditions fundamental to *American society in general* and not peculiar to the slums—were at the core of
the nation's crime and deviance. For Merton the key ingredient to crime was not neighborhood disorganization but the "American Dream"—a message sent to all citizens that
they should strive for social ascent as manifested by economic well-being. The inability
of many Americans to achieve this goal—after all, there must be some losers in the race
for success—had dire consequences for the society.

As might be recalled, the main theorists in the Chicago school hailed from small
rural towns. In comparison to their stable communities of origin, the city struck them
as disorganized; it was a short step to trace crime to this social disruption. Ironically,
Merton grew up in a city slum but was not drawn to disorganization theory. Born in
1910, his experiences taught him that poor neighborhoods were diverse, complex
social spaces. Although impoverishment, gangs, and wayward behavior were present,

there also were libraries, good people, and the possibility
own biography demonstrated (Cullen & Messner, 2007). I
if "sociologists did more probing," they "would get away f
slum environment is wholly and solely conducive to de
Cullen & Messner, 2007, p. 16).

In this chapter, we begin by reviewing Merton's theo
two sources of crime and deviance: anomie and strain.
spective was applied by Albert Cohen and by Richard C
explanation of gang delinquency. Their books proved to be classic
the development of American criminology from the 1960s onward. We then move to a
more contemporary time in which we show how the two strands of Merton's theory
have been revitalized in the influential writings of Robert Agnew on general strain theory and of Steven Messner and Richard Rosenfeld on institutional-anomie theory. As
in other chapters, we end with a discussion of the policy implications of this major
school of criminological thought.

Merton's Strain Theory

In 1938, Robert K. Merton published "Social Structure and Anomie." The original article extended only 10 pages—other elaborations by Merton (1957, 1964, 1968, 1995)
followed—but it succeeded in defining an approach that captured the imagination of
criminologists. As we will see, Merton's paradigm became particularly influential during the 1960s, shaping both theory and policy in important ways. Even today, his theorizing occupies a prominent place in criminological writings (Adler & Laufer, 1995;
Bernard, 1984; Cullen, 1984; Messner, 1988; Passas & Agnew, 1997). Thus, we begin by
reviewing Merton's main theoretical assertions and then follow this discussion by considering the context that shaped strain theory over the years.

AMERICA AS A CRIMINOGENIC SOCIETY

The Chicago school believed that the roots of crime were embedded predominantly
in one area of American society—city slums—and that people became criminal by
learning deviant cultural values. Although Merton never rejected this formulation, he
outlined a very different social process—one involving conformity to conventional
cultural values—that he believed produced high rates of crime and deviance.

Structurally Induced Strain. The United States, in Merton's eyes, is an unusual society,
not simply because American culture places an extraordinary emphasis on economic
success but also because this goal is universal—held up for all to want and achieve.
Poor people are not taught to be satisfied with their lot but rather are instructed to pursue the "American dream." Through hard work, it is said, even the lowliest among us
can rise from rags to riches.

widespread aspiration for success, however, has an ironic and unanticipated quence. Merton (1968) cautioned that the "cardinal American virtue, 'ambition,'" mately "promotes a cardinal American vice, 'deviant behavior'" (p. 200). But why should the desire for social mobility lead to deviance? The problem, Merton observed, is that the social structure limits access to the goal of success through legitimate means (e.g., college education, corporate employment, family connections). Members of the lower class are particularly burdened because they start far behind in the race for success and must be exceptionally talented or fortunate to catch up. The disjunction between what the culture extols (universal striving for success) and what the social structure makes possible (limited legitimate opportunities), therefore, places large segments of the American population in the strain-engendering position of desiring a goal that they cannot reach through conventional means. This situation, Merton concluded, is not without important social consequences: It "produces intense pressure for deviation" (p. 199).

Typology of Adaptations. Merton (1968) proposed that different ways existed for people to resolve the strains generated from the inability to attain success. To conceptualize these possible responses, he developed his classic typology that, as Table 4.1 shows, outlined five possible modes of adaptation.

Merton realized that most people, even if they found their social ascent limited, did not deviate. Instead, the modal response was for people to conform, to continue to ascribe to the cultural success goal, and to believe in the legitimacy of the conventional or institutionalized means through which success was to be attained. But for many others, the strain of their situation proved to be intolerable. Because the disjunction between means and goal was the source of their problem, a requisite for alleviating strain was changing their cultural goal and/or withdrawing their allegiance to institutionalized means. In following either or both courses, however, they were deviating from norms prescribing what should be desired (success) or how this should be achieved (legitimate means such as education or employment).

Table 4.1   Merton's Typology of Modes of Individual Adaptation

Mode of Adaptation	Culture Goals	Institutionalized Means
I. Conformity	+	+
II. Innovation	+	−
III. Ritualism	−	+
IV. Retreatism	−	−
V. Rebellion	±	±

SOURCE: Merton (1968, p. 140). Reprinted with permission of The Free Press, an imprint of Simon & Schuster, from *Social Theory and Social Structure*, revised and enlarged edition, by Robert K. Merton. Copyright © 1967, 1968 by Robert K. Merton. Copyright renewed © 1985 by Robert K. Merton. All rights reserved.

NOTE: Plus sign (+) signifies acceptance; minus sign (−) signifies rejection; plus and minus sign (±) signifies rejection of prevailing values and substitution of new values.

Thus, Merton (1957, 1968) delineated four deviant modes of adaptation. He believed that much criminal behavior could be categorized as *innovation* because this adaptation encompasses those who continue to embrace pecuniary success as a worthy end but turn to illegitimate means when they find their legitimate prospects for economic gain blocked. The behavior of the robber barons, white-collar criminals, and scientists who report discoveries based on fraudulent research are examples of how the intense desire for success can produce innovation among the more affluent. Even so, this adaptation appears to be prevalent particularly in the lower strata. Faced with the "absence of realistic opportunities for advancement," the disadvantaged are especially vulnerable to the "promises of power and high income from organized vice, rackets, and crime" (Merton, 1968, p. 199).

By contrast, *ritualists* maintain outward conformity to the norms governing institutionalized means. They mitigate their strain, however, by scaling down their aspirations to the point where these ends can be reached comfortably. Despite cultural mandates to pursue the goal of success, they are content to avoid taking risks and to live within the confines of their daily routines.

Retreatists make a more dramatic response. Strained by expectations of social ascent through conventional lifestyles, they relinquish allegiance to both the cultural success goal and the norms prescribing acceptable ways of climbing the economic ladder. These are people who "are in society but not of it," and they escape society's requirements through various deviant means—alcoholism, drug addiction, psychosis, vagrancy, and so on (Merton, 1968, p. 207). Suicide, of course, is the ultimate retreat.

Finally, Merton described as *rebellious* citizens who not only reject but also wish to change the existing system. Alienated from prevailing ends and normative standards, they propose to substitute a new set of goals and means. In American society, an example of a rebel might be a socialist who argues for group success rather than individual success and for norms mandating the distribution of wealth equally and according to need rather than unequally and according to the outcome of ruthless competition.

Anomie. Because much of Merton's analysis detailed the social sources of strains potent enough to generate high rates of nonconformity, scholars often have referred to his perspective as *strain theory* (Empey, 1982; Hirschi, 1969; Kornhauser, 1978; Vold & Bernard, 1986). But Merton did not simply identify why individuals might face strains that prompt them to see a deviant adaptation, including crime. Indeed, we can recall that Merton (1938) titled his classic article "Social Structure and Anomie"—not "Social Structure and Strain." What role does *anomie* play in the genesis of crime?

Merton borrowed the notion of anomie—normlessness or deregulation, as it usually is defined—from Émile Durkheim, the French sociologist. In his classic work, *Suicide,* Durkheim (1897/1951) used the concept to describe a social condition in which institutionalized norms lost their power to regulate human needs and action. He argued further that as Western society modernized, a great emphasis was placed on "achieving industrial prosperity" without corresponding attention to restraining people's appetites for success. This development, he observed, had left the economic sphere in a "chronic state" of anomie. People now were free, if not encouraged, to seek seemingly limitless economic success. But disquieting consequences befell those who succumbed to those temptations. "Overweening ambition," Durkheim warned, "always exceeds the results obtained, great as they may be, since there is no warning to pause here. Nothing gives satisfaction, and all this agitation, is uninterruptedly maintained without appeasement" (p. 60). For many, suicide posed the only means of escape from the pain of "being thrown back." We will return to Durkheim's theorizing in Chapter 5.

Merton did not buy Durkheim's framework whole cloth, but he did borrow selectively from it (Cullen, 1984; Vold & Bernard, 1986). Most important, perhaps, he learned that institutionalized norms will weaken—anomie will take hold—in societies placing an intense value on economic success. When this occurs, the pursuit of success no longer is guided by normative standards of right and wrong; instead, "the sole significant question becomes: Which of the available procedures is most efficient in netting the culturally approved value?" (Merton, 1968, pp. 189, 211). The Wall Street insider trading and banking scandals provide examples of how the widespread preoccupation with amassing fortunes results in a breakdown of institutionalized norms—anomie sets in—and fosters the unbridled pursuit of pecuniary rewards.

We also should add that innovative conduct becomes especially prevalent as anomie intensifies. In contrast to ritualism (or to conformity), this adaptation requires an ability to relinquish commitment to institutionalized means in favor of illegitimate means. Fluctuations in levels of anomie, whether over time or within certain sectors of society at any given time, can be expected to determine not only overall rates of deviance but also rates of particular kinds of deviance—including crime, the prototypical innovative response.

Anomie and deviance, moreover, are mutually reinforcing. The weakening of institutionalized norms initially allows a limited number of people to violate socially approved standards. But such deviance, once completed successfully and observed by others, poses a concrete challenge to the norms' legitimacy. This process, Merton (1968) noted, "enlarges the extent of anomie within the system" (p. 234), and this in turn heightens the chance that waywardness will become more pervasive.

The escalating use of marijuana perhaps illustrates this phenomenon. As norms prohibiting its use lost strength during the 1960s—claims of "reefer madness" were ridiculed—increasing numbers of youths experimented with the substance, typically in social situations. This widely observed, if not flaunted, deviant behavior undermined the legitimacy of institutionalized norms, even to the point where in some locales police and courts refused to enforce existing laws and recreational use was decriminalized. Thus, anomie became pervasive, and restraints against "smoking pot" were vitiated greatly—a development that made marijuana use even more pervasive.

Rejecting Individualism. In sum, Merton contended that the very nature of American society generates considerable crime and deviance. The disjunction between the cultural and social structures places many citizens, but particularly the disadvantaged, in the position of desiring unreachable goals. Tremendous strains are engendered that move many people to find deviant ways of resolving this situation. The cultural emphasis on success, moreover, diminishes the power of institutional norms to regulate behavior. As anomie becomes prevalent, people are free to pursue success goals with whatever means are available—legitimate or illegitimate. In this situation, innovation—an adaptation encompassing many forms of crime—becomes possible, if not likely.

Like the Chicago theorists, Merton located the roots of crime and deviance within the very fabric of American society. Again, the Chicago school stressed the criminogenic role of the city and of conformity to a *criminal culture,* whereas Merton stressed the criminogenic role of conformity to the universal and *conventional cultural goal* of pecuniary success. This difference aside, however, both perspectives rejected the notion that crime's origins lay within individuals' minds or bodies.

Indeed, Merton was especially vociferous in his attack on the individualist explanations that prevailed during the 1930s (Merton & Ashley-Montagu, 1940). Most of these theories, he explained, were based on the fallacious premise that the primary impulse for evil lay within human nature. By contrast, Merton (1968) argued for a perspective that "considers socially deviant behavior just as much a product of social structure as conformist behavior" and that "conceives of the social structure as active, as producing fresh motivations which cannot be predicted on the basis of knowledge about man's native drives" (p. 175).

STRAIN THEORY IN CONTEXT

Merton was 28 years old and teaching at Harvard University when "Social Structure and Anomie" was published in 1938. Eventually, he would become a professor at Columbia University (in 1941) and be elected as president of the American Sociological Association. As noted, however, his origins were more modest. He was born in the slums of south Philadelphia and attended college only by winning a scholarship to Temple University and then a graduate assistantship to Harvard (Hunt, 1961; Persell, 1984).

Linking personal biography to a scholar's theorizing involves a degree of speculation. Still, Merton's life seems to mirror the two core features of his paradigm: the significance of the cultural message for all to pursue the American dream and the differential opportunities people had to reach this universal goal.

Thus, Merton's ascent from city slums to elite institutions—from south Philadelphia to Harvard and Columbia—meant that he had lived the American dream. In a very personal way, he also assimilated into the dominant culture. A little-known fact is that Merton, the son of eastern European Jewish immigrants, was born with the name Meyer R. Schkolnick. While a teenager doing a magic act at local events such as birthday parties, he followed the lead of other prominent entertainers in changing his name first to "Robert Merlin" and then, realizing the tackiness of this choice, to "Robert King Merton." He became widely known among friends and at school as "Bob Merton" and chose to retain this Americanized name as he went off to college (Cullen & Messner, 2007).

In short, for Merton, the dominant reality was not ethnic and racial *heterogeneity* and *culture conflict*—ideas highlighted by the Chicago school theorists. For Merton, the defining reality of the United States was cultural *homogeneity* and *universalism*—the fact that Americans shared a dream and an identity. For Merton, there were powerful forces that pushed everyone to become an American and to embrace the national culture of social ascent.

At the same time, it seems reasonable to suggest that Merton's boyhood social context also shaped the theoretical emphasis that he placed on the structural limitations to social mobility. As discussed previously, Merton did not believe that inner-city neighborhoods were fully disorganized and inherently criminogenic (Cullen & Messner, 2007). Still, he was not blind to the limits to mobility that inhered in lower-class origins. Upward mobility certainly was possible—as his own life attested. But as Pfohl (1985) notes, "Most of Merton's slum neighbors did not fare so well," and this was "a lesson of slum life which Merton never forgot" (p. 211). Merton also had experienced the Great Depression and witnessed the consequence of large numbers of people falling to the bottom rungs of society and being deprived of the opportunity to reach what they had been taught to desire.

In contrast to the Chicago school, then, Merton did not believe that life in a slum neighborhood inevitably was criminogenic. In his boyhood community, the families were not all disorganized and criminal traditions were not ever-present. Residents wanted to be Americans and to live the nation's cultural dream. Thus, youngsters were led into crime not so much by life *in the slum*, as the Chicago school claimed, but the denial of the *opportunity to leave the slum*.

Regardless of how Merton came to formulate his paradigm, one point is clear: His article "Social Structure and Anomie" (Merton, 1938) is perhaps the most cited article not only in criminology but in sociology as a whole (Pfohl, 1985). Even so, the article did not receive widespread attention until nearly two decades after its 1938 publication (Pfohl, 1985; see also Cole, 1975). The sudden interest in his work during the late 1950s and early 1960s was prompted in part by the appearance of two important books on juvenile gangs that drew heavily from Merton's theorizing: Cohen's (1955) *Delinquent Boys* and Cloward and Ohlin's (1960) *Delinquency and Opportunity*. We also can point to the effects of Merton's (1957, 1959, 1964, 1968) own renewed interest in elaborating his earlier article.

These observations by themselves, however, do not offer a complete explanation of why, as the United States entered the 1960s, criminologists became so intensely fascinated by this particular way of thinking about crime as opposed to some other set of ideas. The social context of the time, we believe, must be considered as well.

As Charles Murray noted, prior to 1960, poverty was not viewed—in political circles at least—as a major social problem rooted in the very structure of American society (Murray, 1984). By the early part of the 1960s, however, a fundamental transformation in thinking had occurred. An increasing consensus had emerged, Murray (1984) argued, that poverty "was not the just deserts of people who didn't try hard enough. It was produced by conditions that had nothing to do with individual virtue and effort. *Poverty was not the fault of the individual but of the system*" (p. 29, emphasis in original). The civil rights movement, moreover, provided a language for conceptualizing this issue: Minorities and other disadvantaged citizens were being denied "equal opportunity."

This view of the world was embraced increasingly by government officials, journalists, and academics (Murray, 1984). Criminologists also were influenced profoundly by the legitimacy now given to the idea that large segments of the U.S. population were denied access to the American dream. In this context, it becomes understandable why Merton's theory and offshoots, such as the delinquency books by Cohen (1955) and by Cloward and Ohlin (1960), suddenly gained attention. At the core of Merton's paradigm was the lesson that the United States was a society in which all were expected to ascend economically but whose very structure denied equal opportunity to attain this cherished goal. To criminologists of the 1960s, this premise rang true. It made sense that crime and deviance would be a consequence of a system that was to blame for unfairly holding back many of its citizens.

Status Discontent and Delinquency

As indicated previously, the writings of Albert Cohen and of Richard Cloward and Lloyd Ohlin represented important extensions of Merton's deviance approach. Although these scholars offered different variants of strain theory, they shared common themes. First, they investigated how the theory could be applied to the study of juvenile gangs in urban areas. Second, they focused on the origins and effects of delinquent subcultural norms. Third, they drew not only from Merton's structural tradition but also from the Chicago school.

DELINQUENT BOYS

While still an undergraduate at Harvard, Albert K. Cohen enrolled in a senior course instructed by a young professor, Robert K. Merton. A year later, Cohen was off to Indiana University, where he took a seminar run by Edwin H. Sutherland. As might be expected, this fortuitous encounter with two influential and persuasive scholars prompted Cohen to ponder how notions of cultural transmission and structurally induced strain might be reconciled.

Sutherland had convinced Cohen that differential association with a criminal culture would lead youths into legal trouble. Yet Cohen also believed that this thesis of cultural transmission begged more fundamental questions. Where did the criminal culture come from? Why did such subcultures have a specific social distribution, locating themselves in slum areas? Why did the subcultures have a particular normative content? And why did these values persist from generation to generation?

Returning to Harvard for his doctorate, Cohen addressed these questions in his dissertation. A much revised version was published in 1955 carrying the title *Delinquent Boys: The Culture of the Gang*. He began by making several important observations: Delinquent gangs and the subcultural values they embrace are concentrated in urban slums. Moreover, the content of these subcultures not only is supportive of crime but also is "nonutilitarian, malicious, and negativistic" (p. 25). Because slum youths learn and act on the basis of these values, they engage in delinquency that is contemptuous of authority and irrational to conventional citizens. Seemingly, the only guide for their conduct is that they do things for "the hell of it."

To account for these patterns, Cohen (1955) proposed that delinquent subcultures, like all subcultures, arise in response to the special problems that people face. Following Merton's insights, he noted that lower-class youths are disadvantaged in their efforts to be successful and achieve status in conventional institutions. Schools, which embody middle-class values, present a particular obstacle: Poor kids lack the early socialization and resources to compete successfully with their counterparts from more affluent families. Consequently, they are "denied status in the respectable society because they cannot meet the criteria of the respectable status system" (p. 121).

How can these status problems be solved? "The delinquent subculture," Cohen contended, "deals with these problems by providing criteria of status which these children can meet" (p. 121). In a process approximating a reaction formation, lower-class youths reject the middle-class goals and norms that they have been taught to desire but by which they are judged inadequate. In place of middle-class standards, they substitute a set of oppositional values. If conventional society values ambition, responsibility, rationality, courtesy, control of physical aggression, and respect for authority, then these youths will place a premium on behavior that violates these principles. Accordingly, status will be accorded to compatriots who are truant, flout authority, fight, and vandalize property for "kicks."

In short, Cohen (1955) suggested that the strains of class-based status discontent are conducive to the emergence of subcultural values supportive of delinquency. Lower-class youths, thrown together in high-density urban neighborhoods and saddled with a common problem, find a common solution in embracing values that provide both the chance to gain status and the psychic satisfaction of rejecting respectable values that lie beyond their reach. Because American society continues to present each new generation of urban youths with status problems, a structural basis exists for the persistence of these delinquent norms and the gang organization they nourish. Moreover, once in existence, the subculture assumes a reality of its own. As the Chicago theorists taught, this criminal culture can be transmitted to youths in the neighborhood. Cohen cautioned that even juveniles whose status discontent is insufficient in itself to motivate delinquency can be attracted by the lure of the gang and its offer of friendship, excitement, and protection.

DELINQUENCY AND OPPORTUNITY

Like Cohen's work, Richard Cloward and Lloyd Ohlin's work brought together the traditions of the Chicago school and strain theory. As with Cohen, personal circumstance ostensibly played a role in making possible this attempt at a theoretical merger. Ohlin had studied under Sutherland and later received his doctorate from the University of Chicago. Cloward had been Merton's student at Columbia. Eventually, they became colleagues on Columbia's social work faculty (Laub, 1983) and entered into a collaboration that bore the important fruit of *opportunity theory* (Cloward & Ohlin, 1960; see also Cloward, 1959).

From Merton, Cloward and Ohlin learned that the social structure generates pressures for deviance, pressures experienced most intensely in the lower class. Similar to

Cohen's extension of Merton's work, they argued that slum youths face the problem of lacking the legitimate means—the opportunity—to be successful and earn status. In American society, where success in school and careers is valued and rewarded so greatly, this failure presents a special problem: the strain of status discontentment. "The disparity between what lower-class youths are led to want and what is actually available to them," as Cloward and Ohlin (1960) put it, "is the source of a major problem of adjustment." It causes "intense frustrations," and the "exploration of nonconformist alternatives may be the result" (p. 86).

But Cloward and Ohlin also drew an important lesson from the writings of the Chicago theorists. In reading works such as *The Professional Thief* (Sutherland, 1937) and *The Jack-Roller* (Shaw, 1930), which elucidated how criminal roles are learned through cultural transmission, they were led to the conclusion that people are not free to become any type of criminal or deviant they would like. To become a doctor or lawyer, one must, of course, have access to the requisite legitimate means (e.g., education, financial resources). This logic, they reasoned, could be extended to the criminal world: To become a professional thief or jack-roller, one must have access to the requisite *illegitimate means* (e.g., contact with thieves, residence in a slum neighborhood).

This lesson from the Chicago school helped to resolve a shortcoming of Merton's paradigm. As discussed previously, Merton noted that strain could be adapted to through innovation, ritualism, retreatism, or rebellion. He provided only rudimentary insights, however, on the conditions under which a person would choose one adaptation rather than another. Cloward and Ohlin proposed that the Chicago school furnished an answer to this question of why individuals adapt to strain in one way and not others: The selection of adaptations is regulated by the availability throughout the social structure of illegitimate means. Thus, affluent people might have access to the financial positions needed to embezzle or to embark on insider trading schemes. By contrast, although precluded from white-collar crime, lower-class residents might have access to friends who can help them to rob or who can teach them how to fence stolen goods.

Notably, Cloward and Ohlin believed that the concept of illegitimate means could illuminate why delinquent subcultures existed in slum areas and why they took a particular form. One relevant consideration was that lower-class youths experienced high levels of strain. But this factor in itself explains only why youths might be motivated to violate the law, not why subcultural responses of a particular type emerge. "To account for the development of pressures toward deviance," Cloward and Ohlin cautioned, "does not sufficiently explain why these pressures result in one deviant solution rather than another" (p. 34).

Accordingly, Cloward and Ohlin (1960) proposed that delinquent subcultures could emerge and persist only in areas where enough youths were concentrated to band together and to support one another's alienation from conventional values. They further observed, however, that the type of collective response that the youths could make would be shaped intimately by the neighborhood in which they resided. In organized slum areas, for example, criminal subcultures are possible because older offenders serve as role models for a stable criminal life and train youths in the performance of illegal enterprises (e.g., through fencing stolen merchandise).

In more disorganized neighborhoods, on the other hand, access to such organized criminal apprenticeships is absent. Lacking the opportunity to embark on more lucrative, utilitarian illegal careers, youths turn to violence as a way of establishing a "rep" or social status. Therefore, conditions are ripe for the emergence of a "conflict" or fighting-oriented subculture.

Cloward and Ohlin (1960) also identified a third subcultural form: the "retreatist" or drug-using subculture. These groupings arise when sufficient numbers of youths exist who have been "double failures"—people who have failed to achieve status through either legitimate or illegitimate means. These lower-class juveniles not only have been unsuccessful in conventional settings such as the school but also "have failed to find a place for themselves in criminal or conflict subcultures." As a result, they look to "drugs as a solution" to their "status dilemma" (p. 183).

One final point warrants emphasis. Although Cloward and Ohlin focused their substantive analysis on delinquent subcultures, they believed that their opportunity theory—a consolidation of the cultural transmission and strain traditions—offered a *general* framework for studying crime and deviance (Cloward, 1959; Cloward & Piven, 1979). Again, they were persuaded that Merton had identified a major source of pressures for deviance—denial of legitimate opportunity—but that strain theory was incomplete without a systematic explanation of why people solve their status problems in one way and not another. The issue of the selection of adaptations, they believed, could be answered only by focusing on how the illegitimate opportunity structure regulated access to different forms of crime/deviance for people located at different points in society (Cullen, 1984). Thus, white-collar crime and mugging represent two possible innovative methods to acquire financial resources. Even so, participation in one of these offenses rather than the other is determined not by strain but by social class differentials in the availability of illegitimate opportunities.

The Criminological Legacy of Strain Theory

ASSESSING STRAIN THEORY

As strain theory emerged during the early 1960s as the most prominent criminological explanation, it won considerable attention not only from adherents but also from opponents (Cole, 1975). The perspective's critics developed a variety of lines of attack (for summaries, see Empey, 1982; Pfohl, 1985; Vold & Bernard, 1986).

Some scholars, for example, have questioned whether in a society as diverse as the United States, all citizens ascribed to the goal of pecuniary success. At the very least, possible variations in the degree to which different groups were effectively socialized into the American dream would have to be investigated. Other scholars have disparaged strain theory for assuming that strain and deviance were more prevalent in the lower classes. This class-biased assumption is said to ignore white-collar crime and to convey the impression that lawlessness is exclusively a lower-class problem (but see Merton, 1957, 1968). Still other, more radical scholars have expressed concern over

Merton's failure to offer a broader, more penetrating analysis (Taylor, Walton, & Young, 1973). In this view, Merton succeeded in identifying a contradiction central to U.S. society—an open-class ideology and a restricted class structure—but stopped short of asking why this condition originated and persists unabated. The answer, according to Pfohl (1985), is that "the political-economic structure of capitalism must be seen as a basic source of the contradictions which produce high rates of deviance" (p. 234).

The attempts of Cohen (1955) and Cloward and Ohlin (1960) to explain delinquent subcultures have been criticized as well. Commentators most frequently have questioned whether these theorists described the content of subcultures accurately. Cohen (1955) portrayed delinquents as embracing "nonutilitarian, malicious, and negativistic" values, but some youths' criminality is consumption oriented and ostensibly utilitarian. Similarly, Cloward and Ohlin (1960) delineated three distinct subcultural forms— criminal, conflict, and retreatist—but delinquents appear to mix these activities. "Delinquent boys," Empey (1982) observed, "drink, steal, burglarize, damage property, smoke pot, or even experiment with heroin and pills, but rarely do they limit themselves to any single one of these activities" (p. 250; see also Short & Strodtbeck, 1965).

Although these observations have merit, it is important to distinguish between the theorists' substantive analysis of subcultures (which might be in need of revision) and the general framework that they suggest for the study of crime. Cohen (1955) and Cloward and Ohlin (1960) not only identified potentially important sources of youthful disaffiliation from conventional norms but also raised the critical theoretical question of why certain forms of crime (e.g., delinquent gangs, white-collar criminality) are distributed differentially throughout the social structure. Thus, even if these authors' description of juvenile subcultures did not prove to be fully satisfactory, their challenge to criminologists to explain the emergence, persistence, and differential selection of criminal adaptations remains valid.

Lastly, an empirical critique has been leveled against strain theory. Contrary to the perspective's predictions that frustrated ambitions push people outside the law, some researchers have reported that high aspirations are associated with conformity and that low aspirations are associated with delinquent involvement (Hirschi, 1969; Kornhauser, 1978). Bernard's (1984) assessment of available research studies, however, showed that empirical data exist that are supportive of strain theory (see also Burton & Cullen, 1992; Cole, 1975). Moreover, studies that claim to falsify the perspective's premises often use questionable measures of key concepts and do not systematically assess all of the theory's components (Bernard, 1984; Burton & Cullen, 1992; Messner, 1988). For example, most studies rarely assess whether individuals actually experience strain. In this regard, the findings of a community self-report study of adults conducted by Agnew, Cullen, Burton, Evans, and Dunaway (1996) are instructive. In a "new test of classical strain theory," these authors reported that people who express dissatisfaction with how much money they make are, as strain theory would predict, more likely to engage in income-generating crime and drug use.

Despite these considerations, it must be admitted that the popularity of strain theory, which dominated criminologists' attention during the 1960s, has lessened. Some scholars were critical for purely intellectual reasons. But as we have cautioned, the popularity of criminological theories also is shaped, perhaps more profoundly, by changes in the social context that make previously cherished ideas seem odd and make new ideas seem a matter of common sense. Particularly as we discuss the labeling and conflict perspectives (see Chapters 7 and 8, respectively), we will see how events during the late 1960s and early 1970s caused many criminologists to shift their allegiance to theories that emphasized the role of the state and of power in defining what is crime and why people engage in it.

All of this is not to imply, however, that strain theory was relegated to the criminological scrap heap; quite the contrary. Although the paradigm's appeal has diminished, it remains required reading for criminologists and serves as a standard against which newer theories often are compared and tested (Burton & Cullen, 1992). More important, over the past decade or so, a renewed interest in strain theory has cropped up (see, e.g., Adler & Laufer, 1995; Agnew, 1992, 2006a, 2006b; Cullen, 1988; Messner, 1988; Passas, 1990; Passas & Agnew, 1997; Rosenfeld, 1989). Particularly significant are efforts that build on and revise the insights of traditional strain theory. Two promising lines of inquiry warrant our attention: Agnew's general strain theory and Messner and Rosenfeld's analysis of crime and the American dream.

AGNEW'S GENERAL STRAIN THEORY

Robert Agnew suggested that theorizing in the Mertonian school of strain theory is not so much wrong as it is limited. According to Agnew (1992), Merton and status frustration theorists identified *one type of criminogenic strain*: "relationships in which others prevent the individual from achieving positively valued goals" (p. 50), such as economic success and status in a high school. However, there may be other kinds of negative relations or situations that create strain and prompt people to break the law. The time has come, Agnew observed, to move beyond Merton's paradigm, which he termed classic strain theory, and to explore these other sources of criminogenic strain. In short, the challenge is to develop a *general strain theory* of crime.

While a doctoral student at the University of North Carolina, Agnew (2011) was interested in studying the sources of creativity. This project did not come to fruition, because no accessible data set on creativity existed, leaving Agnew with the challenge of how to complete his dissertation. "With each passing week," he has recently recollected, "my level of strain increased. To paraphrase Albert Cohen (1955), I had a problem and was in the market for a solution" (Agnew, 2011, p. 140). A fellow graduate student alerted him to the Youth in Transition survey, which contained, among other variables, measures of delinquency.

With data now in hand, Agnew vigorously delved into the delinquency literature and, despite criticisms of the perspective, was persuaded that a broader version of strain theory held promise. His work was informed by his knowledge of social psychology. Classic strain theory, he felt, focused on a kind of strain—the inability to attain future economic success—that was too distant or far in the future to affect youths' behavior. For Agnew (2011), the "major strains conducive to crime seemed to be more immediate in nature, such as being physically abused by a peer or having a serious argument with a family member. This insight was supported by the research in social psychology, including the stress and aggression research" (p. 141). This line of inquiry eventually led to his developing a "revised strain theory" (Agnew, 1985), which was the prelude to his publication of his general strain theory in 1992. His most systematic presentation of the theory subsequently appeared in *Pressured Into Crime: An Overview of General Strain Theory* (Agnew, 2006a; see also Agnew, 2006b).

Types of Strain. Again, Agnew argued that Merton's classic strain theory identified one category of strain, which involved being blocked from desired goals. Unlike Merton, however, Agnew did not focus exclusively on economic goals. He expanded this type of strain to include blockage from any positively valued goal. This might include not attaining economic success, but it also might entail not making a sports team at school, not achieving status in one's peer group, or not securing a date for a prom dance. Beyond this innovation, he proposed two additional sources of strain (see Table 4.2).

First, Agnew (1992) argued that strain can be generated from the "actual or anticipated removal (loss) of positively valued stimuli from an individual" (p. 57). This strain might occur, for example, when parents take away privileges (e.g., use of the family automobile), when a student is kicked off an athletic team, when a dating relationship is terminated, when an individual's job is eliminated, or when a loved one dies. In these situations, people might take drugs to manage the stress, or they might resort to illegal means to replace what was taken away (e.g., steal a car) or to seek revenge against those who caused their strain (e.g., assault a parent, coach, or boss).

Second, strain can be induced by the "actual or anticipated presentation of negative or noxious stimuli" (Agnew, 1992, p. 58). These adverse situations might include exposure to a sexually or physically abusive relationship, living in a family wracked by conflict, attending a dangerous school, and being supervised by a boss who is unfair or harassing. As Agnew (1992) pointed out, crime and delinquency may take place when people respond to this adversity by seeking to escape (e.g., running away from home),

Table 4.2   Agnew's General Strain Theory: Types of Strain

Types of Strain	How Individuals Are Treated	Types of Strain Most Likely to Lead to Crime
1. Strain as the failure to achieve positively valued goals (traditional strain).	→ Individuals are unable to reach their goals.	1. The strain is seen as unjust. 2. The strain is high in magnitude.
2. Strain as the removal of positively valued stimuli from the individual.	→ Individuals lose something they value.	3. The strain is caused by or associated with low social control.
3. Strain as the presentation of negative stimuli.	→ Individuals are treated in a negative manner by others.	4. The strain creates some pressure or incentive to engage in criminal coping.

to eliminate or obtain revenge against the source of the stress (e.g., murder an abuser), or to dull the psychic pain by taking drugs.

Adapting to Strain. Agnew assumed that, in general, the higher the dose of strain that a person experiences, the greater the likelihood of the person being engaged in crime or in some form of deviance. Even so, theorists in the classic strain theory tradition realized that strain is not related to crime in an ironclad way. Once people are under strain, they may or may not adapt to this state through criminal acts. (Recall Merton's typology of adaptations in Table 4.1.) A complete theory of crime, therefore, must include not only an explanation of what causes strain *but also an explanation of what causes people under strain to respond through criminal conduct.* This insight also is found in the early work of Cloward and Ohlin (1960), who argued that access to different types of illegitimate means shapes the kind of adaptations people make to strain (see also Cullen, 1984).

Thus, Agnew (1992) set out to demarcate the "constraints to non-delinquent and delinquent coping"(pp. 66–74)—that is, the variables that "condition" the response to strain. He identified a range of factors that diminish the risk of a criminal adaptation, such as the availability of other goals to substitute for blocked goals, individual coping resources (e.g., self-efficacy, intelligence), the delivery of social support from others, the fear of the consequences of legal punishments, the presence of strong social bonds, and the denial of access to illegitimate means. Other factors that foster the predisposition to criminality increase the likelihood of crime. These would include, for example, low self-control, prior criminal learning experiences (e.g., association with delinquent friends), the internalization of antisocial beliefs, and the tendency to blame others for being in strain-inducing predicaments. Note that Agnew borrowed many of his conditioning variables from other criminological theories (e.g., bonds, controls). Whereas these other theories would argue that these factors have a *direct* effect on crime, general strain theory contends that they increase criminal behavior only when they occur in *conjunction with strain.* For this reason, the conditioning variables often are measured in empirical tests with an interaction term (e.g., Strain × Number of Delinquent Friends).

Finally, Agnew included *emotions* within general strain theory. In his view, "negative emotions . . . create pressure for corrective action; individuals feel bad and want to do something about it" (Agnew, 2006a, p. 104). He focused, in particular, on the emotion of anger. General strain theory predicts that when strain elicits anger, crime (especially violent crime) is more likely to occur.

The Empirical Status of General Strain Theory. Because Agnew stated general strain theory in a very clear way, the perspective has been subjected to an increasing number of empirical tests (see, e.g., Agnew & White, 1992; Aseltine, Gore, & Gordon, 2000; Brezina, 1996; Hoffman & Miller, 1998; Hoffman & Su, 1997; Katz, 2000; Mazerolle, 1998; Mazerolle, Burton, Cullen, Evans, & Payne, 2000; Mazerolle & Maahs, 2000; Mazerolle & Piquero, 1997; Paternoster & Mazerolle, 1994; see also Kubrin et al., 2009). What does this research tell us about the perspective's empirical support? Three conclusions are possible at this time.

First, although the results are not consistent for every type of strain (an issue we will return to shortly), there is consistent empirical evidence that exposure to strain increases the likelihood of criminal offending (Agnew 2006a, 2006b, 2009). Second, although some positive findings emerge, studies provide less support for the idea that adaptations to strain are conditioned by a range of other factors. This finding may occur because the methodology of using an interaction term (i.e., Strain × Conditioning Variable) is too crude to capture the complex way in which strain is conditioned by individual and social factors. But another possibility is that the conditioning variables identified by Agnew mainly have direct effects on criminal behavior rather than coming into play mainly when a person is under strain. For example, low self-control may cause crime by itself—that is, regardless of whether strain is present or absent. In fact, this is the position that control theorists would take (Gottfredson & Hirschi, 1990; see also Hirschi, 1969). Third, there is some evidence that the combination of strain and anger increases the risk of criminal conduct. What remains to be clarified, however, is whether strain creates anger, which then leads to crime, or whether people who are angry are more likely to create strain in their lives, which then leads to crime. Of course, both of these causal links to crime are possible (Mazerolle et al., 2000; Mazerolle & Piquero, 1997).

We should note that most of the tests of general strain theory are drawn from self-report studies of youths who attend school and live with their families. Further support for this perspective, however, comes from Bill McCarthy and John Hagan's research on delinquency among youths who have left home and are living on the streets (Hagan & McCarthy, 1997; McCarthy & Hagan, 1992). Consistent with Agnew's work, their analysis reveals how adverse or noxious conditions can result in delinquent involvement both directly and indirectly. They observed that youths who leave home often live on "mean streets," where they experience hunger, unemployment, and a lack of shelter. In a direct attempt to alleviate these adverse situations, they steal food and obtain money by pilfering property and by prostitution. Other adverse conditions influence delinquency indirectly by causing youths to take to the streets in the first place. Thus, youths who find themselves in homes where they are sexually and physically abused are more likely to escape this adversity by leaving home—only to find themselves facing the pressures for crime inherent in the deprivations of street life.

Elaborating General Strain Theory. As noted previously, a number of studies have found support for the proposition that exposure to strain increases the risk of criminal involvement. The difficulty, however, is that these studies often have measured a wide variety of strains using a wide variety of measures. Although Agnew stated his perspective parsimoniously—he identifies three major types of strain—each category of strain covers numerous subtypes of strain. Take, for example, the category of the presentation of noxious stimuli, which might involve diverse strains such as exposure to parental conflict, sexual abuse by family members or others, victimization by peers, conflict with friends, unpleasant school experiences, poor working conditions, and noisy and crowded living conditions. For general strain theory to make sense, it must address the issue of which strains—among the thousands that might conceivably be studied—are criminogenic and which are not. Indeed, it is essential for future research to focus on this issue.

Notably, Agnew (2001a) took up the challenge of "specifying the strains most likely to lead to crime and delinquency" (p. 1). He listed four factors that increase the likelihood that strain will prompt a criminal adaptation (see Table 4.2). Notably, this categorization of strain is likely to guide research on general strain theory in the time ahead (see also Agnew 2006a, 2006b).

First, "the strain is seen as unjust" (p. 6). When individuals perceive that the strain they are feeling is due to unfair treatment, they are likely to become angry. As noted previously, general strain theory contends that anger increases the risk of offending. Second, the strain is "high in magnitude" (p. 9). When under severe strain, it is difficult to ignore the strain, keep one's emotions under control, and resolve the strain in legal ways. To the extent that crime can relieve the strain, its immediate benefits may outweigh the more distant and uncertain costs that such conduct might elicit from the law. Third, "the strain is caused by or associated with low social control" (p. 11). For example, if a juvenile is rejected by a parent, then two things occur: This act causes strain while it simultaneously lowers the level of social control over the juvenile by reducing his or her attachment to the parent. Fourth, "the strain creates some pressure or incentive to engage in criminal coping" (p. 12). For example, a youth might experience strain if victimized by a delinquent peer. A failure to respond with violence might lead to more victimization and more strain. In the end, fighting back and getting revenge—potentially a form of "criminal coping"—might be seen as the only realistic option open to the youth.

CRIME AND THE AMERICAN DREAM: INSTITUTIONAL-ANOMIE THEORY

Steven Messner and Richard Rosenfeld argued that much is to be gained by revisiting and revising Merton's paradigm of social structure and anomie. Messner and Rosenfeld (1994) noted that the United States has a higher rate of serious crime than any other industrial nation (see also Currie, 1985, 2009). Why is this so? They doubted that the rampant lawlessness could be traced to individual traits such as Americans having a disproportionate number of defective personalities or bodies. Instead, taking a macro-level perspective, they argued for the need to discern what is distinctive about the very culture and structure of American society. Their perspective typically is referred to as *institutional-anomie theory*.

When Messner and Rosenfeld first became friends in the 1980s, Merton's theory was in decline, with the anomie portion of the paradigm almost completely neglected and relegated to the criminological dustbin (Cole, 1975). However, Messner had been a student and then a colleague of Merton at Columbia University, whereas Rosenfeld had undertaken a dissertation examining inequality and crime that drew on anomie theory and had been schooled in the sociology of Talcott Parsons, one of Merton's mentors at Harvard. Unlike most other scholars of their generation, they were thus sympathetic to the anomie perspective and believed that it had untapped explanatory power. A fortuitous meeting in the early 1980s led to an intellectual merger that later resulted in their classic 1994 book, *Crime and the American Dream* (Rosenfeld & Messner, 2011).

Messner and Rosenfeld believed that criminology had often forgotten the basic lessons taught in introductory sociology. The core components of the social system were culture, which prescribed what individuals should and should not do, and social structure, which includes the positions people hold and the roles they play. Culture (the American dream) and social structure (differential opportunity) were central to Merton's anomie theory. In addition, however, Messner and Rosenfeld highlighted the importance of social institutions—an insight drawn from Talcott Parsons (1951):

> *Social institutions* link culture and social structure together in the context of the basic social functions any society must carry out in order to survive, including adaptation to the environment (economy), collective goal attainment (polity), social integration (legal system), and the maintenance of the society's fundamental normative patterns (family, religion). (Rosenfeld & Messner, 2011, p. 122, emphasis in original)

In their view, Merton's anomie theory had focused on only one social institution, the economy. It did not examine how social institutions are interrelated and how culture can cause one institutional sphere, especially the economy, to be overemphasized and to cause problems in other social institutions (e.g., the family). When this imbalance occurs, they would argue, high crime rates result. Because their perspective incorporated this relationship among institutions into anomie theory, it came to be called *institutional-anomie theory.* The specifics of their approach follow.

The American Dream and Anomie. The special contribution of Merton's paradigm, Messner and Rosenfeld contended, is that it identifies the central role that the American dream plays in generating criminal behavior (see also Messner, 1988; Rosenfeld, 1989). Messner and Rosenfeld (1994) defined the American dream as "a commitment to the goal of material success, to be pursued by everyone in society under conditions of open, individual competition" (p. 69). Like Merton, they noted that the American dream's strong cultural prescription for each individual to achieve monetary success, regardless of personal circumstance or the competition, erodes the power of norms to regulate the use of legitimate or legal means to obtain this success. In short, the American dream fosters anomie or the breakdown of normative control.

In this anomic context, there is a tendency for people to use the "technically most efficient means" to achieve desired goals. On the positive side, the prevailing normative flexibility creates space for innovation and for moving beyond outmoded ways of doing things. But there is a darker side as well: The most efficient means to monetary gain often is to break the law, to rob with a gun, or to defraud the stock market through insider trading. Thus, anomie is criminogenic, and widespread anomie—as exists in American society—creates widespread lawlessness.

Institutional Balance of Power. Although Messner and Rosenfeld embraced Merton's analysis of American culture, they found his institutional analysis lacking. As may be recalled, Merton emphasized the role of America's stratification system (the economic social institution) in restricting opportunities for advancement and, therefore, in creating criminogenic strains for individuals and anomie in society. Messner and Rosenfeld (1994) believed, however, that Merton's approach misses what is distinctive about structural arrangements in the United States: the extent to which the "institutional balance of power is tilted toward the economy" (p. 76).

A capitalist economic system is conducive to producing individual pursuit of self-interest, competition, and innovation, but other developed nations with capitalist economies are not wracked by high rates of serious crimes. What is different about the United States? Messner and Rosenfeld answered that it is the economic institution that dominates other social noneconomic institutions—the family, the educational system, politics, and so on—far more than in other countries. Indeed, U.S. social institutions are arranged to be subservient to and to support the economy. Education, for example, is seen as a means to higher paying employment; students who major in esoteric academic fields merely to learn or to foster personal development risk being ridiculed as "wasting their time" and "hurting their futures." Family values are touted frequently, but corporate executives are expected to uproot their families if promotions require moving to other locations. And politicians' reelection frequently hinges on little more than the strength of the nation's economy.

In other developed nations, however, social institutions counterbalance the capitalist economy's demand to pursue success at all costs and thus exert greater control over people's conduct. As Messner and Rosenfeld (1994) pointed out, these social institutions "inculcate beliefs, values, and commitments other than those of the marketplace" (p. 86), and they tend to demand a level of involvement that constrains behavior. In the United States, however, noneconomic institutions are less successful in socializing, securing the allegiance of, and controlling the citizenry. Instead, "the cultural message that comes through with greatest force is the one most compatible with the logic of the economy: the competitive, individualistic, and materialistic message of the American dream" (p. 86). The result is a context conducive to the spread of anomie and, in turn, of lawlessness.

In summary, Messner and Rosenfeld (1994, 2001) argued that the United States' high rate of serious crime is caused by the nation's distinctive, mutually reinforcing culture and institutional structure. The American dream serves as a powerful cultural force that generates anomie by motivating the pursuit of money "through any means necessary." It not only fosters but also receives continued legitimacy from the dominance of the economic institution. The intersection of culture and structure, therefore, creates a society in which anomie is pervasive and controls are weak. Predictably, the United States' crime problem seems intractable and immune to the many "wars" waged against it.

The Empirical Status of Institutional-Anomie Theory. Messner and Rosenfeld presented a persuasive perspective, but to date it has received a growing, but still relatively limited, body of empirical tests. In large part, the theory is difficult to assess because its main causal variables—the American dream, anomie, the institutional balance of power, and the efficacy of informal control exercised by societal institutions—are difficult to measure at the macro or societal level, especially within a single data set. The studies that do exist, however, provide qualified support for institutional-anomie theory (Baumer & Gustafson, 2007; Chamlin & Cochran, 1995; Cullen, Parboteeah, & Hoegl, 2004; Maume & Lee, 2003; Messner & Rosenfeld, 1997; Piquero & Piquero, 1998; for summaries of the extant research, see Kubrin et al., 2009; Messner & Rosenfeld, 2006).

In particular, there appears to be evidence that crime rates are lower in societies or other geographic areas in which the vitality and support for noneconomic institutions, such as the family, are more pronounced. In this regard, a meta-analysis of research by Pratt and Cullen (2005) revealed that the strength of noneconomic institutions is associated with lower rates of crime across macro-level units of analysis. The evidence is more controversial, however, when scholars have assessed whether the United States has a distinctive "American dream"—what has been called "American exceptionalism." This research examines the responses to questions on surveys conducted across nations. The approach is to see whether, when compared to those in other areas of the world, Americans are more likely to say that they value things like material wealth, competition, and private ownership over government control. In a review of the existing studies, Messner and Rosenfeld (2006, p. 143) reported that there are at least "mixed results" showing that U.S. respondents are exceptional in their strong support for values consistent with the American dream. Clearly, further comparative research, using questions systematically designed to capture all elements of the American dream, will be required to settle this issue.

The Market Economy and Crime. Finally, we should note that in an independent theoretical analysis, Elliott Currie offered a similar explanation of the high rate of serious violent crime in America. In 1985, Currie had published his classic book, *Confronting Crime: An American Challenge* (see Cullen, 2010). Taking what might be called a critical or left realist perspective (see Chapter 9), he counteracted the conservative view that crime was due to defects in individuals and to the failure to get tough by imprisoning more offenders (see Chapter 12). A central thrust to his analysis of high rates of violent crime was that the United States suffered from a high rate of economic inequality, which in turn disrupted other social institutions, such as families and communities. Currie subsequently elaborated this perspective by detailing how America's distinctive "market economy" dominated other parts of society in ways that proved criminogenic (Currie, 1997; see also Currie, 2009). This core insight to his theory thus was similar to the core insight of Messner and Rosenfeld's institutional-anomie theory.

Currie (1997) argued that the United States is characterized by an extreme form of capitalism, a "market economy" in which "the pursuit of personal economic gain becomes increasingly the dominant organizing principle of social life" (pp. 152–153). Echoing Messner and Rosenfeld's concept of institutional balance of power, Currie noted that in the United States, "market principles, instead of being confined to some parts of the *economy* and appropriately buffered and restrained by other social institutions and norms, come to suffuse the whole social fabric—and to undercut and overwhelm other principles that have historically sustained individuals, families, and communities" (p. 152, emphasis in original). He proceeded to contend—again, much like Messner and Rosenfeld—that the United States is "unique" in the "extent to which the principles of the market society have driven America's social and economic development and shaped its national culture" (p. 152).

According to Currie (1997), the dominance of the market economy fosters high rates of crime in at least seven ways:

(1) the progressive destruction of livelihood; (2) the growth of extremes of economic inequality and material deprivation; (3) the withdrawal of public services and supports, especially for families and children; (4) the erosion of informal and communal networks of mutual support, supervision, and care; (5) the spread of materialistic, neglectful, and "hard" culture; (6) the unregulated marketing of the technology of violence (i.e., guns); and (7) the weakening of social and political alternatives. (p. 154)

In this "toxic brew" of conditions, the power of the market economy is so strong that concern for productivity and profits outstrips the concern for the needs of many people who struggle under this system. America's market economy is particularly criminogenic because it "plunges large segments of its population into extreme deprivation—while simultaneously both withdrawing public provision of support and chipping away relentlessly at the informal support networks that might cushion these disadvantages" (Currie, 1997, p. 167). In this context, citizens are urged to embrace a "competitive consumerism" in which individual success and materialism are held in esteem and the losers in this competition are left to deal with their failures and frustrations on their own and with lethal weapons easily obtained. At the same time, the cultural ideology of the market economy disdains the kinds of policy initiatives—commonly found in other Western industrial nations—that might provide disadvantaged Americans with the opportunity to earn living wages and support their families. Instead of social interventions, individuals are largely left to fend for themselves. In the end, the main government program embraced to address the problem of violent crime is to build more and more prisons to house the endless casualties of the nation's unbalanced market economy.

THE FUTURE OF STRAIN THEORY

It is difficult to determine whether Agnew's general strain theory and Messner and Rosenfeld's analysis of crime and the American dream are harbingers of strain theory's reemergence as a dominant criminological theory. At a minimum, the intrinsic scholarly merit of these works will contribute to the revitalization of strain theory. Many criminologists will read these revisions of strain theory, and more than a few will continue to publish articles testing the various propositions put forth by Agnew and by Messner and Rosenfeld. In 21st-century America, however, it also is possible that the prevailing social context will be conducive to renewed interest in the strain perspective. Two conditions seem particularly relevant.

First, in response to the excessive individualism and emphasis on greed during the 1980s and 1990s, there has been a continuing cultural self-examination in the United States. Although many different voices are contributing to this discussion, they share the common theme that the core elements of the American dream—the pursuit of individual self-interest, competitiveness, and materialism—neither bring personal satisfaction nor create a good society. Instead, there is a call for Americans to think more about community, about the obligations of citizenship, and about helping others (Bellah, Madsen,

Sullivan, Swidler, & Tipton, 1985, 1991; Coles, 1993; Etzioni, 1993; I
1988, 2001; Wuthnow, 1991). Even notions of "compassionate conse
theme of George W. Bush's first presidential campaign—emphasize i
self-interest and the neglect of one's fellow citizens (see also Olasky, .
the American dream comes under critical scrutiny—as its darker si
the relevance of strain theory is potentially heightened.

Second, we also may be in a period where it is more difficult to ig.   ...p...-
ity of U.S. society in a range of social ills, including crime. During the 1980s, the strong
conservative ideology of the Reagan era made it fashionable to blame social problems,
including crime, on individuals and not on structural arrangements. But the Los
Angeles riot of 1992, triggered by the acquittal of police officers charged with beating
Rodney King, was a vivid reminder that the neglect of underlying problems eventually
results in a day of reckoning (see also Currie, 1993). A similar lesson can be drawn
from the Cincinnati Riot of 2001. Similarly, Hurricane Katrina not only left New
Orleans in ruins but also laid bare the poverty and racial inequality that placed the
lives of some citizens, but not others, in jeopardy. Writing during the mid-1990s,
Messner and Rosenfeld (1994) made this point more broadly. They noted:

> Given the growing awareness of vexing contemporary social problems—such as home-
> lessness, the urban underclass, persistent economic stagnation accompanied by glaring
> social inequalities, and urban decay in general—explanations of social behavior cast in
> terms of fundamental characteristics of society, rather than individual deficiencies, are
> likely once again to "make sense" to many criminologists. (pp. 14–15)

Even though the U.S. economy boomed during the latter part of the 1990s, the
awareness that some people were not participating in the technological revolution and
were being left behind—largely concentrated in segregated inner-city neighborhoods—
did not slip fully from the national consciousness. This understanding that the nation's
market economy can thrust even good people into financial ruin and stop hard-work-
ing folks from securing employment has been vividly—and in many instances heart-
breakingly illustrated by the Great Recession that started in 2008. To the extent that
this awareness of prevailing inequalities remains strong, strain theory—with its focus
on the criminogenic costs of blocking opportunities for advancement and of placing
people in adverse social circumstances—will receive its fair share of adherents in the
criminological community.

The Consequences of Theory: Policy Implications

EXPAND OPPORTUNITIES

If denial of opportunity generates criminogenic strains, then logic demands that
the solution to crime lies in expanding legitimate opportunities. Most broadly, therefore, strain theory justifies programs that attempt to provide the disadvantaged with
educational resources (e.g., Head Start), job training, and equal access to occupations.

The perspective also would support efforts to introduce into prisons rehabilitation programs that allow offenders to earn educational degrees and to acquire marketable employment skills.

Apart from these general policy implications, it is noteworthy that strain theory has served as the basis for a variety of delinquency prevention programs (Empey & Erickson, 1972; Empey & Lubeck, 1971), the most famous of which was Mobilization for Youth (Empey, 1982; Moynihan, 1969; Pfohl, 1985). MFY, as the program was known, was based directly on Cloward and Ohlin's opportunity theory. The two scholars, in fact, played an integral role in seeing their ideas put into practice.

In 1959, Cloward and Ohlin had been asked by a coalition of settlement houses located in New York's Lower East Side community to develop a theoretical framework for a proposal to secure government funds to provide youths with social services (Laub, 1983). *Delinquency and Opportunity* (Cloward & Ohlin, 1960) was one by-product of this collaboration. The other product was a 617-page report carrying the title *A Proposal for the Prevention and Control of Delinquency by Expanding Opportunities* (Empey, 1982, p. 241).

During different social times, this proposal might have found a home on a dusty shelf, but during the early 1960s it offered a blueprint for social engineering that made eminent sense. The Kennedy administration had taken office recently with hopes of creating a "New Frontier of equal opportunity" and was ready to fulfill a long-standing family commitment to dealing with the problems of young Americans (Pfohl, 1985, pp. 224–225). In 1961, the administration created the President's Committee on Juvenile Delinquency and Youth Crime. David Hackett headed the committee.

In looking for promising strategies to address delinquency problems, Hackett learned of Cloward and Ohlin's work. As Pfohl (1985) noted, their opportunity theory "resonated well with the liberal domestic politics of John F. Kennedy" (p. 224) and particularly with the president's call for equal opportunity. Indeed, the fit between strain theory and the prevailing political context was so close that Ohlin was invited to Washington, D.C., to assume a Health, Education, and Welfare post and to assist in formulating delinquency policy. By May 1962, MFY had received a grant of $12.5 million, with more than half of that amount coming from the federal government. Cloward, meanwhile, was chosen as MFY's director of research.

MFY drew heavily from Merton's strain theory, setting up programs that extended youths' educational and employment support. This reform also drew insights from the Chicago school, particularly Shaw and McKay's admonition that community organization was a prerequisite for delinquency prevention. But one philosophical difference existed between MFY and Shaw's CAP. Although both programs were committed to community self-help, the leaders of MFY also felt the need to change the political structures that sustained inequities in opportunity. They agreed, for example, that the problem with employment was not simply that minorities lacked skills but rather that they were excluded from union apprenticeships, and they agreed that the problem of poor educational opportunities was a matter not simply of youths' lacking books in the home but also of policies that assigned the newest and least talented teachers to schools in slum neighborhoods. What was needed to overcome such formidable barriers to opportunity, therefore, was not community organization but rather community action that attacked entrenched political interests. Accordingly, MFY promoted boycotts against schools, protests against welfare policies, rent strikes against slum landlords, lawsuits to ensure poor people's rights, and voter registration.

Commentators have disputed the wisdom of moving from the more focused agenda of providing youths with educational and employment training to a broader attempt to reform the opportunity structure by empowering slum residents (Moynihan, 1969). The choice of strategy soon brought MFY into a political struggle with city officials dismayed by threats of lawsuits and other efforts by the poor to get their share of the pie. In the ensuing conflict, MFY was subjected to investigations by the Federal Bureau of Investigation on suspicions of misappropriated funds and to claims by the *New York Daily News* that the staff was infested with "Commies and Commie sympathizers." Although they were exonerated of these kinds of charges, the swirl of controversy led to the resignation of key MFY leaders, including Cloward, and ultimately to the disappearance of much of the program (Liska, 1981, p. 52). In Washington, meanwhile, the President's Committee on Juvenile Delinquency and Youth Crime was allowed to wither away. "Influential members of Congress made it clear," Empey (1982) noted, "that the mandate of the President's Committee was to reduce delinquency, not to reform urban society or to try out sociological theories on American youths" (p. 243).

Thus, attempts to reduce delinquency by fundamentally changing the nature of the opportunity structure met with much opposition. But should MFY be judged as a failure? A less ambitious and confrontational reform might have succeeded in gaining more sustained political acceptance and support for programs. Two qualifications, however, must be added.

First, MFY must be credited for its attempt to attack the root causes of crime that most criminal reforms leave untouched. MFY's failure to avoid boisterous and powerful opposition says less about the correctness of the approach taken and more about the political interests that buttress America's structure of inequality. Second, although MFY's impact on delinquency has not been established definitively (Liska, 1981), it did serve as a model for similar community action programs across the nation. Daniel Moynihan (1969), a noted critic of the program, nonetheless gave MFY the credit it is due:

> MFY did lose in its battle with city hall, but it set a pattern for the community action programs that were to spring up across the land at the very moment its own came under fire. . . . Preschool education, legal aid for the poor (not just to defend them but to serve as plaintiffs), a theory of community organization, an emphasis on research and evaluation, and most especially the insistence on the involvement of the poor—all these were the legacy of Mobilization for Youth. It was no small achievement. (p. 123)

TAMING THE AMERICAN DREAM

A key feature of attempts to reduce crime by expanding opportunities is that they implicitly take the American dream—people's striving for individual material success—as legitimate. For Messner and Rosenfeld (2001), however, the American dream and the goals it inspires individuals to pursue are key ingredients in the nation's high rates of serious crime. Indeed, attempts to expand opportunities—the prescription for reducing crime derived from traditional strain theory—may have the unanticipated consequence of intensifying the very cultural beliefs that promote crime. As Messner and Rosenfeld (2001) remind us:

> The criminogenic tendencies of the American Dream derive from its *exaggerated* emphasis on monetary success and its resistance to the limits on the means for the pursuit of success. Any significant lessening of the criminogenic consequences of the dominant culture thus requires the taming of its strong materialistic pressures and the creation of a greater receptivity to socially imposed restraints. (p. 108, emphasis in original)

Taming the American dream, however, may be "better said than done." There is no simple program that can be implemented to make Americans less interested in economic success. Rather, true social change will be necessary—a daunting and, some would say, improbable task in the sense that it asks Americans not to be Americans (but see Putnam, 2000).

The logic of institutional-anomie theory suggests that reducing the power of the American dream would necessitate policies that strengthen other institutions. This might involve, for example, workplace regulations that give adults more time to spend with their families and support schools. Or it might mean social welfare policies that help people to form families and that give everyone a measure of material security. But it also would be important, according to Messner and Rosenfeld (2001), to pursue "cultural regeneration" directly. That is, it is essential to discredit money as the chief currency of a person's success and instead to propose that "parenting, 'spousing,' teaching, learning, and serving the community" become valued "ends in themselves" (p. 108). Such an appeal might appear utopian. Even so, as Messner and Rosenfeld (2001) have cautioned, it seems unlikely that the United States will achieve meaningful and sustained reductions in crime "in the absence of a cultural reorientation that encompasses an enhanced emphasis on the importance of mutual support and collective obligations and a decreased emphasis on individual rights, interests, and privileges" (p. 109).

Conclusion

Over the past two chapters, we have seen how two paradigms—the Chicago school and Merton's strain theory—represent early, yet bold and influential, efforts to show how the very fabric of American society—its slums and the contradictions between its cultural prescriptions and social structure—generates high rates of crime. They rejected as simplistic, if not as incorrect, previous theories that had sought to locate the causes of crime within individuals. Instead, they warned that the social organization of society constrains what people learn to become and what they might be pressured into doing. This is a lesson that cannot be overlooked casually, for it suggests, in a sense, that a society gets the crime it deserves.

The Chicago school and strain theory also illustrate the themes of this book. First, changes in the social context made each of these perspectives make sense to a good number of criminologists. For the Chicago school, the rapid growth and increasing diversity of urban society gave legitimacy to a theory that linked crime to these social transformations. For strain theory, the emergence of equal opportunity as a sociopolitical agenda provided an ideal context in which opportunity theory could win followers.

Second, the history of these perspectives shows how criminological theory can direct, or at least justify, criminal justice policy. CAP and MFY cannot be understood apart from the theories that provided the logic for their development and were convincing enough to prompt financial support. The writings of Shaw and McKay and of Cloward and Ohlin, therefore, clearly made a difference. Ideas, as we have seen, have consequences.

Walter C. Reckless
1899–1988
Ohio State University
Author of Containment Theory

5

Society as Insulation

The Origins of Control Theory

T he most important step in the solution to a problem, theoretical or otherwise, lies in asking the right questions about it. The proper formulation of the problem often may make the answer obvious. At the very least, asking the right question in the right way will contribute greatly to a solution. Asking the wrong question, or asking the right question in the wrong way, will result in no progress at all, although there may be an illusion of getting somewhere.

Unfortunately, the first step also usually is the most difficult one. Not only does it require a thorough understanding of the nature of the situation with which one is faced, but it also demands the capacity to think about it in new and creative ways. Because one is not likely to ask questions about aspects of the situation that one takes for granted, different criminologists have been led to see very different questions as the key to an understanding of crime and delinquency. What one takes for granted needs no explanation, but what one takes for granted depends on assumptions—in this case, assumptions about human nature and social order.

Most criminologists have taken conformity for granted as part of the natural order of things and have concentrated on trying to explain the "crime problem." As we have seen, they have found their explanations in spirits and demons, in theories tracing the nonconformity to individual factors such as biological abnormalities or personality defects, or in theories tracing the nonconformity to social factors such as social disorganization, subcultural traditions, and inequality of opportunity, all of these being factors presumed to operate so as to distort the natural order of conformity. But is conformity really the natural order of things? To what extent should it be taken for granted?

By the time a person is a certain age, he or she speaks a certain language shared by others, drives a car in general obedience to traffic regulations, avoids urination or defecation in public, and in most other ways "goes along." All this tends to be taken for granted, but the evidence indicates that it is not at all "natural" (Davis, 1948). In fact, great effort is expended by parents, teachers, and the individual involved in a concerted attempt to produce these results. Viewed in this way, all of this conformity is a striking thing much in need of explanation. That is the focus of *control theory*, which takes the position that because conformity cannot be taken for granted, *nonconformity such as crime and delinquency is to be expected when social controls are less than completely effective.*

In this sense, control theory is not so much a theory of deviance as a theory of conformity. It does not ask the question, "Why do people commit crimes and acts of delinquency?" Rather, it suggests that crime and delinquency are going to occur unless people conform to all of the social demands placed on them and then asks, "Why do people conform?" That is, if crime is gratifying—fun, exciting, physically enjoyable, emotionally satisfying, and/or materially rewarding—why don't individuals just break the law? From this perspective on human nature and social order, the potential gratifications offered by crime and delinquency will be resisted only when sociocultural controls are operating effectively to prevent such behavior.

Control theory has been at the center of American criminology for the better part of a century—so much so that we devote two chapters to this perspective. Much of its appeal is the simplicity of its main theoretical premise: When controls are present, crime does not occur; when controls are absent, crime is possible and often does occur. Except for the fact that control and crime can be measured independently and the strength of the relationship assessed empirically, there is a tautological quality to this thesis: The very existence of crime seems as persuasive evidence that controls have been rendered ineffective.

Although the organizing premise of control theory is simple—no control permits crime—the perspectives that fall under the umbrella of the control paradigm offer a complex view of how control is linked to criminal conduct. In this chapter, we explore the thinking of early control theorists, selecting those scholars whose work still resonates today. These include Albert Reiss's insights on personal and social controls, F. Ivan Nye's work on family controls, Walter Reckless's influential containment theory, Gresham Sykes and David Matza's revelations about techniques of neutralization, and Matza's extension of that collaboration into his theory of delinquency and drift.

In Chapter 6, we chiefly explore the work of Travis Hirschi, arguably the most influential criminologist of the past half century. Hirschi developed two dominant control perspectives: first his social bond theory and then, in conjunction with Michael Gottfredson, self-control theory. We also consider some more recent variants of control theory that present more complex ideas on how control is implicated in crime and delinquency: Hagan's power–control theory, Tittle's control balance theory, and

Colvin's differential coercion theory. Finally, at the end of Chapter 6, we review the policy implications of control theory.

Before embarking on this excursion across control theory, however, we first examine some forerunners of modern control theory. As in Chapters 3 and 4, here we stress the significance of the work of Durkheim and the Chicago school. Both bodies of work remain rich sources of ideas about both human nature and social order. It is not surprising to find that they also contributed much to the reconsideration of the basic assumptions about the relationship between human nature and social order. This marked the development of control theory.

Forerunners of Control Theory

DURKHEIM'S ANOMIE THEORY

The origins of contemporary control theories of crime and delinquency are to be found in part in the work of French sociologist Émile Durkheim, in the same body of theory that inspired Merton's analysis of anomie as a source of crime (see Chapter 4). Durkheim's work was a product of the late 19th century, a period that had seen dramatic social change in the wake of the industrial revolution. He described anomie not simply as "normlessness" but as the more or less complete collapse of social solidarity itself, the destruction of the fundamental bonds uniting individuals in a collective social order so that each person is forced to go it alone. Technological change had combined with the rise of capitalism, and the old world of agrarian society, in which farmers and herders lived simply in face-to-face relationships with interests in common, was rapidly giving way to a more complex, urban, technologically sophisticated social system. The sense of community was being eroded; the large extended family consisting of many relatives working together was being torn apart and replaced with the new nuclear family of parents alone with their children, and the pace of life was accelerating with an increasing division of labor separating different individuals into occupational specialties.

Describing London in 1844, Karl Marx's friend and collaborator, Friedrich Engels, painted the picture as follows:

> The restless and noisy activity of the crowded streets is highly distasteful and is surely abhorrent to human nature itself. Hundreds of thousands of men and women drawn from all classes and ranks of society pack the streets of London. . . . Yet they rush past each other as if they had nothing in common. They are tacitly agreed on one thing only—that everyone should keep to the right of the pavement so as not to collide with the stream of people moving in the opposite direction. No one even thinks of sparing a glance for his neighbors in the streets . . . . We know well enough that this isolation of the individual—this narrow-minded egotism—is everywhere the fundamental principle of modern society. (cited in Josephson & Josephson, 1962, p. 32)

As we will see in Chapter 8, Marx and Engels saw this situation as a consequence of the underlying economic shifts that had generated capitalism. Writing in France later in the century, Durkheim disagreed. For him, the moral order was more fundamental than the economic order. The idea of social solidarity can be said to represent almost a religion to Durkheim (1933):

> Everything which is a source of solidarity is moral, everything which forces man to take account of other men is moral, everything which forces him to regulate his conduct through something other than the striving of his ego is moral, and morality is as solid as these ties are numerous and strong. (p. 398)

The Importance of Integration and Regulation. In Durkheim's (1897/1951) view, social solidarity was maintained by two distinct sets of social functions: those involving integration and those involving regulation. Integration was described as a state of cohesion amounting to a common "faith" sustained by collective beliefs and practices leading to strong social bonds and the subordination of self to a common cause. For Durkheim, collective activity was what gave purpose and meaning to life. When integrative functions failed, the "collective force of society" was weakened, "mutual moral support" was eroded, and there was a "relaxation of social bonds" leading to extreme individualism (pp. 209–214).

Whereas Durkheim (1897/1951) saw integration as the sum of various social forces of *attraction* that drew people together, *regulation* was considered to be the sum of those forces of *constraint* that bound individuals to norms. Durkheim argued that the constraining regulative functions become more important in an urban society with a complex division of labor. Here different individuals may be attracted to a common goal and may be very willing to submit to the authority of the social system, but their efforts must be coordinated properly if society is to function smoothly. It is especially important that the regulative norms "deliver the goods"—that they maintain their legitimacy by demonstrated effectiveness. Even if an individual wished to work with others toward a common social purpose, he or she might turn in a deviant direction if the norms regulating the common effort were perceived as unnecessary, overly burdensome, or otherwise questionable.

The Nature of "Man." Durkheim's (1897/1951) viewpoint was affected deeply by his conception of human nature. According to his homo duplex conception, any person was a blend of two aspects. On the one side, there was the social self or the aspect of self that looks to society and is a product of socialization and cultivation of human potentials—the "civilized" member of a community. On the other side, there was the egoistic self or the primal self that is incomplete without society and that is full of impulses knowing no natural limits. This conception of the primal aspect is somewhat similar to Freud's notion of the id. In Durkheim's view, conditions of social solidarity based on highly developed functions of social integration and social regulation allowed the more primal self to become fully humanized in a life shared with others on a moral common ground. One implication is that unless such social solidarity is developed and maintained, we may expect crime and delinquency.

THE INFLUENCE OF THE CHICAGO SCHOOL

As described in Chapter 3, the social upheaval that so concerned Durkheim in France was repeated even more dramatically in the United States a few decades later. During the early years of the 20th century, and especially after World War I, the forces of technological change, increased industrialization combined with the rise of the large corporation, and rapid urbanization were accompanied by massive waves of immigration. Chapter 3 already has described the development of the Chicago school with its emphasis on immigration and especially urbanization as the major forces of social change that were leading to crime through social disorganization. But the Chicago school mined a rich vein of ideas about human nature and the social order, ideas that led some criminological theorists not toward a search for criminogenic forces in society that might be pulling or pressuring people away from their normal conformity but rather toward control theories that refused to take conformity for granted as the natural order of things.

Like some of the criminological theories discussed earlier, these control theories also were influenced by the social disorganization perspective, some explicitly and some implicitly. Although the accent was somewhat different among the various control theorists, two related Chicago school themes remained central. The first had to do with the interpretations of the nature of *human nature*. The second had to do with the nature of *community*.

Conceptions of Human Nature. As to human nature, the Chicago school developed the line of thought that explained the self as a blending of a primal self and a social self, but the focus was on showing just how the social self, so important to Durkheim, was formed. Although not at the University of Chicago himself, Charles H. Cooley can be considered part of "what came to be known as the Chicago school of social psychology" (Hinkle & Hinkle, 1954, p. 30). Cooley's work influenced George Herbert Mead of the University of Chicago, and Mead carried forward this and other contributions. Both Cooley and Mead considered the common notion of imitation insufficient to explain social behavior and sought to understand the process by which something like the primal self became the social self.

Cooley (1922) pointed out that the human offspring is dependent on other humans in the family setting for a prolonged period. The family was treated as an example of what Cooley called a *primary group* in which interaction is of an intimate face-to-face character leading to a "we-feeling" or sense of belonging and identification with the group. The child would not even develop a sense of self without the feedback provided by others serving as a mirror. According to Cooley's concept of the "looking-glass self," the child develops a concept of who he or she "really is" by imagining how he or she appears to others and how others interpret and evaluate what they perceive and then by forming a sense of self based on that process (Cooley, 1902, 1909). Without interaction in primary groups, a person would not be fully human. For Cooley (1909), human nature itself, defined as "those sentiments and impulses that are human in being superior to those of lower animals," was essentially the same throughout the world because the intense experience in primary groups was considered to be basically the same everywhere (p. 28).

Mead's (1934) social psychology was reminiscent of Durkheim's *homo duplex* in dividing the individual into an "I" and a "me." Mead did little to describe the "I" except to suggest that this represents a process of fundamental awareness that becomes focused in different ways, leading to the development of the social self or "me." The focusing was said to occur through a process of "taking the role of the other" and seeing things from that perspective, with certain "significant others" being especially important and society representing a "generalized other" (pp. 82–92). It was through this process that socialization occurred.

If successful, such socialization was considered to lead to personally integrated social beings who would see the world not through the narrow and unstable perspective of the shifting "I" (to the extent that the "I" can be said to be capable of organized perception at all) but rather through a social "me" that included others' perspectives and interests as part of oneself. One implication was that *unsuccessful* socialization might lead to *personal disorganization*—to a self lacking in integration and consistency or to a self that was integrated internally but not integrated into society (or both).

The Study of Community. The second related Chicago school theme, the study of community, was reflected in the work of Park and Burgess and in the studies of the spatial distribution of natural areas and conflicting cultural traditions undertaken by Shaw and McKay (as outlined in Chapter 3). There the stress was on the idea of social disorganization. In addition to this approach, much of the social psychological emphasis in the works of different members of the Chicago school had to do with the possibility of such social disorganization through the collapse of community as a consequence of increasing "social distance" among individuals who refused to get close to one another (Hinkle & Hinkle, 1954). Wirth (1938) described it as a problem of segmentation that separated people and made it impossible for them to relate to one another as total personalities. This theme emphasized the impersonality and anonymity of life in urban industrialized societies in which people in the community did not know or care about one another and preferred it that way. The descriptions sounded like Durkheim's discussion of anomie or Engels's picture of London in 1844, only more so.

Beyond the concern with the apparent decline of community at a municipal or neighborhood level, some working within the tradition of the Chicago school focused attention on an even more ominous trend—the apparent decline in the moral integration of the basic primary groups themselves. Even the family, the most primary of all groups, seemed to be losing its influence over its members. If this proved to be true, then it suggested not only social disorganization but also personal disorganization resulting from fundamental problems in the formation of the personal self. It was out of this context of social transformation and sociological thought that control theory was to emerge in criminology.

Early Control Theories

The concept of social disorganization led different criminological theorists in different directions. As suggested in Chapter 3, ideas of social disorganization could be seen in terms of how it created criminal cultures that, when transmitted and learned, *motivated* crime or in terms of a weakening of social controls that simply makes crime and delinquency more *possible*. The latter approach was taken by early control theorists, such as Albert Reiss and F. Ivan Nye. Again, each of these theorists tended to take the position that crime and delinquency could be *expected* in conditions where controls were not effective.

REISS'S THEORY OF PERSONAL AND SOCIAL CONTROLS

During the late 1940s, Albert J. Reiss completed a doctoral dissertation at the University of Chicago in which he attempted to develop an instrument for the prediction of juvenile delinquency (Reiss, 1949). Two years later, Reiss (1951) summarized part of his project in his article "Delinquency as the Failure of Personal and Social Controls." *Personal control* was defined as "the ability of the individual to refrain from meeting needs in ways which conflict with the norms and rules of the community" (p. 196). *Social control* was defined as "the ability of social groups or institutions to make norms or rules effective" (p. 196). According to Reiss (1951):

> Delinquency results when there is a relative absence of internalized norms and rules governing behavior in conformity with the norms of the social system to which legal penalties are attached, a breakdown in previously established controls, and/or a relative absence of or conflict in social rules or techniques for enforcing such behavior in the social groups or institutions of which the person is a member. (p. 196)

This approach clearly was influenced by the social disorganization tradition, but it had a control theory emphasis.

Reiss (1951) maintained that conformity might result either from the individual's *acceptance* of rules and roles or from mere *submission* to them. Such a distinction ran parallel to Durkheim's division of social solidarity forces into those representing *integration* and those representing *regulation*. Personal controls were said to include both

> (a) mature ego ideals or non-delinquent social roles, i.e., internalized controls of social groups governing behavior in conformity with non-delinquent group expectations and
> (b) appropriate and flexible rational controls over behavior which permits conscious guidance of action in accordance with non-delinquent group expectations. (Reiss, 1951, p. 203)

Considered from the "perspective of the person," social control was held to lie "in the acceptance of or submission to the authority of the institution and the reinforcement of existing personal controls by institutional controls" (Reiss, 1951, p. 201). Considered "from the standpoint of the group," it was said to lie "in the nature and strength of the norms of the institutions and the effectiveness of the institutional rules in obtaining behavior in conformity with the norms" (p. 201).

In an obvious reference to Chicago school work in the tradition of Shaw and McKay, Reiss (1951) took pains to stress that "this formulation is not in contradiction with the formulations which view certain types of delinquency as a consequence of social control in the delinquent gang" (pp. 196–197). He was not concerned with the way in which delinquency might be fostered through cultural transmission of norms that conflicted with those of the larger society; rather, he was concerned with underlying processes that seemed (1) to occur *prior* to any later processes such as conflicting cultural transmissions and (2) to be *necessary* before any such subsequent processes (whatever they might be) could ever take place. Being both antecedent and necessary, the processes involving loss of personal and social control might be considered more basic. As Reiss himself put it, "The delinquent peer group is here viewed as a *functional consequence* of the failure of personal and social controls" (p. 197, emphasis added).

It is important to remember that Reiss's goal was to develop a prediction instrument. As one reviewer pointed out at the time, Reiss was not seeking to explain what caused delinquency; rather, he was seeking to pin down those factors that had to occur before any such causes could be expected to produce their effects. "The chief concern as a predictor was failure to submit to social controls, the 'why' of such action being less important" (Symons, 1951, p. 208).

Following the tendency of the Chicago school tradition that stressed the *social psychological* dimension, Reiss (1951) began with the assumption that "primary groups are the basic institutions for the development of personal controls and the exercise of social control over the child" (p. 198). It followed that "delinquency and delinquent recidivism may be viewed as a consequence of the failure of primary groups to provide the child with appropriate non-delinquent roles and to exercise social control over the child so these roles are accepted or submitted to in accordance with needs" (p. 198). The key primary groups were identified as the family, the neighborhood, and the school.

Speaking of the family, for example, Reiss (1951) maintained that there is "social control over the child's behavior when the family milieu is structured so that the child *identifies* with family members . . . and *accepts* the norms" (pp. 198–199, emphasis added). On the other hand, acceptance of or submission to such social control may decline if the family fails to "meet the needs of its members" and "provide for members' needs through the purchase of material goods and services" (p. 198). Control also might be lost if the family exercised either "over-control or under-control of" the child's behavior (p. 199).

NYE'S FAMILY-FOCUSED THEORY OF SOCIAL CONTROLS

One of the leading figures in the sociology of the family, F. Ivan Nye, was a product of Michigan State University and very much under the influence of the Chicago school. During the late 1950s, he set forth a more systematic version of control theory, making explicit the formulation of the problem as one of explaining conformity rather than nonconformity. According to Nye (1958), "It is our position, therefore, that in general behavior prescribed as delinquent or criminal need not be explained in any positive sense, since it usually results in quicker and easier achievement of goals than the normative behavior" (p. 5). Thus, the problem for the theorist was not to find an explanation for delinquent or criminal behavior but rather to explain why delinquent and criminal behavior is not more common.

In this view, it was not necessary to find some "positive" factor(s) of a biological, psychological, or sociological nature that operates to "cause" crime and delinquency. The problem was to locate the social control factor(s) that inhibited such nonconformity, the implication being that when the factor(s) was operating ineffectively for some reason, crime and delinquency became a *possibility* available to an individual. Nye's theoretical and research focus was on adolescents, and he considered the family to be the most important agent of social control over them. As the major primary group representing society, the family could generate *direct control, internalized control, indirect control, and control through alternative means of need satisfaction*. The same patterns of social control could be generated by other social institutions, although perhaps less powerfully.

These four modes of social control also were reminiscent of Durkheim's concepts of a combination of integration and regulation. Direct control was considered to be imposed on the individual by external forces such as parents, teachers, and the police through direct restraints accompanied by punishment for violation. Internalized control was considered to occur when the individual regulated his or her own behavior, even in the absence of direct external regulation, through some process such as "conscience" or superego capable of restraining egoistic impulses. Indirect control had to do with the extent of affection and identification integrating the individual with authority figures in general and with parents in particular. Such integration might serve to keep the person in line when regulation through either direct control or internalized self-control was minimal. Finally, a social system that made available various means of achieving satisfaction rather than demanding that everyone pursue exactly the same goal in exactly the same manner was held to exert social control by "delivering the goods" in such a variety of legitimate ways that the temptation to nonconformity was reduced.

Although these different modes of social control were held to operate somewhat independently, with one being more important in a particular situation and another in a different context, Nye (1958) pointed out that they were mutually reinforcing. The sense of identification with and affection for parents should, to use Reiss's (1951) terms, facilitate both an *acceptance* of and a *submission* to their direct controls. The same integration into the family unit should result in deeper internalization of parental expectations and regulations through an internal conscience that regulated behavior from within the individual. The fact that the family prepared the individual for the pursuit of alternative goals in life through a variety of means and that the larger society made this possible was a major force for the integration of the individual into the social order. Thus, integration and regulation combined to reinforce one another and to reduce the likelihood of nonconforming behavior such as crime and delinquency.

Reckless's Containment Theory

Walter C. Reckless also was raised intellectually in the tradition of the Chicago school. He would come to offer one of the most elaborate and, for some time, influential control theories. He would explore not only different sources of control but also a problem that was ignored by Shaw and McKay: not why many youths in disorganized areas became delinquent, but why some youths did not. Later scholars would use the term *resiliency* to describe youngsters who, despite facing an array of criminogenic risk factors, nonetheless resist crime and pursue conventional lives (see, e.g., Turner, Hartman, Exum, & Cullen, 2007).

Reckless obtained his doctorate at the University of Chicago during the mid-1920s, at the point of theoretical dominance for the Chicago school. Consistent with the Chicago school tradition, Reckless worked with Robert Park and Ernest Burgess. He would eventually undertake a dissertation on prostitution. Huff and Scarpitti (2011) explain Reckless's movement into criminology:

> By then, he was working in a roadhouse playing his violin and observing the activities going on around him: illegal drinking, gambling, prostitution, and various petty rackets. For a fledging sociologist interested in crime, this was fertile ground for the development of idea about illegal behavior. (p. 278)

Although this cannot be verified, Reckless relayed (to Cullen) that Al Capone would come into the establishment and put a large sum of money on the bar. He then would lock the doors and say that nobody could leave until enough shots of alcohol were consumed to exhaust the money. Reckless said that on these nights, he would have to play his violin into the early hours of the morning.

Beyond these experiences, Reckless also was influenced by ideas in the Chicago school about social psychology and their relevance for criminology. He asserted during the early 1940s that the central problem lay in explaining "differential responses" (1943, p. 51). By this he meant that criminology ought to pursue a search for "self-factors" that would explain why some individuals succumbed to social pressures leading to crime and delinquency, whereas others remained relatively law abiding in the same circumstances (see also Cullen, 1984). Partly as a result of his search for self-factors during the 1940s and 1950s, Reckless (1961) was able to present what he called his *containment theory* in some detail at the beginning of the 1960s.

THE SOCIAL PSYCHOLOGY OF THE SELF

Like Durkheim and the members of the Chicago school in general, Reckless argued that the great social transformation from life in fairly simple, integrated, agrarian societies to life in complex, technologically sophisticated, highly industrialized urban environments placed a different set of pressures on the individual and the social order. As he revised his theory during the 1960s, Reckless (1967) became more explicit about the way in which the historical transformation represented a "new pitch" in social psychological terms:

> The new situation or "pitch" may be described as follows. In a fluid, mobile society which has emphasized freedom of action for its individuals, the person is able to soar

like a balloon without the ballast of social relationships. He can readily aggrandize himself at the expense of others. His society does not easily contain him. He no longer fits into expected roles. . . . He plays his major themes in life without agreed-upon ground rules. (p. 21)

Turning specifically to an examination of "crime in the modern world," Reckless focused attention on the "individualization of the self" (pp. 10–12).

As an entity in personal development, the self seems to have had a sort of natural history. It appears to be relatively unimportant in primitive, isolated societies in which individuals very seldom break away from the undifferentiated uniformity of kin, village, and tribe. When human society in the course of evolution becomes differentiated, develops an increasing division of labor, and presents alternate choices to its members, the selves of certain persons begin to individualize and take on separate and distinct identities, departing from the mass of the people. (p. 11)

PUSHES AND PULLS

As a control theorist, Reckless did not purport to offer a theory of crime causation. Indeed, he suggested that a variety of factors (e.g., biophysical forces, psychological pressures, social conditions such as poverty) might "push" a person toward crime or delinquency and that other factors (e.g., illegitimate opportunities) might "pull" one toward misbehavior. He recognized that the leading sociological theories in particular seemed to have effectively analyzed many of the central pushes and pulls. One of his more graphic analogies compared the etiology of crime and delinquency to that of malaria, insisting that the sociological theories stressing social structural forces be complemented by social psychological concepts emphasizing differential response.

Sociological theories that do not account for a self factor to explain differential response are akin to explaining malaria on the basis of the lady mosquito, the swamp, or the lack of screens on houses. But everybody does not get malaria, even under conditions of extreme exposure. Some get a light touch. Some get a very severe attack. Some remain relatively immune. The resistance is within the person—his blood, his chemistry, his differential immunity. (Reckless, 1967, p. 469)

Reckless's containment theory was meant to explain why *in spite of* the various criminogenic pushes and pulls, *whatever they may be,* conformity remains the general state of affairs. He argued that to commit crime or delinquency requires the individual to break through a combination of outer containment and inner containment that together tend to insulate the person from both the pushes and the pulls. With rare exceptions, only when these powerful containing forces were weakened, could deviance occur. And even then it was not assured, given that containment theory was considered a *risk theory* dealing in probabilities. Every weakening of containment was seen as tending to increase the odds for nonconformity by opening a breach in the armor provided by external social control and internal self-control.

Reckless's attraction to a control theory called containment has two other potential sources. First, he was influenced by the work on personal and social controls by Albert Reiss (1951), whose theory was discussed earlier. Similar to Reckless, Reiss also was trained at the University of Chicago. Second, as relayed to one of the book's authors (Cullen), he attributed the origins of his theory to his travels in the Middle East when he still contemplated studying archaeology. He was struck by how regulated or "contained" social behavior was and thus by the lack of outward displays of deviance. These insights, he explained, would later help to shape his thinking on crime (see also Huff & Scarpitti, 2011).

FACTORS IN OUTER CONTAINMENT

The listing of factors involved in outer containment differed somewhat in various statements of containment theory as Reckless (1967) attempted to refine his theoretical position. He pointed out that the key factors binding the individual to the group might vary across different types of societies. Concentrating on "the external containment model for modern, urban, industrial, mobile society," he stressed (1) reasonable limits, (2) meaningful roles and activities, and (3) "several complementary variables such as reinforcement by groups and significant supportive relationships, acceptance, [and] the creation of a sense of belonging and identity" (pp. 470–471). Durkheimian themes were clear here, although not spelled out in detail. For example, although the containing limits reflected effective regulation, the latter set of variables "can also be called incorporation or integration of the individual" (p. 471). Translated into Durkheimian terms, it is clear that the containing force of "meaningful roles and activities" involved a combination of regulation and integration.

> Groups, organizations, associations, [and] bureaucracies, to operate or stay in existence . . . , must expect reasonable conformity. . . . If a group or organization can get its members to internalize their rules, it would be doing an excellent job of containing. If a group or organization can get its members to comply (although they have not internalized the regulations), they are doing an excellent job. If they can minimize the number of infractions or hold the violations to tolerable proportions (where members are effective, if not enthusiastic, conformists), they are doing a fair job. (Reckless, 1967, p. 470)

FACTORS IN INNER CONTAINMENT

For Reckless, however, the emphasis was on inner containment. In contemporary society, the individual who at one point might be operating in a context of powerful outer containment provided by regulating limits, meaningful roles, and a sense of integration into a particular family, organization, or community might in a short time be operating in another context with few regulations, meaningless activities, and a sense of alienation. Inner containment, on the other hand, would tend to control the individual to some extent no matter how the external environment changed. Reckless (1967) identified the key factors here as including self-concept, goal orientation, frustration tolerance, and norm retention.

The importance assigned to *self-concept* echoed Cooley's insistence on the significance of the "looking-glass self." As indicated in Chapter 3, members of the Chicago school, such as Shaw and McKay, focused attention on the manner in which certain natural areas of high delinquency tended to produce "bad boys." As a control theorist concerned with differential response, Reckless asked the *opposite* question: Why were there still so many "good boys" in these "swamps" of high delinquency?

Reckless had served on the sociology faculty of Vanderbilt University from 1924 until 1940, when he moved to The Ohio State University. Starting in 1955 with Simon Dinitz and others, he initiated a series of studies examining the question of resilience—of how boys can be good in delinquency areas (Huff & Scarpitti, 2011). Reckless and his associates concluded that such boys were "insulated" by "favorable self-concepts" (Dinitz, Reckless, & Kay, 1958; Dinitz, Scarpitti, & Reckless, 1962; Reckless, 1967; Reckless & Dinitz, 1967; Reckless, Dinitz, & Kay, 1957; Reckless, Dinitz, & Murray, 1956; Scarpitti, Murray, Dinitz, & Reckless, 1960). These studies suggested that an image of oneself as a law-abiding person of a sort not headed for trouble served to keep potential delinquents in relative conformity in spite of the pushes and pulls described previously. The research suggested that parents were the most influential sources of favorable self-concepts, with teachers and other authority figures also having some influence.

Reckless (1967) maintained that inner containment also was greatly dependent on *goal orientation,* defined as a sense of direction in life involving an orientation toward legitimate goals and an aspiration level "synchronized with approved and realistically obtainable goals" (p. 476). This approach ran directly opposite to the strain theory notion that social aspirations tended to become a major source of crime and delinquency because they were so frequently frustrated due to inadequate opportunity to achieve the goals in question. Instead, containment theory treated such goal orientation as providing a sense of direction that would keep the individual on the straight and narrow path of conformity. Such a perspective implied either (1) the assumption that opportunities actually were more widely available than was assumed by strain theory so that reasonable success goals were indeed "realistically obtainable," (2) the assumption that realistic goal orientations would involve a scaling down of aspirations on the part of many individuals, or (3) both of these assumptions. These insights concerning the controlling role of goal orientation anticipated Hirschi's (1969) discussion of the social bond of "commitment" (see Chapter 6).

Considering *frustration tolerance* as a major factor in inner containment, containment theory accepted the possibility that the control of biophysical urges toward deviance may be very frustrating and that contemporary society, indeed, may generate considerable frustration as a result of facts such as differential opportunity. It did not argue against efforts to deal with the sources of crime and delinquency by movement toward greater equality of opportunity, but it suggested that part of the differential response to familial, economic, political, and sexual frustrations can be accounted for by the fact that different individuals have developed different capacities for coping with frustration and that contemporary individualism generally is characterized by low frustration tolerance and consequent lack of self-control. Thus, Reckless (1967) maintained that the contemporary individual "develops a very low frustration tolerance to the ordinary upsets, failures, and disappointments in life" and that this may result in "the inability to exert self-control, to tolerate frustration, to recognize limits, [and] to relate to others" (pp. 20–21). These insights anticipated the discussion of Agnew (1992, 2006b), in his general strain theory, of the factors that condition the response to strain either toward or away from crime.

The fourth key component of inner containment, *norm retention,* referred to the "adherence to, commitment to, acceptance of, identification with, legitimation of, [and] defense of values, norms, laws, codes, institutions, and customs" (Reckless, 1967, p. 476). Although the emphasis on goal orientation stressed the integration of the individual through the containing power of direction toward legitimate ends, the emphasis on norm retention stressed the integration of the individual through identification with acceptable means. But for containment theory, the key problem here was not norm retention but rather *norm erosion* or understanding the processes by which this containing factor sometimes was eroded to allow for the possibility of crime and delinquency. Norm erosion was described as including "alienation from, emancipation from, withdrawal of legitimacy from, and neutralization of *formerly internalized* ethics, morals, laws, and values" (p. 476, emphasis added). These insights anticipated Hirschi's (1969, p. 26) concept of "belief" and his claim that the weakening of belief in "the moral validity of norms" frees individuals to offend.

SUMMARY

Reckless's theory is based on the simple insight that crime will occur when containment weakens or is absent. As we have seen, however, Reckless fleshed out this central premise through his discussion of sources of criminal motivation and types of containment. To make the complexities of his theory more understandable, we have provided a summary of the key components of containment theory in Table 5.1.

We also want to emphasize that Reckless was influenced by a challenge that faced many scholars of his day: How is order possible as society is rapidly modernized? Thus, it is apparent that containment theory followed the Chicago school approach in a way that was very Durkheimian. It took a historical focus, seeing what apparently was an increase in crime and delinquency as a product of the modern world. Unlike strain theories, it did not stress economic inequality and argue that crime and delinquency must be dealt with primarily through a liberal political agenda stressing equality of opportunity. Like the Durkheimian tradition, it regarded the moral order as more fundamental than the economic order and concerned itself, to a considerable extent, with what it took to be the problem of the individual in a complex society who increasingly is cast adrift with boundless desires, little capacity to tolerate denial, and no real sense of direction or commitment to the traditional rules of social life.

Table 5.1   Summary of Reckless's Containment Theory

Key Explanatory Concepts	Definition
Sources of Criminal Motivation	
Pushes	Factors that propel or motivate offenders toward crime, including biophysical forces, psychological pressures, and social conditions such as poverty.
Pulls	Factors for crime that entice individuals to offend, such as the presence of illegitimate opportunities or peers who offend. Differential association and subculture theories would be "pull" explanations.
Types of Containment	
Outer containment	Factors within an organized group that serve to reinforce conventional behavior, encourage internalization of rules, and offer supportive relationships. Not found in socially disorganized neighborhoods.
Inner containment	Internal factors that insulate, or allow individuals to resist, the pushes and pulls toward crime that they encounter.
Components of Inner Containment	
Self-concept	Positive view of self as a law-abiding citizen. Anticipated the concept of "resilience."
Goal orientation	Realistic view or aspirations that success goals are attainable. Anticipated Hirschi's concept of "commitment."
Frustration tolerance	Self-control to cope with failures and problems in life. Anticipated Agnew's concept of coping or "conditioning" variables.
Norm retention and erosion	Acceptance or belief in conventional values, laws, customs, and ways of behaving. Weakening or erosion of normative commitment can lead to crime. Anticipated Hirschi's concept of "belief."

Sykes and Matza: Neutralization and Drift Theory

As we have seen, control theorists such as Reckless considered many of the sociological theories of crime and delinquency to be overly deterministic. If these

problems were caused by slums, criminal traditions, lack of economic opportunity, and the like, then why was it that many (if not most) of those suffering from such pressures in fact did not become criminals or delinquent at all? During the late 1950s, Gresham Sykes and David Matza had turned attention to a related issue. If the social pressures causing delinquency were so powerful, then why was it that even the worst of delinquents seemed to be fairly conventional people, actually conforming in so many other ways? And why was it that although they continued to live in areas of crime and delinquency and continued to be faced with a lack of economic opportunity, most of them did *not* continue law-violating behavior beyond a certain age but rather settled down in law-abiding lives? Could it be that the delinquency that took up so small a part of their lives for so short a time really was some sort of aberration—a temporary, albeit occasionally dramatic, quirk rather than a basic characteristic?

TECHNIQUES OF NEUTRALIZATION

According to Sykes and Matza (1957), the major theories of the day overemphasized the difference between delinquents and nondelinquents. Subcultural theories, for example, assumed that juveniles learned an alternative, criminal value system that made delinquency a natural choice. In contrast, Sykes and Matza argued that delinquents retained a commitment to conventional society and its standards of behavior; they knew right from wrong. This observation, however, led to a theoretical puzzle: How was it possible for youths to violate the conventional morality if they, in fact, had not rejected it—if they remained essentially integrated into the society and generally respectful of its regulations?

Sykes and Matza argued that delinquency would be possible if youths could escape the control that conventional society had over them. They explained that this was possible because part of the process through which one learned the conventional social norms consisted of the learning of excuses or *techniques of neutralization* by which those norms could be temporarily suspended and their controlling effects neutralized. When this occurred, youths were free to commit certain delinquent acts in particular circumstances without the necessity of rejecting the norms themselves. For example, students might believe that cheating on a test is wrong and not engage in this practice. But if they felt that a professor gave an unfair test, they might be able to justify cheating in this one case "because they had no choice."

Sykes and Matza (1957) listed five specific techniques of neutralization: (1) denial of responsibility, (2) denial of injury, (3) denial of the victim, (4) condemnation of the condemners, and (5) appeal to higher loyalties (for a summary, see Table 5.2). *Denial of responsibility* is exemplified by the excuse that "a lot of the trouble I get into is not my fault." *Denial of injury* is exemplified by excuses such as "they've got so much, they'll never miss it." An excuse such as "I only steal from drunks [or 'queers,' 'rednecks,' 'outsiders,' etc.]" represents *denial of the victim. Condemnation of the condemners* amounts to an insistence that certain people have no right to condemn certain violations because, for example, "they're worse than we are." Finally, "we have to do it to protect our turf size" represents excuses by *appeal to higher loyalties*. Sykes and Matza argued that it was the use of these techniques of neutralization that allowed delinquents to occasionally violate laws that they actually accepted and obeyed most of the time.

For a half century, Sykes and Matza's research has continued to have a major influence in criminology. Thus, qualitative and empirical studies support the conclusion that neutralizations facilitate offending, the original list of five techniques has been elaborated to include other neutralizations, and the perspective has been applied to explain offense types ranging from street crime to white-collar crime (Maruna & Copes, 2005). At the same time, further development of this approach will require integrating Sykes and Matza's views into the evolving literature within cognitive social psychology on how people use excuses and other beliefs when making the decision to break the law (Maruna & Copes, 2005).

Table 5.2 Summary of Sykes and Matza's Techniques of Neutralization

Technique	Neutralization	Slogan
Denial of responsibility	Forces beyond my control made me break the law (e.g., my friends pressured me; my parents abused me).	"I didn't mean it."
Denial of injury	My delinquent act did not injure anyone (e.g., it was just a prank; I was just borrowing the money I took and was going to put it back).	"I didn't really hurt anybody."
Denial of the victim	The victimization was deserved; thus it was not wrong given the circumstances (e.g., I cheated because the teacher gives unfair tests; we beat up the kid because he knows he has no right to be on our turf).	"They had it coming to them."
Condemnation of the condemners	It is okay to break the law because those running society are all corrupt and on the take. They are just hypocrites to criticize what I do (e.g., look at all the "respectable" people committing white-collar frauds).	"Everyone's picking on me."
Appeal to higher loyalties	I was obligated to break the law or I would lose my integrity and morality (e.g., I jumped into the fight to protect my friend; I lied to the police because it is wrong to snitch).	"I didn't do it for myself."

DRIFT THEORY

This explanation of delinquency in terms of techniques of neutralization represented a control perspective in that their use was treated as neutralizing the existing social control, thereby *allowing the possibility* of delinquency. In fact, Ball (1966) argued that the concept of techniques of neutralization is not sufficiently developed to constitute a theory in its own right and might be integrated into Reckless's more systematic containment theory by considering the various techniques as examples of the more general process of norm erosion posited by Reckless.

In any case, Matza (1964) himself developed the notion further during the mid-1960s, arguing that delinquents in general were no more committed to their delinquency than to conventional enterprises but that the delinquency was a matter of "drift" facilitated by the existence of a "subterranean convergence" between their own techniques of neutralization and certain ideologies of the authorities who represented the official moral order. By this he meant that the authorities themselves often excused violations by blaming parents, citing provocation on the part of the victims, or accepting explanations defining the infractions as involving self-defense or "accident" in a way that reinforced the norm neutralization of the juveniles. In this double-edged way, the controlling power of conventional norms might either be weakened by all sorts of qualifications built into them or be eroded through time as attempts to apply them are met with objections, excuses, and vacillations.

Still, neutralization merely makes delinquency *possible*. In Matza's (1964) language, "Those who have been granted the potentiality for freedom through the loosening of social controls but who lack the position, capacity, or inclination to become agents in their own behalf, I call drifters, and it is in this category that I place the juvenile delinquent" (p. 29). Dealing with the argument against control theory, which holds that "delinquency . . . cannot be assumed to be a potentiality of human nature which automatically erupts when the lid is off" (Cohen & Short, 1958, p. 30) but must involve at least some triggering factors, Matza made a concession that many other control theorists would not accept: He agreed that because delinquency may involve unfamiliar and dangerous behaviors, something more than loss of control is necessary to explain it. He held that the triggering factors consisted of a combination of *preparation* and *desperation*.

For Matza, preparation involved a process by which the person discovered that a given infraction could be pulled off by *someone,* that the individual had the ability to do it *himself or herself,* and that fear or *apprehension* could be managed. Even if a delinquent infraction had become possible, it might not occur unless someone learned that it was possible, felt confident that he or she could do it, and was courageous or stupid enough to minimize the dangers. Otherwise the possibility would fail to become a reality. As for the element of desperation, Matza argued that the central force there was a profound sense of fatalism, a feeling that the self was overwhelmed, with a consequent need to violate rules of the system to reassert individuality. Unlike Reckless, Matza did not criticize such a need for individuality as often amounting to a misplaced egoism.

Control Theory in Context

THE CONTEXT OF THE 1950s

When Reiss (1951) published his article treating delinquency as a failure of personal and social controls at the beginning of the 1950s, the social disorganization themes of the Chicago school as developed during the Roaring Twenties still were in fashion, having been reinforced by the experiences of the Great Depression, World War II, and the rapid social change following the war. But almost as if Americans collectively were exhausted by these upheavals and were determined to have peace and quiet, the 1950s became a time of relative social conformity, at least on the surface. Television shows such as *Lassie* and *Leave It to Beaver* were almost caricatures of the era, with the (stereotypical) middle-class American father coming home from the office to greet the (stereotypical) happy (if slightly mischievous) children at their lemonade stand before inquiring of his (stereotypical) wife (clad in her stereotypical housedress and apron), "What's for dinner, honey?" Conformity, conventionality, and complacency were the order of the day.

Given this pervasive atmosphere, approaches such as Nye's theory tracing delinquency to departures from the model of the ideal family with all its conventional controls over the young seemed almost like common sense. Social critics were more alarmed by the atmosphere of stifling overconformity than by concern over general social disorganization. By the close of the 1950s and even into the early 1960s, Americans were being criticized as a "nation of sheep" (Lederer, 1961) in which the mother was trapped at home in a set of stereotypes about women amounting to a "feminine mystique" (Friedan, 1963) and the father went off to work, only to submit meekly as an "organization man" (Whyte, 1957). Despite the fascination with the apparent wildness of the big-city gangs, the young in general were being criticized as a mass of pop culture conformists, and traditionally unruly college students were being castigated as so lacking in any willingness to question things as to constitute a "silent generation" (Starr, 1985, p. 238). Throughout the 1950s, society as a whole seemed to be sleeping through the American dream.

Of course, not everyone was satisfied. As indicated in Chapter 4, there was concern in some quarters about continuing social inequality and differential opportunity, as evidenced by the renewal of interest in Merton's (1938) anomie theory. The general public was fascinated with and strangely alarmed by the apparently senseless behavior of highly publicized adolescent male working-class gangs in the large cities, exotic groups that in some strange way failed to fit into the dream. Cohen's (1955) delinquent subculture theory represented an effort to explain some of this, although Sykes and Matza (1957) maintained that delinquents were not as different as they might appear. By and large, opportunity theory fit the 1950s, stressing as it did the importance of everyone having access to full participation in the American way of life.

The tendency to stress the strain theory aspect of Merton's approach tells a great deal about the preoccupations of the time. Although Merton himself used Durkheim's term and described the way in which the amorality coming with intense emphasis on economic success weakened social control through a de-emphasis on traditional normative standards of right and wrong, it was the idea of a discontinuity between success goals and the institutionalized means for their achievement that captured most of the attention. Many paid little attention to the powerful Durkheimian theme of moral erosion in their excitement over the theme that traced much of the nonconformity that did exist to a lack of access to legitimate opportunities to achieve those conventional economic aspirations so central to the dream.

Although the potentials for social change were seething under the surface of all this apparent tranquility and already were beginning to be felt here and there, this general atmosphere still prevailed in large part when Reckless (1961) first set forth his containment theory at the beginning of the 1960s. The formulation had somewhat less impact than might have been expected from a major theoretical statement by one of the nation's leading criminologists. In some ways, his containment theory seemed to reflect more of the social climate of the post–World War I years of the 1920s, when Reckless was a graduate student at the University of Chicago and concerns over social disorganization filled the air, than of the complacent 1950s. But the onslaught of the 1960s was to make control theory much more popular.

THE CONTEXT OF THE 1960s

The massive social changes that took place in America during the 1960s, including the rising tide of protest over discrimination, the development of a youth counterculture, and the reaction to the war in Vietnam, not only shook American society but also had a considerable impact on criminological theory. Although Martin Luther King, Jr.'s Southern Christian Leadership Conference had begun its work during the middle 1950s, the civil rights movement had not exploded into public consciousness until its rapid acceleration with the dramatic lunch counter sit-in by African American students in Greensboro, North Carolina, in 1960 (Starr, 1985; Zinn, 1964). The coming of the civil rights movement shattered much of the complacency of the 1950s, and its reliance on a nonviolent strategy of civil disobedience entailing the deliberate violation of segregation laws represented a direct repudiation of conventional social control in the name of a higher morality. But this was only the beginning.

The so-called Beats, representing part of the avant-garde literary world of the 1950s, had made a special point of defying conventionality by flaunting drug use, and the latter part of the decade had seen the birth of a strange new brand of rebellious music called "rock and roll." Even so, neither of these had shaken the establishment. But between 1960 and 1966, the number of U.S. students enrolling in college more than doubled, topping 7.3 million (Starr, 1985). Early during the 1960s, widely publicized student protests at respectable and prestigious institutions such as the University of California, Berkeley, attacked the impersonality and complacency of the bureaucracies of higher education and demanded a place for individuality and greater freedom of expression. This developing counterculture tended to reject many of the conventional aspects of the American dream, including the characteristic overconformity, the model of the middle-class family, and the ruthless pursuit of material success. The search for an "authentic self" became a priority. "The perception of the banality of existence became the core of the youth revolt of the 1960s" (Aronowitz, 1973, p. 331).

Changes such as these, as well as dramatic events including the assassination of President Kennedy in 1963 and growing uneasiness over the slow escalation of military involvement in Southeast Asia, meant that Matza's theory positing drift associated with a decline in moral consensus and a combination of preparation and desperation could find a more receptive audience. As the decade moved on, the continuing drama of the civil rights movement, the coming of militant feminism, the Vietnam War protests, the appearance of the hippies, the advocacy of psychedelic drug use on the part of respected Harvard University professors, and a host of other dramatic social and cultural shifts seemed to many to signal the complete collapse of personal and social control.

By the middle of the 1960s, the civil rights movement had taken on a different and more radical tone. The movement for Black separatism, demanding that several states be set apart for African Americans within the United States, had gained considerable strength. The Black Muslim movement had led many African Americans to reject Christianity as a "White man's religion," to change their names, and to embrace a variation of the Islamic faith. For many African Americans, much of the machinery of personal and social control seemed to be designed to ensure their continued social and emotional bondage.

As the escalation of U.S. involvement in Vietnam accelerated under President Johnson, protests became more frequent and more violent. News media increasingly referred to the nation as being "torn apart" by the war. Social consensus appeared to have evaporated. The 1950s began to seem like an era in a very different society.

As for the feminist movement of the 1960s, it represented in large part an attack on the revered institution of the conventional family as a deceptive front for the political, economic, and sexual oppression of women. The counterculture of hippies and others, which grew up among the new breed of college students and spread to include both students and nonstudents as well as both youths and adults, seemed to mock the staid buttoned-down mentality of the 1950s. Many clean-cut, middle-class young men took to beards, sandals, beads, and shoulder-length hair and previously demure young ladies adopted similar attire, except for the beards. Drugs, especially marijuana, seemed to be everywhere among them, and they were most united in their rejection of the middle-class preachments regarding conventional family life and dedication to the economic rat race.

To many observers, the 1960s seemed to be characterized by loss of self-control on the part of the individual and of social control on the part of organized religion, the family, educational institutions, the economic order, and the political state. It sometimes appeared that everything was being called into question and that all of the conventional institutions were crumbling. The times were ripe for acceptance of a perspective linking crime to the breakdown of control if it could be formulated in appropriate theoretical terms. At the close of the 1960s, Travis Hirschi, himself working in the midst of the social unrest at Berkeley, took up this challenge. Hirschi (1969) set forth a highly refined version of control theory—his "social bond" perspective—that became one of the most influential criminological theories during the last quarter of the 20th century. It is to the theoretical contributions of Hirschi that we next turn.

Travis Hirschi
1935–
University of Arizona
Author of Social Bond Theory

6

The Complexity
of Control

Hirschi's Two Theories and Beyond

T ravis Hirschi has dominated control theory for four decades. His influence today
is undiminished and likely will continue for years, if not decades, to come (see,
e.g., Britt & Gottfredson, 2003; Gottfredson, 2006; Kempf, 1993; Pratt & Cullen, 2000).
Beyond the sheer scholarly talent manifested in his writings, what accounts for
Hirschi's enduring influence on criminological theory? Three interrelated considerations appear to nourish the appeal of his thinking.

First, Hirschi's theories are stated parsimoniously. This means that his theory's core
propositions are easily understood (e.g., the lack of social bonds or of self-control
increases criminal involvement). Second, Hirschi is combative and thus controversial.
He stakes out a theoretical position and then argues that alternative perspectives are
wrong. Hirschi (1983) has long been antagonistic to attempts to integrate theories.
Good theories, he believes, have assumptions and an internal consistency that make
them incompatible with other approaches. Attempts to mix them together result in
fuzzy conceptual frameworks and inhibit the growth of the individual theories. Third,
because Hirschi's theories are parsimoniously stated and make claims that other theories are wrong, they are ideal to test empirically. One (but not the only) reason that
theories flourish is that they are able to provide scholars with opportunities to conduct research and gain publications—the very accomplishment that allows for tenure

and career advancement (Cole, 1975). Hirschi's theorizing has thus been a rich resource that criminologists have mined for numerous publications (Gottfredson, 2006; Kempf, 1993; Pratt & Cullen, 2000; Sampson & Laub, 1993). There is little evidence that this vein of research ideas will soon run dry.

This is not to say that Hirschi's theorizing has been universally popular. His frameworks are bold—critics would say pretentious (Geis, 2000)—because they claim to be "general theories" that explain crime across types of crime and types of people. Hirschi also has shown little interest in race, class, and gender inequalities that others—especially those from more critical perspectives—see as fundamental to any explanation of crime (see, e.g., Miller & Burack, 1993). Regardless of their merits, these critiques have done little to dim Hirschi's influence; if anything, the controversy has sparked further research (see, e.g., Blackwell & Piquero, 2005).

In his career, Hirschi's thinking has evolved rather substantially. In fact, he has proposed two related but ultimately competing theories. The first perspective, *social bond theory*, was presented in 1969 in his book *Causes of Delinquency*. The second perspective, *self-control theory*, was presented in 1990 in his book *A General Theory of Crime*—a work he coauthored with Michael Gottfredson. In this chapter, we review each theory and also attempt to show how they are best considered rival theoretical perspectives.

Hirschi's pervasive influence, however, should not mask the theoretical contributions of other scholars who have focused on the role of control in crime causation. In many ways, these alternative theories are richer in that they explore more carefully how control is shaped by context and can have diverse consequences. In short, they illuminate the *complexity of control*. Accordingly, we review three important contemporary control theories: Hagan's *power–control theory*, Tittle's *control balance theory*, and Colvin's *coercion theory*.

Hirschi's First Theory: Social Bonds and Delinquency

The central premise of Hirschi's first theory is that delinquency arises when social bonds are weak or absent. By itself, this proposition seems rather technical and not something that would stir much theoretical controversy. But Hirschi's intent was not simply to identify another variety of control theory—his *social bond theory*—but in doing so to challenge the two major paradigms of his day: Sutherland's differential association theory (which he termed cultural deviance theory) and Merton's strain theory. His goal was to start a theoretical fight; he succeeded (see also Kornhauser, 1978).

Again, as a control theorist, he argued that these two perspectives asked the wrong theoretical question: Why are people motivated to commit crimes? For differential association theory, the answer was that youths are enveloped by a deviant culture that they learn in interaction with others. This positive learning—that is, learning to value crime—is what moves them to break the law. For strain theory, the blockage of goals creates a frustration that is the engine that drives individuals into crime. Hirschi, however, asserted that these theories were explaining something that did not require explanation—*motivation*. If humans would by their natures seek the easy and immediate gratifications inherent in crime, then they did not need to learn to want to commit crimes or be driven into crime by unbearable strains in order to break the law. In effect, such criminal cultural values and strains were redundant and thus did not explain who would be a delinquent and who would not be a delinquent.

For Hirschi, of course, the proper theoretical question was: Why *don't* people break the law? What differentiate offenders from non-offenders are the factors that *restrain* people from acting on their wayward impulses. The theoretical task thus was to identify the nature of the social controls that regulate when crime occurs. Hirschi (1969) called these controls "social bonds."

HIRSCHI'S FORERUNNERS

Hirschi's (1969) theoretical position was expressed with special clarity through a critique of alternative perspectives, an exploration of the differences between his own formulation and those of his predecessors in the development of control theory, and an examination of his position in the light of empirical data. A review of his approach may provide a clear example not only of the way in which criminological theories reflect contexts of time and place but also of the manner in which they are shaped by special considerations aside from the orientations of the particular theorists. Hirschi himself showed an appreciation of this and unusual candor in admitting just how certain considerations affected the way in which he set forth his position. These included the nature of the data available to him and the current unpopularity of one specific tradition through which he might have expressed his ideas—the social disorganization theme of the Chicago school. Hirschi pointed out the significance of the data factor in an interview:

> Control theory, as I stated it, cannot really be understood unless one takes into account the fact that I was attached to a particular method of research. When I was working on the theory, I knew that my data were going to be survey data; therefore, I knew that I was going to have mainly the perceptions, attitudes, and values of individuals as reported by them. . . . Had I data on other people or on the structure of the community, I would have had to state the theory *in a quite different way*. (quoted in Bartollas, 1985, p. 190, emphasis added)

The problem with the social disorganization theme was that it had lost a great deal of its popularity as an explanation of social problems in general and of crime in particular (Rubington & Weinberg, 1971). Like the older concept of *social pathology*, it had come under intense criticism as a matter of vague generalities masking a lack of value neutrality. As Clinard (1957) commented:

> There are a number of objections to this frame of reference. (1) Disorganization is too subjective and vague a concept for analyzing a general society. . . . (2) Social disorganization implies the disruption of a previously existing condition of organization, a situation which generally cannot be established. . . . (3) Social disorganization is usually thought of as something "bad," and what is "bad" is often the value judgment of the

observer and the members of his social class or other social groups. . . . (4) The existence of forms of deviant behavior does not necessarily constitute a major threat to the central values of a society. . . . (5) What seems like disorganization actually may often be highly organized systems of competing norms. . . . (6) Finally, as several sociologists have suggested, it is possible that a variety of subcultures may contribute, through their diversity, to the unity or integration of a society rather than weakening it by constituting a situation of social disorganization. (p. 41)

These and other criticisms of the concept of social disorganization were widely accepted during the 1960s, and Hirschi was equally candid in describing how he deliberately avoided linking his theory to the social disorganization tradition because of its unpopularity at the time:

For example, I was aware at the time I wrote my theory that it was well within the social disorganization tradition. I knew that, but you have to remember the status of social disorganization as a concept in the middle 1960s when I was writing. I felt I was swimming against the current in stating a social control theory at the individual level. Had I tried to sell social disorganization at the same time, I would have been in deep trouble. So I shied away from that tradition. As a result, I did not give social disorganization its due. I went back to Durkheim and Hobbes and ignored an entire American tradition that was directly relevant to what I was saying. But I was aware of it and took comfort in it. I said the same things the social disorganization people had said, but since they had fallen into disfavor I had to disassociate myself from them. Further, as Ruth Kornhauser so acutely points out, social disorganization theories had been associated with the cultural tradition. That was the tradition I was working hardest against; so in that sense, I have compromised my own position or I would have introduced a lot of debate I didn't want to get into had I dealt explicitly with social disorganization theory. Now, with people like Kornhauser on my side and social disorganization back in vogue, I would emphasize my roots in this illustrious tradition. (quoted in Bartollas, 1985, p. 190)

So, Hirschi (1969) was especially careful to avoid working explicitly out of the social disorganization tradition and to ground his position instead in the thought of other forerunners such as Durkheim and Hobbes. Concentrating on a search for the essential variables providing control through bonds to conventional society, he developed his own position and presented a body of systematic research in support of it. This combination seemed to represent the epitome of tightly reasoned and empirically grounded control theory.

HIRSCHI'S SOCIOLOGICAL PERSPECTIVE

The control theorists examined to this point tended to distinguish between control exerted from sources external to the individual and control exerted from within the individual. Indeed, Reckless (1961) argued that the individual is so isolated in contemporary society—so free to move from one context of external control to another or even to escape from most of it—that internal control is the more basic factor in conformity. Hirschi's position was much more sociological in nature. The characteristics that other control theorists took to be aspects of the personality were considered by Hirschi to be factors sustained by ongoing social relationships that he termed *social bonds*.

Other control theorists gave great weight to the notion of *internalization*, the process by which social norms are taken so deeply into the self as to become a fundamental part of the personality structure. Citing Wrong's (1961) critique of this "oversocialized conception of man," Hirschi (1969, p. 4) insisted instead that what seems to be deeply rooted internalization of social expectations actually is much too superficial to guarantee conformity. First, he rescued Durkheim from the strain theorists. Pointing out that "because Merton traces his intellectual history to Durkheim, strain theories are often called 'anomie' theories," Hirschi (1969) showed that "actually, Durkheim's theory is one of the purest examples of control theory" (p. 3). Next he turned to the question posed by Hobbes—"Why do men obey the rules of society?"—observing that an assumption of internalization commonly was used as a means of avoiding Hobbes's own conclusion that conformity was based essentially on fear. Here there was an echo of Reiss's distinction between conformity resulting from acceptance of the rules and conformity based on mere *submission*.

This willingness to acknowledge that conformity might be based on simple submission to the forces of social regulation without internalized acceptance of the norms became fairly clear in Hirschi's first citation of Durkheim as a control theorist. Hirschi (1969) remarked that "both anomie and egoism are conditions of 'deregulation,' and the 'aberrant' behavior that follows is an automatic consequence of such deregulation" (p. 3); he made this remark without noting Durkheim's complementary stress on integration as a factor in conformity. Turning immediately to Hobbes, he put the matter with stark clarity as follows:

> Although the Hobbesian question is granted a central place in the history of sociological theory, few have accepted the Hobbesian answer. . . . It is not so, the sociologist argued: There is more to conformity than fear. Man has an "attitude of respect" toward the rules of society; he "internalizes the norms." Since man has a conscience, he is not free simply to calculate the costs of illegal or deviant behavior. . . .
>
> Having thus established that man is a moral animal who desires to obey the rules, the sociologist was then faced with the problem of explaining his deviance. (p. 5)

For Hirschi (1969), then, the problem of explaining deviance was a false problem based on the mistaken assumption that people are fundamentally moral as a result of having internalized norms during socialization. He insisted, however, that it was an "oversimplification" to say that "strain theory assumes a moral man while control theory assumes an amoral man" because the latter "merely assumes variation in morality: For some men, considerations of morality are important, for others they are not" (p. 11). Unlike Matza, who believed that it was necessary to suggest forces of preparation and desperation as a way of explaining why the mere loss of control might result in delinquency, Hirschi suggested no motivational factors and simply noted that loss of control sets the individual free to calculate the costs of crime. "Because his perspective allows him to free some men from moral sensitivities, the control theorist is likely to shift to a second line of social control—to the rational calculational component in conformity and deviation" (p. 11).

WHY SOCIAL CONTROL MATTERS

Again, Hirschi was a control theorist who believed that the key issue is to explain why people, who are all motivated to seek immediate gratification in the easiest way possible, refrain from doing so. Why don't they do it? Why don't they commit crimes to get what they want?

Here is another way of putting this issue. Hirschi believed that each potential criminal act has benefits and consequences (also called costs). In his view, most people in society see pretty much the same benefits in crime, because such acts allow them to get what they want (e.g., take something), feel something they like (e.g., get high on drugs), or stop something they find unpleasant (e.g., hit someone aggravating them) (Marcus, 2004). Importantly, Hirschi never showed *empirically* that people see the same benefits in crime; he just assumed this to be the case. However, this is a key assumption.

If most people see crime as having the same benefits—that is, if crime is gratifying or tempting to most people—then it logically follows that most people are *equally motivated to offend*. As noted, other criminological theories dispute the equal motivation thesis; in fact, they spend a great amount of time explaining why some individuals are more motivated to offend than others (e.g., they are under more strain; they have internalized criminal values). But if we assume, as did Hirschi, that motivation to offend is universal, then motivation cannot explain who is or is not a criminal.

Think of this issue methodologically: an independent variable—in this case motivation to offend—has to *vary* if it is to explain *variation in the dependent variable*—in this case involvement in criminal conduct. What, then, does vary? Of course, it is social control. Thus, for Hirschi, *variation in the strength of social control is what explains variation in the extent to which people engage in crime.*

Importantly, Hirschi did not just emphasize control but rather *social* control. As noted, he set forth a sociological theory of crime. Control did not reside in some psychological trait or permanently entrenched set of beliefs. Instead, for Hirschi, the control resides in a person's *ties to conventional society*—to its adult members (parents, teachers), its institutions (family, school), and its beliefs (laws, normative standards). The control thus lies in a person's *relationship* to society. Hirschi called these different kinds of ties or relationships *social bonds*. He identified four social bonds: attachment, commitment, involvement, and belief; these are discussed below in detail.

For Hirschi, *variation in social bonds thus explains variation in crime.* The stronger the bond, the more likely criminal enticements will be controlled and that conformity will ensue; the weaker the bond, the more likely individuals will succumb to their desires and break the law. This returns us, then, to the question, "Why don't they do it?" The answer should be clear: People do not engage in crime—they do not act on their desire for gratification—because they are stopped from doing so by their social bonds. In short, social bonds control their attraction to illegal temptations and ensure their conformity. Much like a dam holding back floodwaters, social bonds keep individuals safe from crime. But if the dam cracks or breaks, then criminal motivations can flood these individuals and no barrier exists to prevent them from offending.

Importantly, the stability of the social bond is not a given. The social bond remains strong only so long as it is nourished by interaction with conventional others. If youngsters become distant from parents, give up on going to college and caring about grades, or are cut from sports teams, their bonds can attenuate. And if bonds weaken, crime can take place. Because bonds can vary in strength across time—for example, weaker in the teenage years, stronger before and after—people can move into and out of illegal conduct. Adult offenders might desist from crime if they enter a quality marriage or get a good job. In short, the presence and strength of social bonds can explain *change in offending*. As we will see in Chapter 15, this insight forms the basis of Sampson and Laub's (1993) life-course theory, which uses variation in social bonds to explain why people enter and desist from crime across various points in their lives.

Finally, in his social bond model, Hirschi rejected the view of the classical school of criminology and of later rational choice theorists (see Chapter 13) that crime is simply due to a weighing of costs and benefits. As noted, Hirschi saw the choice of crime as involving costs (or consequences) and benefits (or gratifications). But he differed from the classical school in two important ways. First, he did not see the benefits of crime as varying across individuals, as the classical school would, but rather as easily available to everyone.

Second, the cost of crime was not, as the classical school implied, mainly a matter of legal sanctions, such as imprisonment. Although a simplification, the classical school and some rational choice theorists implicitly suggest that people faced with the decision to commit a crime calculate (1) how much money they will get versus (2) their chances of being arrested and sent to jail. For Hirschi, such a position truncates reality. Indeed, Hirschi's delinquents in *Causes of Delinquency* made choices but did so within a rich social environment populated with parents, teachers, homework, grades, school activities, and so on. Only when youths were loosened from the ties that bound them to the conventional order were they free to choose to pursue the benefits crime had to offer. Again, this attention to relationships within social institutions is what made Hirschi's theory fundamentally sociological.

THE FOUR SOCIAL BONDS

What, then, are the bonds that form the basis of Hirschi's delinquency theory? They include attachment, commitment, involvement, and belief. They are summarized in Table 6.1.

The Social Bond of Attachment. In Hirschi's delinquency theory, *attachment* is the emotional closeness that youths have with adults, with parents typically being the most important. This closeness involves intimate communication, "affectional identification" with

Table 6.1   Summary of Hirschi's Social Bond Theory of Delinquency

Social Bond	Nature of the Social Bond	Nature of Social Control: Why Don't They Do It?
Attachment	Emotional closeness to others, especially parents.	Indirect control—closeness leads youths to care about parents' opinions, including their disapproval of "bad behavior." Youths do not offend because they not want to disappoint their parents (or others to whom they are attached, such as teachers).
Commitment	High educational and occupational aspirations and good grades in school.	Stake in conformity makes the cost of crime too high. This is thus the rational component of the social bond.
Involvement	Participation in conventional activities, including homework, work, sports, school activities, and other recreational pursuits.	Lack of unstructured or leisure time limits opportunities to offend.
Belief	An embrace of the moral validity of the law and of other conventional norms (e.g., school rules).	Moral beliefs restrain impulses to offend; conversely, crime occurs when such conventional beliefs are weakened.

parents (i.e., wanting to be like their parents), and a sense that parents know what they are doing and where they are. This bond is rooted in the extent to which children spend time with parents and "interact with them on a personal basis" (Hirschi, 1969, p. 94).

When close to parents, youngsters care about their opinions and do not wish to disappoint them. As a result, parents are able to exercise *indirect control*. Direct control is when parents supervise their offspring while in their presence (e.g., discipline them for misconduct). Indirect control, however, occurs when children are not in the same location—that is, are physically separated from parents. Hirschi (1969) also referred to this as "virtual supervision." Where, then, does the control come from? According to Hirschi, youngsters refrain from offending because their attachment makes parents psychologically present. They do not skip school, vandalize, or take drugs because, as the saying goes, "my parents would kill me." As Hirschi (1969) stated:

> So-called "direct control" is not, except as a limiting case, of much substantive or theoretical importance. The important consideration is whether the parent is psychologically present when temptation to commit crime appears. If, in the situation of temptation, no thought is given to parental reaction, the child is to this extent free to commit the act. (p. 88)

The Social Bond of Commitment. Commitment involves youths' stake in conformity (Briar & Piliavin, 1965). Because they invest so much in school success, for example, they would not want to "blow their future" by doing something wrong. This is the rational component of the social bond because commitment is part of a cost-benefit calculation. Those highly committed would find delinquency irrational to commit. They are thus controlled by this consideration.

Commitment was defined not in terms of a surrender of self-interest but rather as *the degree to which the individual's self-interest has been invested* in a given set of activities. For Hirschi (1969), this was the "rational component of conformity," essentially a matter of the rational calculation of potential gains and losses, so that the individual contemplating a deviant act "must consider the costs of this deviant behavior, the risk he runs of losing the investment he has made in conventional behavior" (p. 20). In this sense, a youth who has invested much time and energy in conforming to the expectations of parents and teachers, working hard, and perhaps graduating with honors has a tighter bond with society because he or she has a powerful "stake in conformity" and much to lose by getting out of line.

Of course, "in order for such a built-in system of regulation to be effective, actors in the system must perceive the connections between deviation and reward and must value the rewards society proposes to withhold as punishment for deviation" (p. 162). Hirschi went on to point out that "the stance taken toward aspirations here is virtually opposite to that taken in strain theories" because in control theory "such aspirations are viewed as constraints on delinquency" (p. 162). Although strain theory tended to see high aspirations as leading to frustration and consequent deviance, Hirschi, like Reckless, maintained that the opposite was true: Legitimate aspirations gave a "stake in conformity" that tied the individual to the conventional social order, at least when he or she had invested in the pursuit of such goals instead of merely wishing for them.

The Social Bond of Involvement. Involvement is, in effect, another way of proposing that denial of access to criminal opportunities makes delinquency less likely. Discussions of Hirschi's theory do not typically frame involvement in terms of opportunity, but it is useful to do so. He is pointing to the fact that structured conventional activities takes away chances to offend. This insight is now common in environmental or opportunity theories of crime (see Chapter 13).

Thus, as for involvement as a factor in social control, Hirschi (1969) did not stress the psychological theme of emotional entanglement; rather, he stressed the sociological observation that "many persons undoubtedly owe a life of virtue to a lack of opportunity to do otherwise" (p. 21). Acknowledging that the old thesis "idle hands are the devil's workshop" and the commonsense suggestion that delinquency could be prevented by keeping young people busy and off the streets had so far found little support in research, he went on to examine the possibility that involvement defined in terms of sheer *amount of time and energy devoted to a given set of activities* might represent a key factor in social control. Aside from findings such as those relating time spent on homework to extent of delinquency, however, Hirschi's data failed to lend much support to the hypothesis that involvement, as he conceptualized it, represented a variable crucial to preventing wayward behavior.

The Social Bond of Belief. Hirschi's (1969) use of the term *belief* also was much more sociological than psychological. He did not use the term to indicate deeply held convictions; rather, he used it to suggest approbation in the sense of assent to certain values and norms with some degree of approval. Used in this way, beliefs are seen not as profoundly internalized personal creeds but rather as impressions and opinions that are highly dependent on constant social reinforcement. If the degree of approbation is slight, then the belief becomes a matter of simple assent—of a willingness to submit and "go along," at least for the present. If the degree of approbation is greater, then it may amount to a belief to which the individual gives eager approval and wholehearted cooperation. The point is that such beliefs were not taken to be inner states independent of circumstances; instead, they were taken to be somewhat precarious moral positions much in need of social support based on the ongoing attachment to conventional social systems described earlier.

Hirschi (1969) was careful to point out that he was not accepting that approach to control theory in which "beliefs are treated as mere words that mean little or nothing if the other forms of control are missing" (p. 24). He was equally careful to reject the other extreme represented by Sykes and Matza's (1957) insistence that delinquents "believe" in the conventional morality to the extent that techniques of neutralization become necessary before violations can occur. Hirschi (1969) took the position that individuals differ considerably in the depth and power of their beliefs and that this variation is dependent on the degree of attachment to systems representing the beliefs in question. As he put it, "The chain of causation is thus from attachment to parents, through concern for the approval of persons in positions of authority, to belief that the rules of society are binding on one's conduct" (p. 200). In this view, it is not that people lack consciences or that they are, in truth, totally amoral beings who simply babble on about how much they think they "believe in" things. Rather, it is that "attachment to a system and belief in the moral validity of its rules are not independent" (p. 200); what is called *belief* depends on the strength of attachment and will decline with it.

Thus, belief is best seen as *the extent to which adolescents embrace the moral validity of the law and other conventional normative standards*. For Hirschi, youths do not commit crime because, as Sutherland or Akers would suggest, they have learned delinquent values supportive of stealing or fighting. Rather, they know right from wrong. It is just that for wayward juveniles, laws and rules do not have their allegiance. Whereas conforming kids obey the law because they respect it and see it as legitimate, delinquent kids have no belief in the moral validity of such standards; thus belief in the law is too weak to control their desires to gratify their needs through illegal means. "Control theorists," in Hirschi's (1969) words, are "in agreement on one point (the point which makes them control theorists): delinquency is not caused by beliefs that require delinquency but is rather made possible by the absence of (effective) beliefs that forbid delinquency" (p. 198).

This issue of belief is critical in understanding the long-standing feud between differential association/social learning theory and control theory. Recall that Travis Hirschi and Ronald Akers both started their careers on the same sociology faculty at the University of Washington. They have had cordial personal relations for the past four decades. Nonetheless, during these years they have been arch rivals theoretically (Akers, 2011). The crux of the argument is that Akers asserts that beliefs or definitions favorable to crime can be learned and lead people to offend. That is, individuals offend because they are *socialized to embrace criminal cultural* beliefs. By contrast, Hirschi asserts that such "cultural deviance" is a myth (see also Kornhauser, 1978). He denies that any *positive learning* is needed to commit crime. That is, youths—or adults—do not need to learn *criminal beliefs* and skills because (1) as gratification-seeking beings, the motivation to offend already exists within them and (2) crime is easy to commit and thus no acquisition of special skills is required.

Instead, Hirschi contends that crime occurs when people are *not socialized properly into conventional beliefs*. For Hirschi, criminals do not live in some isolated, self-contained criminal subculture where they learn a different way of seeing the world that requires conformity to crime. Rather, they grow up within the dominant society where, from early in life onward, they have received the message from parents, teachers, and clergy that breaking the law is wrong. Hirschi thus observes that criminals violate laws they believe in. What he means by this is that offenders know that crime is wrong because they have been socialized into the dominant culture. Why, then, do they break the law? It is because their socialization has been defective. As a result, their belief in the moral validity of laws or rules is weak or "attenuated." And when bonds are weak, criminal conduct becomes possible.

So, let us put this matter simply. For social learning theorists such as members of the Chicago school and Akers, people go into crime because they *learn criminal beliefs* that define such acts as required, good, or permissible. For Hirschi, people go into crime because they *fail to internalize conventional beliefs* to the degree needed to control them from succumbing to the seductions of vice, violence, or thievery.

ASSESSING SOCIAL BOND THEORY

Since its publication in 1969 in *Causes of Delinquency,* Hirschi's social bond theory has been one of the most, if not *the* most, tested theories in the field of criminology (Kempf, 1993). Results from the myriad of studies are difficult to interpret. As Kempf (1993) pointed out, the existing research has been characterized by diverse and at times weak measures of the four social bonds and by inconsistent findings. Still, the fairest assessment is that there is evidence that the presence of social bonds is inversely related to delinquency and to adult crime (Gottfredson, 2006; Sampson & Laub, 1993). According to Akers and Sellers's (2004, p. 122) review of the existing research, the magnitude of this relationship between bonds and offending appears to range "from moderate to low" (see also Kubrin et al., 2009).

These findings thus suggest that social bonds are implicated in crime, but that they are not the sole cause of offending. This observation prompts us to ask what might be missing from Hirschi's control perspective. One potential limitation to his theory is that it was based on the assumption that humans are naturally self-interested and thus need no special motivation to break the law. Again, as a control theorist Hirschi asserted that the key theoretical problem was to explain what restrains people from acting on their natural inclination to gratify their desires by stealing, hitting, driving fast, becoming inebriated, and other such wrongdoing. However, it seems unlikely that all individuals are, in fact, *equally motivated* to commit crimes. And if not, then a complete theory must include factors—such as exposure to strain and the learning of criminal definitions—that make some people more strongly predisposed or "motivated" to offend than other people.

Another limitation to Hirschi's perspective is his failure to explore how social bonds are potentially affected by the larger social forces in American society. Theorists may legitimately choose to restrict their focus to a limited problem. In this case, Hirschi chose to focus on the emergence of bonds within a youth's immediate or proximate social context of the family and school. Still, in staking out this explanatory territory, Hirschi did not explore how the formation of social bonds is affected by factors such as changing gender roles, neighborhood disorganization, enduring racial inequality, the deterioration of the urban industrial economy (for an example of a more contextualized use of social bond theory, see Sampson & Laub, 1994). Despite having an analytical elegance, his theory thus tends to be detached—to stand at arm's length—from many of the pressing realities of American society. Notably, this contrasts sharply with Shaw and McKay who tried to illuminate how the breakdown of control was shaped by large social forces (e.g., immigration, urbanization) that were placing communities at risk of social disorganization.

Finally, Hirschi argued that social bond theory applied equally to African Americans and to Whites. Based on limited data analyses, he concluded that unjust deprivation from racial discrimination—a factor that strain theorists Cloward and Ohlin (1960) linked to delinquency—was not criminogenic for minorities. As a result, there was no need to develop a theory that was race-specific—that is, that explored whether some experiences might have a unique crime-inducing effect for African Americans but not for Whites. According to Hirschi (1969, p. 79–80), "there is no reason to believe that the causes of crime among Negroes are different from those among whites. . . . It follows . . . that we need not study Negro boys to determine the causes of their delinquency."

Recently, however, James Unnever and colleagues revisited the Richmond Youth Project data set, generously supplied to them by Hirschi, which was the basis for Hirschi's dissertation and, eventually, for *Causes of Delinquency* (Unnever, Cullen, Mathers, McClure, & Allison, 2009). This data set was originally collected in 1964 under the direction of Alan B. Wilson. Then a graduate student, Hirschi rose from an unpaid assistant to the project's deputy director. He was allowed to place questions on the survey and to use the data for his dissertation (Laub, 2002). As is common practice in criminology, Hirschi included in his empirical analysis only those variables that seemed to measure the theories he was assessing (strain, cultural deviance, and control). Unnever et al.'s inspection of the Richmond Youth Project survey instrument, however, revealed a range of other questions that more directly measured perceived racial discrimination by African American boys in the sample. More noteworthy, when they reanalyzed Hirschi's data, they found that perceived racial discrimination was a robust predictor of delinquent involvement whose effects rivaled those of the social bond measures.

This discovery is important for two reasons. First, if Hirschi had expanded his investigation to include these items in the mid-1960s, the future of criminology might have been quite different. Hirschi might have concluded that, while social bonds have similar effects across race, perceived and real racial discrimination is a distinctive risk factor experienced by African American boys. Given Hirschi's stature, a whole line of inquiry might have been undertaken, in the context of the civil rights movement, that explored how the racial animus faced by minorities places a special criminogenic burden on them.

Second, Unnever et al.'s findings are not idiosyncratic. There is now a small but growing body of research showing that perceived racial discrimination leads to delinquency and other problems among African Americans (Agnew, 2006b; Unnever et al., 2009; see also Brody et al., 2006; Gibbons, Gerrard, Cleveland, Wills, & Brody, 2004; Simons, Chen, Stewart, & Brody, 2003; Simons et al., 2006). This is an issue that criminologists—whether control theorist or not—need to explore in the time ahead (see, more generally, Gabbidon, 2007).

Hirschi's Second Theory: Self-Control and Crime

As we have seen, Hirschi's *social bond theory* has continued to be a major paradigm since it was set forth in 1969 in *Causes of Delinquency*. A little over two decades later, however, Hirschi joined with Michael Gottfredson to set forth a related but different control theory: *self-control theory*. This perspective created considerable controversy and generated considerable research on its central premise that self-control had "general effects"—that is, that it was the key causal factor in crime and deviance across an individual's life and across social groups. For this reason, they claimed to have set forth a "general theory of crime."

Gottfredson's first contact with Hirschi came as an undergraduate at the University of California at Davis, when he took Hirschi's course on juvenile delinquency. Gottfredson would later pursue his doctorate at, and be invited to join the faculty of, the University at Albany (then known as the State University of New York at Albany). Hirschi also moved to Albany's School of Criminal Justice, where Gottfredson and he developed a working relationship. This collaboration would continue at the University of Arizona where they would both serve on the faculty (Gottfredson, 2011).

In the sections that follow, we first review the central ideas of self-control theory. We then assess the theory's empirical status and conceptual challenges. Finally, we consider how, despite sharing some common views, Hirschi's two control theories—social bond and self-control—represent incompatible, if not competing, explanations of crime. This discussion includes Hirschi's largely problematic effort to reconcile the constructs of self-control and social bonds through his revised social control theory.

SELF-CONTROL AND CRIME

In *A General Theory of Crime*, Michael Gottfredson and Travis Hirschi crafted an explanation of crime that departs significantly from Hirschi's earlier work. As noted previously, social bond theory rejected the attempt to explain crime through internalized control. Instead, taking a distinctively sociological approach, Hirschi (1969) emphasized that control is sustained by individuals' continuing relationship with the conventional order—by their bonds to family, school, work, everyday activities, and beliefs. By contrast, Gottfredson and Hirschi (1990) abandoned the idea that continuing social bonds insulate against illegal involvement in favor of the proposition that self-control, internalized early in life, determines who will fall prey to the seductions of crime.

It is perhaps instructive that Gottfredson and Hirschi's conception of control as a permanent internal state rather than as an ongoing sociological product reflects a broader trend in criminological theory, encouraged by the context of the 1980s, to revitalize individual theories of crime (see Chapter 12). In any case, Gottfredson and Hirschi contended that their approach was formulated for a more important reason: It explains what we know about the nature of crime.

Gottfredson and Hirschi claimed that much criminological theorizing pays virtually no attention to the facts about the nature of crime uncovered by empirical research: Crime provides short-term gratification such as excitement, small amounts of money, and relief from situational aggravations. People who engage in crime also engage in analogous behaviors that furnish short-term gratification such as smoking, substance abuse, speeding in automobiles, gambling, and irresponsible sexual behaviors. Criminals do not plan their conduct. Their crimes are not specialized or sophisticated but rather are responses to whatever easy illegal opportunities present themselves. Similarly, offenders fail in social domains—school, work, marriage, and so on—that require planning, sustained effort, and delayed gratification. Finally, and importantly, involvement in crime appears to be stable: Children manifesting behavioral problems tend to grow into juvenile delinquents and eventually into adult offenders.

Gottfredson and Hirschi derived their theory from these facts about crime. If crime and analogous behaviors provide easy gratification, then why does not everyone engage in these acts? Logic suggests that something must be restraining law-abiding citizens from taking advantage of the ubiquitous temptations they confront in their daily routines. Furthermore, if crime is stable and its roots are planted early in life, then this restraint must be inculcated during childhood and be capable of operating across time and social situations.

Gottfredson and Hirschi (1990) proposed that self-control is the restraint that allows people to resist crime and other short-term gratification. Because the path toward or away from crime starts early in life, they contended further that the inculcation of self-control depends on the quality of parenting during a child's early years. Those with the misfortune of having parents who are neglectful and ineffectual in child rearing will "tend to be impulsive, insensitive, physical (as opposed to mental), risk-taking, short-sighted, and nonverbal, and they will tend therefore to engage in criminal and analogous acts" (p. 90). In fact, their futures are bleak. Lacking self-control, they not only will be attracted to crime but also are likely to fail in and drop out of school, to lose jobs, and to be unable to sustain meaningful intimate relationships. By contrast, children with parents with enough caring and resources to supervise and punish their misconduct will develop the self-control needed to resist the easy temptations offered by crime and to sustain the hard work necessary to succeed in school, work, and marriage.

ASSESSING SELF-CONTROL THEORY

Empirical tests of Gottfredson and Hirschi's (1990) perspective generally support the theory's conclusion that low self-control is related to criminal involvement

(Brownfield & Sorenson, 1993; Burton, Cullen, Evans, & Dunaway, 1994; Chapple, 2005; Chapple & Hope, 2003; Evans, Cullen, Burton, Dunaway, & Benson, 1997; Grasmick, Tittle, Bursik, & Arneklev, 1993; Keane, Maxim, & Teevan, 1993; Nagin & Paternoster, 1993; Sellers, 1999; Vazsonyi, Pickering, Junger, & Hessing, 2001; Ward, Gibson, Boman, & Leite, 2010; Wood, Pfefferbaum, & Arneklev, 1993; for a more general assessment, see Goode, 2008). Indeed, a meta-analysis of the existing empirical literature found that self-control is an important predictor of crime (Pratt & Cullen, 2000). In a narrative review of over 15 years of research, Gottfredson (2006) has put the matter more force-fully (see also Britt & Gottfredson, 2003; Kubrin et al., 2009):

> In the context of the theory, however, the claims for self-control are quite strong. As a general cause, it should predict rate differences everywhere, for all crime, delin-quencies and related behaviors, for all times, among all groups and countries. . . . A very large number of high quality empirical studies published since the theory was developed now, in the aggregate, provide very significant support for these strong claims. (pp. 83–84)

Given the existing level of empirical support, it seems likely that this perspective will continue to influence criminological thinking in the time ahead. It is clear that Gottfredson and Hirschi have identified a factor in self-control that has wide-ranging effects. At the same time, self-control theory appears to be overstated in places.

Thus, although self-control explains variation in criminal involvement, this does not mean that causes identified by rival theoretical models, such as differential association, are unimportant (Baron, 2003; Brownfield & Sorenson, 1993; Burton et al., 1994; Nagin & Paternoster, 1993; Pratt & Cullen, 2000). In fact, in a study of middle-school students, Unnever, Cullen, and Agnew (2006) found that low self-control and aggressive attitudes not only both independently predict delinquency but also have a significant *interactive* effect on violent and nonviolent offending. It appears that both low self-control and atti-tudes supportive of aggression (a social learning theory variable) are criminogenic risk factors (Andrews & Bonta, 2003). Similarly, it seems unlikely that individual differences in self-control and misconduct, established early in life, will always remain stable across the life-course; change is possible. In research we will revisit in Chapter 15, Robert Sampson and John Laub (1993) presented longitudinal data showing that adult social bonds, such as stable employment and cohesive marriages, can redirect offenders into a pathway to conformity well beyond their childhood years. Furthermore, the relation-ship among self-control, crime, and analogous behaviors is potentially problematic. It is doubtful that criminal and analogous (or deviant) behaviors will be strongly correlated among all offenders—including, for example, white-collar criminals who have evi-denced delayed gratification in acquiring high-status occupational positions (Benson & Moore, 1992). And some (but not all) evidence suggests that, contrary to theoretical pre-dictions, self-control is not strongly related to all types of analogous behaviors, such as smoking (Arneklev, Grasmick, Tittle, & Bursik, 1993), or to all forms of crime, such as intimate violence (Sellers, 1999; cf. Gottfredson, 2006).

Another thesis that has merit but is likely overstated is Gottfredson and Hirschi's contention that ineffective parenting—the failure of parents to care enough to moni-tor, recognize, and punish wayward conduct—is the chief source of low self-control. A number of studies have been conducted that support the impact of parenting on levels of self-control in the predicted direction (for a review, see Cullen, Unnever, Wright, & Beaver, 2008). But research also shows that the origins of self-control are likely more complicated than Gottfredson and Hirschi theorized. One study revealed, for example, that the exertion of parental controls might decrease self-control among girls but increase it among boys in non-patriarchal homes (Blackwell & Piquero, 2005). There also is evidence that, beyond parental management, levels of self-control are increased by effective school socialization and decreased by adverse neighborhood conditions (Pratt, Turner, & Piquero, 2004; Turner, Piquero, & Pratt, 2005). Even more important, research suggests that parents may affect levels of self-control less by their parenting styles and more by genetic transmission (Wright & Beaver, 2005; see also Unnever, Cullen, & Pratt, 2003). These findings are consistent with studies in psychology reporting that personality traits—including those similar to the construct of low self-control (e.g., impulsivity)—are modestly influenced by parenting but have approximately half their variance attributable to heredity (Harris, 1995, 1998).

Beyond contentions that might be overstated, it also is relevant to consider what might be missing from self-control theory. In this regard, Gottfredson and Hirschi (1990) failed to resolve a hidden inconsistency in their thinking. On the one hand, they implied that social class is an unimportant correlate of crime and that crime is found across the class structure. On the other hand, the logic of their model seemingly predicts a strong correlation between class and crime. Their image of offenders is that of people who are social failures. Lacking self-control, offenders do poorly in school and in the job market. They inevitably slide into the lower class. Furthermore, offenders can be expected to be inadequate parents themselves, passing on low self-control and economic disadvantage to their offspring. Over generations, then, crime should be concentrated increasingly in the bottom rungs of society. In the future, this inconsistency will require systematic attention.

Relatedly, although Gottfredson and Hirschi may have identified a crucial link in the chain of conditions causing crime, they remained silent on the larger structural conditions that might affect family well-being, the ability to deliver quality parenting, and the inculcation of self-control. Currie (1985) called this omission the "fallacy of autonomy—the belief that what goes on inside the family can usefully be separated from the forces that affect it from outside: the larger social context in which families are embedded for better or for worse" (p. 185). A more complete understanding of crime, therefore, would place parents and children within the context of a changing American society. In particular, it seems essential to examine the structural forces and government policies that have shredded the social fabric of many inner-city neighborhoods, impeded the development of stable and nurturing families, and placed many youths at risk for early involvement in crime (Currie, 1985, 1989, 1993; Panel on High-Risk Youth, 1993).

Finally, Tittle, Ward, and Grasmick (2004, p. 166) have illuminated the "conceptual incompleteness" of self-control theory. They proposed that the construct of self-control is not a single trait or predisposition but rather involves two elements: the *capacity for self-control* and the *desire for self-control*. Gottfredson and Hirschi have largely theorized about how people differ in their capacity or ability to exercise self-control. Tittle et al. suggested, however, that individuals may also vary in their interest in exercising self-restraint. Although a beginning study, Tittle et al. (2004) present evidence drawn from a community survey of Oklahoma City adults showing that self-control capacity and desire can have independent and interactive effects on forms of misbehavior (see also Cochran, Aleska, & Chamlin, 2006).

In developing the concept of desire for self-control, Tittle et al. (2004) made an effort to bring motivation—what Hirschi had always taken for granted—back into control theory. They noted that one type of motivation is the desire to commit a crime; this is the kind of motivation that traditional theories, such as strain and social learning approaches, try to explain. By contrast, Tittle et al. contended that the desire for self-control was a qualitatively distinct kind of motivation; it was the motivation to resist the lure of offending. These two types of motivation—the desire to offend and the desire to exercise restraint—are not two ends of the same continuum. Rather, Tittle et al. asserted that they are likely competing motivational forces whose comparative strength may determine whether a criminal act occurs. This fresh view of the complexity of motivation is a line of inquiry that future research might profitably investigate.

SELF-CONTROL AND SOCIAL BONDS

As noted, Hirschi set forth two of the most important control models of crime: his early work on social bond theory and his later work, with Gottfredson, on self-control theory. With only limited exceptions (Gottfredson, 2006; Laub, 2002), Hirschi and Gottfredson refrained from detailing precisely how the two perspectives converged and diverged. Thus, in *A General Theory of Crime* there is no attempt to explain the limitations of social bond theory and why they believed that self-control theory would advance our understanding of criminality. Thus, as Laub (2002, p. ix) points out, the "field has been struggling to reconcile" the two perspectives (see, e.g., Longshore, Chang, & Messina, 2005; Taylor, 2001). Below, we add our insights to this controversy (see also Table 6.2).

Hirschi based both perspectives on the notion that the motivation to deviate was rooted in the natural human inclination to pursue immediate gratification in the easiest way possible and without regard for others. Thus, for both theories, the key factor separating wayward and conforming people was whether the *controls* existed to *restrain* them from acting on these impulses. How, then, did Hirschi's control theories differ from one another? As suggested previously, the distinguishing feature was the *source* of the control—social bonds in one case, self-control in the other.

This difference is consequential. Indeed, it is so fundamental that it makes Hirschi's models *rival theoretical perspectives* (cf. Hirschi & Gottfredson, 1995, to Sampson & Laub, 1995). Self-control theory is a sociological explanation in the sense that the effectiveness of early parenting is held to determine the level of self-control that children develop. After this point, however, Hirschi and Gottfredson's perspective becomes a *theory of stable individual differences*. Across the life course, the level of self-control will influence virtually every aspect of a person's life, from involvement in crime to success in all institutional domains (e.g., family, school, employment, marriage). By contrast,

Table 6.2 Hirschi's Two Theories of Crime

Dimension of the Theory	Gottfredson and Hirschi's Self-Control Theory	Hirschi's Social Bond Theory
Nature of control	Self-control	Social bonds
Type of control	Internal	Social: due to quality of relationships to society
Stability of control	Established in childhood; individual differences in self-control persist throughout life	Control may change across life as the strength of the social bonds change
Relationship of bonds to crime	Spurious; quality of bonds and level of crime both caused by level of self-control	Causes crime; quality of bonds determines level of crime

Hirschi's original social bond theory is more of a pure sociological theory. The development of social bonds is not limited to childhood; rather, bonds are potentially formed at any age. The theory implies that when bonds are formed, they will restrain deviant motivations and prevent criminal involvement (Sampson & Laub, 1993).

Hirschi's two control theories diverge, then, on a critical point. His second perspective argues that *social bonds have no influence on criminal involvement*. Instead, the relationship between social bonds and crime is *spurious* (Evans et al., 1997). For example, take the social bond of attachment to parents. Social bond theory would contend that the bond of attachment reduces delinquency. By contrast, self-control theory would argue that children high in self-control are more likely to be attached to parents and to avoid delinquency, whereas children low in self-control have difficulty in forming attachments and are free to break the law. Therefore, attachment and delinquency are related only because both are caused by a third underlying factor—self-control.

Phrased differently, Hirschi changed his mind over the years. He once thought that social bonds were the main determinant of crime. Later, however, he (and Gottfredson) came to believe that social bonds were merely a manifestation of a person's level of self-control and thus had no independent causal relationship to criminal involvement. This issue will surface in Chapter 15, where we revisit Gottfredson and Hirschi and explore the implications of their theory for life-course perspectives, especially the work of Sampson and Laub (1993).

HIRCHI'S REVISED SOCIAL CONTROL THEORY

In 2004, Hirschi reflected on his social bond and self-control perspectives. He was confronted with two theoretical difficulties. The first we have just mentioned: his failure to reconcile his two control theories that seemed fundamentally at odds with one another. The second involved the tendency of scholars to interpret self-control theory as a "trait" explanation of criminal conduct. It is unclear whether these relatively brief thoughts published in a forum not widely read by criminologists will define a third avenue of social control theory (see also Gottfredson, 2006). Even so, his essay—which may comprise his final thoughts on this subject—is worthy of consideration.

The concept of the social bond presented two interrelated problems for Hirschi who, writing with Gottfredson in *A General Theory of Crime* (1990), developed the construct of self-control. First, because social bonds are unstable over time, changes in their strength cause changes in criminal involvement across the life course. Second, as Sampson and Laub (1993) would point out, the strength of the social bond at any given time is affected by the quality of the relationships in which offenders are involved. The bond is a two-way street; it involves the offender (or potential offender) and those with whom the individual interacts. Again, because relationships both form and end, social bonds are a source not only of continuity but also of change in behavior. These assumptions that bonds are unstable and relationship-based are inconsistent with the premises of self-control theory, which sees control as stable and internal.

Hirschi (2004) now rejects the instability thesis and asserts that social bonds are stable. He does this by asserting that the "source and strength of 'bonds' is almost exclusively within the person reporting or displaying them" (p. 544). Thus, attachment is not based on the quality of the relationship between a parent and youngster but rather resides mainly in the youth's mind. It is not clear what might occur when parent–child attachments are compromised (e.g., divorce, child abuse), but this possibility recedes to a minor consideration in Hirschi's revised social control theory. In any event, if the social bond is now stable and internal, how does it differ from self-control? It does not, says Hirschi (2004); "they are the same thing" (p. 543). In this way, social bond theory is now "saved" and becomes identical to self-control theory.

But in his revised perspective, self-control also changes its character. In its original statement, those lacking self-control were portrayed as impulsive, risk taking, insensitive to the needs of others, and unable to defer gratification. These "elements of self-control" (Gottfredson & Hirschi, 1990, p. 89) were hypothesized to constitute a single propensity. For most scholars, this meant, in effect, that self-control was akin to a stable personality trait that individuals carried with them across situations and across the life course.

Hirschi (2004), however, was uncomfortable with his self-control theory being transformed into a psychological trait explanation of crime. In his view, the problem with linking a bad trait to bad behavior is that it omits the way in which individuals make choices. People are not simply bundles of impulses that make delayed gratification difficult. Rather, there is a rational or cognitive process that intervenes between propensity and behavior. People think and then act. Of course, there are individual differences in how people think. Some pay attention to consequences whereas others do not (see also Gottfredson, 2011).

As Hirschi (2004) observed, his theory seemed to suggest—or was interpreted to suggest—that "offenders act as they do because they are what they are (impulsive, hot-headed, selfish, physical risk takers), whereas nonoffenders are, well, none of these" (p. 542). Put another way, self-control was a trait that either prevented or permitted the gratification-seeking side of human nature to seek fulfillment. But for Hirschi, control theory was not meant to strip people of human agency. This was Hirschi's long-standing criticism of positivism in which bad traits were said to equal bad behavior, with the actor somehow disappearing from the equation. Hirschi had meant to retain the classical school idea of people as rational actors—as seeing the world, assessing options, and then acting in their self-interest (Gottfredson & Hirschi, 1990). This feature of his self-control theory, however, was lost in its translation and testing by criminologists. In his 2004 essay, Hirschi intended to reinsert the actor and agency back into the crime equation. Thus, he asserted that "self-control involves cognitive evaluation of competing interests—an idea central to control theories. The theory requires an explanatory mechanism that retains elements of cognizance and rational choice" (p. 542; see also Marcus, 2004).

Thus, in his revised social control theory, Hirschi (2004) redefines self-control as "*the tendency to consider the full range of potential costs of a particular act*" (p. 543, emphasis in original). In short, some people refrain from crime and analogous deviant behaviors because they are able to see the diverse consequences such conduct will have. They do so, Hirschi observes, in large part because they have something to lose—attachments, commitments, involvements, and beliefs they cherish. Social bonds are the costs they weigh that inhibit offending. Other individuals, those with low self-control, think little about consequences and hence are free to pursue immediate gratification. Weak in social bonds, there is little about their lives that inhibits going into crime.

Hirschi's revised theory is provocative and has earned some empirical support (Piquero & Bouffard, 2007). Even so, it suffers from two major shortcomings. First, Hirschi provides no clear explanation of the origins of social bonds. By implication, he seems to be saying that social bonds are not established through social relationships but rather reflect a youngster's internal orientation. Very early on, some children are cooperative and eager to please whereas others are lazy and inattentive (Hirschi, 2004, p. 544). If we have bonds, it is because we are the architects, through the choices we make, of these ties to the conventional order. Apparently, individuals differ in their natural capacity to establish social bonds, which in turn are the costs that make self-control more likely. The causal scheme seems to be that self-control creates social bonds, which in turn creates self-control. Second, Hirschi simply asserts that social bonds are stable and thus the "same" as self-control. But these two constructs—social bonds and self-control—cannot be made the same by theoretical fiat.

Indeed, in the end, the stability or instability of social bonds, where they come from, and the effects that they have independent of self-control are *empirical questions*. In this regard, the work of Sampson and Laub (1993), whose age-graded social bond theory we discuss more fully in Chapter 15, casts doubt on the central claims of Hirschi's revised social control theory.

The Complexity of Control

Although Travis Hirschi has been the dominant figure in contemporary control theory, other scholars have also explored the way in which social controls are related to criminal behavior. At its core, control theories traditionally have linked conformity to the presence of control and crime to the absence of control. More recent perspectives, however, have illuminated that social control is a complex phenomenon that may have *differential effects* depending on its quality, its magnitude, and the context in which it is applied. In the sections below, we review three prominent theories that explore the conditions under which control not only restrains offending but also might well prove criminogenic. These include John Hagan's power–control theory, Charles Tittle's control balance theory, and Mark Colvin's coercion theory.

HAGAN'S POWER–CONTROL THEORY

Gender and Delinquency. John Hagan's power–control theory shares common aspects with Gottfredson and Hirschi's perspective. First, Hagan (1989) contended that delinquency is more likely when a person has a preference for taking risks, an orientation that Gottfredson and Hirschi saw as central to a lack of self-control. Second, both approaches believe that personal orientations, whether risk taking or self-control, are established by the nature of parenting. In short, families are incubators for or prophylactics against criminal involvement.

At this juncture, the two theories diverge. For Gottfredson and Hirschi, parenting is either good or bad, and this determines whether self-control is or is not inculcated. For Hagan, the critical issue is how the balance of power between parents affects the nature of parenting and, in turn, risk preferences and crime. That is, power relations between husbands and wives shape how children are controlled (hence *power–control* theory).

Hagan contended that in patriarchal families, parents exercise greater control over female children than over male children. The family, in effect, tries to reproduce gender relations in the next generation. Daughters are socialized to be feminine and to value domesticity—in short, to prepare for their futures as homemakers. Sons are encouraged to develop boldness and to experience the world—in short, to prepare for their futures as breadwinners. The result is that boys have stronger preferences for risk taking that, in turn, increase their involvement in delinquency.

In egalitarian families, however, parents supervise female and male children more similarly. "In other words," observed Hagan (1989), "as mothers gain power relative to husbands, daughters gain freedom relative to sons" (p. 157). Again, parents tend to reproduce themselves. Daughters—not just sons—are seen as potentially entering the occupational arena and as being equal partners in future relationships. Unlike girls in patriarchal families, they are not socialized as fully into the "cult of domesticity" and are given more latitude to engage in risky activities. The result is that daughters' and sons' risk preferences become more alike and, therefore, their rate of involvement in delinquency converges.

Assessing Power–Control Theory. Although power–control theory has not been without its critics and may be in need of some qualification, the perspective is amassing a fair amount of empirical support as a useful theory of delinquency (Blackwell, 2000; Blackwell & Piquero, 2005; Grasmick, Hagan, Blackwell, & Arneklev, 1996; Hagan, 1989; Hagan, Gillis, & Simpson, 1990; Hagan & Kay, 1990; Hill & Atkinson, 1988; Jensen & Thompson, 1990; McCarthy, Hagan, & Woodward, 1999; Singer & Levine, 1988).

Furthermore, the theory advances criminological thinking by illuminating the need to consider how gender-based power relations in society influence parental control and, ultimately, delinquent involvement.

Several considerations, however, have yet to be addressed systematically by power–control theory. First, perhaps the theory's principal limitation is that it remains largely silent on how other structural conditions affect the nature and effectiveness of parenting. In particular, the theory must address the intersection of class and gender and must comment more clearly on how other types of power relationships in society affect crime. For example, the theory is unclear about how delinquency is affected by the parenting practices of single mothers within the context of impoverished communities. Second, the perspective originally was developed more as an explanation of "common" delinquent behavior than as an explanation of chronic and/or serious offending (Hagan, 1989, p. 160). But if a theory cannot account for the kinds of crime that most concern criminologists and policy makers, then its significance is decreased commensurately. Third, although empirical support for the theory exists, most studies have not tested the theory versus competing theories such as social learning theory and theories of individual differences (e.g., low self-control theory). Unless power–control theory enters into such a theoretical "competition," it will be unclear whether the effects of its variables are real or spurious (i.e., their effects will disappear once the effects of other theories' variables are taken into account in statistical tests).

TITTLE'S CONTROL BALANCE THEORY

Theoretical Propositions. Control theories generally focus on the factors that restrain or "control" the behavior of individuals. They do not consider the control exercised by *these individuals* over their social environment. Charles Tittle, however, made the innovative insight that people are not only *objects* of control but also *agents* of control (Tittle, 1995, 2000). In his *control balance* theory of crime and deviance, he argued that each person has a certain amount of control that he or she is under and a certain amount of control that he or she exerts.

For some individuals, the relative amount of control is in balance; others suffer from a control *deficit,* and still others experience a control *surplus.* Control balance tends to be associated with conformity, and control imbalance tends to be associated with deviance. "The central premise of the theory," observed Tittle (1995), is that "the amount of control to which an individual is subject, relative to the amount of control he or she can exercise, determines the probability of deviance occurring as well as the type of deviance likely to occur" (p. 135). He called this the *control ratio.*

If Tittle merely offered the thesis that control imbalance is criminogenic, then his theory would be parsimonious and easily understood. But for Tittle, the causal process of wayward conduct is complex and contingent on the intersection of an array of factors. Tittle's embrace of complexity is a double-edged sword: He sought to capture— not ignore—the multifaceted conditions that prompt misconduct, but his theory involves so many variables that interact in so many ways that it is difficult to test. Not surprisingly, compared to competing perspectives such as Agnew's general strain theory and Gottfredson and Hirschi's self-control theory, empirical research on control balance theory is limited. Studies published thus far, however, do furnish some supportive evidence for the theory (see, e.g., Baron & Forde, 2007; Piquero & Hickman, 1999; Tittle, 2004, p. 396).

Tittle's theory begins by exploring why individuals become *predisposed* to develop a motivation to deviate. The potential for such a predisposition lies in human nature because we are creatures that have a strong urge for autonomy—that is, a proclivity to escape the control that others wish to impose on us. This desire for autonomy is made even more salient when people are blocked from attaining goals they are seeking and when their control ratios are unbalanced. The convergence of these factors—autonomy, goal blockage, and control imbalance—fosters a "state of readiness to experience motivation for deviant behavior" (Tittle, 2000, pp. 319–320).

This predisposition can develop into a clear deviant *motivation* when two conditions transpire. First, the person must "become acutely aware of his [or her] control imbalance and realize that deviant behavior can change that imbalance either by overcoming a deficit or by extending a surplus" (Tittle, 2000, p. 320). That is, the functionality or payoff of deviance must become apparent. Second, the person must be provoked to experience a "negative emotion"—"a feeling of being debased, humiliated, or denigrated that intensifies the thought that deviance is a possible response to the provocations" (p. 320). Again, deviance is functional in this situation because it allows the person to rectify the attempt to degrade him or her.

Once deviant motivation has emerged, deviant behavior still might not occur. For one thing, a person must have the *opportunity* to engage in a given act. *Constraints* also must be overcome. These might involve situational risks (e.g., getting caught) or an individual's moral inhibitions, level of self-control, or social bonds. Also salient is the control balance ratio, which can shape whether deviant behavior will occur and, if so, what kind will occur.

Tittle sought to have a "general" theory and thus to explain all forms of deviation. He proposed a typology of deviance in which seven behavioral categories are arranged on a continuum. At the midpoint of this continuum lies conformity, which is said to correspond to a situation where there is control balance. On the left side of the continuum, which he labeled *repression,* are three categories, each of which involves a control deficit. Extreme repression yields *submission,* moderate repression yields *defiance,* and marginal repression yields *predation.* On the right side of the continuum, which he called *autonomy,* are three categories that involve a control surplus. Maximum autonomy yields *decadence,* medium autonomy yields *plunder,* and minimum autonomy yields *exploitation.*

Criminologists would mostly be interested in the category of *predation.* Tittle contended that serious forms of crime would occur among people with small deficits in control. When deficits are limited, the individual may well judge that a criminal act might be successful in erasing the control imbalance that he or she is experiencing. A youngster's use of violence, for example, could change his or her control ratio and cause other juveniles to leave the youngster alone. If faced with a larger control deficit, however, a person might merely submit or perhaps engage in less serious forms of deviance, such as vandalism, that show defiance but do not elicit costly actions from those capable of exerting control over the person. Control surpluses generally free people to engage in a range of deviant acts without consequences. Tittle noted that many forms of corporate and white-collar crimes are due to such control surpluses (see Piquero & Piquero, 2006).

More recently, Tittle (2004) has replaced his typology of deviance with a continuum along which deviant acts, including crime, can be placed. Thus, he argues that any deviant act can be rated as to its degree of *control balance desirability*. This construct involves two factors. First, deviant acts vary in their "likely long-range effectiveness" in altering a person's "control imbalance." Second, deviant acts vary in the degree to which committing them requires that a person "is directly involved with a victim or an object that is affected by the deviance" (Tittle, 2004, p. 405). Long-range effectiveness is desirable because it means that the problem of a control imbalance is resolved, thus making further action unnecessary. Avoiding direct involvement with a victim is desirable because distance and impersonality lessen the chance that the person will be subjected to countervailing reactions. For example, if a worker were to assault a boss who humiliated the individual, the assault would have low control balance desirability. Why? Because the worker would only temporarily alter his or her control imbalance vis-à-vis the boss. There also would likely be a reaction by the boss that might ultimately make the control deficit worse (e.g., hit the worker back; fire the worker; have the worker arrested). In any case, the challenge that awaits is to explore how the core elements of Tittle's theory predict when acts of varying levels of control balance desirability will be committed (see Tittle, 2004).

Assessing Control Balance Theory. Again, Tittle presented a fascinating theory that revises the core proposition of previous control theories. Traditional control theories link crime to a *breakdown or lack of control.* Although Tittle would agree with this thesis (to a degree), he made the poignant suggestion that *too much control,* which places a person in a control deficit, also may be a cause of crime (see also Sherman, 1993). His insight that crime can function to restore a sense of control is consistent with other theories that emphasize the role of criminal behavior in resolving problems, such as relieving strain, proving masculinity, and defending self-respect through defiance.

Tittle's theory, however, has some potential weaknesses. First, Tittle can be admired for attempting to develop a theory that not only demarcates the causes of crime and deviance but also links these causal elements to the nature of the phenomena being explained (i.e., first his typology deviance, which then was replaced by his continuum of control balance desirability). But this is likely to be an unprofitable line of inquiry. It seems nearly impossible, for example, to measure what the control balance desirability would be for the endless acts that are seen as being deviant, let alone criminal—especially since the desirability of an act could vary by a host of situational factors (thus, threatening a bothersome boss with assault might be more or less desirable depending on how big the perpetrator was and how scared the boss was). Even if strides could be made in measuring the control balance desirability of acts, scholars are unlikely to find this complex and tedious task an attractive use of their time. In the end, Tittle's theory will find more adherents if he abandons attempts to define and measure deviance and instead concentrates his attention on the conditions under which control balance leads to crime as opposed to other outcomes. Notably, this is the strategy of virtually every other theory.

Second, his emphasis on autonomy as the wellspring of human motivation seems unnecessarily limited. Why not consider, for example, the desire for self-gratification, a drive that many other control theories claim is universal and central to the motivation to deviate?

Third, Tittle relegated the main causal variables from other theories, such as self-control, social bonds, and social learning, to the secondary role of *constraints* or *contingencies*. A competing theorist such as Ronald Akers, for example, would argue instead that social learning has important "main effects" or independent influences on criminal behavior apart from any of the processes outlined by Tittle. As with other newer theories, the challenge is to pit control balance theory in an empirical contest versus other theories. Only in this way will it be possible to assess whether, relative to other known predictors of crime, the control ratio factors identified by Tittle exert strong or weak effects on criminal behavior.

COLVIN'S DIFFERENTIAL COERCION THEORY

Theoretical Propositions. From birth onward, people are exposed to varying levels of coercion—an experience that is consequential. Thus, some people have the good fortune of living in social environments where compliance is secured largely through noncoercive means. Other individuals, however, encounter coercive environments across the life course. Often, this coercion is harsh and erratic, a combination that creates strong criminal predispositions and fosters chronic offending. According to Mark Colvin:

> Chronic criminals are made, not born. They emerge from a developmental process that is punctuated by recurring erratic episodes of coercion. They become both the recipients and the perpetrators of coercion, entrapped in a dynamic that propels them along a pathway toward chronic criminality. (Colvin, 2000, p. 1)

Colvin (2000) defined *interpersonal coercion* as the "threat of force and intimidation aimed at creating compliance through fear" (p. 5). Such coercion may involve physical punishments or the withdrawal of love and support. People also may face *impersonal coercion,* which is "pressure arising from structural arrangements and circumstances that seem beyond individual control such as economic and social pressure caused by unemployment, poverty, or competition among businesses or other groups" (p. 5). Frequently, these two forms of coercion intersect, with those subject to interpersonal coercion living in environments most affected by impersonal forms of coercion.

Colvin called his perspective *differential coercion theory* because, in his view, people vary in the extent to which they are exposed to coercion. This is much like Sutherland used the term *differential association* to refer to variation in exposure to sources of criminal learning. Colvin's use of this term also is similar to Regoli and Hewitt's (1997) *differential oppression theory*. In any case, Colvin argued that controls aimed at securing compliance—aimed at getting people to obey social norms—vary along two dimensions. First, the controls can be either *coercive* or *noncoercive*. Second, the controls can be applied in a way that is either *consistent* or *erratic*. Consistent noncoercive controls are most likely to create psychologically healthy youths who are unlikely to break the law. The most problematic combination, however, is when control is exercised in a coercive and erratic fashion. This form of differential coercion, according to Colvin, produces chronic criminality.

In Colvin's model, coercive and erratic control produces many of the factors that other theorists believe cause crime. In this sense, it is an "integrated" theory. Still, the primary causal factor in this theory—differential coercion—is unique. In any event, harsh and inconsistent coercion creates a sense of unfairness and "anger" (general strain theory), "weak or alienated social bonds" (social bond theory), "coercive modeling" (social learning theory), "perceived control deficits with feelings of debasement" (control balance theory), and "low self-control" (Gottfredson & Hirschi's general theory) (Colvin, 2000, p. 43). In combination, these factors create a strong overall predisposition for chronic involvement in crime. These forces, which Colvin called social psychological deficits, also foster within individuals a *coercive ideation*. Here people have a worldview that coercion can best be overcome by acting coercively in return. Harboring such thinking, they risk becoming involved in predatory behaviors, using violence or the threat of violence to control their environments.

Colvin (2000, p. 87) indicated that the causes of chronic criminality are both intergenerational and developmental. The process begins with parents who come from coercive backgrounds, are employed in coercive workplaces, and are buffeted by impersonal coercive forces (e.g., economic recessions, poverty, racism, harsh living conditions). Such parents then reproduce themselves, so to speak, by using coercive and erratic child-rearing techniques. Their social psychological deficits and coercive ideation are thus transmitted to their children. In turn, these youngsters enter social environments—school, peer groups, and so on—where they experience harsh controls, further reinforcing their deficits and coercive thinking orientations. As they move into early adulthood, they tend to be employed in the secondary labor market, which fails to lift them out of poverty and exposes them to coercive working conditions. Often, they are ensnared in the criminal justice system, where they experience more coercive treatment. These factors across the life course continually nourish coercive ideation and criminal predispositions, thereby placing these individuals at risk for chronic criminality. Eventually, these offenders will reproduce their experiences in a subsequent generation of youths.

Breaking this cycle, Colvin contended, will require a "theory-driven response." On a broad level, Colvin favors creating a less coercive society in which people's human needs are given priority by government policies. Moving in this direction requires the political will to implement a range of supportive social programs (see also Colvin, Cullen, & Vander Ven, 2002; Cullen, Wright, & Chamlin, 1999; Currie, 1998b). These might include, for example, programs to help individuals facing crises such as joblessness and homelessness, programs to help parents raise children more effectively, universal Head Start programs, early intervention programs with youths at risk for crime, more commitment to public education, efforts to make work environments less harsh and more democratic, and a criminal justice system that stresses crime prevention, fairness, restoration, and rehabilitation.

Assessing Differential Coercion Theory. In *Crime and Coercion,* Colvin (2000) marshals evidence supporting the various links on his causal model. Empirical tests of this perspective, however, are needed. Even so, beginning evidence across different social contexts exists that is supportive of coercion theory. Thus, based on a sample of 2,472 middle-school students, Unnever, Colvin, and Cullen (2004) provide evidence consistent with core propositions of differential coercion theory. They found that as predicted by the theory, exposure to coercive environments increased self-reported delinquency and that these effects were mediated by social-psychological deficits. Similarly, based on a sample of 300 homeless street youths ages 16 to 24 in Toronto, Baron (2009) discovered that a multidimensional measure of coercion predicted involvement in violent offenses. In line with Colvin's theory, the direct effect of coercion on violence was partially mediated by low self-control, anger, coercive modeling, and coercive ideation.

The Consequences of Theory: Policy Implications

Unlike some of the other theories examined in this book, control theory has tended to reinforce the sorts of prevention and intervention efforts that have been around for decades and that to many have become a matter of "common sense" (Empey, 1982, p. 268). It is worth noting, however, that the Durkheimian heritage has emphasized prevention through the strengthening of the institutions of socialization rather than through a policy of deterrence relying primarily on fear of getting caught. And it is important to stress that control theory has suggested that regulation of the individual must come through policies fostering integration into the social order rather than through policies of isolation and punishment. That is, these perspectives teach us that attempts to reduce crime by "get tough" laws and harsh penalties—that is, through *state punitive control*—are unlikely to be effective because they do little to establish any self-control and social bonds that insulate against offending.

   The control theories we have examined have less to say about the prevention and control of professional crime, organized crime, and corporate and white-collar crime than about prevention and control of juvenile delinquency or ordinary street crime. With respect to prevention, they provide considerable support for programs to strengthen families, particularly with respect to effective child rearing. These efforts often are labeled *early intervention programs.*

   These programs often target parent–child attachment for improvement, because weakened bonds are a risk factor for misconduct. To the extent that such programs focus on development of self-control, they stress the need for policies that assist the family in inculcating the favorable self-concepts, impulse control, and frustration tolerance that can keep people out of trouble even in situations of weak external control.

School programs also have been developed that specifically target the need to establish youngsters' bonds to school who are at risk of academic failure and a lack of educational commitment. A range of successful results have been reported (Catalano, Arthur, Hawkins, Berglund, & Olson, 1998; Hawkins & Herrenkohl, 2003; Losel & Bender, 2003). For example, the Seattle Social Development Project was "designed specifically to prevent antisocial behaviour by promoting academic achievement and commitment to schooling during the elementary grades" (Hawkins & Herrenkohl, 2003, p. 268). This intervention used multiple methods to improve the ability of students to learn and to solve problems without resorting to anger and aggression, of teachers to manage behavior and instruct effectively, and of parents to support their children's learning. The intervention increased school achievement and commitment and in turn reduced both the initiation of delinquency and violence in the teenage years (Hawkins & Herrenkohl, 2003).

Control theory suggests prevention and reintegration policies moving adults into stable social networks of employment and community activities, but less has been done here than with programs for youths. Control theory also suggests a search for policies capable of demonstrating the payoff of hard work toward conventional goals to draw both adolescents and adults into positions of personal commitment in which there is too much of a stake in conformity to lose by a return to delinquency or crime. Unfortunately, such programs tended to become casualties of the shifting political climate of the 1980s and 1990s and the popularity of "get tough" policies. More recently, however, there has been an inmate "reentry movement" that is illuminating the consequences of severing the ties of offenders from families, community, and work. Part of this agenda is thus the call to explore ways to foster social bonds that promote prosocial behavior (see, e.g., Travis, 2005).

Control theories are most impressive to the extent that a person accepts the larger social structure and conventional middle-class values as things to be taken for granted. For control theory, the systems that are to accomplish the regulation of the individuals at risk for crime and delinquency through their integration almost always are systems defined in conventional middle-class terms. Interestingly enough, this seems true even of approaches such as Hagan's power–control theory, in which the egalitarian family really is the new ideal of the middle class, although it is true that this approach faces up to the possibility that the freedom gained by daughters in such families actually may represent a weakening of control. The larger question is the following: What if all of these systems themselves (e.g., families, schools), or at least some of them, are part of the problem rather than the solution? Integration into a "bad" system may be worse than no integration at all. As we will see in Chapters 7 and 8, some criminological theorists have argued that the major problems are indeed located in the conventional systems themselves and that these systems, rather than the particular criminals or delinquents, are the major sources of crime and delinquency. Labeling theorists insist that the conventional systems tend to aggravate crime and delinquency by overreacting to minor nonconformity, whereas conflict theorists maintain that these systems really are covert instruments of oppression masquerading as helpful agencies of desirable socialization.

Even if one is convinced that conventional institutions, such as the middle-class family and the school, do not create major problems through a tendency to reject and stigmatize those who do not fit into them well and are not instruments of social oppression, it does not follow that successful integration and regulation functions will solve crime and delinquency problems. Speaking of impoverished African Americans, for example, Empey (1982) remarked: "It is not merely that underclass children are sometimes in conflict with their parents, or that their academic achievement is low, but that they are caught in an economic and political system in which they are superfluous" (p. 299). If the basic problem is located in the larger sociopolitical order, then how can policies focused on strengthening families and schools do much to prevent and control crime and delinquency in the long run? Is the sociopolitical order to be accepted as it is, in which case policy must be directed toward inculcating values that support it? Or must more basic policies be developed that challenge some of the assumptions on which the conventional order is based? The labeling theorists and conflict theorists to be examined in Chapters 7 and 8 have directed our attention in just that direction.

Conclusion

It is instructive to note that the control theories that focus on explaining juvenile delinquency tend to locate control influences primarily in the family and secondarily in the school and that those theories that focus on adult crime tend to put greater emphasis on inner factors such as self-concept and self-control. Although juveniles might be influenced less by internal factors than by social forces such as peer pressures and parental control, internal factors might be more important in raising the odds of conformity on the part of the adult out of school and away from family supervision. In short, the adult is more on his or her own in the world, complete with a character structure that presumably is crystallized more fully. The relative mix and impact of internal and external control across the life course is an issue that warrants further investigation.

Even though they vary somewhat in their stress on particular forces of integration and regulation and in their attachment to a social disorganization perspective, the theories covered in the past two chapters share certain similarities beyond the fact that they all take the control perspective. Reckless's stress on *goal orientation* sounds very much like the emphasis that Hirschi placed on *commitment* in his earlier sociologically oriented social bond theory. Both theories consider legitimate aspirations to be a crucial factor in insulating a potential deviant from nonconformity. Reckless's concept of *norm erosion* as the obverse of norm retention is closely related to Sykes and Matza's concept of *techniques of neutralization* in such a way that the latter can easily be subsumed as one aspect of the former (Ball, 1966). In much the same way, Reiss's distinction between conformity as a consequence of *acceptance* and conformity as a consequence of *submission* is quite similar to Nye's distinction between *internalized control* and *direct control*. Gottfredson and Hirschi's psychologically oriented *self-control* theory shows considerable similarity to Reckless's notion of *inner containment,* albeit with a much more systematic theory of how such control may fail to develop. Hagan's power–control theory reads in some ways like an update of Nye's perspective in terms of U.S. families during the late 1980s as compared to the more patriarchal families of the 1950s. There also is an overlap between Tittle's concept of a *control deficit* and Colvin's focus on *coercion.* And with a few contemporary exceptions (e.g., the works of Hagan and Colvin), these theories share a Durkheimian stamp and remain largely silent on issues of how power and inequality influence the quality and impact of social control. These issues will occupy our attention in the chapters ahead.

7

The Irony of State Intervention

Labeling Theory

John Braithwaite
1951–
Australian National University
Author of Reintegrative Shaming Theory

W hen people violate the law, we assume that the state's most prudent response is to make every effort to apprehend the culprits and process them through the criminal justice system. Informing this assumption is the belief that state intervention reduces crime, whether by scaring offenders straight, by rehabilitating them, or by incapacitating them so that they no longer are free to roam the streets victimizing citizens. Scholars embracing the labeling theory of crime, however, attack this line of reasoning vigorously. They caution that, rather than diminishing criminal involvement, state intervention—labeling and reacting to offenders as "criminals" and "ex-felons"—can have the unanticipated and ironic consequence of deepening the very behavior it was meant to halt.

Thus, labeling theorists argue that the criminal justice system not only is limited in its capacity to restrain unlawful conduct but also is a major factor in anchoring people in criminal careers. Pulling people into the system makes matters worse, not better. This contention takes on importance when we consider the width of the net cast by criminal justice officials. On any given day, approximately 2.4 million Americans are in jails or in state or federal prisons, and about 5 million more are on probation or parole supervision and living in communities. For minorities, moreover, the presence of state intervention in their lives is particularly extensive. Nearly one in three African

American males between 20 and 29 years of age is under some form of control by the criminal justice system (Mauer, 1999; see also Pattillo, Weiman, & Western, 2004). In an analysis of a cohort born between 1965 and 1969, Western (2006) has shown the special vulnerability to incarceration of the most disadvantaged minorities. Thus, among African American men who are high school dropouts, he calculated that 58.9%—nearly 6 in 10—will spend some time in prison during their lifetime; the comparable figure for White dropouts is 11.2% (Western, 2006; see also Wacquant, 2001, 2009).

In this chapter, we examine why labeling theorists believe that state intervention is dangerously criminogenic, particularly in light of the faith that current policy makers have in the power of prisons to solve the crime problem. We also consider why, during the 1960s and early 1970s, labeling theory (or, as it also is known, the "societal reaction" approach) grew rapidly in popularity and markedly influenced criminal justice policy. Later, we consider some recent theoretical extensions of the labeling perspective. First, however, we discuss how scholars in this school of thought rooted their work in a revisionist view of what crime is and of how its very nature is tied inextricably to the nature of societal reaction.

The Social Construction of Crime

Before the advent of labeling theory, most criminologists were content to define crime as "behavior that violates criminal laws." This definition was useful in guiding inquiry and in setting rough boundaries for criminology as a field of study. Too often, however, the easy acceptance of this definition led criminologists to take for granted that they knew what crime was and could get on with the business of finding its causes either in offenders or in their environments. This reification of conceptual definitions blinded many scholars from seeing that, as a socially constructed phenomenon, what is or is not "criminal" changes over time, across societies, and even from one situation to the next. Without this insight, scholars failed to explore the social circumstances that determine which behaviors are made criminal, why some people have the label of criminal applied to them, and what consequences exist for those bearing a criminal label.

Labeling theorists sought to correct this oversight. As a starting point, they urged criminologists to surrender the idea that behaviors are somehow inherently criminal or deviant. To be sure, behavior such as killing or raping another person is injurious by nature. Even so, what makes an act criminal is not the harm it incurs but rather whether this label is conferred on the act by the state. Thus, it is the nature of the societal reaction and the reality it constructs—not the immutable nature of the act per se—that determines whether a crime has occurred (Becker, 1963; Erikson, 1966). Pfohl's (1985) discussion of whether killing is "naturally deviant" illustrates the essence of this position nicely:

Homicide is a way of categorizing the act of killing, such that taking another's life is viewed as totally reprehensible and devoid of any redeeming social justification. Some types of killing are categorized as homicide. Others are not. What differs is not the behavior but the manner in which reactions to that behavior are socially organized. The behavior is essentially the same: killing a police officer or killing by a police officer; stabbing an old lady in the back or stabbing the unsuspecting wartime enemy; a Black slave shooting a White master or a White master lynching a Black slave; being run over by a drunken driver or slowly dying a painful cancer death caused by a polluting factory. Each is a type of killing. Some are labeled homicide. Others are excused, justified, or viewed, as in the case of dangerous industrial pollution, as environmental risks, necessary for the health of our economy, if not our bodies. The form and content of what is seen as homicide thus varies with social context and circumstance. This is hardly the characteristic of something which can be considered naturally or universally deviant. (p. 284)

Armed with this vision of crime as socially constructed, labeling theorists argued that criminologists could ill afford to neglect the nature and effects of societal reaction, particularly when the state was the labeling agent. One important area they investigated was the origins of criminal labels or categories. Howard Becker (1963), for example, explored how the commissioner of the treasury department's Federal Bureau of Narcotics served as a "moral entrepreneur" who led a campaign to outlaw marijuana through the Marijuana Tax Act of 1937. In Becker's view, this campaign, marked by attempts to arouse the public by claims that smoking pot caused youths to lose control and commit senseless crimes, was undertaken to advance the bureau's organizational interests. The bureau's success in securing passage of the act, Becker concluded, resulted in "the creation of a new fragment of the moral constitution of society, its code of right and wrong" (p. 145; see also Galliher & Walker, 1977).

Numerous other scholars provided explanations of attempts to criminalize other forms of behavior. Anthony Platt (1969) studied how, at the turn of the 20th century, affluent women "invented delinquency" through their successful campaign to create a court exclusively for juveniles. With the establishment of this court, juveniles were treated as a separate class of offenders, and the state was granted the power to intervene not only when youths committed criminal acts but also when they showed signs of a profligate lifestyle—acts such as truancy and promiscuity that became known as "status offenses." This movement, Platt observed, was class biased because it was directed primarily at "saving" lower-class youths, reaffirmed middle-class values, and left unaddressed the structural roots of poverty.

In a like vein, Stephan Pfohl (1977) investigated the "discovery of child abuse," which prior to the 1960s had largely escaped criminal sanctioning. Pfohl showed how pediatric radiologists, who read X-rays in hospitals, were instrumental in bringing attention to the abuse of children. Their efforts, he claimed, were fueled by the incentive to demonstrate the importance of pediatric radiology and, thereby, to enhance the low prestige of this specialty in the medical community.

Kathleen Tierney (1982) focused on the "creation of the wife-beating problem." She revealed how wife battering did not emerge as a salient social issue deserving of criminal justice intervention until the mid-1970s, when feminist organizations and networks were developed sufficiently to make domestic violence socially visible, to establish victim shelters, and to earn the passage of new laws. Also critical to the battered women's movement was the media, which dramatized wife beating because

> it mixed elements of violence and social relevance . . . [and] provided a focal point for serious media discussion of such issues as feminism, inequality, and family life in the United States—without requiring a sacrifice of the entertainment value, action, and urgency on which the media typically depend. (pp. 213–214)

These and similar analyses showed, therefore, that what the state designated as criminal was not a constant but rather the result of concrete efforts by men and women to construct a different reality—to transform how a particular type of behavior was officially defined. Moreover, it was not simply the extent or harmfulness of the behavior that determined its criminalization. After all, drug use, juvenile waywardness, child abuse, and wife battering had long escaped state criminal intervention despite their pervasiveness and injurious effects. These behaviors were criminalized only when the social context was ripe for change and groups existed that were sufficiently motivated and powerful to bring about legal reform.

But what occurs once a form of behavior is defined as unlawful and a criminal label is created? To whom will this label be applied? The common answer to this question is that those who engage in the proscribed activity will be labeled as criminals. Labeling theorists, however, again were quick to point out that such thinking implicitly assumes that societal reaction can be taken for granted and treated as nonproblematic. It ignores not only that the innocent occasionally are falsely accused but also that only some lawbreakers actually are arrested and processed through the criminal justice system (Becker, 1963). A lawbreaker's behavior, therefore, is only one factor—and perhaps not the most important factor at that—in determining whether a criminal label is conferred.

A number of labeling theory studies illustrate this principle. In one experimental study conducted in Los Angeles, a racially mixed sample of college students, all of whom had perfect driving records during the past year, had Black Panther bumper stickers affixed to their car bumpers. Within hours of the experiment's start, they began to accumulate numerous tickets for traffic violations (e.g., improper lane changes), thereby suggesting that police officers were "labeling" differentially on the basis of the bumper stickers (Heusenstamm, 1975). Other researchers interested in police encounters with juveniles observed that officers' decisions to arrest wayward youths were based less on what laws were violated and more on the juveniles' demeanor—whether they were respectful and cooperative or surly and uncooperative (Piliavin & Briar, 1964).

Still another researcher, William Chambliss (1984), examined one community's societal reaction to two groups of high school boys: a middle-class group he called the "Saints" and a working-class group known as the "Roughnecks." Although the two groups had a rate of delinquency that was "about equal," the "community, the school, and the police react[ed] to the Saints as though they were good, upstanding, nondelinquent youths with bright futures but [reacted] to the Roughnecks as though they were tough, young criminals who were headed for trouble" (pp. 126, 131). Why did this occur? In Chambliss's view, a major reason was the community's lower-class bias, which led police to define the Saints' behavior as pranks and to anticipate that the poorly dressed and poorly mannered Roughnecks were up to no good.

Through these and similar studies, labeling theorists revealed that the nature of state criminal intervention was not simply a matter of an objective response to illegal behavior but rather was shaped intimately by a range of extralegal contingencies (Cullen & Cullen, 1978). Much attention was focused on how criminal justice decision making was influenced by individual characteristics such as race, class, and gender. In addition, however, researchers explored how rates of labeling vary according to the resources available to and political demands placed on police and other criminal justice organizations. This body of research also highlighted how official measures of the extent of crime, such as arrest statistics reported each year by the Federal Bureau of Investigation, depend not only on how many offenses are committed but also on the arrest practices of police. Accordingly, official crime statistics may be inaccurate to the extent that they reflect a systematic bias in enforcement against certain groups (e.g., the urban poor) or fluctuations in the willingness of police to enforce certain laws (e.g., rape).

In sum, labeling theorists elucidated the importance of considering the origins of criminal labels and the circumstances that affected their application. But they did not confine their attention to these concerns. They proceeded to put forward the more controversial proposition that labeling and reacting to people as criminals composed the major source of chronic involvement in illegal activity. They claimed that state intervention created crime rather than halted crime. We consider this line of reasoning next.

Labeling as Criminogenic: Creating Career Criminals

Where should the search for the cause of crime begin? As we have seen, scholars traditionally have argued that the starting point for criminological inquiry should be either individual offenders themselves or the social environments in which they reside. Labeling theorists, however, argued that causal analysis should commence not with offenders and their environs but rather with the societal reaction that *other* people—including state officials—have toward offenders. Again, their contention was based on the belief that labeling and treating lawbreakers as criminals have the unanticipated consequence of creating the very behavior they were meant to prevent.

EARLY STATEMENTS OF LABELING THEORY

The idea that criminal justice intervention can deepen criminality did not originate with the labeling theorists of the 1960s. A number of early criminologists, for example, noted that prisons—a severe form of societal reaction—were breeding grounds for crime. Jeremy Bentham, the classical school theorist, lamented that "an ordinary prison is a school in which wickedness is taught by surer means than can ever be employed for the inculcation of virtue. Weariness, revenge, and want preside over these academies of crime" (cited in Hawkins, 1976, p. 57). In 1911, Lombroso echoed this theme in his observation that "the degrading influences of prison life and contact with vulgar criminals . . . cause criminaloids who have committed their initial offenses with repugnance and hesitation to develop later into habitual criminals" (Lombroso-Ferrero, 1972, pp. 110–111). Willem Bonger, the Dutch Marxist scholar, noted similarly that in imprisoning "young people who have committed merely misdemeanors of minor importance . . . , we are bringing up professional criminals" (Bonger, 1916/1969, p. 118). And Shaw (1930) felt compelled to title the chapter in *The Jack-Roller* on correctional institutions "The House of Corruption."

Although observations such as these anticipated the more developed views of later labeling theorists, Frank Tannenbaum (1938) was perhaps the earliest scholar to state in general terms the principle that state intervention is criminogenic because it "dramatizes evil." "Only some of the children [who break the law] are caught," noted Tannenbaum (1938), "though all may be equally guilty." And this event is not without consequence. The youth is "singled out for specialized treatment" as the "arrest suddenly precipitates a series of institutions, attitudes, and experiences which other children do not share." Now the youth's world is changed fundamentally; people react differently, and the youth starts to reconsider his (or her) identity. "He is made conscious of himself as a different human being than he was before his arrest," Tannenbaum observed. "He becomes classified as a thief, perhaps, and the entire world about him has suddenly become a different place for him and will remain different for the rest of his life" (p. 19). This is particularly true if the youth is placed in prison, for that is where incipient, "uncrystallized," criminal attitudes are "hardened" through the "education" that older offenders provide (pp. 66–81).

In the end, Tannenbaum (1938) cautioned, we would do well to consider the potential consequences before taking the initial step of pulling a juvenile into the criminal justice system:

> The first dramatization of "evil" which separates the child out of his group for specialized treatment plays a greater role in making the criminal than perhaps any other experience. . . . He has been tagged. A new and hitherto nonexistent environment has been precipitated out of him. The process of making the criminal, therefore, is a process of tagging, defining, identifying, segregating, describing, emphasizing, [and] making conscious and self-conscious; it becomes a way of stimulating, suggesting, emphasizing, and evoking the very traits that are complained of. . . . The person becomes the thing he is described as being. (pp. 19–20)

In 1951, Edwin Lemert further formalized these insights when he distinguished between two types of deviance: primary and secondary. *Primary deviance,* he contended, arises from a variety of sociocultural and psychological sources. At this initial point, however, the offender often tries to rationalize the behavior as a temporary aberration or sees it as part of a socially acceptable role. The offender does not conceive of himself or herself as deviant, nor does the offender organize his or her life around this identity (p. 75).

By contrast, *secondary deviance* is precipitated by the responses of others to the initial proscribed conduct. As societal reaction intensifies progressively with each act of primary deviance, the offender becomes stigmatized through "name calling, labeling, or stereotyping" (Lemert, 1951, pp. 76–77). The original sources of waywardness lose their salience as others' reactions emerge as the overriding concern in the person's life and demand to be addressed. Most often, the offender solves this problem by accepting his or her "deviant status" and by organizing his or her "life and identity . . . around the facts of deviance." Accordingly, the offender becomes more, rather than less, embedded in nonconformity. As Lemert (1972) explained:

> Primary deviance is assumed to arise in a wide variety of social, cultural, and psychological contexts, and at best [it] has only marginal implications for the psychic structure of the individual; it does not lead to symbolic reorganization at the level of self-regarding attitudes and social roles. Secondary deviation is deviant behavior, or social roles based upon it, which becomes means of defense, attack, or adaptation to overt and covert problems created by the societal reaction to primary deviation. In effect, the original "causes" of the deviation recede and give way to the central importance of the disapproving, degradational, and isolating reactions of society. (p. 48)

LABELING AS A SELF-FULFILLING PROPHECY

Although Tannenbaum and Lemert stated fairly explicitly the theme that societal reaction can induce waywardness, this view of labeling did not win wide intellectual attention and become an identifiable school of thought in the criminological community until the mid- to late 1960s (Cole, 1975). As will be discussed later, the sudden pervasive appeal of labeling theory can be traced largely to the social context of the 1960s that made it seem plausible that state intervention was the crime problem's cause and not its solution. But another circumstance also was important in contributing to labeling theory's ascendancy: the existence of a group of scholars—Howard Becker, Kai Erikson, and John Kitsuse were perhaps the most influential among them—whose combined writings argued convincingly that societal reaction is integral to the creation of crime and deviance (Becker, 1963; Erikson, 1966; Kitsuse, 1964).

To show how societal reaction brings about more crime, these labeling theorists borrowed Merton's (1968) concept of the "self-fulfilling prophecy." For Merton, "The self-fulfilling prophecy is, in the beginning, a *false* definition of the situation evoking a new behavior which makes the originally false conception come *true*" (p. 477, emphasis in original).

Consistent with this reasoning, labeling scholars argued that most offenders are defined falsely as criminal. In making this claim, they did not mean to imply that offenders do not violate the law or that justice system officials have no basis for intervening in people's lives. Instead, the falseness in definition is tied to the fact that criminal labels, once conferred, do not simply provide a social judgment of the offenders' behavior; they also publicly degrade the offenders' moral character (Garfinkel, 1956). That is, being arrested and processed through the justice system means that citizens not only define the offenders' lawbreaking conduct as bad but also assume that the offenders as *people* are criminal and, as a consequence, are the "type" that soon would be in trouble again. Yet as Lemert's (1951) work suggested, such predictions about personal character and future behavior are likely to be incorrect. Much primary deviance—including initial experiments in crime and delinquency—is not rooted fundamentally in character or lifestyle and thus is likely to be transitory and not stable (Scheff, 1966).

In short, theorists observed that the meaning of the label "criminal" in our society leads citizens to make assumptions about offenders that are wrong or only partially accurate. These assumptions are consequential, moreover, because they shape how people react to offenders. Equipped with false definitions or stereotypes of criminals, citizens treat all offenders as though they were of poor character and likely to recidivate. On one level, these reactions are prudent, if not rational; after all, it seems safer not to chance having one's child associate with the neighborhood's "juvenile delinquent" and not to employ a "convicted thief" to handle one's cash register. Yet on another level, these reactions have the power to set in motion processes that evoke the very behavior that was anticipated—that transform an offender into the very type of criminal that was feared.

But how is the prophecy fulfilled? How are incipient criminals, who might well have gone straight if left to their own devices, turned into chronic offenders? Again, the conferring of a criminal label singles out a person for special treatment. The offender becomes, in Becker's (1963) words, "one who is different from the rest of us, who cannot or will not act as a moral human being and therefore might break other important rules" (p. 34). As a result, being a "criminal" becomes the person's "master status" or controlling public identification (pp. 33–34; see also Hughes, 1945). In social encounters, citizens do not consider the offender's social status as a spouse, as a parent, or perhaps as a worker; they focus, first and foremost, on the fact that they are interacting with a criminal.

Admittedly, this public scrutiny might scare or shame some offenders into conformity. But for other offenders, the constant accentuation of their criminal status and the accompanying social rebuke has the unanticipated consequence of undermining the conforming influences in their lives and of pushing them into criminal careers. Thus, in the face of repeated designation as criminals, offenders are likely to forfeit their self-concepts as conformists or "normal" persons and to increasingly internalize their public definition as deviants. As this identity change takes place, the offenders' self-concepts lose their power to encourage conformity; the pressure to act consistently with their self-concepts now demands breaking the law.

Similarly, people who are stigmatized as criminal often are cut off from previous prosocial relationships. As one's reputation as a reprobate spreads, phone calls are not returned, invitations to social engagements are not extended, friends suddenly no longer can find time to meet, and intimate relationships terminate. One solution to being a social pariah is to seek out those of a like status. Accordingly, conditions are conducive for offenders wearing a criminal label to differentially associate with other lawbreakers, thereby forming criminal subcultural groupings. Such associations are likely to further reinforce antisocial values and to provide a ready supply of partners in crime.

The abrogation of ties to conventional society, labeling theorists warned, is most probable when state intervention involves institutionalization. Imprisonment entails the loss of existing employment and strains family relations to the point where they might not survive. It also mandates that offenders reside in a social setting where contact with other, more hardened criminals is enforced. Education in crime, as Tannenbaum (1938) and other early criminologists noted, is the likely result.

Finally, saddling offenders with an official criminal label, particularly when they have spent time in jail and carry the status of ex-convict, limits their employment opportunities. Through their jail sentences, offenders may "pay back" society for their illegal behavior, but they find it far more difficult to shake their definition as persons of bad character who might fall by the wayside at any time. Therefore, employers see them as poor risks and hesitate to hire them or to place them in positions of trust. Most often, offenders are relegated to low-paying dreary jobs with few prospects for advancement. In this context, crime emerges as a more profitable option and a lure that only the irrational choose to resist.

In sum, labeling theorists asserted that the false definition of offenders as permanently criminal and destined for lives of crime fulfills this very prophecy by evoking societal reactions that make conformity difficult and criminality necessary, if not attractive. The labeling process thus is a powerful criminogenic force that stabilizes participation in illegal roles and turns those marginally involved in crime into chronic or career offenders. It is an especially dangerous source of crime, moreover, because its effects are unanticipated and rarely observed. Indeed, the common response to escalating crime rates is not to minimize societal reaction but rather to arrest and imprison more people. From a labeling perspective, such "get tough" policies ultimately will prove self-defeating, for they will succeed only in subjecting increasing numbers of offenders to a self-fulfilling process that makes probable lives of crime.

ASSESSING LABELING THEORY

Few criminologists would dispute either that labeling theory succeeded in bringing attention to the issue of societal reaction or that this innovative theoretical focus was an important reason behind the perspective's popularity. Even so, labeling theory's central propositions have not escaped considerable critical analysis (Gove, 1975, 1980). As we will see, this assessment suggests the need to temper some of the perspective's boldest claims but also indicates that it would be unwise to discount the insights set forth in the labeling tradition.

One line of criticism came from conflict or radical criminologists. Although they agreed that crime was socially constructed and that labels were differentially applied, they did not believe that labeling theorists went far enough in their analysis. Radical scholars argued that the origins and application of criminal labels were influenced fundamentally by inequities rooted in the very structure of capitalism. As Chapter 8 will show in greater detail, radicals insisted that differences in power determined that the behaviors of the poor, but not those of the rich, would be criminalized. Labeling theorists understood that political interest and social disadvantage influenced societal reaction, but again, they did not make explicit the connection of the criminal justice system to the underlying economic order. As Taylor, Walton, and Young (1973) observed, they failed "to lay bare the structured inequalities in power and interest which underpin processes whereby laws are created and enforced." Thus, they stopped short of exploring "the way in which deviancy and criminality are shaped by society's larger structure of power and institutions" (pp. 168–169).

A very different critique was leveled by criminologists of a traditional positivist bent, who maintained that labeling theory's major tenets wilted when subjected to empirical test. These critics contended—correctly, we believe—that the perspective's popularity had less to do with its empirical adequacy and more to do with its voicing a provocative message that meshed with the social times (Hagan, 1973; Hirschi, 1975). After all, in asserting that state intervention deepened criminality, none of the early labeling theorists had presented hard data supporting this thesis.

Thus, these positivist or empirically oriented criminologists brought data to bear on what were considered labeling theory's two principal propositions: first, that extralegal factors, not behavior alone, shaped who was labeled; and second, that labeling increased criminal involvement.

The Extralegal Factors Proposition. First, the positivist scholars assessed the premise that extralegal factors, such as an offender's race, class, and gender, are more important in regulating criminal justice labeling than are legal factors, such as the seriousness of the illegality or the offender's past record. Contrary to the expectations of the labeling theorists, research studies have found repeatedly that the seriousness of the crime—not the offender's social background—is the largest determinant of labeling by police and court officials (Sampson, 1986b). The critics have concluded that extralegal variables exert only a weak effect on labeling (Gove, 1980; Hirschi, 1975; Tittle, 1975b).

Not all criminologists, however, would agree with this conclusion. Research by Robert Sampson, for example, forced a reconsideration of the critics' sweeping rejection of the idea that official reactions are influenced by a variety of contingencies. Thus, Sampson (1986b) uncovered an "ecological bias" in police control of juveniles. Even when he took into account the seriousness of lawbreaking behavior, police were found to be more likely to make arrests in poor neighborhoods than in more affluent neighborhoods. One possible explanation for this ecological pattern is that police resources are concentrated more heavily in lower-class areas, where it is assumed that anyone encountered is likely to have a character sufficiently disreputable to warrant close surveillance. In any case, Sampson provided convincing evidence of an extralegal circumstance that causes differential selection in the criminal justice system.

The controversy over labeling has resurfaced in the recent debate over "racial profiling." Research is now probing whether African Americans (and other minorities) experience traffic stops because of their race and ethnicity—that is, because they are guilty of "driving while Black (or Brown)." Although there is some conflicting evidence, even when other behavioral factors are taken into account, minority drivers appear to be stopped, cited, searched, arrested, and have force used against them more often. It is unclear whether this disparity is due to individual prejudice on the part of officers or to more institutionalized practices in which officers stop minorities because they believe that they are more likely to be transporting drugs or because they are driving in neighborhoods "where they are out of place and do not belong." Regardless, consistent with labeling theory, whether individuals are subjected to social control and potentially have a criminal label attached to them is determined by more than simple legal factors (Engel & Calnon, 2004; Engel, Calnon, & Bernard, 2002; Novak, 2004).

State Intervention Is Criminogenic Proposition. Labeling theory's second major proposition is that state intervention through the justice system causes stable or career criminality. Phrased in these terms, however, the proposition is difficult to sustain. As critics point out, many offenders become deeply involved in crime *before* coming to the attention of criminal justice officials (Mankoff, 1971). Chronic delinquency, for example, seems far more tied to the criminogenic effects of spending years growing up in a slum neighborhood than to being arrested and hauled into juvenile court as a teenager. Similarly, we know that offenders become extensively involved in illegalities such as corporate crime, political corruption, wife battering, and sexual abuse without ever being subjected to criminal sanctioning.

Over the last quarter of the 1990s, tests of labeling's causal effects yielded mixed results (Bazemore, 1985; Klein, 1986; Morash, 1982; Palamara, Cullen, & Gersten, 1986; Shannon, 1982; Thomas & Bishop, 1984; Ward & Tittle, 1993; Kubrin et al., 2009). This resulted in two responses by criminologists. One involved rejecting labeling theory, the other extending the perspective.

The first response was to conclude that criminal justice labeling has "no effects" (Hirschi, 1975, p. 198), thus making the perspective irrelevant as a theory of criminal behavior. If this view is accurate, then it means that state intervention neither deepens criminality (as the labeling theorists have claimed) nor deters criminality (as advocates of punishment have asserted) (Thomas & Bishop, 1984). It also suggests that the sources of why people experiment with crime, become chronic offenders, or recidivate after incarceration are to be found not in the operation of the criminal justice system but rather in the social forces that impinge on offenders in their everyday lives (as traditional criminologists have argued). Phrased differently, people become criminals because they live in disorganized areas, lack controls, learn criminal values, and are under strain. Only after they become deeply involved in crime are they arrested and imprisoned. State intervention is thus a reaction to criminal involvement, not a cause of it. Labeling effects are negligible.

The second response was to suggest that the studies' mixed results may occur because the effect of labeling varies under different circumstances. It may well be that labeling's overall effect is unclear because researchers have yet to disentangle the conditions under which contact with the criminal justice system increases or diminishes commitment to crime (Palamara et al., 1986; Tittle, 1975a). At present, research has addressed in only a rudimentary way how vulnerability to criminal labels might vary by factors such as individual sociodemographic characteristics, stage in a criminal career, family strength, and neighborhood context. Until these empirical issues are settled, general statements about labeling effects remain premature (see also Paternoster & Iovanni, 1989).

Indeed, the complexity of labeling can be seen in experimental studies assessing how police officers' mandatory arrest of batterers influences subsequent episodes of domestic violence (Sherman, 1992). Although the finding and its interpretation still must be viewed tentatively, the research suggests that the impact of arrest—or of labeling—varies according to whether batterers are employed. Perhaps because their economic well-being and "stake in conformity" are threatened, those who are working are less likely to recidivate after arrest. By contrast, arrest appears to escalate incidents of abuse among batterers who are unemployed and, therefore, have weak bonds to conventional society (Sherman, 1992).

Notably, these kinds of insights helped to move scholars to develop theories that explored the differential effects of labeling. In this regard, Braithwaite's (1989) shaming theory and Sherman's (1993) defiance theory are presented later in this chapter. They focus in particular on how the quality of the state intervention—whether it is reintegrative or stigmatizing and disrespectful—can evoke either conformity or greater criminality.

In recent years, it is notable that labeling theory is enjoying a resurgence of interest and growing empirical support. These works do not claim that state intervention is the major cause of crime, but they do propose that it is a criminogenic risk factor—that is, that it is one factor that contributes to continued criminal involvement.

Thus, in a recent meta-analysis of 29 experimental studies, Petrosino, Turpin-Petrosino, and Guckenburg (2010) calculated that juvenile justice processing has no crime control effect and that "almost all the results are in the negative direction" (p. 6). In short, arresting and processing juveniles, as opposed to diverting them out of the system or into services, increased delinquent involvement. Similar results have been found using data from the longitudinal Rochester Youth Development Study. Analyses revealed that official intervention (arrest, involvement in the juvenile justice system) results in greater delinquency in the short term (mainly by increasing involvement in deviant groups, such as gangs) and, in the longer term, in greater crime in early adulthood (mainly by decreasing educational achievement and employment (Bernburg & Krohn, 2003; Bernburg, Krohn, & Rivera, 2006). Similar iatrogenic results of juvenile justice processing were reported by Gatti, Tremblay, and Vitaro (2009) using a sample of 779 disadvantaged Montreal youths.

Chiricos, Barrick, Bales, and Bontrager (2007) examined the fate of more than 95,000 adult men and women convicted of a felony and facing probation. This is a unique study because the labeling condition was created by a Florida law that allows judges not to assign a formal felony label to guilty offenders. In essence, the felony label is withheld:

> The consequence of this unique labeling event is that offenders who are equivalent in terms of factual guilt can either be labeled a convicted felon or not. For those offenders who have adjudication withheld—about half of the felony probationers in Florida in recent years—no civil rights are lost and such individuals may legitimately say on employment applications and elsewhere that a felony conviction did not occur. For those offenders who *are* formally adjudicated, all of the structural impediments of being a convicted offender are possible. (Chiricos et al., 2007, p. 548, emphasis in original)

In support of labeling theory, Chiricos et al. found that "being adjudicated a felon significantly and substantially increases the likelihood of recidivism in comparison with those who have had adjudication withheld" (p. 570). Differential labeling effects also were detected, with Whites, women, and those with no prior conviction before age 30 most affected by the assignment of a felony status. As Chiricos et al. note, these groups typically are less at risk of recidivating. If so, then this result is consistent with the labeling theory notion that state intervention applied to "primary deviants" is likely to be especially criminogenic.

Research on the effects of imprisonment also lends support to labeling theory. It is a remarkable oversight by criminologists that, despite having an inmate population in the United States of approximately 2.4 million, studies on how the prison experience affects recidivism are relatively few in number and often methodologically suspect (Nagin, Cullen, & Jonson, 2009). Still, useful investigations have been undertaken (see, e.g., Chen & Shapiro, 2007; Nieuwbeerta, Nagin, & Blokland, 2009; Petersilia & Turner, 1986; Sampson & Laub, 1993; Smith, 2006; Spohn & Holleran, 2002). They allow for three general conclusions, the first of which is most firmly established and important (Nagin et al., 2009; see also Gendreau, Goggin, Cullen, & Andrews, 2000; Jonson, 2010; Lipsey & Cullen, 2007; Villettaz, Killias, & Zoder, 2006).

First, overall, a custodial sanction versus a noncustodial sanction either has a null effect or is criminogenic, especially for low-risk offenders. Second, the longer the time spent in prison, the more likely it is that an offender will recidivate (however, see Jonson, 2010). Third, the harsher the prison living conditions are, the higher the rate of reoffending is likely to be. Notably, these findings are contrary to the predictions of conservative criminologists favoring "get tough" policies and of specific deterrence theorists (see Chapters 12 and 13). In both instances, these scholars link more punishment, especially imprisonment, with less recidivism.

Again, showing that incarceration has an effect on recidivism is not the same as contending that it is the main factor in explaining stable involvement in crime; other factors matter and matter more than confinement. At the same time, given that millions of Americans spend some time in prison during their lives (Pattillo et al., 2004), even a modest criminogenic effect of the prison sanction could have a meaningful impact on public safety.

Beyond crime control policy, the recent research on labeling theory leads to two further points. First, the consistent findings favoring labeling theory suggest that the role of state intervention in causing crime cannot be dismissed. As it becomes more common to conduct longitudinal studies that follow offenders across their lives, it would be inexcusable to ignore the consequences of a key social experience of career offenders: arrest and incarceration. Life-course theories should thus strive to incorporate imprisonment into their models (for an example, see Sampson & Laub, 1993). Further, the challenge ahead for labeling theorists is to specify more carefully how different types of state intervention impact the lives of different types of offenders, especially those low and high in their risk of recidivating. Societal reaction is a complex process, and its effects have yet to be fully unpackaged so that can be truly understood (see, more generally, Link, Cullen, Frank, & Wozniak, 1987). We return to this issue later in the chapter when we discuss contemporary extensions of labeling theory.

LABELING THEORY IN CONTEXT

A central theme informing this book has been that changes in society expose people to new experiences, which in turn prompt them to think differently about many issues, including crime. The 1960s were just such a period when social change gripped the United States and caused citizens and criminologists alike to take stock of their previous assumptions about criminal behavior (Sykes, 1974). We saw in Chapter 5, for example, that the tumultuous context of the 1960s sensitized some scholars to the importance of controls in constraining human conduct. In Chapter 8, we will see how that decade's events radicalized other scholars and led them to assert that crime and criminal justice were intimately shaped by the conflict and inequities inherent in capitalism. More relevant to our present concerns, we can recall that the 1960s also proved to be fertile ground for the growth of labeling theory (Cole, 1975).

Why did many criminologists suddenly embrace the notion that state intervention through the criminal justice system was the principal cause of the crime problem? In the absence of strong empirical evidence, why did labeling propositions, voiced earlier by Tannenbaum (1938) and Lemert (1951), suddenly strike a chord and seem sensible to so many scholars?

The key to answering these questions, we believe, lies in understanding how the prevailing context led many people to lose trust or confidence in the government. During the early 1960s, optimism ran high. As noted in Chapter 4, the Kennedy administration instilled the expectation that a "New Frontier of equal opportunity" was within reach; a "Great Society" was possible that would eliminate poverty and its associated ills such as crime. Moreover, this agenda reaffirmed the Progressives' belief that the government should play a central role through social programs in effecting this change. The state could be trusted to do good.

But as the 1960s unfolded, this optimism declined, eventually turning to despair as the bold promises made at the decade's start went unfulfilled (Bayer, 1981; Empey, 1979). Thus, the civil rights movement not only laid bare the existence of pernicious patterns of racism, sexism, and class inequality but also revealed the inability, if not the unwillingness, of government officials to address these long-standing injustices. The war in Vietnam raised other concerns. Although the use of U.S. troops was justified as necessary to protect democracy, this policy lost its moral value as many citizens perceived the United States as merely propping up a corrupt regime. More disquieting, however, was the government's response to political protest. The United States witnessed not only demonstrators being chased and beaten by police but also students being gunned down at Kent State University. The Attica riot, in which troopers storming the prison fatally wounded 29 inmates and 10 guards being held hostage, confirmed the state's proclivity to abuse its power in the suppression of insurgency. The state's moral bankruptcy seemed complete with the disclosure of the Watergate scandal, which showed that corruption not only penetrated but also pervaded the government's highest echelons (Cullen & Gilbert, 1982).

In short, the state faced what commentators have called a "legitimacy crisis" (Friedrichs, 1979) or a "confidence gap" (Lipset & Schneider, 1983); citizens no longer trusted the motives or competence of government officials. Such feelings spread and intensified as the 1960s advanced and turned into the 1970s, and they created a context ripe for harvesting the ideas of labeling theorists who blamed the state for the crime problem. Due to their social experiences, it now made sense to many criminologists, policy makers, and members of the public that government officials would label the disadvantaged more than the advantaged and would operate prisons, such as Attica, that drove offenders deeper into crime. Accordingly, labeling theory won wide support and offered a stiff challenge to traditional theories of crime. The stage was set, moreover, for a reexamination of existing crime control policies.

The Consequences of Theory: Policy Implications

Labeling theory, Empey (1982) observed, "had a profound impact on social policy" (p. 409). As the ranks of the perspective's proponents swelled, an increasingly loud warning was sounded that pulling offenders into the criminal justice system only exacerbated the crime problem. The prescription for policy change was eminently logical and straightforward: If state intervention causes crime, then steps should be taken to limit it (Schur, 1973).

But how might this be accomplished? As Empey (1982) noted, labeling theorists embraced four policies that promised to reduce the intrusion of the state into offenders' lives: decriminalization, diversion, due process, and deinstitutionalization. These four reforms were implemented to different degrees and with uneven consequences. Even so, the agenda identified by labeling theory to this day remains an important vision of the direction that criminal justice policy should take.

DECRIMINALIZATION

Labeling theorists insisted that the "overreach of the criminal law" constituted a critical public policy problem (Morris & Hawkins, 1970, p. 2; Schur, 1965; Schur & Bedeau, 1974). Thus, the criminal justice system traditionally has been used to control not only threats to life and property but also a range of "victimless crimes" (e.g., public drunkenness, drug use, gambling, pornography) and juvenile status offenses (e.g., truancy, promiscuity). The morality of these behaviors might be open to debate, theorists admitted, but using the criminal law as a means of control is an "unwarranted extension [that] is expensive, ineffective, and criminogenic" (Morris & Hawkins, 1970, p. 2).

Edwin Schur, for example, argued that the criminalization of victimless deviance, such as drug use, creates crime in various ways (Schur, 1965; Schur & Bedeau, 1974). First, the mere existence of the laws turns those who participate in the behavior into candidates for arrest and criminal justice processing. Second, it often drives them to commit related offenses, such as when drug addicts rob to support their habits. Third, by prohibiting the legal acquisition of desired goods and services, criminalization creates a lucrative illicit market, the operation of which fuels the coffers of organized crime. Finally, the existence of such illicit exchanges fosters strong incentives for the corruption of law enforcement officials, who are enticed through payoffs to "look the other way."

Accordingly, labeling theorists argued for the prudent use of decriminalization—the removal of many forms of conduct from the scope of the criminal law. This policy might involve outright legalization or treating the acts much like traffic violations (e.g., speeding) for which penalties are limited to minor fines. In any case, the goal was to limit the law's reach and thus to reduce the extent to which people were labeled and treated as criminals.

The policy of decriminalization evoked much debate and encouraged some significant legal changes. Abortion was legalized by a U.S. Supreme Court decision, possession of small amounts of marijuana frequently was reduced to a minor violation, the criminal status of pornographic material was left to local communities to decide, forms of gambling (e.g., state-run lotteries and casinos) were legalized, and status offenses were made the concern of social welfare agencies. These changes did not occur across all states, however, and other forms of behavior remained illegal. Especially noteworthy, the most controversial of the labeling theorists' policy proposals—the call to decriminalize all forms of drug use—fell on deaf ears. Indeed, as we have seen during recent years, an enormous campaign has been launched to stop the flow of drugs and to place those participating in this illicit market behind bars (Currie, 1993; Inciardi, 1986).

DIVERSION

Given that laws exist and offenders come to the attention of law enforcement officials, how should the criminal justice system respond? Labeling theorists had a ready answer: diversion. For juveniles, as Empey (1982) indicated, this policy might entail taking youths from the province of the juvenile court and placing them under the auspices of "youth service bureaus, welfare agencies, or special schools" (p. 410). For adults, it might involve releasing them to privately run mental health agencies, community substance abuse programs, or government-sponsored job training classes. Diversion also might involve the substitution of a less severe intervention such as when offenders are "diverted" from prison and instead placed in the community under "intensive probation supervision" or under "home incarceration" (Ball, Huff, & Lilly, 1988; Binder & Geis, 1984; Latessa, 1987).

Diversion programs became widespread during the past two decades, a development that can be traced, at least in part, to the persuasive writings of labeling theorists (Klein, 1979). The current inmate crowding problem, moreover, is furnishing fresh incentives for jurisdictions to establish diversion programs that will help to empty their prisons and jails. The popularity of diversion, however, has given labeling theorists little cause for celebration.

Originally conceived as an alternative to involvement in the criminal justice system or to incarceration, diversion programs most often have functioned as add-ons to the system. That is, participants in programs have not been those who would have stayed in the system or gone to jail but rather those who previously would have been released, fined, or perhaps given suspended sentences. Ironically, the very policy suggested by labeling theorists to lessen state intervention has had the effect of increasing that intervention: Diversion has "widened the net" of state control by creating a "system with an even greater reach" (Klein, 1979, p. 184; see also Binder & Geis, 1984; Frazier & Cochran, 1986).

DUE PROCESS

Labeling theorists also were quick to join the mounting due process movement, which sought to extend to offenders legal protections (e.g., right to an attorney, right not to be searched illegally). As Empey (1982) pointed out, although labeling theory did not prompt the concern for offender rights, the perspective and the concern for due process had a common source: "Both were part of the growing distrust of governmental and other institutions in the 1960s" (p. 410).

Labeling scholars' call for expanding due process was tied up with their critique of the "rehabilitative ideal." As noted in Chapters 2 and 3, reforms during the Progressive era had provided criminal justice officials with enormous discretion to effect the individualized treatment of offenders. Such discretionary powers were unbridled in the juvenile court, where the state was trusted to "save children" by acting as a "kindly parent" (Platt, 1969; Rothman, 1978, 1980). Labeling theorists, however, accused state officials of abusing this trust. Individualized treatment, they claimed, was merely a euphemism for judicial decisions that discriminated against the powerless and for parole board decisions that denied release to inmates who dared to resist the coercive control of correctional officials.

The solution to this situation was clear. Schur (1973) urged that "individualized justice *must give way to a return to the rule of law*" (p. 169, emphasis in original). The worst abuses must be curbed by an extension of constitutional protections, particularly to juveniles who had been blindly left in the hands of the state. Labeling theorists, moreover, embraced what amounted to the principles of the classical school: Punishments should be prescribed by law, and sentences should be determinate. Accordingly, discretionary abuse would be eliminated: Judges would be forced to sentence according to written codes and not according to whim, and determinate sentences with set release dates would replace parole decision making.

Labeling theorists hoped that these policies would result in shorter and more equitable sentences and thus would reduce the extent and worst effects of state intervention. This blind faith in the rule of law, however, has proven to be a mixed blessing. On the one hand, due process has provided offenders with needed protections against state abuse of discretion. On the other hand, it remains unclear whether the corresponding attack on rehabilitation has succeeded in creating a system that is less committed to interventionist policies and more committed to humanistic ideals (Cullen &

.t, 1982). As we will see in Chapter 12, recent trends in criminal justice policy do
. furnish reason for optimism on this point.

DEINSTITUTIONALIZATION

Finally, labeling theorists took special pains to detail the criminogenic effects of incarceration and to vigorously advocate the policy of lessening prison populations through deinstitutionalization. The time had come, they insisted, for a moratorium on prison construction and for the move to a system that corrected its wayward members in the community.

This proposal received a stunning test in 1972 when Jerome Miller, commissioner of Massachusetts's Department of Youth Services, took the bold action of closing the state's major juvenile facilities and placing youths in community programs (Empey, 1982; Miller, 1991). Only a small number of youths remained in secure detention (Klein, 1979). Significantly, a subsequent evaluation revealed that recidivism rates were only slightly higher after the institutions were emptied. More instructive, the researchers found that in those sections of the state where the reform was pursued enthusiastically through the creation of "a large number of diverse program options so that the special needs of each youth could be more nearly met," the recidivism rates were lower than before Miller's deinstitutionalization policy was implemented. The researchers termed these results "dramatic" (Miller & Ohlin, 1985, p. 70; see also "Deinstitutionalization," 1975).

One might have expected that these empirical results would cause policy makers in other states to consider the wisdom of making a community response to crime rather than an institutional one. But as we have noted, policy decisions are based less on research and more on what seems sensible and politically feasible. During the last quarter of the 20th century, the tenor of American society changed and new ways of thinking about crime emerged (J. Q. Wilson, 1975). Not surprisingly, policies have reflected this change in thinking, as we have abandoned the idea of deinstitutionalization and chosen instead to incarcerate offenders in unprecedented numbers.

Extending Labeling Theory

During recent decades, advocates of "get tough" criminal justice policies have shown remarkable hubris in making grand and typically unsupported claims for the ability of punishment to deter offenders. They have endorsed efforts to increase surveillance on offenders in the community (e.g., electronic monitoring, intensive supervision) and have endorsed efforts to increase the harshness of imprisonment (e.g., longer sentences; more ascetic living conditions) (Irwin, 2005; Irwin & Austin, 1994; Pattillo et al., 2004; Whitman, 2003). Of course, such attempts to heighten state intervention in the lives of offenders are antithetical to labeling theory, which would predict that these policies would only increase criminal behavior.

Contemporary criminologists often question the wisdom of this massive attempt to inflict more control and pain on offenders (Clear, 1994; Currie, 1998b; Pattillo et al., 2004). Other criminologists, however, recognize that *under certain circumstances,* criminal justice sanctions might reduce recidivism—a possibility fully discounted by labeling theory. At the same time, they also assert that such punishments, as typically applied in the criminal justice system, are likely either to have no effect or, consistent with labeling theory, to amplify criminal involvement. The key issue is not simply whether a sanction is applied but also the *quality* of the sanction—what actually happens to an offender during the criminal justice process. As Sherman (2000) noted, "The major failing of the science of sanction effects has been the assumption that all sanctions were alike in quality, varying only in quantity" (p. 6). Notably, two important attempts have been made to develop a theory of how the quality of sanctioning affects reoffending: Braithwaite's (1989) *theory of shame and reintegration* and Sherman's (1993) *defiance theory.* These perspectives occupy our attention next.

In addition, Rose and Clear (1998) have called attention to the way in which mass incarceration can have macro-level or community-level effects on crime. In so doing, they extend labeling theory by showing how, beyond the effects on individuals, state sanctions can have the unanticipated consequences of creating criminogenic conditions within neighborhoods. Their *coerced mobility theory* is also presented below.

BRAITHWAITE'S THEORY OF SHAMING AND CRIME

In *Crime, Shame, and Reintegration,* John Braithwaite took up the issue of the conditions under which societal reaction increases crime (as labeling theorists contend) or decreases crime (as advocates of punishment predict). Legal violations evoke formal attempts by the state and informal efforts by intimates and community members to control the misconduct. Central to social control is what Braithwaite (1989) called *shaming,* which he defined as "all processes of expressing disapproval which have the intention or effect of invoking remorse in the person being shamed and/or condemnation by others who become aware of the shaming" (p. 9).

Shaming comes in two varieties—reintegrative and disintegrative—and each has a different impact on recidivism. Consistent with labeling theory, Braithwaite (1989) argued that *disintegrative* shaming stigmatizes and excludes, thereby creating a "class of outcasts" (p. 55). The offender not only is castigated for his or her wrongdoing but also is branded as a criminal who is beyond forgiveness and unworthy of restoration to membership in the community. As labeling theory warns, the result is further entrenchment in crime. The offender is denied employment and other legitimate opportunities to bond with conventional society and, consequently, joins with other outcasts in creating and participating in criminal subcultures.

But shaming also can be *reintegrative.* In these instances, an illegal act initially evokes community disapproval but then is followed by attempts "to reintegrate the offender back into the community of law-abiding or respectable citizens through words or gestures of forgiveness or ceremonies to decertify the offender as deviant" (Braithwaite, 1989, pp. 100–101). There is a stick followed by a carrot—condemnation followed by community responses aimed at binding the offender to the social order. In this case, shaming has two faces: It makes certain that the inappropriateness of the misconduct is known to the offender and all observers, and it presents an opportunity to restore the offender to membership in the group. This combination, Braithwaite argued, reduces crime by exerting greater control over offenders and by not setting in motion the criminogenic processes caused by stigmatization and social exclusion.

Braithwaite (1989) extended labeling theory not only by delineating types of shaming or societal reaction but also by observing that the underlying social context determines the degree to which shaming will be reintegrative or disintegrative. In communitarian societies such as Japan, "individuals are densely enmeshed in interdependencies which have the special qualities of mutual help and trust" (p. 100). As might be expected, in this context shaming is reintegrative and produces low crime rates.

In the United States, however, communitarianism is weakened by urbanization, racial and ethnic heterogeneity, extensive residential mobility, and a strong ideology of individualism. As a consequence, America lacks the cultural and institutional basis that would encourage seeing offenders as part of an interdependent community. Thus, social control has a strong disintegrative quality: Lawbreakers are marked indelibly as ex-offenders and are provided with few avenues to reestablish full societal membership. This stigmatization creates a class of outcasts who individually turn to crime and who collectively develop criminal subcultures and illegal opportunity structures. The result, Braithwaite claimed, is that America is burdened with a high rate of lawlessness.

In summary, Braithwaite enriched labeling theory by illuminating not only that shaming (or labeling) varies in its nature and effects but also why this variation ultimately is contingent on the society in which shaming takes place. The empirical adequacy of Braithwaite's contentions remains to be convincingly demonstrated. Braithwaite, Ahmed, and Braithwaite (2006) have marshaled evidence from an array of quantitative and qualitative sources that is consistent with the theory (e.g., studies of parenting, corporate regulation, and restorative justice programs) (see also Braithwaite, 2002; Makkai & Braithwaite, 1991). Studies that employ surveys to measure reintegrative shaming and then assess its impact on self-reported delinquency—the methodological approach most often used to test criminological theories—are still in short supply (Ahmed & Braithwaite, 2004; Hay, 2001; Tittle, Bratton, & Gertz, 2003). This research has yielded promising but mixed results. One finding of potential consequence is that measures of reintegration and of shaming tend to have main (or independent) effects on outcomes. As Braithwaite et al. (2006) have observed, the "need to break down different elements of reintegrative shaming to see which are theoretically crucial and which are not should be an exciting challenge to criminologists in the survey research tradition" (p. 410).

SHERMAN'S DEFIANCE THEORY

Sherman (1993) began with the observation that labeling theory "does not account for the many examples of sanctions reducing crime" (p. 457). At the same time, he realized that there also are many examples in which sanctions increase crime. Given these seemingly opposed realities, Sherman noted that there is a pressing need for a theory of the criminal sanction that will address the question: "Under what conditions does each type of criminal sanction reduce, increase, or have no effect on future crimes?" (p. 445).

Sherman's (1993) central concept is that of *defiance,* which he defined as the "net increase in the prevalence, incidence, or seriousness of future offending against a sanctioning community caused by a proud, shameless reaction to the administration of a criminal sanction" (p. 459). A key insight is that when offenders are treated unfairly or with disrespect by police officers and/or the court, or when they perceive such mistreatment, they are likely to act defiantly. In such cases, criminal sanctions are not given legitimacy by offenders and are incapable of bringing about their intended effect of reducing crime. If anything, by provoking defiance, they cause offenders to assert their anger and autonomy by flouting the law and recidivating. There is some empirical evidence to support this thesis (Paternoster, Brame, Bachman, & Sherman, 1997; Sherman, 2000).

But Sherman understood that defiance does not inevitably follow from unjust treatment—or from what Braithwaite would call stigmatizing shaming. Three factors, in particular, are seen as increasing the risk that disrespect and unfairness will prompt increased offending. First, when offenders have few social bonds to the community, there is little to restrain their defiance and arising criminal inclinations. Second, consistent with Braithwaite, offenders are more likely to be defiant when they perceive the sanction as stigmatizing not their actions but rather the offenders personally. Third, when offenders deny or refuse to acknowledge the stigmatizing shame that has been imposed on them, they are more likely to respond with pride and use crime to exact revenge on conventional society. Empirical tests of these propositions remain in short supply. The dearth of research is likely because existing data sets do not contain measures of the theory's key components. The studies that have been undertaken show mixed support for the perspective (Bouffard & Piquero, 2010).

Regardless, the value of Sherman's work is that it shows that criminal sanctions can backfire and, through processes such as defiance, create the kind of self-fulfilling prophecy first identified by labeling theorists. The theory's implications are especially disquieting because many offenders targeted for searches and arrest by police are youths who come from "street families" (recall Anderson's [1999] "code of the street" discussed in Chapter 3). As Sherman (2000) noted, these adolescents are likely to be treated with little respect and might act in ways that will provoke harsh responses from police officers. They also are likely to have weak social bonds and a "code" that accords little legitimacy to conventional attempts to "disrespect" them. This might be one reason why police interactions with inner-city minority youths are potentially volatile: The conditions are conducive to defiance. More generally, Sherman's theory would predict that unless considerable attention is paid to the quality of relationships between inner-city youths and criminal justice representatives (e.g., police, judges, probation officers), attempts to sanction these youngsters might well prove to be counterproductive, fostering defiance and not deterrence.

ROSE AND CLEAR'S COERCED MOBILITY THEORY

Labeling theory has been primarily a theory of how state punishment has the unanticipated consequence of increasing, if not stabilizing, the criminality of *individuals.*

Dina Rose and Todd Clear (1998), however, elevate the perspective from individuals to the *community* (for a more complete discussion, see Clear, 2007). Their project is to explore what happens when the government adopts a policy of mass incarceration that disproportionately removes young, minority males from inner cities. As Rose and Clear (1998, pp. 450–451) point out, it is estimated that nearly 1 in 10 African American, underclass males between the ages of 26 and 30 are currently imprisoned, and that nearly 3 in 10 Black men will spend time in a state or federal prison during their lifetime (see also Mauer, 1999; Tonry, 1995). They conceptualize this incarceration as a form of *coerced mobility*—a practice that regularly takes large numbers of males out of inner-city communities for prolonged absences (see also Clear, 2002).

As noted, labeling theory is counterintuitive, because it suggests that efforts to achieve social control can backfire in unexpected ways (Hagan, 1973). In this case, Rose and Clear theorize that the mass coerced mobility of minority males may well have the unanticipated consequence of increasing, rather than decreasing, a community's crime rate. This thesis is counterintuitive because it would seem that locking up predatory criminals would make inner-city neighborhoods safer. In fact, DiIulio (1994, p. 23) has claimed that incarcerating such offenders is, in essence, a form of social justice that would "save Black lives." Rose and Clear (1998, p. 441) are not naïve to this reality; they realize that it is hard to comprehend how it can "be bad for neighborhood life to remove people who are committing crimes in those very neighborhoods." Still, when the incarceration is *massive and concentrated* in vulnerable communities, then it may become a *macro-level* force that undermines existing social institutions in such a way as to produce more, rather than less, social disorganization and conditions conducive to crime. Theoretically, they are suggesting that incarceration might have a feedback loop that causes disorganization; if so, then their thesis is a notable extension of social disorganization theory, calling attention to an important source of a neighborhood's ability to achieve organization and social control.

For Rose and Clear (1998), offenders are community liabilities (e.g., victimizers, often less than ideal parents), but they also are community assets (e.g., producers of income, supporters of families, parents, members of social networks). Their incarceration thus lessens liabilities (as is often recognized), but it also depletes the community of assets (as is often not recognized). The impact on neighborhoods will differ by the area's affluence and existing level of social organization. In stable working-class and middle-class communities, the incarceration of a limited number of resident-offenders is likely a social benefit. But in underclass neighborhoods, the overuse of prisons—high levels of incarceration year after year after year—risks depleting the area of resources it desperately needs and of weakening core social institutions. Incarceration thus has differential impacts depending on its magnitude and on the nature of the neighborhood.

Consider, for example, the impact of incarceration on family stability. Offenders typically earn money both legally and illegally and, although not always reliable providers, are a source of income for their partners and children. When they are incarcerated, family life is often disrupted, which contributes to social disorganization. Lacking income, families frequently change addresses, move into different school districts, and go on public assistance. Mothers have less time to supervise their children. And new males, not the children's father, may enter the household, creating more instability. On a broader level, the coerced mobility of offenders to prison means that a substantial portion of the community's male population spends years out of the labor market and builds little human capital. Their job prospects upon reentry are limited (Holzer, Raphael, & Stoll, 2004). As a result, the "marriage pool" of males in the area—consisting of ex-prisoners with few marketable skills—is unattractive, leading women to forgo marriage but not necessarily motherhood (Wilson, 1987). A high concentration of single-headed households thus prevails, a condition that fosters social disorganization and crime (Sampson & Groves, 1989; Wilson, 1987).

After the coerced mobility of prison, the offenders tend to return to their neighborhoods more of a liability than when they left. Not only do they face limited opportunities to secure legitimate work, but also they import from prison cultural values and networks supportive of crime; these challenge the ability of a community to socialize its youth into a common, prosocial value system. Because it is so common among young males in the area, being sent to prison might, in fact, lose its stigma. The legitimacy of the government, especially the criminal justice system, is thus called into question. Again, these countervailing forces, all nourished by mass imprisonment, serve to weaken convention institutions.

Rose and Clear thus offer a potentially important advance of labeling theory, exploring how formal control can produce crime by undermining informal control. In their words:

> We argue that state social controls, which typically are directed at individual behavior, have important secondary effects on family and neighborhood structures. These, in turn, impede the neighborhood's capacity for social control. Thus, at the ecological level, the side effects of policies intended to fight crime by controlling individual criminals may exacerbate problems that lead to crime in the first place. (Rose & Clear, 1998, p. 441)

Thus far, empirical research on their theory of coerced mobility is in short supply and has yielded mixed results (Clear, Rose, Waring, & Scully, 2003; Lynch & Sabol, 2004). The most formidable challenge to the theory is research showing that, due mostly to the effects of incapacitation, incarceration is inversely related to crime rates (Lynch & Sabol, 2004; Pratt & Cullen, 2005; Spelman, 2000). These findings, however, may be misleading. They do not show, for example, whether greater crime savings might be achieved if money devoted to mass imprisonment were allocated to criminal justice sanctions that were community-based and strategically oriented to building human capital in offenders. Further, although imprisonment might achieve contemporaneous or short-term reductions in crime rates, this does not mean that its overuse might not still be a factor in worsening levels of social disorganization and contributing to the emergence of the next generation of offenders—youngsters who will eventually follow their fathers and brothers into prison.

POLICY IMPLICATIONS: RESTORATIVE JUSTICE AND PRISONER REENTRY

Consistent with labeling theory, the three contemporary perspectives reviewed above call attention to the way in which the quality of the societal reaction to offenders—including their arrest, sanctioning (especially imprisonment), and stigmatizing social exclusion—may have the unanticipated consequence of deepening people's criminality. From the viewpoint of Braithwaite, Sherman, and Rose and Clear, the challenge is to find ways to blunt the negative, criminogenic effects of the sanctions typically imposed on offenders. Two recent policy developments are informed by this line of theorizing: restorative justice and prison reentry programs.

Restorative Justice. Perhaps the most significant recent development in criminal justice in the United States and other nations is "restorative justice" (Bazemore & Walgrave, 1999; Braithwaite, 1998, 1999, 2002; Hahn, 1998; Harris, 1998; Sullivan & Tifft, 2006; Van Ness & Strong, 1997). In traditional criminal justice, the state acts—presumably on the victim's behalf—to sanction offenders. The intent of the sanction often is to exact a measure of "just deserts" for the victim and larger community by inflicting some sort of discomfort on the offender. Restorative justice, however, rejects the logic that equates the state's harming of an offender with victims' receiving any meaningful sense of justice. In fact, the risk is that the state's action will only increase the overall amount of harm being inflicted without achieving anything of value.

As an alternative, advocates of restorative justice suggest that the guiding principle of the criminal sanction should be to *decrease harm* by *restoring* (1) the victim to his or her prior unharmed status and (2) the offender to the community. Instead of a traditional trial in which the state is an adversary prosecuting defendants, such advocates favor a victim–offender conference in which the state functions more as a mediator. In such a conference, which often is attended by family members and interested members of the community, the actions of the offender are condemned or shamed, and the offender is encouraged to take responsibility, express remorse, and apologize to the victim. Offender accountability is integral to the proceedings. Thus, efforts are made to design plans for how the offender will compensate the victim, thereby restoring the victim by undoing the harm that was experienced (e.g., restitution). Community service also might be required. In exchange, the goal is to reintegrate the offender into the community, providing the supports that are necessary to accomplish this end. Throughout this process, the offender is treated fairly and is respected; the offender's actions, but not him or her personally, are shamed; and the offender is brought into a context where it is possible to be restored to society without facing the continuing stigma of being an "ex-con."

The restorative justice movement has gained strength from a number of sources, ranging from evangelical Christians and victim rights advocates to peacemaking and feminist criminologists (Immarigeon & Daly, 1997). Still, Braithwaite's (1998, 1999, 2002) shaming theory has furnished a significant intellectual justification for restorative justice. In Braithwaite's terms, restorative justice is built on the premise of "reintegrative" shaming rather than "stigmatizing" shaming. His perspective is critical of state-centered punishment because its quality typically is disintegrative and crime inducing. By contrast, Braithwaite's reintegrative approach seeks to shame the crime but not the criminal and to find ways of reattaching the offender to conventional society. In a similar way, Sherman's defiance theory would be receptive to restorative justice. Because offenders are treated with respect and fairness, his model would predict that restorative justice would be far less likely than traditional criminal justice sanctions to foster defiance and increased criminality.

It remains to be seen, however, whether restorative justice is an approach capable not only of increasing justice but also of reducing offenders' criminal involvement (compare Braithwaite, 2002, with Levrant, Cullen, Fulton, & Wozniak, 1999). Because restorative justice programs are mostly directed toward minor offenders, their impact on serious chronic offenders is open to question. Studies assessing these interventions, especially randomized experimental program evaluations, are in short supply. There are some very promising results but also conflicting findings (Braithwaite, 2002; Kurki, 2000; McGarrell & Hipple, 2007; Schiff, 1999; Shapland et al., 2008; Sherman & Strang, 2007; Strang & Sherman, 2006). Furthermore, when positive findings are forthcoming, it is often difficult to determine whether reductions in recidivism are due to the restorative features of the program or to other services, such as counseling, that were provided to offenders (Bonta, Wallace-Capretta, & Rooney, 1998).

Perhaps the most systematic assessment of the effectiveness of restorative justice programs has been provided by Bonta, Jesseman, Rugge, and Cormier (2006; see also Gendreau & Goggin, 2000; Latimer, Dowden, & Muise, 2005; Lipsey, 2009). They conducted a meta-analysis of 39 studies evaluating the impact on recidivism of restorative justice interventions (hereinafter referred to as RJIs). They reached three conclusions (all quotes to follow from Bonta et al., 2006, p. 117).

First, the effects of RJIs are "relatively small, but they are significant" and larger in more recent studies (see also Lipsey, 2009). Second, when RJIs are court-ordered, they have no effect on recidivism; those that are conducted in a "non-coercive environment and that attempt to involve victims and community members in a collaborative manner" achieve the largest reductions in reoffending. Consistent with the work of Braithwaite (2002) and Sherman (1993), this finding indicates that a less stigmatizing sanctioning process may be more effective with offenders.

Third, RJIs "appear to be more effective with low-risk offenders" than with "high-risk offenders." When focused on high-risk offenders—that is, those with a high probability of recidivating—it appears that RJIs may not be sufficient, by themselves, to counteract the strong criminal tendencies of these individuals. For this group, it might be necessary to combine an RJI with rehabilitation programs that target known criminogenic risk factors for change and that have been shown to decrease reoffending among high-risk criminals (Bonta, Wallace-Capretta, Rooney, & McAnoy, 2002; see also Cullen & Gendreau, 2000; Gendreau, Smith, & French, 2006; Levrant et al., 1999). It should be noted, however, that Sherman and Strang (2007) dispute this finding. Their review of studies leads them to conclude that restorative justice *"seems to reduce crime more effectively with more rather than less serious crimes"* (p. 8, emphasis in original). Future research on the differential effects of RJI's by risk level thus seems in order.

Prisoner Reentry Programs. In the first half of the 1900s, inmates were allowed to leave an institution on parole only if they had a place to live (usually with their family) and a job awaiting them. As Simon (1993) notes, this system of "industrial parole" broke down due to the confluence of two factors: rising prison populations and the deterioration of the economies in the nation's inner cities that made the guarantee of a job implausible. In "post-industrial parole," the goal became more to manage offender reentry—whether that was through a treatment model that emphasized the provision of services or, more recently, through a policing model that emphasized surveillance and the threat of reincarceration.

For several decades, most correctional observers did not give priority to the disquieting reality that offenders reentering society face an array of daunting challenges (Irwin, 2005)—which predictably lead to high recidivism rates. As Petersilia (2003) points out, "more than two-thirds of those released from prison will be rearrested; nearly half will be returned to jail or prison for a new crime or technical violation; and about a quarter will be returned to prison for a new crime conviction in the three years following their release" (p. 153). These figures are not new; they have remained fairly stable since the mid-1960s (Petersilia, 2003). Still, as prison populations jumped sevenfold since the early 1970s, it became increasingly difficult to ignore the sheer number of inmates "coming home" each year. As Travis (2005, p. xx) notes, the "reality of mass incarceration translates into a reality of reentry." It is estimated that each year in the United States, well over 650,000 offenders leave prison and return to society.

As scholars shifted their attention to this issue, they soon detailed a correctional situation that, as labeling and related theories would predict, almost certainly stabilized, rather than "knifed off," criminal behavior. Many offenders leave prison with their criminogenic needs untreated or worsened by their stay behind bars, with tenuous ties to their families, with no place to live, and with no driver's license or identification. Their job prospects are dismal, given that they have learned few marketable job skills while institutionalized and, as ex-offenders, will have difficulty being hired; if employed, they will earn low wages and work in unpleasant settings (Bushway, Stoll, & Weiman, 2007; Pager, 2007). They will be stripped of many civil rights, including the right to vote. Further, in the 1990s, federal and state legislatures moved to impose more restrictions on offenders, especially those convicted of drug and violent crimes. Enacting "legislation to cut offenders off from the remnants of the welfare state" (Travis, 2002, p. 23), they barred offenders from welfare assistance and food stamps, living in public housing, receiving loans for higher education, and the right to hold a driver's license (see Irwin, 2005; Manza & Uggen, 2006; Mauer & Chesney-Lind, 2002; Pager, 2003; Pager & Quillian, 2005; Pattillo et al., 2004; Petersilia, 2003; Travis, 2005).

Within the last decade or so, there has been a new realization that failing to address prisoner reentry and pursuing a policy of stigmatizing reintegration exacerbate recidivism and pose a threat to public safety (Taxman, Young, Byrne, Holsinger, & Anspach, 2002; Travis, 2005; Western, 2006). During this time, for example, Ohio's corrections director, Reginald Wilkinson, created "The Ohio Plan for Productive Offender Reentry and Recidivism Reduction" (Petersilia, 2003), and communities in eight states around the nation implemented Reentry Partnership Initiatives that involved the "formation of a partnership between criminal justice, social service, and community groups to develop and implement a reentry process" (Taxman et al., 2002). Models or principles of how to develop an effective reentry program also are emerging, generally emphasizing the need (1) to start reentry preparation while offenders are in prison, (2) to focus on the challenges and crises that are faced immediately upon release (e.g., food, shelter, job), and (3) to provide treatment services and support to facilitate long-term community reintegration (Petersilia, 2003; Taxman et al., 2002; Travis, 2005). It remains to be seen if reentry programs will be well designed and be capable of reducing recidivism, but prisoner reentry has emerged from the shadows and promises to be an important correctional policy issue for the foreseeable future (Bushway et al., 2007; Listwan, Cullen, & Latessa, 2006).

Conclusion

Labeling theory's distinctive focus on societal reaction succeeded in sensitizing criminologists to the important insights that the criminal nature of behavior is socially constructed by the response to it and that a variety of factors can shape who comes to bear a criminal label. The perspective also forced consideration of the possibility that state intervention can have the ironic, unanticipated consequence of causing the very conduct—lawlessness—that it is meant to suppress. In light of conflicting findings, empirical research has yet to definitively confirm this causal thesis. But recent research is increasingly showing that state intervention, especially the use of imprisonment, contributes to entrenching certain kinds of offenders in criminal careers. At the least, labeling theory provides an important reminder that the effects of criminal justice sanctions are complex and may contradict what common sense would dictate. As contemporary criminologists in this tradition remind us, the quality of the sanction that is imposed—what we do to and with offenders—is potentially consequential. This warning assumes significance when we consider policy makers' repeated assurances that the panacea for the crime problem can be found in widening the reach of the criminal law and in the enormous expansion of prison populations.

William J. Chambliss
1933–
George Washington University
Author of Conflict Theory

8

Social Power and the Construction of Crime

Conflict Theory

A s Chapter 7 showed, theories purporting to explain crime by locating its sources in biological, psychological, or social factors associated with the offender have tended to ignore the way in which "crime" is produced and aggravated by the reaction to the real or imagined attributes or behavior of those who are being labeled offenders. Labeling theory tried to correct this oversight, going into great detail in an effort to explain the labeling process and its consequences. But as the previous chapter also indicated, some criminological theorists did not believe that labeling theory went far enough. The labeling theorists showed some appreciation for the way in which political interests and political power affected social reaction, but they did little to explore these deeper issues.

What determines whether a label will stick? What determines the extent to which those labeled are punished? The nature of the labeling may depend on differences between those labeled and those doing the labeling. Whether labels can be made to stick and the extent to which those labeled can be punished may essentially depend on who has power. Theories that focus attention on struggles between individuals and/or groups in terms of power differentials fall into the general category of *conflict theory*.

Some conflict theories try to search for the sources of the apparent conflicts. Some seek to elucidate the basic principles by which conflict evolves. Others try to develop a theoretical foundation for eliminating the conflict. Still others try to do all of this and more.

In this chapter, we examine some of the leading criminological conflict theories, beginning with conflict theory in general as found in the pioneering work of Marx and Engels and in the later work of Simmel. Then we examine Bonger's attempt to develop a Marxist theory of crime, Sutherland's and Sellin's focus on the relationship between culture conflict and crime, and Vold's effort to build a criminological conflict theory on the tradition exemplified by Simmel. Because contemporary American criminological conflict theory, like other criminological control theories, drew much of its inspiration from the turmoil of the 1960s, it is necessary to reexamine the 1960s from another point of view before proceeding to a consideration of influential conflict theories such as those of Turk, Chambliss, and Quinney. Finally, we consider some of the consequences within criminology as well as the policy implications of conflict theory.

Forerunners of Conflict Theory

MARX AND ENGELS: CAPITALISM AND CRIME

As indicated in Chapter 5, Karl Marx and Friedrich Engels already were expressing concern over the apparent decline in social solidarity by the middle of the 1800s, preceding Durkheim by several decades. For Marx and Engels, as well as for Durkheim, crime was to some extent a symptom of this decline and would diminish (although Durkheim made it clear that it never would disappear) if social solidarity could be regained. They differed in their analysis of the source of the erosion of solidarity and their prescription for its restoration.

Durkheim saw the situation as a moral problem and argued that the social solidarity of the future would depend on an effective combination of controls through modes of social integration and social regulation that could operate in tune with the new division of labor that had been created by industrialization. Marx and Engels saw the problem in economic terms, denounced the new division of labor as the unjust exploitation of one social class by another, and insisted that social solidarity could be regained only with the overthrow of capitalism itself. They proposed revolution followed by a period of socialism. Because in their view the political state existed essentially as a mechanism for the perpetuation of capitalism, it was deemed historically inevitable that the state would then "wither away," leaving a society based on the true brotherhood and sisterhood of communism (Marx & Engels, 1848/1992).

Marx himself had very little to say about crime. Those criminological theorists who cite him as a major forerunner tend to extrapolate from his general approach, to cite his collaborator Engels's more directly relevant writings on the assumption that Marx agreed, or to do both by citing Marx in general with specific quotations from Engels to nail down the point. Certainly, Marx and Engels stressed differences in interests and in power much more than did Durkheim. For them, conflict was inherent in the nature of social arrangements under capitalism, for it was capitalism that generated the vast differences in interests and capitalism that gave the few at the top so much power over the many at the bottom. Above all, their theoretical approach was action oriented; they were less concerned with the pure understanding of social problems than with changing things for what they considered the better.

Although the work of Marx and Engels was extremely complex, certain basic propositions stand out (Turner, 1978). First there was the proposition that conflict of interests between different groups will be increased by inequality in the distribution of scarce resources (e.g., food, clothing, shelter). A second proposition was that those receiving less of the needed resources would question the legitimacy of the arrangement as they became aware of the nature of the "raw deal" they were getting. The third proposition was that these groups then would be more likely to organize and to bring the conflict out into the open, after which there would be polarization and violence leading to the redistribution of the scarce resources in such a way that they would be shared by everyone. Capitalism was considered to be at the root of the conflict because it was taken to be the source of the unjust inequality. In this view, greater integration and regulation simply would tend to perpetuate an unjust economic system. The way in which to solve the problem of collapsing social solidarity was not to find some new sources of faith in the social order or some more effective means of regulating its members but rather to destroy capitalism and build toward the one just form of social solidarity—communism.

SIMMEL: FORMS OF CONFLICT

Like Marx, Georg Simmel was a German intellectual with a deep interest in social theory. On the other hand, Simmel was a contemporary of Durkheim, doing his theoretical work a few decades later than Marx and concerning himself with a search for precise intellectual understanding of abstract laws governing human interaction rather than with changing the world. Simmel was an exponent of "sociological formalism" (Martindale, 1960, p. 233). He was concerned not so much with the changing *content* of social life as with its recurring *forms* or patterns. Speaking of Simmel, Wolff (1964) remarked that his approach represented a "preponderance of the logical over the normative" (p. xviii). Simmel's interest was not in particular conflicts but rather in conflict in general, and he was less concerned with normative questions such as the justice of a particular outcome than with the abstract logic of conflict itself.

Although Simmel was deeply interested in conflict as a form common in social life, he was not as preoccupied with it as was Marx, seeing it as a normal part of life and as one form of interaction among others, some of which were operating in different and even opposite directions in an integrated social system (Turner, 1978). Conflict was regarded not as a problem necessarily calling for solutions or even leading to change but rather as a typical aspect of social order that often actually contributed to that order. Of equal importance is that whereas Marx focused on the *causes* of conflict and sought to find means for their elimination, Simmel focused on the *consequences* of conflict, with little interest in its sources and great interest in the complex formal patterns through which it developed.

BONGER: CAPITALISM AND CRIME

Early during the 20th century, the conflict perspective of Marx and Engels was applied specifically to criminological theory by the Dutch criminologist Willem Bonger. As we will see later, the turbulence of the 1960s brought renewed interest in Marxist theory and in the work of Bonger (1916/1969). In the introduction to an abridged edition of Bonger's major work published near the end of the 1960s, the American criminologist and conflict theorist Austin T. Turk spoke of him appreciatively as a man who "combined a passion to alleviate human misery with an equal passion for scientific research" (Turk, 1969b, p. 3). Like Marx and Engels, and unlike Freud and Durkheim, Bonger believed that the human was innately social. If so, then crime would have to be traced to an unfavorable environment that distorted human nature. Bonger held that just such an unfavorable environment had been generated by the rise of capitalism.

Under capitalism, according to Bonger, there had arisen a sharp division between the rulers and the ruled that originated not in innate differences between them but rather in the economic system itself. In such an unfavorable environment in which people were pitted against each other in the economic struggle, in which the individual was encouraged to seek pleasure by any means possible without regard for others, and in which the search required money, human nature was distorted into an intense "egoism" that made people more capable of committing crimes against one another. Thus, like the control theorists as far back as Durkheim, Bonger traced crime in part to individual egoism. Unlike them, however, Bonger took the Marxist position that the decline of social integration and the rise of extremely disruptive individualism could be traced to capitalism. Such egoism never could be reduced by social controls that bound the individual more closely to society, for society under capitalism was itself the very *source* of the egoism.

Bonger traced much crime to the poverty generated by capitalism, both directly because crime among the subordinate class sometimes was necessary for survival and indirectly because the sense of injustice in a world where many had next to nothing while a few had nearly everything was held to demoralize the individual and stifle the social instincts. At the same time, however, he recognized that the more powerful bourgeoisie also committed crimes. He traced this to the opportunities that came with power and the decline of morality that came with capitalism. Crime was seen as a product of an economic system that fostered a greedy, egoistic, "look out for number one" mentality while at the same time making the rich richer and the poor poorer.

Long before the labeling theorists, Bonger stressed that although it certainly was true that crime fell into the category of immoral actions, definitions of morality varied. Indeed, he went further, insisting that the source of the prevailing definitions and of their variations could be found in the interests of the powerful. In Bonger's view, behaviors were defined as crimes when they significantly threatened the interests of the powerful, and "hardly any act is punished if it does not injure the interests of the dominant class" (Turk, 1969b, p. 9). Thus, Bonger took note of the statistics showing a lower crime rate among the bourgeoisie and traced this to the fact that the legal system "tends to legalize the egoistic actions of the bourgeoisie and to penalize those of the proletariat" (p. 10). Given Bonger's theory of the causes of crime, his conclusion that the abolition of capitalism and the redistribution of wealth and power would restore a favorable environment and eliminate crime followed almost entirely as a matter of political logic.

SUTHERLAND AND SELLIN: CULTURE CONFLICT AND CRIME

In criminological theory, Edwin H. Sutherland is best known for his differential association theory discussed in Chapter 3. As indicated in that chapter, the concept of differential association was built on the concept of differential social organization, which was itself an attempt to move away from the value judgments entailed in describing situations organized differently as representing social disorganization. The concept of differential social organization represented one type of conflict perspective. Inherent in the notion was the assumption that society did not rest on complete consensus but rather was made up of different segments with conflicting cultural patterns. Individual criminal activity could be explained by assuming that a person whose associations were dominated by relationships with those in a less law-abiding segment of society would tend to learn criminal techniques and develop criminal orientations.

Sutherland already had begun systematic research into the crimes of the wealthy and powerful by the mid-1920s. During the 1930s, he pioneered the study of white-collar criminality, publishing a groundbreaking article at the beginning of the 1940s (Sutherland, 1940) and a widely cited book on the subject by the end of the decade (Sutherland, 1949). In these works, he called attention to the fact that powerful economic interests such as the huge corporations that had arisen during the 20th century represented a segment of society whose organization and policy made them "habitual criminals," although their wealth and political power protected them from prosecution in the criminal courts. As was pointed out in Chapter 3, he explained the participation of *individuals* in the white-collar crimes of these corporations in terms of their *differential association* by way of immersion in the criminal patterns of the business.

During the depression of the 1930s, another criminologist, Thorsten Sellin, best known today for his studies of capital punishment and his efforts at developing crime indexes, argued for a broader definition of crime and a less conventional approach to criminological theory. Sellin (1938) stressed the problem of "culture conflict" as a source of crime, maintaining that different groups learn different "conduct norms" and that the conduct norms of one group might clash with those of another. As for which conduct norms would become a part of the criminal law, Sellin held that

> the conduct which the state denotes as criminal is, of course, that deemed injurious to society or, in the last analysis, to those who wield the political power within that society and therefore control the legislative, judicial, and executive functions which are the external manifestations of authority. (p. 3)

VOLD: CONFLICT AND CRIME

Near the close of the 1950s, George B. Vold set forth the most extensive and detailed treatment of criminological theory from a conflict perspective yet seen in criminology. Citing Simmel, Vold (1958) agreed that conflict should be regarded not as abnormal but rather as a fundamental social form characteristic of social life in general. He argued, "As social interaction processes grind their way through varying kinds of uneasy adjustment to a more or less stable equilibrium of *balanced forces in opposition,* the resulting condition of relative stability is what is usually called social order or social organization" (p. 204, emphasis added). In this perspective, social order was not presumed to rest entirely on consensus; rather, it was presumed to rest in part on the stability resulting from a balance of power among the various conflicting forces that compose society. The conflict between groups was analyzed in Simmelian terms as potentially making a *positive contribution* to the strengthening of the different groups because participation in the struggle resulted in group esprit de corps and in-group solidarity. According to this analysis, it was perfectly normal for groups in a complex society to come into conflict as their interests clashed, and "politics, as it flourishes in a democracy, is primarily a matter of finding practical compromises between antagonistic groups in the community at large" (p. 208).

As to the nature of legal compromises, Vold (1958) cited Sutherland's work from the 1920s, noting that "those who produce legislative majorities win control over the police power and dominate the policies that decide who is likely to be involved in violation of the law" (p. 209). Vold considered politics to be the art of compromise and insisted that "the principle of compromise from positions of strength operates at every stage of this conflict process" (p. 209). His own theoretical interest in the *forms* rather than the *content* of such conflict may be seen in his comparison of the delinquent gangs' patterns to those formed by conscientious objectors during time of war. Both were analyzed as identical patterns or social forms in which group ideology existed in conflict with established authority. Like Simmel, Vold was concerned with the similarity of the forms of conflict rather than with normative distinctions such as the nature of the morality involved.

Vold (1958) pointed out that much crime was of an obviously political nature. He included crimes resulting from protest movements aimed at political reform, noting that "a successful revolution makes criminals out of the government officials previously in power, and an unsuccessful revolution makes its leaders into traitors subject to immediate execution" (p. 214). His formal analysis avoided the normative and stuck to the logical. Regardless of who was "right" and who was "wrong," those who lost were the "criminals." In this sense, their (formal) "crime" consisted of losing.

Vold (1958) also gave attention to crimes involving conflicts between unions and management and between different unions themselves. In a statement portending what was to come in the civil rights movement in the United States during the 1960s and elsewhere in the world during the 1980s, he pointed out that "numerous kinds of crimes result from the clashes incidental to attempts to change or to upset the caste system of racial segregation in various parts of the world, notably in the United States and in the Union of South Africa" (p. 217). Vold undertook an intensive analysis of organized crime and white-collar crime as examples of groups organizing themselves in the pursuit of their interests, examining the strategies and tactics employed in the conflict.

Theory in Context: The Turmoil of the 1960s

Although the various forerunners of conflict theory discussed previously had anticipated much of what was to follow, it was not until the social upheavals of the 1960s that criminological conflict theory came into its own. The 1960s represented a major turning point for criminology. As we have seen already, it was within the context of those times that both control theory and labeling theory developed from theoretical seeds planted earlier. The same was true for conflict theory. Although control theory reacted by stressing the tenuous nature of complex society under conditions of rapid social change and insisted that crime and delinquency tended to spread with any significant weakening of forces containing the individual, conflict theory highlighted the newly revealed patterns of social division and questioned the legitimacy of the motives, strategies, and tactics of those in power. Although labeling theory exposed the way in which crime was a social construction of moral entrepreneurs and others in a position to influence the definitions developed by the political state and sometimes pointed to class bias in the labeling process, conflict theory was much more explicit about the connection between the criminal justice system and the underlying economic order, sometimes condemning the state itself.

Seeking to explain the rise of criminological conflict theory (which he termed *critical theory*) during the 1960s, Sykes (1974) pointed to three factors of special importance. First, there was the impact of the war in Vietnam on American society. Second, there was the growth of the counterculture. Third, there was the rising political protest over discrimination, particularly racial discrimination, and the use of the police power of the state to suppress political dissent—associated issues that had been smoldering and threatening to break into the open since World War II.

As the war in Vietnam escalated, doubts deepened not only about the wisdom of governmental policy but also about the fundamental motives and credibility of those in power. Reasons given for the escalating involvement seemed to many to be rather farfetched, and doubts were increased by discoveries of *disinformation*—a bureaucratic term for governmental lies. Protest marches spread. Armed troops fired on and killed apparently peaceful protesters on college campuses. The conscientious objectors about whom Vold had written so coolly were everywhere, voluntarily choosing to become "criminals" by leaving for Canada or burning their draft cards in heated public protests.

As indicated in Chapter 5, the developing counterculture represented in large part a repudiation of middle-class standards. Countercultural behavior dramatized a fundamental conflict in values. Millions were engaging in a variety of activities that they considered harmless but that the legal system regarded as criminal offenses—so-called victimless crimes such as fornication, vagrancy, and illegal drug use. The latter had in fact become not only a way of "getting high" but also a symbolic political protest.

As Chapter 5 emphasized, along with the impact of the war in Vietnam and the growth of the counterculture came the rise of social movements aimed at eliminating discrimination and demanding an end to the suppression of political dissent. The feminist movement spoke more and more radically in terms of the political, economic, and sexual oppression of women. The gay community became more politicized, organizing to resist labeling and discrimination. Underlying social conflict came more and more into the open as the civil rights movement seemed to make it clear that African Americans were not going to gain social equality without civil disobedience and that the only way of dealing with the unjust laws enforcing segregation was to violate them en masse—for the protesters to become "criminals" and spend some time in jail as the price for their beliefs.

At the same time, a society that had tolerated the McCarthyism of the 1950s was upset to learn that some of its most revered symbols of law and order, such as the Federal Bureau of Investigation, had become involved in the dissemination of internal disinformation aimed at destroying political opposition and had fallen into the use of illegal tactics in dealing with citizens seeking to express legitimate grievances. It often appeared that the agents of the political state, whether southern sheriffs threatening African American schoolchildren with police dogs or members of Congress harassing as radicals or Communists those who seriously questioned the political order, really were moral criminals disguised as legally constituted authorities. The impression of society as composed of groups in conflict, with the legal system tending to brand as criminal significant threats to the interests of the powerful, became more and more prevalent, providing fertile ground for the development of conflict theory.

Varieties of Conflict Theory

The new criminological conflict theory went much further than criminologists had taken it before, appealing to Marx and Bonger as well as to Simmel. As Gibbons and Garabedian (1974) pointed out in their own examination of the conflict theory movement (there termed *radical criminology*), "Sutherland certainly did not characterize the seventy corporations that he studied as 'exploiters of the people'; instead, he stopped far short of that sort of condemnation" (p. 51). Similarly, Sellin may have noted that the law is defined and applied in the interests of dominant groups, but he did not denounce them in the manner of Marx. As for Vold's analysis of the state of affairs, what had seemed like a bold outline of the conflict perspective during the late 1950s now struck some as in need of much more elaboration and others as far too abstract, academic, and aloof from the injustices that so angered them.

Sykes (1974) listed four major factors behind the shift. First, there was now a profound skepticism toward any theory that traced crime to something about the individual, including not only the biological and psychological theories but also sociological theories referring to inadequate socialization and the like. Second, there was a marked shift from the assumption that the inadequacies of the criminal justice system were to be traced to incompetent or corrupt individuals or minor organizational flaws to the conclusion that these problems were inherent in the system, either because it was fundamentally out of control or because it had been so designed by powerful interests. Third, the older assumptions that criminal law represented the collective will of the people and that the job of the criminologist was to do theoretical analysis and empirical research and not to deal in normative issues of right and wrong increasingly were rejected as fallacies that made it impossible to ask the basic questions. Finally, as discussed in Chapter 7, it had become clear not only that official crime rate figures did not reflect the amount of criminal behavior actually present in society but also that what they did reflect more often was the labeling behavior of the authorities.

During the 1960s, several criminologists, influenced by a blending of labeling theory with political theory, began to develop their own brand of criminological conflict theory. It is fair to characterize their work as part of a theoretical movement. Here we concentrate on the efforts of three of the leading conflict theorists: Turk, Chambliss, and Quinney. As we will see, they were not working in isolation but rather working in close contact within a particular sociohistorical context. All of them presented the outlines of their own approaches in articles written during the 1960s, and all published book-length volumes in 1969. Their differences were to depend, to a considerable extent, on influences from the various theoretical forerunners discussed previously.

TURK: THE CRIMINALIZATION PROCESS

Completing his graduate work at the beginning of the 1960s, Austin T. Turk had come to be increasingly interested in the culture conflict perspective of criminologists such as Sellin and in the emerging labeling theory. As with any theorist, his perspective was influenced by his life experiences. Turk (1987) himself put it this way:

> Growing up as a working class boy in a small segregated Georgia town, I learned early that life is neither easy nor just for most folks; that irrationality and contradiction are very much part (maybe the biggest part) of social reality; that access to resources and opportunities [has] no necessary association with ability or character; that the meaning of justice in theory is debatable and of justice in practice [is] manipulable; and that whatever degree[s] of freedom, equality, brotherhood, or security exist in a society are hard-won and tenuous. (p. 3)

By the end of the 1960s, Turk (1969a) had presented a complete statement of his own brand of conflict theory in *Criminality and the Legal Order,* quoting at length from Sutherland in the introduction and citing the more recent efforts of Vold and Dahrendorf. It was this effort to build on Dahrendorf's perspective, along with his Simmelian-Voldian approach that treated conflict not as some abnormality but rather as a fundamental social form, that distinguished Turk's theoretical contribution. For Turk, recognition of social conflict as a basic fact of life represented simple realism rather than any particular tendency toward cynicism.

Although also a conflict theorist, Dahrendorf (1958, 1968) disagreed with Marx on the question of inequality. Instead of tracing inequality back to an unjust economic system, Dahrendorf located the source in *power* differences, more specifically in differences in *authority* or power that had been accepted as legitimate. Unlike Marx, who had argued for the abolition of inequality, Dahrendorf took the position that because cultural norms always exist and have to depend on sanctions for their enforcement, some people *must* have more power than others to make the sanctions stick. In Dahrendorf's view, it was not the economic inequality resulting from capitalism that produced social inequality; rather, inequality was an inescapable fact because the basic units of society *necessarily* involved dominance-subjection relationships. Thus, the idea of eliminating inequality was treated as a utopian dream.

Turk (1969a) seemed to have been persuaded of the essential relativity of crime in much the same way as the labeling theorists. For him, the theoretical problem of explaining crime lay not in explaining varieties of behavior, for these may or may not be crimes depending on time and place. Instead, the problem lay in explaining "criminalization," the process of "assignment of criminal status to individuals" (p. xi), which results in the production of "criminality" (p. 1). Dahrendorf's influence was clear in Turk's definition of the study of "criminality" as "the study of relations between the statuses and roles of legal *authorities . . .* and those *subjects*—acceptors or resisters but not makers of such law creating, interpreting, and enforcing decisions" (p. 1, emphasis in original).

Such an approach was deemed necessary if the criminologist was to explain "facts" such as variations in crime rates or to develop better methods of dealing with "criminals." Turk (1969a) stressed that assignment of criminal status to an individual may have less connection with the behavior of that person than with his or her relationship to the authorities. "Indeed, criminal status may be ascribed to persons because of real or fancied *attributes,* because of what they *are* rather than what they *do,* and justified by reference to real or imagined or fabricated behavior" (pp. 9–10, emphasis in original). From this point of view, even if the criminologist eventually could succeed in explaining the *behavior* of criminals, such an achievement would not help in accounting for their *criminality,* which had more to do with the behavior of the authorities in control of the criminalization process.

Central to the concept of *authority* is its accepted legitimacy; authority differs from raw power because it is regarded as legitimate power, the use of which is accepted by those subject to it. Any theorist concerned with the relationship between subjects and authorities must investigate the basis for such acceptance. Like some of the control theorists, Turk (1969a) rejected the argument that acceptance of authority must be the result of internalization. He maintained that acceptance could be explained as a consequence of people learning the *roles* assigned to the *statuses* they occupied and simply acquiescing and going along as a matter of routine. Some have the status of authorities; others play the part of subjects. "The legality of norms is defined solely by the words and behavior of authorities" to which subjects will tend to defer (p. 51).

Like Simmel and Vold, Turk (1969a) was concerned with the logical *consequences* of the fact that some people had authority over others, not with the sources of this authority or whether it was just or unjust according to some normative conception of justice. In fact, logical consistency might be expected to force anyone who was convinced that concepts such as crime were a matter of labeling relative to time and place to a similar position with respect to concepts such as justice. In any case, Turk focused on the logical consequences of authority relationships, holding that "how authorities come to be authorities is irrelevant" to such an analysis (p. 51).

Because of Turk's (1969a) argument that the assignment of a criminal status would have to be justified by the authorities "by reference to real or imagined or fabricated behavior" (pp. 9–10) that was held to represent a violation of legal norms, he found it important to make a formal distinction between two types of legal norms: cultural norms and social norms. The first he defined as those set forth in symbolic terms such as words—as norms dealing with *what is expected*. The second he identified as those found in patterns of actual behavior—in terms of what is being done rather than *what is being said*. Turk pointed out that the cultural norms and social norms in a given situation may or may not correspond.

According to Turk (1969a), a satisfactory theory accounting for the assignment of criminal status would include

> a statement of the conditions under which cultural and social differences between authorities and subjects will probably result in conflict, the conditions under which criminalization will probably occur in the course of conflict, and the conditions under which the degree of deprivation associated with becoming a criminal will probably be greater or lesser. (p. 53)

Like Simmel, he proceeded to examine the nature of these conditions through a series of formal logical propositions.

Thus, Turk argued that, given that cultural and social norms might not agree, the existence of a difference between authorities and subjects in their evaluation of a particular attribute (e.g., past membership in a radical political organization) or a particular act (e.g., marijuana use) logically implies four situational possibilities. Each logical possibility carries a different conflict potential. The conflict probability would be highest, for example, in the "high-high" situation where there was (1) high congruence between the cultural norms preached by the authorities and their actual behavior patterns and (2) similarly high congruence between the cultural evaluation of a particular attribute or act and the actual possession of that attribute or commission of that act on the part of the subjects. If both sides not only *hold* different standards but also act *in accordance* with them, then there is no room for compromise. For example, in a situation where the authorities not only say "smoking marijuana is wrong" but also act to stop it, *and* marijuana users not only say "pot is okay" but also insist on using it in spite of the normative clash, the conflict potential would be logically greatest.

On the other hand, Turk's logic suggested that the conflict potential would be lowest in situations where there was neither agreement between authorities' stated cultural norms and their actual behavioral norms nor agreement between the cultural norms and the social norms of subjects. In such situations, the preachments would clash, but because neither side practices what it preaches in any case, the probability of conflict would be low. Why should they fight over words when neither side lives by them anyhow?

A third logical possibility was described as one in which authorities' talk and behavior were highly congruent, although there was little, if any, agreement between the words and actions of subjects. Such a situation would fall somewhere in the middle in terms of Turk's formal logic of conflict potential, but with a somewhat higher conflict potential than that logically inherent in the fourth and final possibility. The latter situation was described as one in which the attribute or act, as described in the announced cultural norm, happened to be in close agreement among subjects, whereas the cultural norm preached by the authorities actually had little relationship to their behavior. According to Turk, the third possibility would entail somewhat more conflict potential than would the fourth because the authorities would be less likely to tolerate norms different from their own when their cultural norms were reinforced by their social norms.

According to Turk's analysis, the logic of the relationships between cultural and social norms is complicated by additional formal propositions. Under the assumption that an individual who has group support is going to be more resistant to efforts to change him or her, Turk concluded that the probability of conflict with authorities grows with the extent to which those having the illegal attributes or engaging in the illegal activities are *organized*. Under the assumption that the more sophisticated norm resisters would be better at avoiding open conflict through clever tactical maneuvers (e.g., pretending to submit while secretly continuing as before), he concluded that the probability of conflict increases as the authorities confront norm resisters who are less *sophisticated*. The logical possibilities resulting from the combination of the two variables of organization and sophistication were set forth as follows: (1) organized and unsophisticated, (2) unorganized and unsophisticated, (3) organized and sophisticated, and (4) unorganized and sophisticated. Conflict odds were attached to each combination.

Proceeding through formal analysis of the four formally distinct possibilities, Turk reached certain conclusions. First, he concluded that conflict between authorities and subjects is most probable where the latter are *highly organized and relatively unsophisticated* (e.g., delinquent gangs). He then concluded that the odds of such conflict declined to the extent that the subjects involved are *unorganized and unsophisticated* (e.g., skid row transients) and still further to the extent that they are *organized and sophisticated* (e.g., syndicate criminals). It followed logically that the lowest probability of conflict would be associated with a situation in which the norm-resisting subjects are *unorganized and sophisticated* (e.g., professional con artists). These formal deductions may or may not match empirical reality, but they have a certain logical consistency, and Turk encouraged research designed to assess their actual empirical validity.

As for the authorities themselves, Turk pointed out that they must be *organized* or, by definition, they would not be the authorities; instead, they would be some sort of illegitimate mob. He concluded that the probability of conflict between these authorities and subjects resisting their norms would be greatest where the authorities were *least sophisticated* in the use of power. Interestingly enough, Turk's (1969a) logic carried him to the same conclusion as that reached by the control theorist Hirschi at the same time—that the probability of conflict was affected by the "nature of the bonds between authorities and subjects" (p. 61). He concluded that "where subjects are strongly identified with the authorities and generally agree in moral evaluations, an announced norm may be accepted in a 'father knows best' spirit" (p. 61). But where Hirschi stressed these bonds as central, Turk devoted much less attention to them, considering society to be less a matter of Durkheimian bonds and more a matter of constant Simmelian conflict working itself out over time. We must remember, of course, that Hirschi was focusing theoretical attention on juvenile delinquency, whereas Turk was devoting considerable attention to organized crime, political crime, and white-collar crime in general. Accordingly, the relative stress on bonds as compared to conflict potentials is hardly surprising.

For Turk (1969a), an analysis of conflict probabilities was only a first step. The key question was as follows: "Once conflict has begun, what are the conditions affecting the probability that members of the opposition will become criminals . . . , that they will be subjected to less or more severe deprivation?" (p. 64). Part of the answer was traced to the same factors outlined previously that would continue to affect probabilities throughout the criminalization process. Turk concluded, however, that additional variables tended to come into play with actual criminalization.

First, Turk admitted that although the crucial norms defining the conflict were those of the higher authorities, the major factor in the probability of criminalization was likely to be the extent to which the official legal norms agreed with both the cultural norms and the social norms *of those specifically charged with enforcing the legal norms,* especially the police but also prosecutors, judges, and the like. Because of the importance of police discretion and decisions on the spot, he concluded that the extent to which the police agreed with the legal norms they were expected to enforce would have a major effect on the odds of arrest and criminalization. Prosecutors, judges, juries, and others were expected to affect the probabilities somewhat, but ultimately it would be the police as the frontline enforcers who would determine the extent to which norm resisters actually would be defined as criminal.

In Turk's analysis, the *relative power of enforcers and resisters* became still another variable affecting the odds of criminalization. He proposed that the greater the power difference in favor of norm enforcers over resisters, the greater would be the probability of criminalization. He asserted that the reluctance of the enforcers to move against very powerful resisters would keep the criminality of the very powerful low regardless of their behavior. He added the qualification, however, that some of the disapproved behavior of the least powerful also might be ignored if these individuals seemed to pose no threat and "weren't worth the bother."

The final set of variables to which Turk assigned special significance in determining the odds of criminalization had to do with the *realism of conflict moves.* Although in part a matter of the sophistication mentioned previously, success in avoiding or producing criminalization also was regarded as dependent on factors beyond the use of knowledge of others' behavior patterns in manipulating them. Any move by *resisters* was considered to be unrealistic if it (1) increased the *visibility* of the offensive attribute or behavior, thereby increasing the risk that the authorities would be forced to act; (2) increased the *offensiveness* of the attribute or behavior (e.g., emphasizing it, calling attention to additional offensive attributes, violating an even more significant norm of the authorities); (3) increased *consensus* among the various levels of enforcers (e.g., moving from simple opposition to a particular norm or set of norms to a wholesale attack on the system with emphasis on stereotyping of enforcers as brutal, ignorant, and corrupt); or (4) increased the *power difference* in favor of the enforcers (e.g., upsetting the public in such a way that enforcers would be able to get significantly increased resources such as budget increases).

In Turk's analysis, any move by the *authorities* was likely to be unrealistic if it (1) *shifted the basis of their legitimacy away from consensus* toward the "norm of deference" or obligation to obey despite disagreement, which would be likely in the case of power plays; (2) represented a *departure from standard legal procedures,* especially if the shifts were unofficial, sudden, or sharp; (3) *generalized from a particular offensive attribute* so that additional attributes of the opposition also became grounds for criminalization (e.g., if roundups of similar types were used in lieu of a systematic search for a particular offender); (4) *increased the size and power of the opposition* (e.g., by creating martyrs among them so that they gained sympathy and other resources from other segments of society); or (5) *decreased consensus* among the various levels of enforcers.

Turk went much further than this in pointing out additional factors that may be expected to alter the nature and outcome of social conflict. He attempted to provide a logical integration of his propositions to demonstrate how various combinations affect conflict probabilities differently and went into great detail in describing possible indicators that might be used to operationalize his propositions to facilitate research. Thus, he dealt with the means by which legal norms could be classified in practice; with the question of the best measures of key concepts such as normative-legal conflict, the significance of legal norms, relative power, and realism in conflict moves; and with other issues that would have to be solved if his theory were to be tested by data.

CHAMBLISS: CRIME, POWER, AND LEGAL PROCESS

Finishing his own graduate work in the same year (1962) as did Turk, William J. Chambliss had become interested in the development of criminal law, specifically in "sociologically relevant analyses of the relationship between particular laws and the social setting in which these laws emerge, are interpreted, and take form" (Chambliss, 1964, p. 67). Chambliss (1987) also has been candid about the relationship between his theoretical perspective and his life experiences:

> After I graduated from UCLA [University of California, Los Angeles], I hitchhiked across the country again to see my father. It was 1955, and in short order I was drafted into the army and sent to Korea with the Counter Intelligence Corps (CIC). I learned a lot about crime during that period. American and Korean soldiers raped, stole, assaulted, intimidated, and generally terrorized the Koreans. Because they had the power, nothing was done about it. . . . How could crime be understood from the paradigms I learned in psychology and sociology? (pp. 5–6)

Chambliss completed his graduate work at Indiana University before going on to teach at the University of Washington and then to postdoctoral studies at the University of Wisconsin. It is worth noting that (1) both Turk and Quinney, the third of the contemporary criminological conflict theorists to be examined in this chapter, completed their graduate work at the University of Wisconsin in the same year that

Chambliss was completing his at Indiana and that (2) Turk himself joined the faculty at Indiana after completion of his graduate work at Wisconsin. The ties between these institutions were very close, and there was considerable mutual influence:

> Our lives are far more dependent on chance occurrences than we ever want to acknowl-
> edge. A year after I arrived at Washington, the university hired Pierre van den Berghe,
> who was extremely knowledgeable about Marxism. I organized a faculty seminar on the
> sociology of law with Pierre, a philosopher, and two anthropologists. . . . About this
> time, the Russell Sage Foundation decided to support the resurrection of the sociology
> of law. I was awarded a fellowship to study law at the University of Wisconsin.
> (Chambliss, 1987, pp. 3, 7)

Like Turk, Chambliss was impressed by the relativity of crime and the way in which the criminal label seemed to be the product of social conflict. But whereas Turk's work was influenced by a formal Simmelian approach, Chambliss was at first inspired less by the older European tradition of conflict theory than by the American tradition of *legal realism,* the pioneering work of the American legal scholar Jerome Hall, and the results of his own empirical research. Chambliss (1964) had undertaken a study of the development of vagrancy laws in England, concluding that these laws could be traced to vested interests: "There is little question that these statutes were designed for one express purpose: to force laborers . . . to accept employment at a low wage in order to insure the landowner an adequate supply of labor at a price he could afford to pay" (p. 69).

In the same year that Turk (1969a) published *Criminality and the Legal Order,* Chambliss (1969) published *Crime and the Legal Process,* an edited volume consisting of actual empirical research studies of the legal system, tying them together with his own theoretical framework. This soon was followed by a more elaborate presentation, *Law, Order, and Power,* in collaboration with Robert T. Seidman, a law professor at the University of Wisconsin (Chambliss & Seidman, 1971). In the earlier statement, Chambliss (1969) simply identified his position as representing an "interest group" perspective rather than a "value expression" perspective, a distinction that he compared with that seen in "the debate in social science theory between the 'conflict' and 'functional' theorists" (p. 8). His approach was influenced greatly by the American school of legal realism, which concerned itself with the distinction between the "law in the books" and the "law in action." It insisted that the study of abstract legal theory must be complemented by the study of the law as it works itself out in actual practice. It is interesting to note that although the first of these had been the focus of the classical school discussed in Chapter 2, the new attention to the law in action represented the influence of positivism in legal studies. Chambliss hoped to develop a theory of the law in action—a theory based on empirical research.

Early in the second volume, Chambliss and Seidman (1971) sounded an almost Durkheimian theme, asserting that the "first variable in our theoretical model is the relative complexity of the society" (p. 31). They went on, however, to insist that (1) the complexity, which comes with technological development and necessitates more complicated, differentiated, and sophisticated social roles, actually operates to (2) put people *at odds with one another,* thereby (3) requiring *formal institutions* designed to sanction what some consider to be norm violations. This argument was reminiscent of Dahrendorf's, with the major difference being that formal sanctioning institutions were regarded not as inherently necessary to society but rather as a contemporary necessity resulting from increasing social complexity. Their conflict perspective led them to a theory of legal development almost exactly opposite to that expounded in Durkheim's (1964) control theory approach, which maintained that increased societal complexity tended to lead society from an emphasis on "repressive" law to an emphasis on "restitutive" law. Instead, Chambliss and Seidman (1971) argued, "We may formulate, therefore, the following proposition: The lower the level of complexity of a society, the more emphasis will be placed in the dispute-settling process upon reconciliation; the more complex the society, the more emphasis will be placed on rule enforcement" (p. 32).

Chambliss and Seidman (1971) began by arguing that increasing *social complexity* itself tended to call for sanctioning institutions designed to keep order among the conflicting interests. Going further, they maintained that this sanctioning process would become even more pronounced to the degree that the social complexity became a matter of *social stratification,* with some groups having more wealth and power than others: "The more economically stratified a society becomes, the more it becomes necessary for the dominant groups in the society to enforce through coercion the norms of conduct which guarantee their supremacy" (p. 33). Here was something of a Marxian theme, although Chambliss and Seidman stopped short of a clear-cut Marxian position that traced the most serious problems of social stratification to polarization produced by capitalism.

Chambliss and Seidman (1971) also stressed the significance of the fact that the developing sanctions tended to be enforced through *bureaucratic organizations*. In their view, the basis of the sanctioning would be organized in the interests of the dominant groups, but the actual application of the sanctions tended to come through bureaucracies *that had their own interests*. In this sense, the law in action might be expected to reflect a combination of the interests of the powerful and the interests of the bureaucratic organizations created to enforce the rules.

Because the essence of Chambliss's theory was aimed at explaining the law in action in contemporary, complex, industrial societies, we can concentrate on the second and third variables in the theory of the law in action—*social stratification* and *bureaucracy*. In *Crime and the Legal Process,* Chambliss (1969) stressed that "the single most important characteristic of contemporary Anglo-American society influential in shaping the legal order has been the emerging domination of the middle classes" along with "the attempt by [members of] the middle class to impose their own standards and their own view of proper behavior on people whose values differ" (pp. 10–11). In this view, the middle class was coming to represent the conventional morality, but Chambliss emphasized that behind the values lay self-interest.

In *Law, Order, and Power,* Chambliss and Seidman (1971) set forth five fundamental propositions with respect to the relationship between social stratification and the law, beginning with the proposition that (1) *the conditions of one's life affect one's values and norms.* They then asserted that (2) *complex societies are composed of groups with widely different life conditions* and that (3) *complex societies, therefore, are composed of highly disparate and conflicting sets of norms.* As for the relationship between these conflict-ing norms and the law itself, they maintained that (4) *the probability of a given group's having its particular normative system embodied in law is not distributed equally but rather is closely related to the political and economic position of that group.* These propo-sitions, taken together, led the authors to the final proposition that (5) *the higher a group's political or economic position, the greater is the probability that its views will be reflected in the laws* (pp. 473–474).

Most of Chambliss and Seidman's theoretical explanation for the law in action focused on the argument that although the law represents the values and interests of the more powerful elements in the stratification system of complex societies, it is specifically created and enforced by bureaucratic organizations with their own agen-das. In *Crime and the Legal Process,* Chambliss (1969) insisted that "the most salient characteristic of organizational behavior is that the ongoing policies and activities are those designed to maximize rewards and minimize strains for the organization" (p. 84). Furthermore,

> this general principle is reflected in the fact that in the administration of the criminal law, *those persons are arrested, tried, and sentenced who can offer the fewest rewards for nonenforcement of the laws and who can be processed without creating any undue strain* for the organizations which comprise the legal system. (pp. 84–85, emphasis in original)

In Chambliss's view, the criminal justice bureaucracies tended to treat those of lower social class position more harshly for the same offenses committed by middle-class and upper-class people because lower social class people had little to offer in return for lenience and were in no position to fight the system. In addition, he insisted, these bureaucracies tended to ignore or deal leniently with the same offenses when commit-ted by those higher in the stratification hierarchy.

The process by which the goals of bureaucratic efficiency and avoidance of trouble displace the official goal of impartial law enforcement has been termed *goal displacement* or goal substitution. According to Chambliss, a bureaucratic organization might be expected to take the easy way out—the path of least resistance—especially in contexts where (1) the members have little motivation to resist the easy way out, (2) they have a great deal of discretion in how they actually will behave, and (3) adherence to the official goals is not enforced. "It will maximize rewards and minimize strains for the organiza-tion to process those who are politically weak and powerless and to refrain from pro-cessing those who are politically powerful" (Chambliss & Seidman, 1971, p. 269). Thus,

> the failure of the legal system to exploit the potential source of offenses that is offered by middle- and upper-class violators . . . derives instead from the very rational choice on the part of the legal system to pursue those violators [who] the community will reward them for pursuing and to ignore those violators who have the capability of caus-ing trouble for the agencies. (Chambliss, 1969, p. 88)

Speaking of the police, Chambliss and Seidman (1971) presented evidence leading to the conclusion that the police, as a bureaucracy, "act illegally, breaching the norms of due process at every point: in committing brutality, in their searches and seizures, in arrests and interrogation" (p. 391). This illegality takes place not because the police are evil but rather because they are not committed to due process in the first place, they have enormous discretion, and there is little enforcement of due process norms by the public or other agencies of the criminal justice system. As for prosecution following an arrest, Chambliss and Seidman concluded that "how favorable a 'bargain' one can strike with the prosecutor in the pretrial confrontations is a direct function of how politically and economically powerful the defendant is" (p. 412). The alleged safeguards provided by the right to trial by jury were taken to be largely a matter of myth because of the "built-in hazards of the jury trial . . . which exert the greatest pressure on accused persons to plead guilty," leading the powerless to surrender this so-called right in 9 out of 10 cases (p. 444). Finally, Chambliss and Seidman attempted to establish the validity of the proposition that "the tendency and necessity to bureaucratize is far and away the single most important variable in determining the actual day-to-day functioning of the legal system" by demonstrating how, even in sentencing, "institutionalized patterns of discrimination against the poor are inevitable" (pp. 468–469).

For several years prior to his original presentation of his theoretical framework, Chambliss had been involved in a study of the relationship between professional crime and the legal system in a large American city. His experiences while conducting this study clearly influenced his thinking, especially with respect to the impact of bureaucracy on the legal system. Pointing out that although at first glance "it would appear that the professional criminal (be he a thief, gambler, prostitute, or hustler) would have little to offer the law enforcement agencies," he noted that there appeared in practice to be a situation of "symbiosis" or mutual dependence there (Chambliss, 1969, p. 89). He argued that because law enforcement agencies in fact depend on professional criminals for inside information that makes their job easier, they tend to cooperate with these offenders rather than enforcing the law against them. In the later elaboration, Chambliss and Seidman (1971) went into much more detail with respect to the symbiosis involving organized crime, presenting evidence in support of the proposition that "from the standpoint of the sociology of legal systems, the most important aspect of the widespread presence of organized crime . . . is that such organizations are impossible without the cooperation of the legal system" (p. 489).

Chambliss (1969) used this argument to explain some characteristics of the Anglo-American legal system that seem somewhat illogical. As he pointed out, "Ironically, most of the criminal-legal effort is devoted to processing and sanctioning those persons *least* likely to be deterred by legal sanctions" (p. 370, emphasis in original). He cited the use of harsh punishments against drug addicts and capital punishment against murderers as examples of severe sanctioning in exactly those cases where it has little deterrent effect. By contrast, he cited the reluctance to impose stiff sanctions against white-collar criminals and professional criminals as examples of allowing precisely those criminals who do tend to be deterred by sanctions to escape them. According to his argument, such a policy goes directly against the formal logic of deterrence but fits perfectly the bureaucratic logic of demonstrating "effectiveness" by harsh treatment of the powerless while avoiding the organizational strains that would follow from taking on the powerful.

As for Durkheim's control theory argument that law contributed to social solidarity, Chambliss (1969) concluded that "the imposition of legal sanctions is likely to increase community solidarity only when the emergent morality also serves other interests of persons in positions of power in the community" (p. 373). Although admitting that law might *occasionally* contribute to social solidarity, he instead stressed the manner in which the law in action *splintered the community* by labeling and excluding certain of its (powerless) members and the way in which policies such as the death penalty "may play an extremely important role in the general attitude toward the legitimacy of the use of extreme violence to settle disputes" (p. 376). In Chambliss's view, the creation and enforcement of law grew out of social conflict and then tended to *add to and reinforce that conflict.*

By the mid-1970s, however, political events in general and developments in social theory in particular seemed to have combined to produce a significant shift in Chambliss's perspective. The social changes of the 1960s, which had influenced both Turk and Chambliss, had at the same time led a number of important social theorists to return to the Marxist tradition that had lain dormant during the war years of the 1940s and the McCarthyism of the 1950s. These Marxist theorists traced the problem of racial discrimination and the war in Vietnam directly to the economic interests of capitalists. They argued that the countercultural response to the American dream was politically naïve and would be crushed unless its adherents developed a revolutionary socialist mentality. As we will see in greater detail later, the end of the 1960s saw the emergence of a political backlash that threatened not only to block the prospects for reform implicit in Chambliss's critique of the law in action but also to "turn the clock back" to restore the social and political climate of the 1950s. Chambliss apparently concluded that the problems lay deeper than he had suspected, that the Marxists had come closer to the truth, and that his own analysis must be turned in a Marxist direction.

Chambliss's (1975) shift was reflected in nine specific propositions. With respect to the *content and operation of criminal law,* he now asserted that (1) "acts are defined as criminal because it is in the interests of the ruling class to so define them"; that (2) "members of the ruling class will be able to violate the laws with impunity while members of the subject class will be punished"; and that (3) "as capitalist societies industrialize and the gap between the bourgeoisie and the proletariat widens, penal law will expand in an effort to coerce the proletariat into submission" (p. 152). As for the *consequences of crime for society,* he maintained that (1) "crime reduces surplus labor by creating employment not only for the criminals but for law enforcers, welfare workers, professors of criminology, and a horde of people who live off the fact that crime exists"; that (2) "crime diverts the lower classes' attention from the exploitation they experience and directs it toward other members of their own class rather than toward the capitalist class or the economic system"; and that (3) "crime is a reality which exists only as it is created by those in the society whose interests are served by its presence" (pp. 152–153).

Summarizing his position and its implications, Chambliss (1975) argued that (1) "criminal and noncriminal behavior stem from people acting rationally in ways that are compatible with their class position . . . , a reaction to the life conditions of a person's social class"; that (2) "crime varies from society to society depending on the political and economic structures of society"; and that (3) "socialist countries should have much lower rates of crime because the less intense class struggle should reduce the forces leading to and the functions of crime" (p. 153). The similarity to Bonger's position was clear, and the use of terms such as *bourgeoisie, proletariat, exploitation,* and *class struggle* stressed the shift to a more Marxist formulation focusing on capitalism as the problem and socialism as a way of dealing with it.

QUINNEY: SOCIAL REALITY, CAPITALISM, AND CRIME

As pointed out earlier, the third of the criminological conflict theorists to be examined in this chapter, Richard Quinney, also completed graduate work in 1962, entering criminology along with Turk and Chambliss at a time when the ferment of the 1960s already was producing a markedly different social atmosphere. Quinney was to become not only the most prolific of the criminological conflict theorists but also the most controversial, revising his theoretical perspective time after time. Beginning with a position similar in many ways to that of Turk and Chambliss, he was to alter it almost immediately, only to develop a Marxist perspective at about the same time as did Chambliss and then to move in still another direction (Wozniak, 2011; see also Wozniak, Braswell, Vogel, & Blevins, 2008). Quinney explained these theoretical turns as follows:

> I have moved through the various epistemologies and ontologies in the social sciences. After applying one, I have found that another is necessary for incorporating what was excluded from the former, and so on. Also, I have tried to keep my work informed by the latest developments in the philosophy of science. In addition, I have always been a part of the progressive movements of the time. My work is thus an integral part of the social and intellectual changes that are taking place in the larger society, outside of criminology and sociology. One other fact has affected my work in recent years: the search for meaning in my life and in the world. (quoted in Bartollas, 1985, p. 230)

The section to follow thus traces the evolution of Quinney's thinking about crime. Later in the chapter, we will revisit Quinney and his role in founding *peacemaking criminology.*

Constructing the Reality of Crime. In the same year that saw publication of Turk's (1969a) *Criminality and Legal Order* and Chambliss's (1969) *Crime and the Legal Process,* Quinney (1969) set forth his own position in the introduction to an edited volume on the sociology of law called *Crime and Justice in Society.* At this point, Quinney's perspective was similar to that of Turk in some ways and similar to that of Chambliss in others. Like Chambliss, Quinney focused on the sociology of conflicting interests. But he preferred to begin with the *sociological jurisprudence* of Pound (1942) rather than with the "so-called legal realists," arguing that it was Pound who had first made the "call for the study of 'law in action' as distinguished from the study of 'law in the books'" (Quinney, 1969, pp. 22–23). But unlike Pound, who had seen law as operating for the good of society as a whole, Quinney, like Turk (and also citing Dahrendorf), took a position "based on the coercion model of society as opposed to the integrative" (p. 29).

Quinney (1969) defined law as "the creation and interpretation of specialized rules in a politically organized society" (p. 26). He then asserted that "politically organized society is based on an interest structure," that this structure "is characterized by unequal distribution of power and by conflict," and that "law is formulated and administered within the interest structure" (pp. 27–29). Referring to the "politicality of law," Quinney argued that "whenever a law is created or interpreted, the values of some are necessarily assured and the values of others are either ignored or negated" (p. 27). In this view, law was seen as part of the interest structure of society, with changes in the law reflecting changes in the interest structure, and as changing with changes in that structure.

One year later, Quinney (1970a, 1970b) published *The Problem of Crime* and *The Social Reality of Crime*, two volumes in which he presented somewhat different versions of criminological conflict theory. Like both Turk and Chambliss, Quinney was impressed by the relativity of crime. In *The Problem of Crime*, Quinney (1970a) began by taking the position, like Turk, that crime must be considered in relative terms as a "legal status that is assigned to behaviors and persons by authorized others in society," with the criminal defined as "a person who is assigned the status of criminal on the basis of the official judgment that his conduct constitutes a crime" (pp. 6–7). Like both Turk and Chambliss, Quinney also argued that social differentiation and social change tended to produce complex societies with different and often conflicting conduct norms prevailing in different segments. His critique of criminal justice statistics was quite similar to those of Turk and Chambliss. He asserted that "the crucial question is why societies and their agencies report, manufacture, or produce the volume of crime they do" (p. 16), and his analysis of American society as a criminogenic social system drew from the traditions described in earlier chapters—except for the accent on the "politicality of crime" (p. 180). By the politicality of crime, Quinney meant that

> the actions of the criminally defined are not so much the result of inadequate socialization and personality problems as they are conscientious actions taken against something . . . , the only appropriate means for expressing certain thoughts and feelings—and the only possibilities for bringing about social changes. (p. 180)

In his discussion of philosophical principles underlying his approach, however, Quinney diverged considerably from Turk and Chambliss. This distinct difference was to set him apart. Quinney (1970a) pointed out that a number of criminologists had expressed concern over the positivistic conception of "cause," outlined deeper issues beneath this philosophical debate, and concluded that "under the impact of the philosophical implications of modern physics, most physical scientists have abandoned the idea that science is a *copy of reality*" (p. 134, emphasis in original). This amounted to a total rejection of positivism. Drawing on the European tradition of *philosophical idealism,* he took the position that "accordingly, to state the extreme, there is no reality beyond man's conception of it: *reality is a state of mind,*" and he argued that "there is no reason to believe in the objective existence of anything" (pp. 136–138, emphasis in original). As Quinney saw it, the problem was not to understand some reality that stood apart from the observer but rather to formulate ideas that were helpful in terms of one's purposes.

To criminologists untrained in philosophy, this might sound highly implausible or even ridiculous. Nevertheless, philosophical idealism (the theory that the world is a product of mind) has a long and distinguished history, going back to Plato and beyond. It holds that what we take to be the objective world outside us is an image produced by our senses and the thoughts that interpret what they seem to reflect. For example, the world would be different for another creature with different senses and different minds, and no creature's world is any more real than that of another. Labeling theory itself had represented a modest move in this direction, throwing light on crime as a matter of perception and definition. Quinney went a step further, following the lead of the social constructionists in pointing out that social reality in general, as well as deviance or crime in particular, is a matter of changing perceptions and interpretations.

Many variations on some form of philosophical idealism were "blowing in the wind" during the 1960s, from the political commitment to achieving major social change through moral persuasion and nonviolence that Martin Luther King, Jr., had learned from Gandhi to the "far out" activities of the so-called countercultural yippies, who may be taken as an example of the extremes to which this approach can be taken. In some ways, the 1960s seemed to be a time when it appeared possible to *change things* by *redefining them* through the development of a new consciousness. The more extreme version of such "consciousness politics" aimed at "blowing people's minds" is captured in the following summary:

> To blow people's minds was to confront them with a situation that could shatter their cultural assumptions and, perhaps, liberate them from ruling-class images. . . . Without question, the master mindblowers in the New Left were a scruffy band of self-described "anti-intellectual action freaks" who called themselves the Youth International Party (YIP).
>
> Lacking any theory or organization and with a narrow social base, the Yippies concentrated on tactics. They used dramatic irony to reveal absurd contradictions in a social order whose legitimacy depended on the appearance of rationality.
>
> The Yippies' first major prank was to shower dollar bills from the visitors gallery onto the floor of the New York Stock Exchange.
>
> The Yippies appeared naked in church; invaded university classrooms, where they stripped to the waist and French kissed; dressed as Keystone Kops and staged a mock raid on the State University of New York Stony Brook campus to arrest all the whiskey drinkers; planted trees in the center of city streets; dumped soot and smoke bombs in Con Edison's lobby; and called a press conference to demonstrate a drug called "lace," which when squirted at the police made them take their clothes off and make love. (Starr, 1985, pp. 267–270)

Quinney was no yippie, but he clearly was influenced by the pervasive sense of the way in which the taken-for-granted aspects of social life really were a matter of collective definitions with which people went along largely without thinking. His position was laid out in greater detail in *The Social Reality of Crime*. In that work, Quinney (1970b) developed an analysis of the social reality of crime, drawing from Berger and Luckmann (1966), Schutz (1962), and others. Whatever physical reality may be, these theorists had argued that *social reality* consisted of the "meaningful world of everyday life" that was tied together by the fact that "human behavior is *intentional,* has *meaning* for the actors, is *goal oriented,* and takes place with an *awareness* of the consequences" in such a way that the individuals share a collective reality made up of shared meanings and understandings (p. 14, emphasis in original). For Quinney, the theoretical problem lay in the exploration and explanation of the *phenomenological processes* by which this collective meaning is developed and sustained.

The theory set forth in *The Social Reality of Crime* consists of six propositions. Quinney's theory began with a general definition of crime in which it was to be regarded as "a definition of human conduct that is created by authorized agents in a politically organized society." This was followed by the second proposition that "criminal definitions describe behaviors that conflict with the interests of segments of society that have the power to shape public policy" (Quinney, 1970b, pp. 15–16). Here Quinney indicated his indebtedness to both Vold and Turk; thus, he observed that the probability of powerful segments of society formulating criminal definitions becomes greater with an increase in the conflict of interests between the segments of a society, and he insisted that the history of law reflected changes in the interest structure of society.

Quinney's (1970b) third proposition focused on the "law in action," asserting that "criminal definitions are *applied* by the segments of society that have the power to shape the enforcement and administration of criminal law" (p. 18, emphasis added). He argued:

> The probability that criminal definitions will be applied is influenced by such community and organizational factors as (1) community expectations of law enforcement and administration, (2) the visibility and public reporting of offenses, and (3) the occupational organization, ideology, and actions of the legal agents to whom authority to enforce criminal law is delegated. (1970b, pp. 19–20)

In a fourth proposition dealing with the sources of the behavior resulting in the criminal label, he asserted that "behavior patterns are structured in segmentally organized society in relation to criminal definitions, and within this context persons engage in actions that have relative probabilities of being defined as criminal" (p. 20). Taking the position that it is not the quality of the behavior but rather the action taken against it that makes it criminal, Quinney went on to say that "persons in the segments of society whose behavior patterns are not represented in formulating and applying criminal definitions are more likely to act in ways that will be defined as criminal than those in the segments that formulate and apply criminal definitions" (p. 21).

Fifth, Quinney (1970b) argued further that the definitions of crime developed by certain social segments had to be successfully diffused within the overall society before the generally accepted social reality could be altered: "Conceptions of crime are constructed and diffused in the segments of society by various means of communication" (p. 22).

His final proposition summarized the entire theoretical framework: "The social reality of crime is constructed by the formulation and application of criminal definitions, the development of behavior patterns related to criminal definitions, and the construction of criminal conceptions" (p. 23).

The Influence of Marx. Four years later, however, Quinney (1974a, 1974b) published *Criminal Justice in America* and *Critique of the Legal Order,* two volumes that reflected a significant shift to a Marxist approach. He now criticized not only positivism but also the sort of social constructionism and phenomenology that he had used so effectively 4 years earlier, charging that "positivists have regarded law as a natural phenomenon; social constructionists have regarded it relativistically, as one of man's conveniences; and even the phenomenologists, though examining underlying assumptions, have done little to provide or promote an alternative existence" (Quinney, 1974b, p. 15). Like Marx, Quinney seemed to have concluded that the point was not simply to understand social life as a collective construction but also to change it, going on to say that "with a sense of the more authentic life than may be possible for us, I am suggesting that a critical philosophy for understanding the social order should be based on a development of Marxist thought for our age" (p. 15).

This shift in thinking was influenced by his reading of the work of the Frankfurt school of German social theorists including Habermas (1970, 1971) and Marcuse (1960, 1964, 1972), recent American work in the revived Marxist tradition mentioned previously (Baran & Sweezy, 1966; Edwards, Reich, & Weisskopf, 1972; Milibrand, 1969), and his own interpretation of the backlash, particularly the "war on crime" as it had developed during the Johnson and Nixon administrations. Just when Chambliss and Quinney were calling for further change, the backlash appeared, exemplified initially by the crackdown on crime in the Johnson administration and then by the election of Nixon, a political figure who had been rejected twice by the voters during the early 1960s only to be elected president by the end of the decade. While criminologists such as Chambliss and Quinney were crying for *more* reform, the voters were electing a president who promised just the opposite.

It also is apparent that Quinney himself believed, as did many people around him, that the foundations of American life had to be changed if people were to regain a more "authentic existence." Quinney's (1974b) six Marxist propositions read as follows:

(1) American society is based on an *advanced capitalist economy;* (2) the state is organized to serve the interests *of the dominant economic class,* the capitalist ruling class; (3) criminal law is an instrument of the state and ruling class to *maintain and perpetuate the existing social and economic order;* (4) crime control in capitalist society is accomplished through a variety of institutions and agencies *established and administered by a governmental elite,* representing ruling class interests, for the purpose of establishing domestic order; (5) the contradictions of advanced capitalism—the disjunction between existence and essence—require *that the subordinate classes remain oppressed by whatever means necessary,* especially through the coercion and violence of the legal system; and (6) only with the collapse of capitalist society and the creation of a new society, *based on socialist principles,* will there be a solution to the crime problem. (p. 16, emphasis added)

This line of argument followed Engels in asserting that the institution of the political state arises only at a point in the development of society when private property appears and then becomes concentrated in the hands of a few and that the law is "the ultimate means by which the state secures the interests of the ruling class" (Quinney, 1974b, p. 98). Developing the argument contained in *The Social Reality of Crime* in a Marxist direction, Quinney (1974b) maintained that the clever manipulations of the ruling class were obscured by an ideology serving to justify the system and that "manipulating the minds of the people is capitalism's most subtle means of control" (p. 137). He stressed that the sort of socialism advocated was not that of a centralized state bureaucracy such as existed in the (then) Soviet Union and some Eastern European societies but rather a "democratic socialism" based on equality and giving everyone a chance to participate in control over his or her own life (p. 188).

Quinney's thought continued to evolve in reaction to criticisms leveled at Marxist criminology and newer contributions to Marxist theory. Three years later, he published *Class, State, and Crime* (1977), a volume in which he criticized recent theories of justice, arguing that they all were rooted in an implicit acceptance of the current economic order. He now laid great stress on the Marxist argument that capitalism generates a *surplus population* made up of unemployed laborers. The general problem of the capitalist state was seen as providing support for the growth of capitalism while trying to manage the resulting problems by mechanisms such as the welfare state and the criminal justice system. According to Quinney, some members of the surplus population are not co-opted by mechanisms such as the welfare system, especially in view of the fact that capitalism finds it difficult to fund this mechanism adequately. They may adapt to their plight by turning to crime. Quinney observed, "Nearly all crimes among the working class in capitalist society are actually a means of survival, an attempt to exist in a society where survival is not assured by other, collective means" (p. 58).

At the same time, Quinney moved from a position that seemed to suggest that the state was in the hands of a powerful and all-seeing elite to one emphasizing a dialectical concept of social class. Quinney (1977) asserted that "a theory which posits an opposition between an elite (or a 'ruling class') and the 'masses' (or the 'people') fails to provide an adequate understanding of the forces of capitalist society" (p. 64). His new emphasis was closer to a structural Marxism that saw political outcomes as natural results of the dynamics of the economic system than to an instrumental Marxism that saw political strings being pulled by members of a small elite looking out for themselves. Crime was considered in Engels's terms as a "primitive form of insurrection, a response to deprivation and oppression," but one that "in itself is not a satisfactory form of politics" (pp. 98–99). Although there still were signs of the older conception of the politicality of crime, crime was now clearly considered an unsatisfactory form of politics. It was not a sufficiently rational response to oppression unless it succeeded in developing a revolutionary consciousness so that it represented an *informed rebellion* against capitalist conditions.

Class, State, and Crime also presented a typology of crime including *crimes of domination* and *crimes of accommodation and resistance*. Crimes of domination were said to include *crimes of control* (e.g., police brutality), *crimes of the government* (e.g., Watergate-style offenses), and *crimes of economic domination* (e.g., white-collar crime, organized crime). Crimes of accommodation and resistance were said to include *predatory crimes* (e.g., theft) and *personal crimes* (e.g., homicide), which were provoked by the conditions of capitalism, and *crimes of resistance* (e.g., terrorism), which involved the political struggle against the state.

Transcendence and the Search for Transformative Justice. Throughout Quinney's theoretical turns, there was a constant tension between the realms of the subjective and the objective. Whereas *The Social Reality of Crime* had taken a subjectively oriented, phenomenological, or constructionist view of the world, *Critique of the Legal Order* had thrown Quinney into a Marxist tradition of strict *materialism,* a tradition that not only explicitly rejected philosophical idealism but even regarded it as the enemy.

This tension broke through in the second edition of *Class, State, and Crime* (Quinney, 1980), which is heavily theological in content. Indeed, it is interesting that current textbooks tend to remain content with an examination of Quinney's social reality period and his later Marxism without mention of this later development, almost as if it is either embarrassing or irrelevant. It certainly was not irrelevant to Quinney, whose thought tended thereafter to deal less with criminology per se and more with existential philosophy and theology.

In the preface to the second edition of *Class, State, and Crime,* Quinney (1980) argued that "ultimately, the answer to the human predicament is a salvation achieved through the overcoming and healing of the disparity between existence and essence" (p. ix). Although continuing to advocate a socialist solution to the crime problem, he increasingly emphasized the *religious* nature of the goal, going so far as to reject Marxist materialism in favor of the theology of Tillich: "The contemporary capitalist world is caught in what Tillich, going beyond Marx's materialistic analysis of capitalism, calls a *sacred void,* the human predicament on both a spiritual and a socio-political level" (p. 3, emphasis in original). Quinney described the deeper problem as follows:

> Among the vacuous characteristics of present civilization are a mode of production that enslaves workers, an analytic rationalism that saps the vital forces of life and transforms all things (including human beings) into objects of calculation and control, a loss of feeling for the translucence of nature and the sense of history, a demotion of our world to a mere environment, a secularized humanism that cuts us off from our creative sources, a demonic quality to our political state, and a hopelessness about the future. (p. 3)

Returning to the theme of justice and citing the biblical prophets, Quinney asserted that "justice is more than a normative idea; it is charged with the transcendent power of the infinite and the eternal, with the essence of divine revelation" (pp. 30–31). What was necessary, he now argued, was a "prophetic understanding" of reality (p. 40).

Much of the second edition of *Class, State, and Crime* was devoted to discussions of the religious implications of socialism. Quinney (1980) asserted, "The rise of political consciousness in the late stage of capitalism is increasingly accompanied by a consciousness about matters of ultimate concern" (p. 112). He held that Marx had erred in considering religion merely the "opium of the people" and that "a social criticism that does not consider the sacred meaning of our existence systematically excludes the full potential and essence of our being" (p. 199). He held that "prophetic criticism takes place with an awareness of divine involvement in history" (p. 204). Quinney drew heavily from theological writings such as Tillich's work on religious socialism, seeing the effort in this direction as one in which "we hope to recover our wholeness, to heal our estrangement from the source of our being." He concluded, "The socialist struggle in our age is a search for God at the same time that it is a struggle for justice in human society" (p. 204).

Notably, as will be seen later in this chapter, these themes would continue to inform Quinney's writings as he embraced peacemaking criminology. Quinney's life is best understood as an academic and personal journey marked by extensive reading and careful study, self-examination and actualization, and engagement with the world (Wozniak, 2011). As he traveled across time, his concerns involved transcending repressive mind-sets that constrain our thinking and lead to harm in the world. By thinking in new ways—by deconstructing ideologies justifying structures of power, by envisioning fresh ways of reducing suffering among the poor and defenseless—Quinney believed that it becomes possible to transform oneself and the social order in the pursuit of greater justice (Wozniak, 2011; Wozniak et al., 2008). His work thus represents both a self-exploration as he endeavored to grow individually and, through his publications, an attempt to inspire criminologists and others to create a more humane society.

CONFLICT THEORY AND THE CAUSES OF CRIME

During the 1960s, many criminologists had turned away from the search for the causes of crime. As we have seen, control theory focused on the sources of *conformity*, under the assumption that crime and delinquency should be expected when there was a decline in the holding power of the conformity influences. Labeling theory treated crime as a matter of definition, with the source of the definition being the labelers. The early work of the conflict theorists Turk, Chambliss, and Quinney explored the criminalization process with a focus on factors that might explain *the behavior of the authorities* rather than that of the offenders.

Still, criminological conflict theory did have something to say about the causes of offenders' behavior. We already have seen some examples. Turk had referred to cultural and social norms, Chambliss had considered crime as a rational reaction to exploitation, and Quinney had discussed the politicality of crime as involving the use of the only available and appropriate means to express certain thoughts and bring about certain changes. The question of causality, however, became more important in the criminological conflict theory that began to appear with the shift in the social climate at the end of the 1960s and the beginning of the 1970s. Although they mentioned possible sources of offenders' behavior from time to time, the earlier conflict theorists had concentrated on the way in which the traditional search for causes of criminal and delinquent behavior had deflected attention away from the fact that crime was the result of the criminalization of certain behavior by the powerful. Having made the point so powerfully, criminological conflict theory then turned more attention to the sources of the *behavior* that was being criminalized.

Early during the 1970s, for example, Gordon (1971) offered a Marxist economic analysis that traced a great deal of crime to the underlying economic structure of American society. He attempted to show how many crimes represented a rational response to the fact that the economic position of many people was kept in constant danger by the very nature of capitalism. By the middle of the decade, Spitzer (1976) had set forth the argument that capitalism generated both a *surplus population* that consisted essentially of economic outcasts and a series of *internal contradictions* in the institutions developed to maintain capitalist domination. He maintained that members of the surplus population were chronically unemployed outsiders who sometimes turned to deviant behavior including crime and that additional deviance resulted from the tensions in institutions such as schools, which are said to serve youth but really serve the ruling class.

Focusing on the sources of delinquency, Greenberg (1977), for example, argued that theft was one response to a situation in which adolescents in capitalist society were put under heavy pressure to spend considerable money in a consumption-oriented youth culture even as the economic system was eliminating the employment opportunities. This lack of employment opportunities was said to produce considerable anxiety about prospects for achieving secure adult status during a time when institutions such as schools were taking away any sense of independence and subjecting the young people warehoused there to a variety of humiliating experiences. Greenberg maintained that such circumstances tended to produce a deep resentment and a fear of failure that precipitated violent behavior, itself a reflection of a demand for respect.

Early during the 1980s, Colvin and Pauly (1983) attempted to combine control theory with a Marxist approach to social class issues. In their view, capitalist society tended to exert a pattern of "coercive control" over the lower classes, threatening those at or near the bottom with loss of jobs or of any economic assistance unless they completely conformed to the expectations of the powerful. Colvin and Pauly maintained that this pressure produced an "alienative involvement" on the part of those under such oppression, breaking what Hirschi might term their "bonds to society" and increasing the likelihood of criminal activity (see also Colvin, 2000).

A number of other conflict theorists also have attempted to say something about the causes of criminal and delinquent behavior (see, e.g., Currie, 1997), but the focus has remained on the notion of crime as an outcome of definitions imposed as part of the consequences of conflict among various segments of society. Seen in this way, the central theoretical problem still is to understand the nature of social conflict. Some of the more recent developments in criminological conflict theory will be examined in Chapter 9.

Consequences of Conflict Theory

Criminological conflict theory has had notable consequences in terms of subsequent theorizing and rethinking within mainstream criminology but has had relatively little direct impact on social policy except perhaps for its recent metamorphosis in the form of "peacemaking criminology" (Pepinsky & Quinney, 1991). As indicated earlier, and as we will see in greater detail in Chapter 9, the period of social turmoil that gave rise to contemporary criminological conflict theory was followed by a period of exhaustion and a social backlash of conservatism—as if people were trying to pretend that the 1960s had not happened and were determined to recapture the sense of tranquility of the complacent 1950s. Some of the conflict perspective could be integrated into contemporary criminological theory and applied to social policy, but much was rejected. What was accepted and what was rejected depended primarily on whether the formulations called for further social reform in the tradition of some of the earlier theories or demanded social revolution.

Turk's conflict approach had been highly formalized and had treated conflict patterns as essentially inevitable, so it was not to be expected that his theoretical perspective would lead to specific alterations in social policy. It did, however, have considerable impact not only within criminology but also within the more general field of "deviant behavior" studies. The concept of "criminal" as a *status* assigned by the authorities as a result of a process working itself out through conflict probabilities was especially appealing to those with a formal sociological orientation who were interested in the conflict perspective but were put off by the ideological fervor of many Marxists.

As for the conflict perspective presented by Chambliss (1969) in *Crime and the Legal Process* and elaborated in the later work with Seidman, it is worth noting that a number of the policy implications inherent in that approach already had been addressed by the U.S. Supreme Court under Chief Justice Earl Warren. Indeed, the "Warren Court" was in its own way as much a part of the turbulent 1960s as was the civil rights movement, the counterculture, and the Vietnam War protests. Years before Chambliss began his work, the Warren Court had shown considerable appreciation for the legal realists' distinction between the law in the books and the law in action. It had extended the legal rights of convicted offenders as well as suspects and private citizens to provide them with additional protections in their struggles with the bureaucratic agendas of the police, courts, and corrections. Chambliss himself was well aware of this, and he actually dedicated his first book to Warren. It is ironic that it appeared at the time of President Nixon's appointment of the conservative Warren Burger to replace the retiring Warren, which took the court back into a posture that gave much *less* attention to the way in which the law in action might differ from the law in the books.

Chambliss's use of the tradition of legal realism has been of considerable influence within criminology and, like Turk's work, also has been influential within the field of deviant behavior studies. Quinney's early work had essentially the same theoretical implications as Turk's, but any interpretation of its policy implications depends on one's interpretation of philosophical idealism. As noted previously, Quinney later decided that his earlier phenomenological constructionist approach tended to impede efforts to change things. Nevertheless, the earlier position seems to have had more impact on both criminological thought and deviant behavior studies than his later Marxist and theological approaches. The concept of crime as a result of the "social construction of reality" is broader than labeling theory, and the general perspective has become extremely influential in the larger field of "social problems" theory (Ball & Lilly, 1982) as well as in the newer "postmodern criminologies" to be discussed in Chapter 9.

So long as the conflict theories went only one step further than the theories tracing crime to criminogenic elements in society (as discussed in Chapters 3 and 4) or the labeling theories pointing to the relativity of crime (as discussed in Chapter 7), they tended to strengthen the case for policies of social reform suggested by those perspectives. In fact, they gave the appearance of greater political realism. They recognized that the inequalities stressed by the opportunity theories and the stigmatizing processes emphasized by the labeling theories had a great deal to do with the perceived interests of the powerful, and they suggested how the powerful and the bureaucracies representing them might be held more accountable. For example, some of the radicals took an active part in community campaigns to curb police brutality, raise bail for poor defendants, abolish the death penalty, stop the repression of political dissidents, and provide support for prisoners (Greenberg, 1981). Some pushed for policies allowing for greater social diversity without criminalization of those who were different and for more informal community-based policies of conflict accommodation such as arbitration, informal dispute settlement, and conflict resolution through negotiation outside the mechanisms of the political state (Mathiesen, 1974; Pepinsky, 1976; Quinney, 1974b). These efforts were to crystallize into what came to be called "peacemaking criminology" (Pepinsky & Quinney, 1991).

But when they called for the abolition of capitalism, the radical criminologists produced mostly charges of either misunderstanding Marxism (which indicated that they were not to be taken seriously) or understanding it too well (which indicated that they were dangerous revolutionaries of the sort that Senator McCarthy had warned America about during the 1950s). This was generally the fate of the later conflict theorizing of both Chambliss and Quinney. Here it is useful to contrast the Marxist approach with the "peacemaking" approach.

MARXIST APPROACH

Even if society had been ready for further dramatic social change during the late 1970s, the problem of policy impact probably would have remained simply because of the inability or unwillingness of Marxist conflict theorists to provide blueprints for policy. The tendency of Marxists was to condemn capitalism and insist that individuals equipped with a revolutionary consciousness would be able to work their way through policy changes as they arose. They generally held that it was too early to try to spell out exactly what law and criminal justice (if such continued to exist) would look like under democratic socialism. Although some were willing to predict, for example, that prisons probably still would be needed (but for many fewer prisoners in a very different setting), others were less inclined to offer blueprints for the future. Although this stance can be understood in terms of the Marxist focus on principles of dialectical logic, it made no sense to contemporary policy makers (and many criminological theorists) who knew and trusted only the instrumental logic of ends-means calculation that insists that the policy makers spell out precisely what they expect to find at the end of the political journey before undertaking it in the first place (Ball, 1978a, 1979).

Although conflict theory had little direct impact on social policy, it had a great deal of impact within criminology itself. It led to considerable rethinking as to the nature of law, with most textbooks taking a more critical stand than was the case at the beginning of the 1960s. More recent criminological conflict theorists working in the Marxist tradition have moved toward structural Marxism and away from the instrumental Marxism that seemed so close to the notion of a conspiracy of dominance by a tiny elite at the top. This instrumental Marxism had tended to portray the capitalist elite as an omniscient few who knew everything and always pulled the strings at exactly the right moment to ensure that their interests were served. Structuralism locates the basis of social control factors such as law in class *relations* in general rather than asserting that it was entirely within the total conscious control of the capitalists at all times. Structural Marxism bears a resemblance to the formalism of Simmel, Vold, and Turk in that it maintains that social conflict is a matter of the inherent social dynamics of a particular system. It differs by taking a more historically relative position, maintaining that the key to understanding these dynamics lies not in formal logic that deals with social systems *in the abstract* but rather in a dialectical logic that is rooted in an understanding of the structural features of capitalism *in particular*.

At the same time that the Marxist conflict theories were attacking capitalism, they were criticizing conventional criminology. This criticism had consequences, even if only in making it more difficult to be complacent about the field. These criticisms included charges that conventional criminology was itself a part of the capitalist system and tended to support it, thereby contributing to the crime problem rather than to its understanding or its solution. The more radical conflict theorists insisted that conventional criminology had tended to accept the law as given, concentrating on the behavior of the offenders and searching for some pathological source of their behavior in biological, psychological, or social factors. They maintained that this very search tended to exaggerate the notion that the criminals were in some important way "different from the rest of us" when in fact the problem lay in the creation and enforcement of the laws that produced the criminals. Some demanded that crimes also be considered from the perspective of the criminals, using the writings of prisoners as part of an attempt to escape a narrow class-biased point of view.

Some of the radical criminologists leveled heavy criticisms at the efforts of conventional criminology to maintain a value-free "scientific" posture that refused to get involved in political debate or political action, insisting that scientific criminology confine itself to the "facts." This position went back to the development of the positivistic approach discussed in Chapter 2. The critics of positivism charged that conventional criminology, by refusing to take a moral stand, was implicitly accepting the moral ideology provided by the powerful and forced onto the powerless. When mainstream criminologists responded that they were not simply accepting the ideology of the powerful and that they were in fact operating on the basis of broad social consensus, the radicals argued that this public consensus was a misleading mirage based on the power of the elite to shape public opinion (Michalowski & Bolander, 1976; Quinney, 1970b; Reiman, 1979). As such, it seemed to them to represent a "manufactured consensus" (Greenberg, 1981, p. 9). Indeed, some radicals insisted that conventional criminologists seemed to "reproduce the hegemony of the existing relations of property, race, and sexual privilege," either "to regain prestige, to build research empires, or because they truly believe in the ideas of those in positions of power" (Krisberg & Austin, 1978, p. 119).

The radical theorists insisted that even the reforms urged by conventional criminology were simply minor tinkering that tended to support the further survival of a corrupt social system by making it appear that the powerful do care and that progress is being made toward economic and political justice. Meanwhile, they added, conventional criminology was providing the knowledge necessary to detect and control those seen as threats to the powerful. For example, Platt (1969) argued that the original establishment of the juvenile court was not a means of helping youths but rather a technique by which upper-class, Republican, Protestant women extended control over the children of the Catholic and Jewish immigrants. In a similar vein, the concept of rehabilitation was criticized as a tool of political oppression that justified prolonged and invasive tinkering with the minds and bodies of prisoners under the guise of assisting them (Smith & Fried, 1974; Wright, 1973).

PEACEMAKING CRIMINOLOGY

Perhaps nothing exemplifies the way in which the context of the times conditions the development of criminological theory so clearly as the evolution in the thinking of Quinney (1969, 1970a, 1970b, 1974a, 1974b, 1977, 1980), especially as he moved into the more recent phases of his theorizing (Pepinsky & Quinney, 1991). As he made clear in the earlier quotation, Quinney "moved through the various epistemologies and ontologies in the social sciences," always finding that another approach was "necessary for incorporating what was excluded from the former, and so on," trying to keep his work "informed by the latest developments in the philosophy of science" (in Bartollas, 1985, p. 230). He also made clear that this search has been conditioned by "the social and intellectual changes that are taking place in the larger society, outside of criminology and sociology" (p. 230). Eventually, Quinney came to join those who stressed an approach to social conflict reminiscent of Gandhi. Instead of following the Marxist path that calls for a revolutionary overthrow of capitalism, he embraced the "peacemaking" approach that tries to accommodate conflict through various means of conflict resolution such as moral suasion and informal negotiation (Wozniak, 2003). It is interesting to note that this approach, which often is scorned as totally impractical by those who favor confrontation, actually may be much more effective in the long run (Fuller, 1998; Fuller & Wozniak, 2006; Wozniak et al., 2008).

Peacemaking criminology accepts the notion that conflict is at the root of crime, but it advocates a policy response that refuses to escalate this conflict in favor of policies of conciliation and mediation (Fuller, 1998; Fuller & Wozniak, 2006). According to Pepinsky (1999), peacemaking is one of two ways of approaching social control, with the other being "warmaking." Whereas warmaking leads to more and more distrust, as well as efforts to secure more and more power over the opponent or even to destroy the opponent, peacemaking aims to build trust and a sense of community. Although most associated with the tradition of conflict theory, it actually suggests that conflict can be best resolved by building social bonds among people, much as control theory stresses.

This view is much more within the spiritual traditions of faiths such as Christianity (as exemplified by groups such as the Quakers), Buddhism (as illustrated by the fact that Quinney became a practicing Buddhist), and the Hindu tradition embodied by Gandhi. It also has been placed within certain humanistic traditions, so that conflict theorists who wish to stress the earlier, more humanistic side of Marx also are able to connect the Marxist tradition to peacemaking criminology (Anderson, 1991). As Pepinsky (1999) pointed out, peacemaking criminology reflects the position of many marginalized people who realize that they cannot obtain their goals by overpowering the opposition. Thus, it also has been seen as close to certain elements of the feminist criminology that will be described in Chapter 10.

Peacemaking criminology reflects an assumption that "two wrongs do not make a right" or that a "war on crime" will only tend to make society itself more violent. Very sensitive to many of the issues raised by labeling theory, it represents an effort to step back from the alienating competitive sense of the other as a stranger to recapture a sense of common humanity and mutual community. Although it stresses development of social bonds, the bonds that it wishes to create run deeper than those suggested by much control theory. Pepinsky (1999) pointed to Christie's (1981) observation that the more we come to really know someone, the more we lose our capacity to inflict pain on that person. Peacemaking criminology also stresses that the more we know ourselves, especially the uglier side of ourselves, the less likely we are to project it onto others (Pepinsky & Quinney, 1991).

In this view, the violence endemic in contemporary society is manifested by the homicide committed by the criminal, on the one hand, and the death penalty administered by the state, on the other. Each expresses the same underlying set of social premises. The violence of the state reflects the same underlying values as the violence of the criminal—a willingness to use violent means to deal with opposition.

It is not surprising that peacemaking criminology focuses on developing means of building trust even in the midst of conflict or that it stresses construction of a "social fabric of mutual love, respect, and concern" (Pepinsky, 1999, p. 59). In this way, it leads to social policies such as the recent "restorative justice" movement (Consedine, 1995; Fuller & Wozniak, 2006). Recall that restorative justice was described in the last chapter. The goal of this approach is to reduce harm by restoring victims, the communities, and the offender from the damage a crime has caused. The role of the state is not to inflict more pain and to seek vengeance but to create a context in which reconciliation can occur. Offenders must show accountability and personal responsibility, but they also receive support, reintegration, and, in some cases, forgiveness. The purpose of this approach is to do justice but in ways that connect people together, not separate them (Fuller & Wozniak, 2006).

As might be expected, peacemaking criminology has been strongly criticized for utopian thinking and for the suggestion that power can be countered by something other than power. Gibbons (1994) pointed out that the sorts of humanistic policies advocated by peacemaking criminologists have a long history in reformist criminology and that they have not developed a theory that would suggest how to achieve the larger structural changes necessary to build close human ties, spiritual understanding, and genuine caring within a large and inherently impersonal society (see also Fuller, 1998; Fuller & Wozniak, 2006; Wozniak et al., 2008).

Conclusion

One does not have to accept radical conflict theory and peacemaking criminology to realize that it is easy to be taken in by the assumptions of the times. Indeed, the present book itself represents an effort to show how thinking about crime is intimately shaped by the nature of social context. Complacency has no place in criminology; both theories are anything but complacent. Whatever their defects, criminological conflict theory and peacemaking criminology certainly have succeeded in providing for a broadened reorientation and an increased sensitivity to issues previously overlooked or treated only in passing (Thomas & Hepburn, 1983).

New Directions
in Critical Theory

Ian Taylor
1945–2001
University of Durham, UK
Thoughtful left realist critic

The social context that nourished the development of conflict theory, with its challenging of social reality and of existing structures of inequality, soon shifted in a conservative direction. This social transformation would be conducive to a criminology that was less oppositional and less questioning of existing arrangements and, as in the case of conservative criminology, antagonistic to criticisms of traditional values and relationships of power. Even so, the scholarly tradition initiated by conflict theorists has persisted—albeit at times on the margins of criminology. In Chapter 10, we review a significant line of inquiry in this general paradigm—feminist criminology—that called attention to the neglect of gender and of patriarchy in criminological theorizing. In this chapter, we examine the most important "new directions" in critical criminology that have emerged in recent times. As will be apparent, critical theory has remained a vital perspective, rich in its variety and attracting prominent adherents from several countries, including among others Australia, Canada, the United Kingdom, and the United States.

In the pages to follow, we explore the significant history of "new criminology" in shaping the development and contemporary status of critical criminology. We also consider left realism, the European approach of abolitionism, anarchist criminology, cultural criminology, convict criminology, and consumer criminology. Each of these perspectives is an example of what has been identified as "postmodern criminology," which in turn is part of what has been termed postmodern social thought (Milovanovic, 1995). Although general postmodern thought is at first a little bit confusing, it is an important intellectual advance that merits our attention because it has had, and is continuing to have, an impact on criminology. It represents a broad and complex philosophical shift away from the traditional Enlightenment emphasis on discovering the natural and social world through the scientific method. We hasten to add that the postmodern criminological work discussed here is not exhaustive. It is a diverse group of developing critical perspectives that include chaos theory, discourse analysis, topology theory, peacemaking criminology (see Chapter 8), constitutive theory, convict criminology, and anarchic criminology. Before discussing the concept of "modern," we examine the definition of "critical criminology."

Put simply, "there is no single critical criminology. Rather, there are critical criminologies that have different origins, that use different methods, and that have diverse political beliefs" (DeKeseredy, in press). At heart, nonetheless, they share a perspective that asserts that the major sources of crime stem from the fact that unequal class, race/ethnic, and gender relations do in fact control society. This point is nicely captured in the words of Friedrichs (2009): "The unequal distribution of power or of material resources within contemporary societies provides a unifying point of departure for all strains of critical criminology" (quoted in DeKeseredy, in press). Compared to what some critical criminologists pejoratively call orthodox, conventional, mainstream, or liberal progressive criminologists, these criminologists generally reject official/legal definitions and measurements of crime. Crime fighting policies that emphasize such things as "zero tolerance" policing, three-strike sentencing, and private prisons—solutions that ignore major structural and cultural changes in society as essential to crime reduction—are also rejected by critical criminologists (DeKeseredy, in press; Lynch & Stretesky, 2006). Whereas conventional criminologists often claim to be value-neutral scientific experts, critical criminologists disavow this position as ideologically naïve and prefer to see themselves as more inclined to be politically active and committed to having their work reduce pain and suffering (DeKeseredy, in press).

During the last four decades, these shared assumptions and commitments— among others—have generated rich theoretical and research-based contributions. As a result, critical criminology now rivals mainstream criminology as a perspective that shapes thinking in the field. Indeed, it is a paradigm that arguably shares "an equal partnership with conventional criminology" (Lilly, 2010). We turn now to the concept of "modern."

Modernity and Postmodernity

In its most concise meaning, *modern* refers to a form of thought or philosophy that developed during the Enlightenment of the 18th and 19th centuries. It emphasized, among several important ideas, that the social world contained a "natural" order that could be discovered by the scientific method. With its emphasis on the manipulation of variables, systematic observation, and measurement and application of findings (policy), modern thought claimed that problems such as crime could be "discovered" and "solved." Once problems were discovered and solved, it was argued, the human condition would experience progress.

For approximately 200 years, all fields of scholarly and policy interests experienced the power and attraction of this perspective. Politicians and scholars alike claimed that the scientific method, through state intervention, could almost guarantee reductions, if not the control and elimination, of most social problems, especially crime. Over time, however, it became clear to some observers that the scientific method was a dangerous two-edged sword. It could help to relieve human pain, but it also could contribute to the infliction of enormous human suffering.

The very tangible technological advances of the industrial revolution that freed much of humankind from labor-intensive struggles for survival, for example, were celebrated with great fanfare and hopes for the future by the 1889 erection of the soaring Eiffel Tower for the Paris Exhibition. But by the early part of the 20th century, the context of Western Europe and North America had shifted radically with World War I. By this time gravity-defying technological developments had led to the invention of airplanes and the promise of rapid travel in the sky. Unfortunately, they were used by various nations—including the United States, Britain, France, Russia, and Germany—to drop bombs on and cause near unimaginable death and destruction for military troops and civilians. One important result was that people from many different walks of life, especially in Europe, began to damn technology, including planes, as inventions of the devil while, at the same time, seriously questioning the idea that progress resulted from science in the hands of the state.

Postmodern thought, according to some observers, began in earnest in Europe after World War II. It contains many forms of theorizing, some of which are more skeptical about the promises and assumptions of the scientific method than others. Fundamentally, it rejects the Enlightenment belief in scientific rationality and state intervention as the main vehicles to knowledge and progress (Barak, 1994). In general terms, postmodernism argues that the modern social world and its rules for behavior, including definitions of crime and law, are arbitrary linguistic constructions. These, it is argued, have resulted in encrusted notions about certainty, truth, and power (DiCristina, 1995). In fact, for postmodernists, truth is not absolute, and scientific inquiry fails to fully reveal reality. "In other words, positivist [deterministic] science offers a way of understanding, but it does not provide the sole method for comprehending human behavior and social phenomena" (Arrigo & Bersot, 2010, p. 41).

Furthermore, this logic has created false hierarchies and divisions within the social order that are divisive and repressive. Among these linguistic constructions are racial and sexual categories of approved and disapproved people and behavior. Many postmodern theorists argue, accordingly, that these creations should be debunked or "destructed" because they are in fact not much more than elitist constructions or claims (Arrigo & Williams, 2006; Milovanovic, 2002).

Another part of the postmodern argument rests on the observation that modernism, along with its emphasis on traditional scientific logic, directs efforts to fixing or changing individuals or instructions while neglecting the larger picture of the society as a whole. This means that official ideas and policies about the rightfulness or normalcy of the social order are left in place. Some postmodernists argue that these ideas should be replaced with approaches that are more relevant to the current era. For example, alternative methods and epistemologies for learning about the complexities and truths of the social world should be developed.

Consistency in the meaning of the term *postmodern* within the social sciences beyond these two distinctions is complex and at first examination off-putting. Writing on this and related issues in the late 1980s, Featherstone (1988) implied that any reference to the term exposed one to the risk of being accused of jumping on a bandwagon that perpetuated a shallow and meaningless intellectual fad. Thus, although fashionable, the term was nonetheless "irritatingly elusive to define" (p. 195). While open to the charge that the term was invented by theorists in hopes of creating a movement that would advance their careers, today the term has gained wide intellectual capital and usage.

Postmodern Criminological Thought: The End of Grand Narratives?

From this perspective, crime is not simply a violation of formal law or an objective fact that can be discovered by using the scientific method. The reasoning behind this position on crime is relatively simple. According to postmodernists today, there are no agreed-upon true "grand narratives" or stories of reality that people live by such as the idea that democracy will lead to universal happiness or the Marxian argument that the collapse of capitalism would lead to utopia. For postmodern criminologists generally, crimes are linguistic constructions made by official institutions. Laws and official claims about right and wrong, therefore, are structures of domination that have led to increased repression rather than to liberty (Schwartz & Friedrichs, 1994, p. 224). By the 1960s and 1970s, nowhere was this argument made more forcefully than in critiquing the state's application of the scientific method in the areas of crime and justice. Time and again, critical observers argued that the state's law and order efforts at correcting individual behavior were directed at those who were least able to resist the official language of the state. It is no surprise, therefore, that it is with a sense of irony that postmodern criminologists point out that the very core values and the material foundations of society that generate crime are left in place when the state attempts to solve the very problems they generate.

According to Henry and Milovanovic (2005, pp. 1245–1249)—two of the leading postmodernist criminologists—postmodernism did not reach criminology until the late 1980s. According to them, postmodernism today consists of a cluster of ideas including the following: truth is unknowable; rational thought is merely one way of thinking, and not necessarily a superior way; knowledge is not cumulative; facts are only social constructions that are supported claims to truth that themselves are expressions or reflections of a discourse—a way of talking about things; and criticism assumes an alternative truth that can be found by continuously attempting to expose or reveal the assumptions on which claims of knowledge are based.

Although Henry and Milovanovic's summation may be correct, other postmodernists are less certain about the term's definition. Indeed, it has a multiplicity of meanings, including a historical epoch, aesthetic orientations, and antisocial movements. In the face of this definitional quagmire, Ferrell (1998) argued that criminology and criminal justice postmodernism is best understood in terms of what it *opposes*. It stands against—opposes—"the intellectual and legal machinery of modernism" and the conventional "forms of legality, illegality, and crime that criminology conventionally investigates" (p. 63). By advocating this position, postmodernism attempts to expose and repudiate modern law and the state as "a system of coordinated control found on economic and social inequality and perpetuated through coercion and cultural manipulation" (p. 64).

This point is illustrated by considering how some postmodern feminist scholars are concerned about female role formations and what they term "gendered" subjects (Arrigo & Milovanovic, 2009). The status of the feminine in society, for example, is defined in masculine terms that are inherently "malestream" and misogynous. This means that the identities of women and minorities are reduced and repressed, particularly through criminal, legal, and correctional practices are defined in masculine terms (Arrigo & Milovanovic 2009).

In essence, this opposition is a "double negation" of modern criminology and of modern constructions of law and justice. The weaknesses and limitations of these "official accounts" stem from the fact that they rest upon "grand" or "meta narratives" that claim the existence of true and universal knowledge.

In less abstract terms, Arrigo (2003) suggests that postmodern criminological thought is based on three key language-based propositions that are remarkably similar to the basic tenets of the social construction of reality and the social construction of crime perspectives (see Chapters 7 and 8 on labeling theory and conflict theory).

1. *The centrality of language:* Reality and our social actions are shaped by the written and spoken word. Language is not neutral but contains values and assumptions that define who we are, our interaction with others and the institutions of society, and our participation in them.

2. *Partial knowledge and provisional truth:* The meanings and actions of daily life are structured by language that is not neutral. Inescapably, most if not all forms of our understanding are limited and incomplete. "Knowledge" and "truth" are unquestionably partial and provisional.

3. *Deconstruction, difference, and possibility:* Postmodern methods emphasize the deconstruction of the written and spoken word in order to expose implicit and hidden assumptions located in specific narratives (e.g., the hidden and privileged political messages in certain court decisions). "Decoding" texts (language written or spoken) exposes how particular "truth claims" are accepted while others are dismissed or ignored. Postmodern criminological theory encompasses articulating the differences in language of all the voices involved in social interaction.

Looking Back at Early British and European Influences

With hindsight and the passage of nearly four decades since critical criminological perspectives in Britain and Europe started to be developed, we can now more clearly understand and assess these theoretical contributions. Today, we can now look at what was called the "new criminology," left realism, and abolitionism, and we can see that they were part of postmodern critiques grounded in the grand narrative of Marxism. While much changed in their current forms, these three perspectives nonetheless have been some of the most fertile in modern criminological theory. More specifically, during the 1970s and early 1980s, they were concerned with the social, economic, and political significance of crime. We turn to this time period in order to contextualize these theories. Next we examine the more recent contributions of cultural and convict criminology.

BACKGROUND: THE NEW CRIMINOLOGY

Known during this period by other terms—including Marxist, materialistic, dialectical, radical, social, and critical criminology—the new criminology experienced strong popularity and support on both sides of the Atlantic. Of particular concern here is the British version of the new criminology because of the "exceptional success of radical criminology in this country" (Young, 1988, p. 159). (See Chapter 8 for a discussion that focuses primarily on U.S. theorists' work on social power and the construction of crime.) Of special importance beyond its appeal in Britain was the fact that it not only had a significant impact on academic criminology more than four decades ago but also remains an important influence today, albeit outside of conventional criminology. Unlike so many criminological theories that have enjoyed relatively brief moments of popularity and then faded away, the new criminology had a brief period of decline and experienced a resurgence of interest and influence (Walton & Young, 1998).

Early on, it was much influenced by what Young (1988) identified as the "impact of the West Coast labeling theory centering around Howard Becker that set the creaking chariot of radical criminology off on its course [in Britain]" (p. 163). According to Young, this involved a transformation of U.S. ideas that, rather like popular music, play back to the United States the culture of Europe. But this development did not involve the simple translation of U.S. ideas to Britain. It was a transposition of ideas that had to travel "a considerable distance politically, culturally, and indeed in terms of the contours of crime itself" (p. 163).

More than this specific influence was at work. During the late 1960s, British criminology was at a crossroads because traditional positivism—with its emphasis on crude and simple biological, psychological, and sociological determinism—was in crisis. The central problem was that, unlike what traditional criminological explanations of crime would have predicted, a "wholesale improvement in social conditions resulted not in a drop in crime but [rather] the reverse" (Young, 1988, p. 159). Confounding as this problem was for conventional deterministic explanations of crime, it was not the only social abnormality that needed to be explained and corrected. It was part and parcel of a number of social problems challenging the British post–World War II welfare state, including urban housing problems and the destruction of communities, public health issues, and education issues. "Thus, the crisis in criminology did not . . . come out of the blue; it was part . . . of the particular crisis in politics and culture . . . which was refracted in the internal problems of criminology" (p. 160). It is against this background that the new criminology in Britain emerged.

THEORETICAL ARGUMENTS

Published during the early 1970s (and still in print) with a highly complimentary foreword by American Marxist Alvin Gouldner (1973), this perspective's major tome, *The New Criminology* (Taylor, Walton, & Young, 1973), was described as so powerful in its critique of traditional criminology that "it redirects the total structure of technical discourse concerning 'crime' and 'deviance'" (Gouldner, 1973, p. ix). One year later, Currie (1974) praised it as an important document in the "effort to build a more humane criminology" and "probably the most comprehensive critical review of 'the field' that has been produced so far" (p. 133). The content of this perspective was different in important ways from its counterpart in the United States. For example, it often was more intellectually sophisticated and well grounded in criminology literature and continental philosophy than that found in the United States. And unlike the minority position of radical criminology in the United States (Inciardi, 1980), it grew to share equal partnership with conventional criminology.

Central to its early development were its objections to (1) structural functionalism's assumption that the social order was based on a public consensus and (2) traditional criminology's overly deterministic treatment of crime. Writing of its early focus, Cohen (1988) stated, "The initial intellectual shape taken by the break . . . was a systematic attempt to overturn the taken-for-granted assumptions of the positivist criminological tradition" (p. 9). To overturn these assumptions, the first job was to demonstrate that conventional studies of crime were too narrowly entrenched in more general theories and paradigms that assumed that they had a monopoly on the "correct," "scientific," and "deterministic" understanding of human nature and social order. To explain crime as the result of broken homes, for example, was too simple. The task facing the new criminology in part was to successfully challenge this narrow position and to demonstrate that conventional criminology was grounded in ideological constructs central to the policies of the state. The challenge required establishing that "there *was* a debate going on in criminology" over these issues (Young, 1988, p. 161, emphasis in original).

Next, the task was to make crime the central focus of concern, rather than a peripheral topic, for social scientists given that it had been placed in an intellectual ghetto by traditional positivism. In other words, the new criminology claimed that studying crime illuminates both social order and disorder. Crime is indeed not a marginalized topic but rather a central plank of social scientists' explanations of stability and change. For this argument to take hold and change the direction of traditional criminology, the new criminology had to "undo the notion of . . . objectivity" that deterministic "positivism had created in its pursuit of a mistaken scientificity" and to focus on the political nature of crime (Young, 1988, p. 161). This was accomplished by focusing on the political and economic structures and the institutions of capitalism and on how these form the social conditions that generate crime.

For the new criminology, capitalism was an exploitative and alienating social order in which inequality was institutionalized by an elite ruling class. Under these conditions, crime is a rational response to forced social arrangements; it is a by-product of the political economy. Young's position in the mid-1970s (1976)—one of the major contributors to the new criminology—was that: "Central to our concerns is the explanation of law and criminality in terms of the dominant mode of production and the class nature of society" (p. 14).

Under capitalism, criminal law all too often is manipulated to benefit particular interest groups to the detriment of all others. In particular, criminal law is used by the state and the ruling class to secure the survival of the capitalist system (Bohm, 1982). For the new criminology, then, crime was defined as capitalist policies and interests that contribute to human misery and deprive people of their human potential. Here the violation of human rights was of central concern for the definition of crime and it included imperialism, capitalism, racism, and sexism. Although assaulted by more than three decades of nonintervention and the privatization of social services ideology under Prime Ministers Thatcher, Major, and Blair, this focus remains an important part of the British criminological enterprise. In recent years, its appeal has expanded into several other countries and other forms of critical criminology.

CRITIQUE OF THE NEW CRIMINOLOGY

Early on, this perspective was criticized as soundly and thoughtfully as it was praised, sometimes by the same people. One critic was quick to notice that *The New Criminology* was in many ways far too "close to the tradition it supposedly sets out to criticize and replace" (Currie, 1974, p. 133). Three specific problems were identified.

First, although Taylor et al. provided a solid and illuminating critique of the shortcomings of conventional criminology, especially ethnomethodology and conflict theory, they often were sketchy and sometimes just plain wrong. Furthermore, their discussion of the classical school was incomplete, and the critique of positivism was about biological determinism, avoiding other varieties of determinism; accordingly, they gave the misleading impression that criticism of biologism applied equally to psychological and sociological determinism. Nor did their critique of biological determinism go beyond previous criticisms of the same topic. Taylor et al.'s (1973) work also did not live up to their assertion that they would "explain why certain theories . . . survive" (p. 31). Instead, they treated criminological theories as if they existed in a scholarly limbo rather than in wider ideological currents rooted in material conditions of advanced capitalist societies. This oversight not only obscured the connections that conventional criminology has with state intervention into crime and other social problems but also gave credence to the claim that Taylor et al. were themselves conventional criminological theorists because they did not go into a deeper level of critical analysis (Currie, 1974, p. 136).

Second, the writing style in *The New Criminology* was closely akin to that of people with finely tuned interests in the field of criminology. While this might sound like an unfair criticism, for Currie it was an indication that Taylor et al. actually were writing for traditional criminology rather than working to create meaningful social change for the exploited.

The third and most damaging criticism was that, in its rush to offer a more heuristic social explanation of crime than that offered by traditional criminology, the new criminology failed to present a cogent discussion of human nature and the social order. Without this information, it is unclear just what it was offering as an alternative to traditional criminology. Part of the explanation for this weakness is that the new criminology's concentrated critique of biological theories threw out the baby with the bathwater. "A more useful approach is not to reject biological and psychological conceptions altogether, but to integrate them with an overall vision of human nature" (Currie, 1974, p. 138). For these and other reasons *The New Criminology*'s main theoretical assumptions were a bit confused and fundamentally misleading.

Early Left Realism

THE THEORY

Other factors eventually contributed to diminishing the initial praise and impact of the new criminology, not the least of which was the changing political context of Britain with the rise of the "New" Right under Prime Minister Thatcher. In 1979, the New Right succeeded in a Conservative/Tory Party victory in Britain's general election and brought into office a government determined to effect change (Jenkins, 1987). It was committed specifically to ending Britain's domestic and international economic and political decline. Doing this involved making an ideological and political break with the assumptions and rules restraining government under the ideology of social democracy that developed around 1940 and post–World War II.

This represented a major change in British politics. At the heart of social democracy was a four-decade expansion of the functions and responsibilities of the state. The public enterprise sector, for example, was substantially "increased through the nationalization of major public utilities such as gas, electricity, coal, and railroads" (Gamble, 1989, p. 2). In the social arena, this expansion meant that welfare provisions were extended, as were national health care, housing, and education. But by the 1970s, social democracy and its welfare policies came under attack in Britain and in many other countries. The 1979 Conservative victory ushered in a new governmental ideology that used as its major agenda the privatization of government industries and the placing of restrictions on welfare, national health care, and educational support. Over the past three decades, continuing efforts have been made to dismantle the welfare state, although the Conservative government was markedly weakened in popularity by failed social policies, high-level governmental scandals, and the election of the Labor Party in 1997 ("Assault," 1993; "British Scandals," 1994).

As the New Right's governmental policies were being formulated and implemented, radical criminology recognized that its tide had turned. Writing about the impact of radical criminology less than a decade after the publication of *The New Criminology*, critical criminologist Cohen (1981) stated:

> There are more corners and cavities than ten years ago, but for the most part the institutional foundations of British criminology remain intact and unaltered, for the establishment saw the new theories as simply fashion which would eventually pass over or as a few interesting ideas which could be swallowed up without changing the existing paradigm at all. (p. 236)

Indeed, to a large extent this interpretation of the impact of radical criminology in Britain during the early 1980s was not seriously disputed. In a reflective comment about radical criminology's history, Matthews and Young (1986) stated that it had "concentrated on the impact of the state—through the process of labeling—on the criminal at the expense of neglecting the effect of crime upon the victim" (p. 1). In their words, this neglect was improper because radical criminology should have concentrated on "the basic triangle of relations which is the proper subject matter of criminology—the offender, the state, and the victim" (p. 1).

In part as a remedy to this oversight and the claim in some circles that radical criminology was in a state of crisis of its own, radical criminology moved away from the new criminology and developed a different approach to studying crime called *left realism,* a name used because of its emphasis on the *real* aspects of crime. Central to left realism was a strong concern that the new criminology not only had placed perhaps too much emphasis on the state but also had neglected the etiology of crime (Young, 1986). This point cannot be overemphasized because it marked a significant shift in Britain's radical criminology, one not so much *from* theoretical issues as *toward* research and statistical analyses of crime causation and its consequences.

More specifically, left realism was explicitly, although not exclusively, concerned with the origins, nature, and impact of crime in the working class. This emphasis was conducive to creating a research agenda that included "an accurate [study of] victimology" (Young, 1986, p. 23). But this was not just an emphasis on "victims"; it was a broader approach that not only stressed the geography and social dimensions of vulnerable sections of a community but also included studying the "risk rate of vulnerability" in the community. One way of conceptualizing this point is to think of the working class as a victim of crime from all directions. The more vulnerable people are economically and socially, for example, the more likely it is that *both* working-class and white-collar crime will occur against them (p. 23).

One example of left realism's concern for victims is the emphasis it places on feminist perspectives in criminology, a topic neglected by new criminology. With the growing awareness of the problem of rape and other feminist issues, feminists forced a rethinking of criminological theory itself. This development was not lost on left realism, which stresses that crime should be studied as "problems as people experience them" (Young, 1986, p. 24). As Young (1986) stated, "It takes seriously the complaints of women" (p. 24). This should not be interpreted to mean that left realism was and is concerned only about women as crime victims. Its agenda is concerned equally with racism, police brutality, and a number of other "everyday crimes" (Matthews & Young, 1992; Young & Matthews, 1992).

Left realism's emphasis on "everyday crimes"—crimes in the street—did not call for stronger punishment in order to win the war on crime. Rather they advocated minimal sanctions for minor and "victimless crimes"—including minor property crime, soft drug usage, and prostitution—while calling for expanded social control for more harmful crimes, such as industrial pollution and corporate malfeasance (Young, 1991). Nonetheless, Cohen (1986) stated that left realists "by their overall commitment to 'order through law' . . . have retreated too far from the theoretical gains of twenty years ago. Their regression into the assumptions of the standard criminal law of social control criminalisation and punishment is premature" (p. 131).

The uniqueness of its perspective was its strong interest in the class and power dimensions of crime causation and what can be done about it (Young, 1992b). It also represented an effort at synthesizing several theories, including labeling, strain, subcultural, radical Marxism, and some feminist perspectives (Alvi, 2005). As it neared its 20th anniversary, left realism's appeal and contributors were found in a number of countries, including Canada, the United States, Australia, and Britain, and it had generated several criticisms. One issue was whether it has strayed too far from its roots in radical thought, especially radical Marxism, and become "nothing more than liberal theory in a new guise" (Alvi, 2005, p. 933). Another concern was its emphasis on realistic approaches to the causes of crime comes perilously close to advocating punitive control strategies popular with conservatives (Schwartz, 1991, cited in Alvi, 2005), and the fact that it did not gain converts among some of the most influential radical or critical criminologists in England and Wales, including the eminent scholar Stan Cohen, among others.

CONSEQUENCES OF NEW CRIMINOLOGY/LEFT REALISM

The early contributions by the new criminology and realist criminology were not created in a social and political vacuum. Initially, they were part of the emergence of the "new left" in North America and Britain during the late 1960s and early 1970s. It held in its scope of criticism not only the issues discussed here within traditional criminology but also what has become known as the anti-psychiatry movement, prison support groups, campus sit-ins, and community action efforts (Young, 1988). Moreover, during the late 1960s and 1970s, higher education social science in Britain was expanded at an unprecedented level, and sociology/criminology/deviance offerings were developed not only in well-established universities but also in other universities and the expanding polytechnics. The new courses often were "taught . . . by individuals in or around the new radical criminology . . . to sociology students heavily imbued with the new left ideas and practice" (p. 168). Many of these scholars and their students remain in academia and are active critical criminologists.

In addition to leading the attack on traditional positivism, which held sway between post–World War II and the late 1960s, radical and realist criminologies have contributed to a long list of concepts that "are now staples of criminological culture, whatever its political persuasion" (Young, 1988, p. 164). These include a powerful critique of the mechanical determinism associated with biological explanations of crime, the social construction of statistics, emphasis on the endemic rather than a solely class-based nature of crime, and the largely invisible victimization of racist crime, domestic violence against women, and the abuse of children.

Several general social and criminal justice policy implications flow from left realists, including striking a balance between the "crimes of the powerful and the realities of street crime," creating a social justice agenda that emphasizes democratic-based reforms that would provide safe and affordable housing, fulfilling work, child care, universal health care, and adequate transportation (Alvi, 2005, p. 933). Left realists support minimal incarceration, programs that would enhance the likelihood of released prisoners successfully reentering society, and democratized forms of social control such as proactive instead of reactive policing.

The New Criminology Revisited

During the late 1990s, two events occurred that provided an opportunity to reevaluate the impact of the new criminology. The first was what all accounts described at first glance as a historic shift in Britain's politics and the beginning of a new and different political philosophy. The second was the publication of *The New Criminology Revisited* (Walton & Young, 1998).

One year before, on May 1, 1997, 44-year-old Tony Blair was elected with a huge majority as the youngest prime minister since 1812 to the most popular new government in British history, ending two decades of Conservative rule. Once the lead singer for a student band called Ugly Rumors, Blair had campaigned for what was termed the "New Labor" Party and promised a national transition to a "New Britain"—one far different from its bowler-hatted image of divisive classism. Central to his agenda was a rhetoric supporting community inclusiveness and a reforming zeal. Once in office, Blair and his Labor Party quickly focused on the modernization of health care, the reduction of Britain's runaway welfare bill, human rights, globalization, poverty, the devolution of Scotland and Wales, and a more cooperative relationship with the European Union. Most of these items generally had been opposed by Thatcher and her successor, John Major.

By late 1997 and early 1998, however, skeptics were questioning whether Blair and New Labor were any less conservative and hopeful in their efforts to solve social problems (e.g., crime) than were their predecessors. Indeed, some critics argued by the time Blair was in his third term he was the greatest Tory since Thatcher, a criticism given greater credence post-9/11 because of his fondness and support for President George W. Bush and the Iraq War. Since then and the 2005 London subway bombings by terrorists, the Blair government initiated anti-terror legislation and administrative policies and practices that have been strongly criticized as violations of human rights, invasions of privacy, illegal, and unjust (Lea, 2005).

Prior to these policies, Labor's approach to welfare reform, for example, was to cut support for jobless single mothers and to submit the benefits for sick and disabled people to means testing. Blair called these decisions a "tough choice" and dismissed the warnings of harm to the disadvantaged as "scare-mongering." To some critics, New Labor's approach to crime, while initially influenced by left realism, was soon equally as authoritarian, punitive, and conservative as that of the Tories. New policies, for example, included more private prisons, curfews for young people, enhanced use of electronic monitoring ("tagging"), harassing beggars, "zero tolerance," and enhanced automatic sentences for persistent petty offenders (Hanley & Nellis, 2001; Lilly, 2006a, 2006b, 2010; Lilly & Nellis, 2001; Nellis, 2003b, 2003c, 2004, 2006; Nellis & Lilly, 2000, 2004).

Of these policies, privatizing (along with centralizing) public service—including health, education, and criminal justice agencies—increasingly became Blair's touchstone when using "what works" was believed to be his guiding principle. That was replaced with an ideology of "does it introduce competition?" By 2002, for example, he had presided over the opening of eight new adult jails and three juvenile jails, all funded by private efforts; this approach permitted the London-based Centre for Public Services to conclude that Britain has the most privatized criminal justice system in the world. During the same time period Blair and his political supporters attempted to extend privatization—also known as "contestability" and introduced by the Carter Report (Carter, 2004)—to include probation services, an effort that was defeated, perhaps temporarily, in late 2005 and early 2006 (Dean, 2005; Travis, 2006). By late 2005, probation had already been largely transformed from its nearly 100-year-old social work ethos toward a "punishment in the community" ideology that blurred the time-honored distinction "between community and custodial penalties and the creation of 'seamless sentences,' which contain elements of both" (Nellis, 2003a, p. 41).

The effort to privatize probation followed an earlier national centralization development that had brought England's and Wales's prison and probation service together under the rubric of the National Offender Management System [NOMS]. Ostensibly presented and justified by Blair as a way to improve public service standards, it was arguably in fact a way to introduce more technological innovation into criminal justice, especially various forms of electronic monitoring, including GPS real-time tracking (Nellis & Lilly, 2000). The latter strategy of managing offenders was part of government concern about the high percentage of prisoner reentry or resettlement to society failures. It was argued that intense electronic monitoring, coupled with well-planned reentry plans, would reduce reoffending and the cost of traditional incarceration.

England's incarcerated population continued to grow under New Labor as it had under the Conservatives, and it remained the highest per capita prison population—142 per 100,000—in Western Europe. In late 2005, England and Wales's prison population was nearly 75,000, more than a 50% increase since 1994 (Home Office, 2004; Howard League for Penal Reform, 2005). Under Gordon Brown, a member of the Labor Party who was elected as Blair's successor on June 24, 2007, England and Wales's prison population remained high. In 2010, it reached 83,378 (*World Prison Brief,* 2010). More instructive, with the exception of Spain (165 per 100,000), England and Wales's per capita prison population—152 per 100,000—outpaced that of most other prominent Western European nations, including Denmark (66), Finland (67), Norway (70), Sweden (74), Switzerland (76), Ireland (85), Germany (88), Belgium (94), France (96), the Netherlands (100), Portugal (106), and Italy (107) (*World Prison Brief,* 2010). It is unclear, however, what, if any, impact Brown's May 7, 2010, defeat and the new coalition government—the first since World War II—under Conservative David Cameron will have on England and Wale's prison population (Burns, 2010).

New Labor's "tough on crime" policies often failed to attack the causes of crime. Not unlike Clinton's and Bush's support for more police, longer sentences, and expansion of the death penalty, Blair and New Labor soon replaced the Conservatives as the law and order party. It is against these changes that we can examine anew the impact of the 1973 publication of *The New Criminology* and its critical heir apparent, left realism.

Occasioned perhaps more by the long-running popularity of the original *The New Criminology,* and the fact that this volume has been in print continually since 1973, than by the political shifts associated with Blair and New Labor, *The New Criminology Revisited* (Walton & Young, 1998) was an unusual and important development because few, if any, developments in criminology have survived so well as to merit a revisitation. In retrospect, the editors of and contributors to *The New Criminology Revisited* saw the 1973 publication as part of the 1970's wave of radicalism that continued with various degrees of impact into the 21st century. A close reading of this work indicates that most, if not all, of the major points advanced in 1973 were reaffirmed:

- Crime and the processes of criminalization are embedded in the core structures of society, whether it be in its class relations, its patriarchal form, or its inherent authoritarianism. (Walton & Young, 1998, p. vii)

- The sole and precise aim of new criminology is improving the human condition. New criminology has a utopian commitment.

- The new criminology was and still is not committed to corrections as supported by establishment criminology a la administrative criminology. Human behavior does not need "correcting."

- The new criminology is wedded to social change. Its adherents wish to do more than make professional contributions to human knowledge.

- The new criminology aims to deconstruct criminological theories in an attempt to construct a social theory of crime and deviance.

However much the ideology of new criminology generally had not changed during the intervening years, the 1998 reaffirmation of its agenda could not disguise the fact that significant new theoretical developments had occurred. Feminist perspectives had become more developed and central to critical criminology as had postmodernist thought. The Marxist heritage had not been so much abandoned as refined and redefined (Muncie, 1998, p. 227). Instead of maintaining its position as a mega narrative, it became a source for a "set of provisional hypotheses or a frame of conceptual resources/deposits" (Muncie, 1998, p. 227). Ian Taylor's (1998) and Elliott Currie's

(1998b) analysis of the impact of the U.S. and Britain's 1980s and 1990s unleashing of "free market forces" on the increase in property crime in each country and globally was illustrative of reworked Marxist explanations of crime. Although heralded as a vindication of conservatives' social and economic policies, much evidence supports the proposition that it increased property crime, poverty, homelessness, inner-city drug abuse, and violent crime.

In retrospect some of the ideas developed in *The New Criminology Revisited* were forerunners of much that captured the imagination of today's cultural criminologist. Each of these scholars discussed some of the features of late modernity and crime and by doing so they laid some of the theoretical groundwork that can now be appreciated as a bridge between new criminology and cultural criminology. Most notably here are the pieces by Currie (1998a), Lea (1998), Muncie (1998), and Taylor (1998).

Left realism remains the new criminology's critical heir apparent and Britain's major alternative to mainstream criminology. Tierney's 1996 assessment was prescient— left realism continues to offer a "radical alternative to the rhetoric of the right" (Tierney, 1996, p. 233).

Left Realism Today

In recent years, interest in left realism has grown considerably as one of the fields of inquiry under the broader category of critical criminology, despite the fact that some critics pronounced it dead (DeKeseredy, 2010). There are several reasons for this premature and incorrect pronouncement. One reason is found in the fact that prior to the 1980s, critical criminology focused primarily on corporate and white-collar crime and on the importance of understanding the impact class and race have on the administration of criminal justice agencies, while neglecting left realism's call for attention to be directed to the crimes committed by the powerless as well as working-class and female victimization. Ignoring left realism's position on this point misleadingly gave some critics the impression that it had little or nothing to say worth heeding.

Whereas left realists agree that these are important topics that warrant attention, they nonetheless claim that it has been empirically demonstrated that failure to acknowledge and study the crimes of the powerless plays into the hands of conservative politicians who can manufacture ideological support for traditional right-wing law and order policies. Because left realists argue that the crimes of the powerless result largely from inequalities inherent in the social structure, the crimes of the disenfranchised—in their opinion—must first be recognized before an egalitarian society based on social justice principles can develop. For some time, however, this argument was largely ignored because it had less political appeal than focusing on crimes of the powerful and on race and gender issues, regardless of theoretical and empirical accuracy to the contrary. This, too, added to the impression that left realism had a weak critical voice.

Left realists had another important reason for criticizing critical criminology's neglect of the crimes of the powerless, a position that did little to enhance its own stature. They argued that the failure to take working-class crimes and victimization seriously—especially female victimization—helped right-wing groups to dominate control over knowledge about crime and policing. This can be seen, for instance, in the vitriolic discussions on radio and TV talk-shows. In England, during early 2010 in the run up to national elections, the conservative Tory Party was accused of issuing "dodgy crime figures" suggesting that violent crime had increased dramatically under the Labor Party (Watt, 2010). In fact, for some time, early left realism itself was criticized for paying only lip service to gender-related issues, such as the role of broader patriarchal forces. This also contributed to the notion that left realism was dead.

This has changed. Now "the bulk of their theoretical work addresses street crime, draconian means of policing, and violence against women in heterosexual relationships" (DeKeseredy, 2010). In addition,

> left realists [now] conducts local crime victimization surveys, which include quantitative and qualitative questions that elicit data on harms generally considered irrelevant to the police, conservative politicians, and most middle- and upper-class members of the general public. These topics include: male-to-female physical and sexual assaults in adult intimate relationships; sexual harassment of gays, lesbians and people of color in public places; and corporate crime. (DeKeseredy, 2010, p. 548)

There are, however, discernable differences between British and North American left realism, none of which support the notion that this perspective is dead. To the contrary, the differences reflect the sound health of left realism. In general terms, the bulk of British realist policies focus on criminal justice reforms "including democratic control of policing" (DeKeseredy, 2010, p. 549), whereas its counterpart across the Atlantic devotes more attention to anticrime proposals. Left realists in both locations agree theoretically that such policies, including "'hard' police tactics such as stopping and searching people who are publicly drunk only serves to alienate socially, economically, and politically excluded urban communities" (DeKeserdy, 2010, p. 547).

For some time too, left realism seemed to be culture-bound, with most of its work being conducted in Britain, especially its theoretical work on crime in impoverished inner-city communities. This focus left many other areas of criminological concern neglected. To some extent, this is true. Today, however, left realism constructs and tests theories in a number of key problems facing contemporary societies around the world (DeKeseredy, 2010).

Nonetheless, until recently, left realism has given little critical attention to how structural factors, such as the shift from a manufacturing to a service-based economy and what they call the neo-conservative assaults on social services, have negatively impacted today's middle-class youths. Or to how double-digit unemployment on both sides of the Atlantic, home foreclosures, factory slowdowns and closings, and additional economic turmoil (caused by the implementation of economic policies fostered by University of Chicago economist, Milton Friedman), the near elimination of the public sector, deregulation of corporations (including incredibly risky Wall Street and banking practices), and skeletal social spending lead to various forms of social exclusion, gendered crime, and a negative impact on middle-class adolescence (Currie, 2004;

Klein, 2007). For example, low, demeaning, and marginal male employment has been found by Currie (2009) and others to be universally associated with violence against women (see also Basran, Gill, & MacLean, 1995; DeKeseredy, Alvi, Schswartz, & Tomaszewski, 2003). Or how in early 2010 the structure of the U.S. federal legislative process permitted a single senator to prevent the Senate from taking up a House of Representatives measure that provided another 30 days of unemployment aid and extended television broadcasts to a half-million people living in rural areas (Hulse, 2010).

The vitality of left realism today is also found in its rich discussions of what some critics have called the public irrelevance and marginality of orthodox criminology, or what has been labeled with the moniker of "so what? criminology." This is no small matter for mainstream criminology, especially in face of the fact that according to some sources—including a 2005 United States Government Accountability Office (GAO) report—crime is going down and thus is perhaps contributing to making the criminological enterprise an endangered species (Zimring, 2006). The major contours of the current debate were offered by the award-winning and internationally esteemed left realist Elliott Currie at the 2004 annual meeting of the American Society of Criminology (Currie, 2007; Feeley, 2010). The author of *Confronting Crime* (1985), one of the classic criminological works (Cullen, 2010), Currie in essence said that conventional criminology had, despite its accumulated theoretical and empirical heft, distressingly little impact on the course of public policy toward crime and criminal justice. In his view—a variation on a similar position espoused by radical and critical criminologists in the 1960s and 1970s that was soon taken up by others—this situation had developed not because there was so few crime problems to address. Indeed, the situation was just the opposite. The United States

> stands out around the world in both our level of incarceration and our rates of routine violence in the streets and homes. The specter of prison haunts poor communities, wrecking havoc on families, destroying futures and disenfranchising great numbers of minority Americans. Mass incarceration has deepened the historic gaps in life chances between affluent and poor, white and black, and created a vast army of the socially and economically disabled; it fruitlessly sucks up billions of dollars, year after year, that we need for more productive human purposes." (Currie, 2007, pp. 175–176)

The voice of criminology, according to Currie (2007), was never more needed than now. For decades, however, it had become increasingly marginal to the larger public discussions of crime and criminal justice and "decreasingly capable of affecting the thrust of public policy" (p. 176). It was not that criminology had nothing or little of relevance to say about public policy. Rather the problem was that it had become isolated from such debates because criminologists do a lousy job of educating the public about what they in fact know about what to do about crime. Currie identified several reasons for this situation.

At the top of his criticisms is the current tendency in major research universities to define criminological scholarship too narrowly, favoring "original research" and "significant findings" to be published in peer-reviewed journals with obtuse language and incomprehensible mathematical formulas that have limited appeal even to conventional criminologists and even less interest or relevance to public policy (see also Young, 2004). This approach, observed Currie, sacrificed making *sense* of research and *disseminating* it to a "broader and potentially more efficacious audience than ourselves" (Currie, 2007, p. 180).

According to Currie, other root factors that have contributed to the isolation of conventional criminology include a national political shift to the right that has 40 years of success in extolling the virtues of punitive responses—including mass incarceration—to deviance, the acceptance of a kind of predatory individualism as a guiding principle of public life, and a social Darwinian view of social relationships that supports cutting or eliminating social services to solve the problems of isolation and marginality. He argued that criminology should develop rewarding and effective strategies, especially within research universities, that allow criminologists to have vigorous and systematic intervention in the world of social action and social policy (Currie, 2007).

Although Roger Matthews—one of the founders of left realism—agreed with Currie's assessment, he disagreed with his proposed solution. Rather than develop new strategies for criminologists to intervene in crime and criminal justice policy debates, Matthews (2009) proposed engaging in "theoretically informed interventions employing an appropriate methodology" (p. 343). This left realist idea—also known as the "holy trinity" because it incorporates theory, method, and practice—represented a proposal long associated with other radical and critical thinkers. However, one important factor had changed since the time when liberals' major contributions had been formed around rejecting or opposing conservative "get tough on crime" policies. According to Matthews, the change is the fact that conservative criminology itself is experiencing a demise, one that has left liberal criminologists with little more to complain about than "manipulative politicians who play the 'law and order' card, on gullible publics who are charged with becoming more intolerant and punitive" (Matthews, 2009, p. 344).

Matthews offers a refashioned realist criminology that prioritizes the role of theory around concepts such as class, the state, and structure. Using these ideas, he argues, coupled with the recognition that a method of analysis that stresses the *meaning* of crime (instead of trends) to victims and offenders is an essential component that would link theory to effective intervention. This approach, Matthews proposed, would create a left realist criminology that would stand in stark contrast to forms of "positivism and empiricism that are prevalent in [conventional] criminology" (Matthews, 2009, p. 344).

DeKeseredy and Schwartz (2010) agree with Matthews's assessment of the importance of left realism and the fact that it has been rediscovered. They also praise him for attempting to write a coherent critical realist theory that explains crime and punishment in view of the major structural transformations of late modernity, including increasing unemployment, marginal work, and social exclusion. They criticize him, however, for committing the same mistake made by earlier left realists—gender blindness. "In fact," observe DeKeseredy & Schwartz (2010), "the word 'gender' only appears once in the main text of his article and the words 'feminism' and 'patriarchy' are nowhere to be found. From one feminist standpoint, then, his re-fashioned realist criminology is more of the same" (p. 161).

To address this limitation, DeKeseredy and Schwartz offer *a new left realism subcultural theory* that places gender at the forefront. Their main point is that laissez-faire economic policies, such as those advocated by University of Chicago economist Milton Friedman, have caused a relatively "new assault" on workers that has increasingly made North America "categorically unequal" (Massey, 2007, quoted in DeKeseredy & Schwartz, 2010, p. 163). One of the specific disenfranchising economic policies identified involved corporations moving to developing countries "to take advantage of weak environmental and work place safety laws" (p. 163). One result is that a substantial portion of North America's male labor market ends up facing a number of challenges to their masculine identity that puts them at great risk "of teaming up with others to create a subculture that promotes, expresses, and validates masculinity through violent means" where violence against women and male-to-male violence develop as a type of "compensatory masculinity" (Dekeseredy & Schwartz, 2010, pp. 163, 164; see also DeKeseredy, Schwartz, Fagen, & Hall, 2006). DeKeseredy and Schwartz refer to their own research in rural Ohio and other places in North America to support their claim.

In summary, there are a variety of left realist theories. DeKeseredy and Schwartz's and Matthews's are just two of the most recent. Other left realists mix Mertonian, subcultural, and Marxist theory together; others integrate feminism, male peer support theory, and Marxism; and still others focus simultaneously on the relationship between victims, offenders, the community, and the state (DeKeseredy, Alvi, & Schwartz, 2006). More recently, Gibbs (2009) has looked at terrorism through left realist lenses.

The New European Criminology

CONTRIBUTIONS AND CONTEXT

Historically, various schools of thought and scholars beyond the English-speaking countries of North America and Britain influenced the developments in criminology discussed in this book. Unfortunately, today this is recognized too infrequently. The result is a false impression that there is or has been little, if any, significant contribution to criminology from other parts of the world. This is not true.

One reason for the relative failure to recognize contributions from non–English-speaking countries lies in the fact that early contributors to criminology came from disparate languages, contexts, and disciplines. For example, Lombroso, Ferri, and Garofalo were Italian; Gabriel Tarde and Durkheim were French; and Bonger was Dutch. Disciplinary training also was different. Lombroso was a physician, whereas Tarde was a provincial magistrate interested in philosophy and psychology. One result was that a coherent body of literature that could be identified as criminology did not develop smoothly.

Another reason why more recent European contributions to criminology have been neglected in the United States is that traditionally it was an "auxiliary discipline to criminal law" (van Swaaningen, 1999, p. 6). Unfortunately, this allowed criminology to be easily used as a utilitarian tool for social control by the state. With the rise and implementation of Nazi law during World War II, criminology faced an ethical crisis that was not overcome until after the war. This retarded its development as a discipline committed to the rigors of the scientific method. With the help of continental refugees and nascent criminological interests in the United States and Britain, the growth of post–World War II criminology boomed.

In the aftermath of World War II's nationalism gone awry, great international effort was directed to creating a grand unified Europe. This was to be a Europe "where nationalities constantly mix while national borders slowly crumble" (Ruggiero, South, & Taylor, 1998, p. 1). Up to and through the 1960s, the dream held real promise as the economies underpinning efforts to harmonize Europe saw unemployment rates at less than 2%. By the late 1990s and early 21st century, this had changed. The unemployment rate in some European countries rose to more than 12%. This change reflected two powerful developments that were to have a great impact on developing a new European criminology.

First, unemployment and related social problems were an outgrowth of a major shift in the economic market. What until the 1980s had been a strong emphasis in Europe on state-sponsored community welfare shifted to a free market economy more concerned with corporate profits. This change to "market forces" witnessed a withdrawal of efforts to create a harmonized Europe where nationalism was expected to disappear. In its place came a form of liberalism that stressed freedom of choice, nearly unfettered competition, and the wondrous joys of consumerism. A sense of duty and responsibility to others were seen as old-fashioned and as obstacles to the freedom of the market (Ruggiero et al., 1998, p. 4).

Second, combined with the shift to a free market economy, a flood of immigrants fleeing political uncertainties and economic hardships in Eastern Europe and parts of Africa posed another major threat to European harmonization. Competition for profits in the face of an oversupply of workers contributed to a number of predictable problems. Working conditions declined. Intense competition for housing, education, and health care also occurred. In some places, unemployed native-born citizens physically assaulted and sometimes murdered immigrants. According to several observers, the 2005 French riots were the result of youth unemployment among immigrants. Jobless people there under age 26 took between 8 and 11 years "to find a permanent job compared to three to five years in most of Europe" (Gow, 2006). To some observers, the deterioration in working conditions and welfare provisions were necessary in the face of marketplace competition from North America (with its protective North American Free Trade Agreement) and from Asia.

ABOLITIONISM

In part as a result of the crime problems associated with the disruptions to the grand dream of a unified Europe, and in part as a continuation of efforts begun in 1973 by the European Group for the Study of Deviance and Social Control, the late 1990s saw the birth of a "new European criminology" (Ruggiero et al., 1998). One of its objectives, loosely stated, was to maintain an exchange of criminological communications and comparisons across Europe that, rather than contribute to free market liberalism, would contribute to developing a European public sphere that emphasizes a sharing of experiences. A second objective was to develop a European criminological community that would "help develop an understanding of trends and concerns in Europe" (p. 10).

Shortly thereafter, in 2000, the European Society of Criminology (ESC) was established, and one year later it held its first annual meeting in Lausanne. In 2004, the ESC published the first issue of its *European Journal of Criminology*. Today, European criminology is experiencing great vitality and diversity. One example of this vitality and diversity is found in "abolitionism."

There is no concrete definition of abolitionism (Brushett, 2010). To a large degree, this approach is an example of European criminology whose advocates are based in England, Sweden, Denmark, the Netherlands, and Norway. Until recently it was neglected in the United States where it has modest support. Similar to new criminology/left realism, abolitionism's origins sprang from changing contexts, but it has a few closely related points of emphasis. According to one observer, the modern abolitionist perspective was nurtured by critical criminology and birthed in some of the Scandinavian countries during the 1960s (Tierney, 1996, p. 132). One of its major current contributors, however, locates its origin in the anti-slavery movement mentioned earlier in this chapter (van Swaaningen, 1997, p. 116). At this juncture—40 years after it started—abolitionism comes in two forms: general and restricted.

A central tenet of general abolitionism is that punishment never is justified because it is a form of social control that is based on the faulty logic that the infliction of pain will prevent crime. Furthermore, this form of abolitionism argues that the criminal justice system as a whole is a social problem that should be dismantled and replaced with alternative dispute resolution (Brushett, 2010). These concepts reflect not only the idea that traditional incarceration is destructive to individuals and the community but also that justice should be founded on social and economic concerns for all of its victims and victimizers (Knopp, 1991). This point is especially compatible with peacemaking criminology's emphasis on mediation, conciliation, and dispute settlement, as discussed in Chapter 8.

Restricted abolitionism deals with the elimination of specific aspects of the criminal justice system (Brushett, 2010). Prisons, it is argued, are a form of violence that should be destroyed because they reflect "a social ethos of violence and degradation" (Thomas & Boehlefeld, 1991, p. 242) and because they do not prevent recidivism. One of the primary purposes of prisons, it is argued, is to control the least productive members of society. Restricted abolitionists argue that prisons should be replaced, or at least decentralized, by democratic community control and community-based treatment that would emphasize "redress" or "restorative justice."

Whereas the abolition of prisons has not been accomplished in any major industrial nation, there are "several major groups strongly backing penal abolition including Britain's Radical Alternatives to Prisons (RAP), the Norwegian Association for Penal Reform (KROM), the Danish Association for Penal Reform (KRIM), and the Swedish Association for Penal Reform" (Brushett, 2010, p. 2).

CONSEQUENCES OF ABOLITIONISM

Similar to other reformist schools of critical criminology, abolitionism has been criticized for being romantic, for being imprecise, and for lacking a well-grounded theoretical opposition to punishment (Thomas & Boehlefeld, 1991). Or, in the words of Hudson (1998), abolition has a "vision without a strategy" (quoted in Brushett, 2010, p. 2). In particular, it does not have practical plans for dealing with dangerous predatory criminals. Neither does it have a well-articulated research agenda with significant funding. Nevertheless, it has made significant contributions, such as Christie's (1981) eloquent critique of the limitations of conventional punishment and his (1993, 1997, 2001) insightful analysis of crime control as industry—an argument that has had three editions. In the United States, Angela Davis (2003, 2005) lent her critical anti-racism voice to making the case for the abolition of prisons and torture. Nonetheless, "abolitionism is still a perspective that is structured primarily by analogies and metaphors" (Brushett, 2010, p. 2). Abolitionism is but one potent critical criminological perspective.

The voices of critical theorists are highly valued as they continue to question the domain assumptions that criminologists and politicians tend to take for granted when explaining the causes of crime. Their contributions are no less valuable when it comes to questioning responses to crime. Worthy of special note here are the now widely accepted and deeply respected contributions from feminist perspectives on crime (see Chapter 10). Equally significant are the contributions from what was once called "new criminology," from left realism, and from European-based abolitionism.

THE IMPORTANCE OF OTHER VOICES: JOCK YOUNG

While it would be foolhardy to attempt to identify the most influential criminological voices outside of the United States during the last 40 years, it behooves us nonetheless to be circumspect and appreciative of rich contributions from abroad. The initial and continued development of *new criminology,* left realism, and more recently cultural criminology have been greatly influenced by the writing of Jock Young. There is little if any doubt that he has been and continues to be one of the major criminological theorists in the world. In 1998 the American Society of Criminology awarded him its Sellin-Glueck Award for Distinguished International Scholarship, the most prestigious award given in the United States to a non-American criminological scholar. In 2003 the Critical Criminology Division of the American Society of Criminology gave him its Lifetime Achievement Award. Former sociology professor and head of the Centre for Criminology at Middlesex University, England, Young is now the Distinguished Professor of Sociology at the John Jay College of Criminal Justice and a visiting professor of sociology at the University of Kent, U.K.

Reminiscent of Richard Quinney's varied contributions to criminological theory, Young earned his B.Sc., M.Sc., and Ph.D. from the London School of Economics, and in his early career was best known as the coauthor of *The New Criminology.* Much of his other work is equally seminal. His *The Drugtakers* (Young, 1971) is a classic study of drug policy and drug behavior, and *The Exclusive Society* (1999) and his *The Vertigo of Later Modernity* (2007) are considered blistering critiques of social exclusion in contemporary society. By 2003, Young was contributing not only to left reality but to cultural criminology (Hayward & Young, 2004, 2005; Young, 2003; Young & Brotherton, 2005), a development we will review in the section to follow. The central and persistent theme in his intellectual journeys—new criminology, left realism, and cultural criminology—has been those groups marginalized by capitalism. The focus was first present in his highly articulate chapter "Working-Class Criminology" in 1975 and has been reaffirmed ever since.

Cultural Criminology

Cultural criminology is a perspective that has developed in both the United States and the United Kingdom since the mid-1990s with writings having important cross-national impacts. It is a theoretical orientation that is based on the argument that crime and crime control cannot be understood apart from the domain of culture. Whereas criminology has traditionally focused on issues like the structural facilitators of patterns of everyday crime, crimes are in fact cultural in nature. According to this perspective, crimes are constructed out of symbolic interactions among groups and people "and are shaped by ongoing conflicts over their meaning and perceptions" (Ferrell, 2010, p. 249). In 2005, according to American criminologist Jeff Ferrell (2005), cultural criminology built on the "groundbreaking work of Cohen (1971, 1980), Hebdige (1979), and other British cultural theorists" (p. 358). This mutual influence is hardly unprecedented. Thus, in the 1960s, some of the young British criminologists at that time were greatly influenced by several American criminologists/sociologists including Howard Becker (1963), John Kitsuse (1964), Edwin Lemert (1951, 1972), and David Matza (1964, 1969), and "supplemented by the social construction work of writers such as Peter Berger and Thomas Luckman" (Hayward & Young, 2004, p. 261). Later, during the 1970s, many of the same young British developed the "new criminology" at the same time that the closely allied cultural studies movement came of age at the now defunct Birmingham Centre for Contemporary Cultural Studies (Hayward & Young 2004, p. 261; Webster, 2004). The work by Katz (1988), Manning (1998), Young (1999), and Ferrell, Hayward, Morrison, and Presdee (2004a) has influenced cultural criminology on both sides of the Atlantic.

LATE MODERNITY AND GLOBALIZATION: CONTEXTUAL CHANGES

According to Young (2003), the impact of economic and cultural globalization is creating widespread resentment and tensions within the First World and internationally. Although economically and politically beneficial, especially in the short run, globalization nonetheless exacerbates "both relative deprivation and crises of identity"

(Young, 2003, p. 389). The combination of these conditions, Young theorizes, generates a sense of unfairness and humiliation that results in offensive behavior that is *transgressive* and *expressive* instead of rational and instrumental as was argued by new criminology and left realism.

Central to this argument is the idea that in the present period of late modernity, boundaries and categories of behavior and culture associated with the social embeddedness found in traditional and modern societies (i.e., job security, marriage, and community) are blurred and confused. The new "flexibility" demanded by management in the labor market has replaced job security with "easy hire and fire" where in Denmark, for example, a third of the workforce changes jobs each year and the aged work longer (Gow, 2006; Peters, 2006). Whereas Merton (1938, 1957), Felson (1998), and Garland (2001) explain crime by relying on "notions of opportunities, on one side, and lack of control on the other," cultural criminology emphasizes the exact opposite (Young, 2003, pp. 390–391). It focuses on "the sensual nature of crime, the adrenaline rushes of edgework—voluntary illicit risk-taking and the dialectic of fear and pleasure" (Young, 2003, p. 391).

In cultural criminology, the *meaning* of crime and criminality in everyday life for criminals, politicians, criminal justice organizations, the media, and popular culture is *contested*—not agreed upon. According to this perspective, crime does not have the same meaning to those who want to control or prevent, study, and report it in the media or who participate in it. The meaning of crime is rather the result of complex processes "through which illicit subcultures, the mass media, political authorities, criminal justice professionals, and others contest the meaning of crime and criminality" (Ferrell, 2005, p. 358).

One important theoretical implication, which is central to the cultural criminology perspective, is that the meaning of crime is obviously socially constructed and not simply the result of rationally chosen violations of law. The *situated* meanings of crime and illicit subcultures that develop in the clashes or dynamics of social control is of great concern to cultural criminology. To get a handle on this point cultural criminology investigates how the image, style, and the representation of crime and crime control actually occur. On the one hand, political debates about crime and crime control—represented through the media—often include dynamic examples and symbols of mayhem and threat to the public while simultaneously denigrating criminals. Criminals and criminal subcultures, however, may very likely give the same behavior an "alternative meaning, organized around elaborate conventions of appearance, vocabularies of motive, and stylized presentations of self" (Ferrell, 2005, p. 359). One result is that crime in the hands of the media turns into politics as moral entrepreneurs launch campaigns of moral panic over crime and criminals. In fact, according to Ferrell, one of the distinctive features of contemporary society is the constant interplay of the media, crime, and criminal justice that comprises a model of *media loops and spirals* (Ferrell, Hayward, & Young, 2008).

A key dimension of this model is the idea that traditional models of examining the effects of media on crime and subsequent crime, or the accuracy of the media in reporting criminality, are outdated. What is needed, it is argued, is a model "that can account for a world *so saturated* with media technology and media images that distinctions between a crime and its mediated image is often lost. In this world, crime and media are linked by a looping effect in which crime and the image of [it] circle back on one another, together constructing the reality of crime for *participants and the public*" (Ferrell, 2010, p. 251, emphasis added). Examples of media loops and spirals, which are illustrative of this theoretical observation, are found in squad car cameras influencing everyday policing, in juries demanding evidence like what is presented in *CSI*-style television programs, in violence such as a staged encounter of teenage females beating up another female that is recorded and posted on the Web for profit, and in what is called "underground fighting"—fistfights for fun, meaning, and profit (Ferrell et al., 2008; see also Brent & Kraska, 2010). Crime, criminal justice agencies such as the police, and the media are part of an ongoing feedback loop (Ferrell et al., 2008).

Cultural criminology also critiques the methods often used by conventional criminology and offers alternatives. Survey research and statistical analysis of survey research are, according to cultural criminologists, inadequate for understanding the interpretative dimensions—meaning, representation, and emotion—of crime. Conventional criminology methods reduce the behavior of offenders to brain-dead soulless pieces of data that are fit into preset categories that are analyzed with cross-tabulations. Cultural criminology methods stress the human dynamics of surprise, ambiguity, and such things as anger—factors that are often ignored by conventional criminology as well as by the media. "[C]ultural criminologists instead turn to a variety of methods that are informed by an ethnographic sensibility" (Ferrell, 2010, p. 252).

Political figures meanwhile build political capital by marketing imaginary solutions to crime as criminals (and those who romanticize them) create images of themselves as romantic outlaws or outsiders. According to Ferrell (2005), "In all this, cultural criminologists emphasize that crime and crime control are increasingly becoming a public carnival, a media circus, and a hall of mirrors where images mostly reflect each other" (p. 359). Or, in the words of cultural criminologist Mike Presdee, "the creeping criminalization of everyday life provokes transgression rather than conformity" (quoted in Young, 2003, p. 391).

CONSEQUENCES OF CULTURAL CRIMINOLOGY

Cultural criminology is not without theoretical and methodological critics, some of whom have been particularly insightful. Its use of *culture*—its central concept—is illustrative (O'Brien, 2005). Put succinctly, cultural criminology is charged with using a definition of culture that is based more on political rather than analytical motivations. One result is confusion about the meaning of *culture* and *subculture*. In Ferrell's (1996) work about graffiti, for example, *culture* generally refers to the state or the mainstream and its followers. While they are ideologically critiqued as boring, average, and one-dimensional, members of subcultures are celebrated with a minutia of detail depicting them both as somehow morally superior to and more creative than the mainstream (O'Brien, 2005).

At the heart of O'Brien's specific critique of Ferrell and of cultural criminology in general—including Young's efforts (1999, 2003) to revitalize left realism—is what

O'Brien identifies as lack of understanding and engagement with the classic debates on the meaning of culture found in social anthropology (see also Yar & Penna, 2004). The result is a fundamental confusion "about what culture represents in relation to different levels of analysis and a collection of forced definitions intended to deal with the confusion on a case-by-case basis" (O'Brien, 2005, p. 600). Put differently, where is *culture* in the work of cultural criminology if their ethnographic work makes no distinctions between "psychological, economic, political and geographical forces that impinge" on experiencing crime individually or the patterns of crime over time (O'Brien, 2005, p. 605)?

The work by Hall and Winlow (2004) and Hall, Winlow, and Ancrum (2008), however, does address some of these questions. They do so by creating a theoretical perspective that utilizes, among others, Jacques Lacan's (2006) refashioned Freudian psychiatry, Lasch's (1979) work on narcissism, Zizek's (2002, 2006a, 2006b, 2006c, 2007, 2008) philosophy, psychoanalysis and cultural criticism, and Frank's (1997) commentary on hip consumerism and the deep contradictions and ironies of American politics.

A key to understanding this take on cultural criminology is found in Hall and Winlow's observation that whereas cultural criminology has much promise for an analysis of crime and social disorder, it has thus far placed too much emphasis on its perspective being grounded firmly in the economic logic of advanced capitalism. They propose instead an approach that places crime—especially violence—within what they call "the context of increasing instrumentalism in consumer culture and the breakdown of the pseudo-pacification process" (Hall & Winlow, 2004, p. 284). The latter term describes what Hall and Winlow see as the weakening strength of the contemporary culture to hold together the collective social solidarity that followed after the Enlightenment and until recent decades characterized Western capitalistic societies. Today, they argue, the civilizing impact of the Hobbsian social contract with government on the super-ego and diminishing anxiety, volatility, and violence has weakened considerably for some parts of our society.

In fact, they agree with Rose (1969) that the post-1980s "era has seen the 'death of the social' and the fracture of relations of mutual reciprocity that have always existed in one form or another across our diverse anthropological histories" (Hall et al., 2008, p. 6). Traditional collective social bonds, they argue, have been replaced by "what is quite possibly the most complete and pervasive form of atomised competitive individualism yet seen" (p. 6). One result is that the acquisition of cultural commodities by brutal barbaric behavior is in fact an instrumental response to extremely heightened anxiety and insecurity. However, as Hall and Winlow (2004) emphasize, it is not the material aspects of consumer goods that provokes desire and ambition, but rather its social symbolism and its power to identity and meaning in what Bauman has identified as *liquid modernity*.

To elaborate, Hall et al. (2008) are of the opinion that the hoped-for increases in equality, opportunities, and freedoms promised by Britain and the United States during the 1960s and 1970s have generally failed, although, as they note, there has been some progress made, especially in race/ethnic relations. There were also discernible decreases in crime during the 1990s, some of which can be explained by the growth in mass incarceration. But even that did not generate much, if any, impact on various forms of violent crime. Thus, in "many troubled locales blighted by permanent recession, unemployment and burgeoning criminal markets the murder rate is six times the national average" (Hall et al., 2008, p. 3). To explain this paradox, Hall and Winlow argue that changes in the 1980s were not merely another structural adjustment of the economy and labor market. They were instead a radical change in the political economy and culture that allowed the market to exert an unprecedented domination of life. It is in this radically "altered terrain in the 1980s that the most resounding phase of the 'crime explosion' took place, and it presents new challenges for criminological theory" (Hall et al., 2008, p. 4).

For these scholars, cultural criminology is in danger of developing into *culturalism*—an extreme reductionist argument that attempts to explain culture and identity in late or postmodern capitalism by emphasizing the explanatory power of culture at the expense of neglecting political, economic, and historical processes and shifting contexts. The type of crimes they attempt to explain are committed by new barbarians or *anelpis*—"an historically unique section of humanity, which cannot be described as an 'underclass,' in the structural sense because their wage needs have priced them out of the global labor market" (Hall & Winlow, 2004, p. 277). They live in what have been identified as locations of *permanent recession* in old, economically abandoned, micro-communities of the industrial past where the inhabitants experience an extreme sense of status-anxiety and insecurity under advanced consumerism. Unlike other groups who have similar experiences, some members of these groups live with total cynicism and nihilism, without hope, opinion, fear of authority, or realistic expectations. They are the new barbarians at the gate.

Cultural criminology's analysis of media-based images of crime is somewhat less problematic. This tack indeed reveals how images of crime are manufactured and contested, yet it tells very little about crime itself from a positivistic perspective, or *the meaning of crime to its victims*—including, for example, women, children, the aged, and property owners, among others. Its claim, too, that it has "reshape[d] theory and practice of contemporary criminology" and "instigated a quiet revolution in criminology" in "its traditional domains of inquiry" are unsubstantiated (Ferrell, 2005, p. 359).

This said, some of the work in cultural criminology is interesting, thought provoking, and worthy of praise. Matthews's (2005) review of Ferrell et al.'s edited *Cultural Criminology Unleashed* (2004a) is illustrative. According to Matthews's (2005) review, cultural criminology is a refreshing development in part because traditional criminology has become pedantic, tedious, apolitical, and decontextualized. Cultural criminology, on the other hand, is a product of "and a response to changing late-modern conditions and sensibilities" where traditional, rigid, and bound categories of what is crime are too problematic and difficult to sustain (p. 419). The same, he says, applies to criminology where the boundaries between it and other subjects including sociology, politics, and media studies, are blurred. According to cultural criminology, crime and criminology are increasingly subject to cultural influences both in terms of shaping interpersonal relations and in relation to the social production of meaning. In the opinion of Maggie O'Neill (2004), one of the contributors to *Cultural Criminology Unleashed*, the major tasks for cultural criminology are to engage in critical analysis while "simultaneously engaging with the lived experience of the marginalized, employing a range of methods and representations that can be used to challenge accepted social knowledge" (Matthews, 2005, p. 420).

Hallworth's assessment of cultural criminology generally and of *Cultural Criminology Unleashed* strikes a similar chord. Rather than seeing cultural criminology as a paradigm shift in the Kuhnian sense, "it represents a meeting place wherein various members of criminology's wandering tribes have met profitably together" (Hallsworth, 2006, p. 147). Their theoretical interests include micro-sociology of subcultural theory, phenomenology, and post-structuralism. Coupled with and inspired by political engagement and critiques of oppression, the result is a "hybridized movement that is theoretically rich while also politically motivated and committed" (Hallsworth, 2006, p. 147). At the heart of this enterprise is an effort to "uncover the phenomenology of transgression by taking cognizance of the meanings delinquents bring to their actions and interactions" (Hallsworth, 2006, p. 147). Examples of this emphasis are found in chapters on fighting and street-racing. Other chapters, notably Brotherton's (2004, p. 263), use cultural and political economy theories to analyze the transformation of one of the "most notorious street gangs in the United States, the Almighty Latin King and Queen Nation," into a political movement that developed in the face of the state's war on poverty.

Cultural criminology uses the destabilizing conditions of late modernity to study how populations "position themselves in relation to these shifts, not least in their [the state's] demand for more punishment and greater security" (Hallsworth, 2006, p. 148). The enemy for cultural criminology is the state and an administrative criminology that advocates rational choice theory. As Hallsworth notes, however, for cultural criminology to go beyond its current "coming of age," it needs to do more than to treat agents of control solely as negative forces; culture and edgework should be studied with the same diligence, otherwise cultural criminology's political and intellectual impact is weakened. In sum, cultural criminology seeks to dissolve conventional understandings of crime regardless of whether they are specific theories of the institutionalized discipline of criminology itself (Ferrell et al., 2008, p. 5).

Convict Criminology

BACKGROUND: PRIMARILY AN AMERICAN CONTRIBUTION

Though not the first criminal biography or autobiography, Shaw's (1930) classic study of the "jack-roller"—a person who in the 1920s robbed drunks and skid-row bums—was the first in criminology to give this genre any sort of theoretical or social organizational framework. Since then, criminology has developed a long-standing, yet spotty, tradition that links it to the subjective experiences of offenders. While not widely publicized or acclaimed until John Irwin became the first convict to openly use—to much critical acclaim—his criminal experiences to help him enter academe, there have been a number of convicts who became academic criminologists, including Gwynn

Nettler who once taught at the University of California, Santa Barbara. But it was not until the late 1990s that convicts-turned-academics had enough critical mass, energy, and determination to start developing what is now called "convict criminology."

According to two of its leading American contributors, this "perspective grew out of six interrelated movements" (Richards & Ross, 2001; Ross & Richards, 2003b). These include theoretical developments in criminology, writings in victimology and constitutive criminology, the failure of the prisoners' rights movement, the authenticity of insider perspectives, and the growing importance of ethnography. Central to convict criminologists' claims is the argument that while a multiplicity of radical and critical perspectives has developed in criminology during the last 40 years, they too "often have remained the intellectual products of the well meaning yet privileged, with only minimal reference and relevance to the victims of the criminal justice machine" (Ross & Richards, 2003b, p. 2).

Beginning formally in 1997 with a session at the annual meeting of the American Society of Criminology, convict criminology is a much welcomed development in the long-standing tradition of "prisoner biography" (Franklin, 1998). By 2003, convict criminology was self-declared as a new school within criminology. Among its basic tenets was the belief that convicts, ex-convicts, and enlightened academics who critique "existing literature, policies, and practices" would create a realistic paradigm "that promises to challenge the conventional research of the past" (Ross & Richards, 2003b, p. 6). They claim that their work is collaborative; inclusive of established and budding scholars; often mentor-based, with ex-convict criminologist and scholar John Irwin (1970, 1980, 1985; Irwin & Cressey, 1962) as an early leader; cognizant of the fact that criminologists commit crime and engage in deviant and unethical behavior; and issue oriented. These latter points sound very much like some of the central tenets of cultural criminology and its emphasis on the situated *meaning* of crime.

Its humanitarian orientation encompasses a kind of "back to basics" criminology, one that listens to the people on the receiving end of criminal justice while avoiding a criminology that is solely academic or academic–government driven. Thus far, it has not fallen into a "fantasy radical" position like some forms of abolitionism. It is much too realistic for that. Whereas academia has empowered some ex-cons, convict criminology has in turn given voice to prison workers close to the ground in prison administration and prison research. To its advantage, it is not sect-like because it listens to the relevant voices of non-convicts as evidenced by its leading anthology, *Convict Criminology* (Ross & Richards, 2003a).

CONSEQUENCES OF THE "NEW SCHOOL OF CONVICT CRIMINOLOGY"

Ex-con Alan Mobley (2003) perhaps said it best: "Criminology is a curious business, and it is not clear where, or if, the convict criminologist fits in" (p. 223). Although he was writing about convict criminologists, his observations have important implications for evaluating the current state of the "new school of convict criminology." It is not at all clear that convict criminology is doing anything new that has not been done sporadically during the past 45 years with the exception that about half of its current contributors openly embrace their identity as an "ex-con."

Few if any of the issues and policy recommendations that thus far fall under the umbrella of convict criminology are new to criminology and criminal justice, nor is its preference for ethnographic methodologies—the "insiders' perspective." Certainly, in recent years there has been a decline in old style prison ethnographies, perhaps because of prison secrecy and the dangerousness of prisons. It remains to be seen if this "bottom up, inside-out perspective" gives much of a "voice to the millions of men and women convicts and felons," and if, as claimed, it represents a paradigm shift within criminology (Richards & Ross, 2005, p. 235). Criminology, however, has a history of convicts-turned-academics who have used their criminal experiences as the basis for making significant contributions minus the claim that they know the truth or that they were unfairly incarcerated or oppressed. Whether an increasingly larger number of more recent convicts-turned-academics will be as successful remains to be seen.

Thus far, its contributors have not placed their writing within mainstream journals, but they have participated in numerous international, state, and local criminology and criminal justice conferences and published (mostly edited) books. It also remains to be seen if the messages and policy changes that convict criminology wishes to make will be successful, given that much of its literature thus far relies heavily on metaphorical/emotional language rather than on traditional positivist or scientific language—a feature shared and emphasized with cultural criminology (see Ferrell et al., 2004a). The result at this time is an odd mixture.

Richards's 1998 work (pp. 122–144) is illustrative but not necessarily representative. Titled "Critical and Radical Perspectives on Community Punishment: Lessons From the Darkness," the meanings of "lessons" and "darkness" are not directly addressed; the meaning of these words is only implied. Bold observations, though ringing with emotion and some sense of truth, risk the charge of being more provocative than informative. For example, Richards's (1998) claim that "[M]ost people are unable to comprehend and do not want to know about" the degradation of American convicts is asserted without citing a supporting literature, as is his claim that, "Prisons were built to destroy people"—save for a reference to murderer and radical chic celebrity Jack Henry Abbott's (1981) *In the Belly of the Beast: Letters From Prison* (p. 123). Equally provocative is the contention that ethnographic methodologies have special value because they have "penetrated and illuminated the life of real people" (Richards, 1998, p. 131). *Penetrated, illuminated,* and *real people* (as opposed to "unreal" people?) are presented without sufficient explanation. Broad generalizations such as these—as well as "convicts . . . have little interest in legal or social morality," to mention another—are major obstacles to clarity and broad-based acceptance and support (see Richards, 1998, p. 138). In this context, while the "convict story" from convicts' perspectives is interesting, it risks having more in common with some forms of journalism and novels/memoirs by ex-convicts, such as Edward Bunker's (2000; see his *Education of a Felon*), than with academic positivistic criminology.

Whether these potential weaknesses are indicative of all or most of the convict criminology to date—and in the future—is a question that cannot be answered here. Most clearly, this approach is struggling to negotiate a position of critical relevance within what Mobley (2003, p. 223) correctly called a curious—and, we would add, politically charged and sensitive—business. What remains to be determined is whether convict criminology will prove to be a truly new path of critical thinking that offers knowledge that is both unavailable to "outsiders" and generalizable or will become a small and insular paradigm within the discipline.

With these cautionary remarks in mind, it can be said that during its first decade, convict criminology has created what appears to be somewhat of a sustained presence within criminology and the media. At the 61st annual meeting of the American Society of Criminology, convict criminology had six thematic sessions on a variety of topics, including "voices from inside the prison-industrial complex," "the various uses of criminal justice fiction in the classroom," "the politics of punishment and corrections," "convicts, students, and college," "beyond bars, reentering the free world," and "the last convict, criminal careers, and reentry" (American Society of Criminology, 2009, pp. 70, 84, 146, 207, 280, 310). In the same year, two convict criminologists, Ross and Richards (2009), published a self-help book on convicts reentering society that earned media attention on National Public Radio ("'Behind Bars' Amid Bad Times," 2009). The group includes members from Australia, Canada, Finland, New Zealand, Sweden, the United Kingdom, and the United States.

Conclusion

Critical criminology stands outside mainstream criminology and outside the structures of power in society. As we have seen in Chapters 8 and 9, the many varieties of theories within this approach might lead us to conclude that it is better termed critical *criminologies*. That said, the central theme that informs these diverse perspectives is that official, legitimate, and hegemonic realities should not be taken for granted. They often mask self-interest, inequality, and inhumanity. When unraveled, it becomes possible to see that existing realities are not inevitable but are socially and politically constructed. It becomes possible, in short, to imagine different ways in which crime might be understood and addressed. It becomes possible to imagine a very different society.

Freda Adler
1933–
University of Pennsylvania
Author of Liberal Feminist Theory

10

The Gendering of Criminology

Feminist Theory

B etween the first edition of *Criminological Theory*, a collaboration initiated two decades ago, and the current edition, theorizing about crime has undergone major changes, none more dramatic and significant than what we review in this chapter. Our 1989 handling of feminists' perspectives was severely limited; only twice did we even mention "feminism." Feminist perspectives were then just beginning to develop beyond the seminal work of Adler (1975), Simon (1975), and Smart (1976). In the second edition, which appeared in 1995, we undertook a major renovation with regard to women and crime that was reflective of contextual and theoretical developments. Indeed, feminist theories were included as the first half of a new chapter, "New Directions in Critical Theory"—a plan we continued in the third edition, albeit much expanded. Since then, a number of important milestones in feminist criminology have occurred, not the least of which is the fact that the American Society of Criminology's Division on Women and Crime—with its global membership—celebrated its 20th anniversary. However, its membership declined approximately 25% between 2007 and early 2010. Although we are not at all certain why this occurred, we are nonetheless of the opinion that the interest in women and crime theory and research continues to be strong. This expanded chapter reflects these continued developments and explores the important contextual and theoretical advances that elucidate our current understanding of crime and criminology from various gendered perspectives.

The tour across the development of theories about gender and crime is lengthy and reaches into the somewhat distant past. We begin by examining early prefeminist theories. These works had the advantage of focusing on women as offenders—something most other scholars ignored—but they did so in a limited, if not sexist, way, locating the causes of crime in females' sexuality, biology, or pathology. A significant transition in criminology occurred when, in the midst of the second wave of the women's movement in the 1960s and 1970s, theories emerged that linked crime to gender roles, arguing that as women were liberated and gained equality, the nature and amount of crime would equalize with their male counterparts. Similar to the discipline as a whole, theories of female crime then moved farther to the political left—in a "critical" direction. In fact, at this time, there is—according to Daly (2010)—a good deal of affinity and crossover between feminist perspectives in criminology and other critical theories, including cultural criminology among others.

Most important, there was a growing recognition that crimes by and against women were shaped by the gender inequality inherent in patriarchy. In a significant development, scholars observed that gender is only one structure of inequality in society. Accordingly, they argued that a full understanding of female criminality necessitates theoretical and empirical investigations that examined how gender intersects with race and class. Another line of inquiry suggested that men's criminality also was fundamentally a product of gender—that crime was a form of displaying masculinity or of "doing gender." More recently, feminist scholars have explored different ways to insert gender into the criminological enterprise and thus have "gendered" the study of crime. This journey across feminist theory ends with a discussion of postmodernism and, more concretely, with a consideration of the role of feminist ideas in bringing gender into the policy domain.

Background

Before examining contemporary feminist perspectives in criminological theory, we first must recognize that feminism's roots rest in antiquity. Among other origins, feminism has been traced to the Roman Empire, where several women championed emancipatory issues. Cornelia, the "Romans' own 'favorite woman,'" led a reform movement of the plebeians against the patricians (Boulding, 1992, p. 8). Even after social unrest that saw the death of her two sons, Cornelia remained a voice for equality and became an international figure visited by men of affairs seeking her advice. A monument erected to her by the Romans indicates the reverence in which she was held. According to Boulding, Seneca and his mother Helvia both taught the equality of the sexes to their offspring and were promoters of women's emancipation (Boulding, 1992). Although their actions did not constitute a sea change ending the subordination of women and others, neither did emancipatory issues disappear throughout the ages.

The beginning of the first wave of the feminist perspective in the United States conventionally is located during the mid-19th century when the first women's rights convention was held at Seneca Falls, New York, in 1848 (Daly & Chesney-Lind, 1988, p. 497). From this meeting's "Declarations of Sentiments and Resolutions," it was resolved that "the history of mankind is the history of repeated injuries and usurpations on the part of man toward woman, having in direct object the establishment of an absolute tyranny over her" (Miles, 1989, p. 221).

This resolution and the first wave of the feminist perspective were themselves part of another emancipatory movement, the abolition of slavery. In fact, Miles (1989) argued that of "all the causes that fueled the fight for the rights of women, most important was the parallel struggle against the slavery of the southern states in America" (p. 242). This and the feminist struggle, however, reached beyond the United States to include some European countries. But it was from England during the 1840 World Anti-Slavery Congress that abolitionists "imparted their feminist vision to their American sisters" (p. 237). This was a major impetus for the 1848 Seneca Falls convention. Still, as important as the anti-slavery and women's movements were as mid-19th-century emancipatory developments, they were not the only two social struggles.

It is well to remember that the early to mid-19th century witnessed what some have termed the apex of one of the most profound and dramatic transformations in history—the industrial revolution. Begun in the mid-18th century in England, by the mid-19th century the industrial revolution, coupled with the rise of capitalism, had largely changed traditional family and village economies into factory production. One result was the near destruction of what previously had been a valued and necessary "household" partnership between spouses, their offspring, and extended households. Writing from Paris in the mid-19th century, the utopians Marx and Engels (1848/1992) argued in their *Communist Manifesto* that the bourgeoisie had

> pitilessly torn asunder the motley feudal ties that bound man to his "natural superiors," and . . . left remaining no other nexus between man and man than naked self-interest, than callous "cash payment." It . . . drowned the most heavenly ecstasies of religious fervor, of chivalrous enthusiasms, of philistine sentimentalism, in the icy water of egotistical calculation. It has resolved personal worth into exchange value and, in place of the numberless indefeasible chartered freedoms, has set up that single unconscionable freedom—free trade. In one word, for exploitation, veiled by religious and political illusions, it has substituted naked, shameless, direct, brutal exploitation. . . . [It] has reduced the family relation to a mere money relation. (pp. 20–21)

It is against this broader background of human struggle and exploitation that we now examine the feminist perspectives on criminology.

Prefeminist Pioneers and Themes

Until recent years, the criminality of women had a long history of neglect. In both the late 19th and early to mid-20th centuries, however, there was a small group of writings specifically concerned with women and crime. Throughout this early literature, there were different analytical approaches, yet there were many shared assumptions about the nature of women and the crimes they committed (Klein, 1973). These assumptions focused on crime as the result of *individual* physiological or psychological characteristics of women. It also was thought that these characteristics were *universal* to women and that they transcended any historical time frame. Central, too, was the assumption that there is an *inherent nature of women* (Klein, 1973).

The focus on crime as a result of individual characteristics rather than on conditions in the existing social structure generated theoretical and research attention directed toward determining the differences between criminal and noncriminal women. One result was the creation of two distinct classes of women: good women who are not criminal and bad women who are criminal (Klein, 1973). Built into this distinction was another important assumption: Crime resulted from individual choices. Therefore, women were conceptualized as freely choosing to act criminal or noncriminal, void of any influences from the social, economic, and political worlds.

CESARE LOMBROSO

Often called the "father of modern criminology" (see Chapter 2), Cesare Lombroso had a great interest in tracing an overall pattern of evolution in the human species. According to his arguments, evolution accounts for the uneven development of groups. For example, Whites were more advanced than non-Whites, men were more advanced than women, and adults were more advanced than children. In the early 20th century, Lombroso (1903/1920) published *The Female Offender,* in which he described female criminality as an inherent tendency of women who, in effect, had not developed properly into feminine women with moral refinements. He buttressed his argument with physiological evidence that he thought explained why female criminals were biological atavists. Included here were cranial and facial features, moles, height, dark hair, and skin color. Short, dark-haired women with moles and masculine cranial and facial features were good candidates for crime.

Lombroso went further than this, however, by arguing that women also were characterized by physiological immobility, psychological passivity, and amorality featuring a cold and calculating predisposition. Criminal women, then, were in fact more *masculine* than feminine; they could "think like a man," whereas "good women" could not. Criminal women also were thought to be "stronger" than men in some ways. Lombroso observed that criminal women could adjust more easily than men to mental and physical pain. As an example, he argued that criminal women often adjusted so well to prison life that it hardly affected them at all. For Lombroso, criminal women were abnormal.

W. I. THOMAS

A modified emphasis on the physiological explanations of female crime offered by Lombroso appeared in the work of W. I. Thomas. Born in an isolated region of Virginia and a graduate of the University of Tennessee, Thomas wrote his doctoral dissertation in sociology, *On a Difference in the Metabolism of the Sexes,* at the University of Chicago (where he taught his first sociology course in 1886). He wrote two influential books on sexual behavior and society: *Sex and Society* (Thomas, 1907) and *The Unadjusted Girl* (Thomas, 1923). Separated by 16 years between publications, the shift in emphasis in these books moved the discussions of women's behavior to more sophisticated theories that "embrace physiological, psychological, and social-structural factors" (Klein, 1973, p. 11).

In *Sex and Society*, Thomas (1907) began by offering the age-old dichotomy that men and women were fundamentally different. For Thomas, men were destructive of energy and what can flow from it, whereas women stored energy much like plants; women were more motionless and conservative than men. This difference, according to Thomas, had contributed to a relative decline in the structure of women, especially in "civilized" societies. Thomas was ambivalent, however, as to why the stature of women had declined. At one point, he attributed the decline to a lack of superior motor fitness on the part of women. At another point, he argued the decline could be explained by women's loss of sexual freedom. He claimed that women, under the development of monogamy, had to confine their sexual behavior to being wives and mothers and adjust to the fact that they were treated as property controlled by men.

Underlying these arguments of the inferior status of women was a focus primarily on *physiological* issues. Men, for example, had more sexual energy than did women. This allowed men to pursue women for sexual reasons and allowed women, in turn, to exchange sex for domesticity (Klein, 1973). In essence, monogamy and chastity became a form of accommodation to men's basic urges.

In *The Unadjusted Girl,* Thomas (1923) shifted his position on female criminality in two directions. First, he argued that female delinquency was normal under certain circumstances given certain "assumptions about the nature of women" (Klein, 1973, p. 14). Unfortunately, Thomas did not specify the nature of these assumptions.

Second and more important, however, was Thomas's shift in focus from punishment of criminals to *rehabilitation* and *prevention*. This point represented a radical departure from the Lombrosian biological position, which claimed that crime-prone individuals must be locked away or sterilized as a preventive strategy (Klein, 1973). One important implication that followed was Thomas's contention that there was no individual who could not be made to be socially useful (Thomas, 1923). For him, individuals could be socialized to prevent antisocial attitudes even if they had been poorly socialized in slum families or neighborhoods. The key to this strategy lay in how individuals were taught to define their situations. Thomas claimed, for example, that the way in which to prevent women dissatisfied with their conventional sexual roles from committing crime was to socialize them to change their attitudes. In other words, they needed to adjust to their situations by first redefining them as acceptable. This, he explained, was one of the important reasons why middle-class women committed so few crimes; they had been socialized to accept their positions and treasure their chastity as investments (Klein, 1973). Lower-class women, on the other hand, had not been socialized to suppress their need for security and instead committed crimes rather inadvertently out of a desire for excitement and new experiences (Klein, 1973). Sexual behavior in delinquent girls, Thomas maintained, was used as a means for realizing other wishes.

Despite Thomas's shift in focusing attention away from biological determinism and toward rehabilitation, he nevertheless relied on the ever-present dichotomy of the good and bad woman to explain female crime. Bad women exploited men for fulfillment of their desires; good women used sex as a protective measure against the future and uncertainty. Not unlike other men who had attempted to explain female crime, Thomas also underemphasized the importance of economic factors in favor of psychological characteristics.

SIGMUND FREUD

A similar emphasis is found in the work of Sigmund Freud. For him, the position of women was based on explicit biological assumptions about their nature. In essence, for Freud, "anatomy is destiny," and because he believed women's anatomy was inferior to men's anatomy, it was appropriate that women were destined to occupy an inferior social status, including being mothers and wives (Klein, 1973, p. 16). The specific anatomical characteristics that Freud considered were men's and women's sex organs. According to Freud, the inferiority of women's sex organs was recognized universally. Beginning during childhood, for example, male and female children were aware of this distinction, and girls grew up assuming that they had lost their penises as a form of punishment. One central consequence, according to Freud, was that girls developed penis envy and became revengeful, whereas boys came to dread their envy and vengeance.

An additional difference between men and women based on women's anatomical inferiority, according to Freud, was the fact that women also developed an inferiority complex and tried to compensate for it by being exhibitionistic, narcissistic, and well dressed (Klein, 1973). According to Freud, this concern with personal matters helped to explain why women had little sense of justice, scant broad social concerns, and few significant contributions to building civilization. Because they could not sublimate their individual needs, they were concerned with marginal matters. On the other hand, men, as builders of civilization, recognized that individual urges, especially the sex drive, must be repressed to get on with worldly affairs. Otherwise, little of lasting value would be accomplished or produced. Therefore, men were rational, and women were irrational.

In this framework, "the deviant woman is one who is attempting to be a *man*" (Klein, 1973, p. 17, emphasis in original). Female aggression and rebellion, for example, were expressions of longing for a penis, and if not "treated," women would only end up "neurotic." Freudian-based treatment for women, therefore, was intended to have them adjust to appropriate sex roles. Again, as we have seen, the emphasis was placed on changing women to fit into society as defined by men rather than on changing society. To be normal, women had to adjust to and accommodate the glorified duties of wives and mothers at the expense of gender equality. As Klein (1973) observed, one clear implication of Freud's logic is a class bias: "Only upper and middle class women could possibly enjoy lives as sheltered darlings" (p. 18). The lives of poor and Third World women are not so fortunate.

Another important implication is that Freudianism has had a powerful influence on transforming a gender and sexual ideology of proper female behavior and sexuality into a *scientific* framework. As such, Freudianism has been used for decades to maintain female sexual repression, sexual passivity, and the "woman's place" in the nuclear family. Although a very controversial and often discredited theorist, his work is now experiencing a renewed popularity (Adler, 2006, pp. 43–49; Kalb, 2006, pp. 50–51). His early legacy influenced the writing of many scholars including Pollak.

OTTO POLLAK

For Otto Pollak, one of the most influential post–World War II scholars in the field of female crime, female involvement in crime, compared to male involvement, was largely "hidden" from public view (Klein, 1973, p. 21; Pollak, 1950). Pollak advanced this theory based on the idea that women were inherently deceitful because of physiological reasons. Pollak reasoned that because men, unlike women, must achieve erections to perform sex acts, they could not hide their emotions or deny their failure to perform sexually. Women's *physiological* nature, on the other hand, permitted them to hide their emotional involvement in sex to a degree. Therefore, Pollak suggested that women were innately deceitful. When combined with the domestic opportunities they had as maids, nurses, teachers, and homemakers, this deceitful nature permitted them to commit undetectable crimes (Klein, 1973).

Women also were vengeful, according to Pollak, especially during their menstrual periods. During this time, women once again recognized that their "anatomy is destiny" and that any desire to be men was doomed. False accusations, for example, were typical female crimes because they were an outgrowth of their nature and treachery. Shoplifting also was a special type of female crime reflecting the mental disease of kleptomania.

A final factor that Pollak advanced for explaining female hidden crime was chivalry in the criminal justice system. His argument was that although there was no major discrepancy between crime rates for men and those for women, women were treated differentially by the law, thereby keeping the rates of their crimes hidden (Klein, 1973; Pollak, 1950). He failed to consider that female criminals often were poor or were women who had stepped outside of chauvinistic, classist, and racist definitions of women's proper roles.

Unfortunately, Pollak, like many other early theorists, failed to see that, in some instances, female crime could be explained by economic necessity. Instead, psychologically based and physiologically based sexual motives or mental illness often were invoked to explain female economic crimes. Only during the past few decades has this logic been challenged.

The Emergence of New Questions: Bringing Women In

In 1961, criminologist Walter C. Reckless questioned whether any theory of delinquency would be accepted if a criminologist paused to consider whether it applied to women (Reckless, 1961). It was nearly a decade later, however, before Bernard (1969) and Heidensohn (1968) drew attention to the "omission of women from general theories of crime" and "signaled an awakening of criminology from its androcentric slumber" (Daly & Chesney-Lind, 1988, p. 507). Indeed, in hindsight, the discovery of *women* as a conceptual term that could be incorporated into criminological theory has been called "quite literally, pioneering" (Young, 1992b, p. 289).

Two issues were critical at this stage of the development of feminist perspectives in criminology. The first issue was the uncertainty identified by Reckless of whether general theories of crime generated by men to explain crime by men and boys could be applied or generalized to women and girls. The second issue involved social structure and categories of risk. The question here was whether class, race, and age structures found to be the core of criminological theory for males also held for gender. This issue addressed the gender-based ratio question of why women commit less crime than men (Daly & Chesney-Lind, 1988).

But early feminist writing in criminology did not always focus on these problems. Instead, it centered on what Gelsthorpe (1988) called criminologists' "amnesia" of women (p. 98). What followed was a number of critiques on the intellectual and institutional sexism found in explanations of female crime and criminal justice systems (Daly & Chesney-Lind, 1988). Some feminist writings focused on how women were represented or misrepresented in conventional criminological literature. This work, essentially a critique of "accumulated wisdom" about female offenders (in Britain, see Campbell, 1981; Smart, 1976), was of immense importance for demonstrating that theories of criminality developed by and "validated on men had limited relevance for explaining women's crime" (Gelsthorpe, 1988, p. 98). Examples of such work abounded at that time and were easily found in leading criminology texts.

For many generations of American criminologists, no better example existed than the rabbit hole through which so many of them "fell" into the study of crime and justice, Sutherland and Cressey's (1970) *Criminology*. Of the many introductory textbooks on criminology published in the United States so far, none had been "born again" in as many editions as this book (Lilly & Jeffrey, 1979). Even its 11th edition, published in 1992, contained no discussion of a feminist perspective per se on crime. Although recognizing that recent research indicated a narrowing of the sex ratio between men and women for some reported crimes, this relationship was explained partially as a result of modifications in law enforcement practices that witnessed changes in the "somewhat chivalrous treatment they [women] traditionally received" (Sutherland, Cressey, & Luckenbill, 1992, p. 162).

This type of explanation of female crime, although perhaps having some empirical basis, pointed to a wider and more serious shortcoming: Sutherland and other early criminologists had little or no understanding of the social worlds of women and girls. It is not surprising, therefore, that historical explanations of female crime and deviance often focused more on biological forces than on social or economic ones (Daly & Chesney-Lind, 1988; Heidensohn, 1985; Morris, 1987). Nor is it surprising that examinations of juvenile and criminal justice system responses to girls' misbehavior found that they were more concerned with girls' proper sexual behavior than with the sexual behavior of boys (Chesney-Lind, 1973). In fact, there was very little juvenile justice system concern about boys' sexual behavior. This suggested a sexist ideology reflecting "a set of ideas about the place of women in the social order that emerged in the . . . nineteenth century. . . . It . . . placed men in the public sphere . . . and women in the private sphere" (Daly & Chesney-Lind, 1988, p. 509). Here men were found in the paid workforce, politics, and law, whereas women were located as the moral guardians of the household and family life.

The first wave of feminism ended in the United States in 1920 with the ratification of the Nineteenth Amendment to the U.S. Constitution, giving nationwide suffrage to women. Second-wave feminism denounced the domestic or private "sphere as oppressive to women and sought to achieve equality with men in the public sphere" (Daly & Chesney-Lind, 1988, p. 509). This ideological shift contributed to the development of a number of feminist critiques of criminology and to additional questions about equality being raised by feminists.

The Second Wave:
From Women's Emancipation to Patriarchy

WOMEN'S EMANCIPATION AND CRIME

The late 1960s and early 1970s saw the beginning of renewed emphasis on women's issues, a development that is known as the second wave of the women's movement. Born in part out of questions about social, political, and economic equality with men, it has had a profound impact on the nation's social agenda, especially on how women and crime were examined (see Freidan, 1963). Prior to this development, criminologists often reported that women simply committed fewer and different crimes than men. Subsequent to an increased participation of women in the workforce, new explanations of female crime were developed.

During the mid-1970s, two controversial books—Adler's (1975) *Sisters in Crime* and Simon's (1975) *Women and Crime*—proposed ideas about women's criminality based on analyses of female arrest trends of the 1960s and early 1970s. The two books reached somewhat different conclusions, but both were largely "an outgrowth of the unexamined assumption that the emancipation of women resided solely in achieving legal and social equality with men in the public sphere" (Daly & Chesney-Lind, 1988, p. 510). Adler (1975) argued that lifting restrictions on women's opportunities in the marketplace gave them the chance to be as greedy, violent, and crime prone as men. Simon (1975) read the evidence and concluded that women's increasing share of arrests for property crime (she found no increases for violent crime) might be explained by their increased opportunities in the workplace to commit crime. Simon also wondered whether the emancipation of women might encourage law enforcement and courts to be more interested in treating men and women the same.

Both books attracted critical attention, and several scholars called into question many of the issues raised by both Adler and Simon. Steffensmeier (1978, 1980), for example, found that research contradicted the view that women were committing more masculine violent crimes. He reported that arrest rates for women were in fact increasing but that the gap between male and female crime rates was not closing. Rather, because female arrest rates historically had been so much lower than those of males, any small increase in the *absolute* number of female arrests showed up as a *relatively* large percentage increase. But Steffensmeier (1980) did find that there were rate increases for women in the larceny–theft category, a type of crime in which women always have been found in great numbers.

Steffensmeier (1981) also disputed the claim that changes in the occupational structure generated increases in women's white-collar crime. He argued that Simon was wrong to classify larceny, fraud, forgery, and embezzlement by women as traditional white-collar or occupational crime. Steffensmeier argued that women were not committing stock fraud or embezzling large sums of money. Instead, they were being arrested for shoplifting and writing bad checks. For Steffensmeier, these crimes did not fit the white-collar criminal image presented by Simon. Overall, Steffensmeier and Cobb (1981) concluded that any increase in female arrest rates matched the daily activities of women, such as shopping and paying family bills, and not those associated with occupational positions.

This raised an important question: If increased occupational opportunities do not explain increased female crime, then what does? Two suggestions were offered by Steffensmeier. First, there were greater opportunities than in the past for women to commit petty theft and fraud because of a self-service marketplace. With this type of market comes a greater availability of credit for women, which in turn provides new opportunities for old crimes—shoplifting, passing bad checks, and credit card fraud. These new opportunities also are associated with increased security and detection procedures that improve the arrest rates for female offenders. Second, Steffensmeier and Cobb (1981) provided data indicating that law enforcement and court attitudes toward female offenders are changing and that now there is a greater willingness to arrest and prosecute women.

But Steffensmeier and his colleagues were not without their critics. Giordano, Kerbel, and Dudley (1981) argued that women's roles might be changing more gradually than can be measured in the relatively short time periods that Steffensmeier examined. Giordano et al. analyzed police records from 1890 to 1976 and found that female crime patterns differed greatly from those at the turn of the 20th century. During the early years of the 20th century, high percentages of female arrests were for prostitution, in marked contrast to the reasons for women being arrested during the late 20th century.

Herein was a serious problem, as critics noted: the failure to examine whether the trends that early research on female crime associated with the women's movement actually were occurring (Daly & Chesney-Lind, 1988). Were low-income women actually seeking equality with their male counterparts in the public sphere, as Adler argued? Were female arrests for property crime the result of opportunities, as Simon claimed? Or was their reading of the arrest data hampered by a liberal feminist perspective on gender that "ignores class and race differences among women and defines gender either as the possession of masculine or feminine attitudes or as role differences between men and women" (p. 511)? Both Adler (1975) and Simon (1975) ignored the impact of power relations in a patriarchy where the social structure allows men to exercise control over women's labor and sexuality. By focusing on the public sphere's aspects of equality and the opportunities of women compared to those of men, these scholars failed to consider one of the more important features of the dominant social structure—patriarchy.

PATRIARCHY AND CRIME

The "emancipation thesis" had great value in focusing attention on female crime, but it carried insights on female criminality only so far. Following Adler's and Simon's pioneering work, the next generation of feminist criminologists went from focusing on emancipation to patriarchy. As Akers noted in 1994, the major theme in feminist theories then was "the pervasiveness of male dominance in patriarchal society and its impact on crimes committed both by and against women" (p. 175). While this theoretical shift was a departure from the earlier focus on women's liberation, it was not a total departure from criminological theories that focused on power, such as conflict and Marxist theories. The difference was found in what "*type* of power is placed at the center" (p. 175, emphasis in original). Marxist theories of crime, for example, focused on the power of the ruling class, and most conflict theories focused on the conflict between various powerful and powerless groups. For much of feminist theory at this time the focus was on men's power over women.

This emphasis fit nicely with feminists' contributions to our understanding of the nature of the crimes women committed as well as some of the crimes women experienced at the hands of men. The emphasis on power differences between men and women—it was argued—led women into "powerless" types of crime such as prostitution and small-scale fraud (Messerschmidt, 1986). These crimes bring little reward to women because they are marginalized by economic destitution. As such, some female crime is a manifestation of, and helps to reproduce, sexual stratification in society (see also Cloward & Piven, 1979).

Rape, other forms of sexual abuse including date and marital rape, and domestic violence were all explained by patriarchal dominance. Again, these crimes by men— and, therefore, the victimization of women—reflected the ability of men to use their power against women. Unfortunately, relatively little research has tested the notion that patriarchy explains female crime. A major problem here is that patriarchy is difficult to measure as an independent variable, so that its explanatory value can be determined in different settings. Until this occurs, progress in assessing patriarchy theory will be slow. Fortunately for the development of feminist perspectives on crime, attention was given not only to the patriarchal structure of society but to the idea that gender is a social construction rather than merely biological sex. In retrospect it is clear that this shift in emphasis was a major development for feminist perspectives.

Varieties of Feminist Thought

EARLY FEMINIST PERSPECTIVES

There are now several strands of theorizing and research comprising feminist perspectives that can be classified as "early" and "contemporary" feminist perspectives on crime. *Liberal feminism,* not surprisingly, has its foundations in 18th-century and 19th-century ideas of liberty and equality (Jaggar & Rothenberg, 1984). It emphasizes gender socialization as the cause of crime. Male dominance and female subordination are reflections of how each gender is taught to behave socially and culturally. According to this perspective, official crime data show that, for example, men commit more aggressive offenses than do women. Each gender, it is argued, commits crime consistent with role expectations, an important point that was developed more fully later with work that explored the complexities of the intersection of gender, race, class and age (see the next section: *Contemporary Feminist Perspectives*).

Marxist feminists adhere to the idea that the class and gender division of labor combine to determine the social position of men and women. The gender division of labor, however, is viewed as the result of the class division of labor, which is dominated by men. Masculine dominance of women, therefore, is not just an expression of sexism. The criminal victimization of women and the crimes they commit result from the mode of production under capitalism. From this perspective, women's labor in the home and in the marketplace creates profit for capitalists.

Radical feminism, by comparison, sees crime as part of the biological fact that men are born to be aggressive and dominant. Thus, crime is an expression—but not the only one—of men's need to control. Other expressions of this need are sexual slavery, imperialism, rape, and forcing women into motherhood. Rape, according to Brownmiller (1975), is nothing more than an attempt by all men to keep all women in a state of fear (Lilly, 2003, 2004, 2007; Lilly & Marshall, 2000).

Not entirely unlike Marxist feminism, radical feminism argues that women are first subordinated by men into a sexual division of labor that originates in procreation and child care and is "extended into every area of life" (Jaggar, 1983, p. 249). It is argued that various institutions—including the state, employment, and the family—reflect and reinforce this pattern of male dominance. Women's cultures and self-concepts also reflect male dominance and contribute to women's servile status.

Socialist feminism is unique because it attempts to merge Marxist and radical feminism by examining the various connections between patriarchy and capitalism that lead men to crime and women to subordination. In a major statement from this perspective, Messerschmidt (1986) contended that crime results from capitalist exploitation of workers. But men and women have different positions of power in relations of production, with men being more powerful. This gives men more opportunity to commit crime and, at the same time, keeps women relatively subordinate. Thus, women are given less opportunity to benefit from either legitimate or noncrime opportunities. Recent research on women's status and risk of homicide victimization lends support to Marxist and social feminist arguments. Vieraitis, Kovandzic, and Britto (2008) examined the relationship between female homicide rates and various measures of women's status and gender inequality. They used data from cities with populations greater than 100,000 in 2000. They found support for the idea that higher socioeconomic status gives women greater protection from lethal violence.

Unlike some of the other feminist perspectives on crime, social feminism argues that human behavior is shaped more by social forces than by "pre-social givens" (Jaggar, 1983, p. 304). Rather, human behavior is socially constructed and alterable, a theme that is central to more recent and complex gendered explanations of crime.

CONTEMPORARY FEMINIST PERSPECTIVES

Each of the above perspectives has shortcomings that have contributed to the development of additional perspectives, or morphed into an alternative theoretical formation. While not completely separated or divorced from the "traditional" or early feminist perspectives, these efforts have placed gender—especially women—more centrally into their inquiries. They seek to understand crime and gender and how both intersect with race, class, and age.

The Intersection of Race, Class, and Gender

As important as many early contributions were for initial feminist perspectives, they tended to share a common limitation: They implicitly treated women as a monolithic or homogenous unit of analysis. However, scholars began to understand that women did not simply possess "gender roles" and exist in the single structure of inequality of patriarchy. Instead, their structural location and collateral experiences were seen as diverse and potentially complicated. As a result, in moving beyond an exclusive focus on gender, feminist scholars began to argue for the importance of theories and investigations that explore how crime is shaped by the intersection of race, class, and gender. In this regard, several important contributions in criminology merit consideration.

Writing in 1991, Sally Simpson argued that for a decade criminological research had targeted gender as an important indicator of criminal participation and persistence. But as important and insightful as this research had been for understanding gender and crime, Simpson (1991) noted that too often it had focused on contrasts between the criminality of males and that of females. More specifically, she argued that the past "research on gender and crime supported an incorrect portrait of violent criminality as primarily a lower-class phenomenon . . . disproportionately enacted by young males" (p. 115). This picture was only partially accurate, according to Simpson, because it did not address the complex interactive effect of gender, race, and class.

After first acknowledging the difficulty of addressing these interactive effects by relying on official statistics, surveys, and self-reported instruments, Simpson illustrated that African American females indeed have higher rates of homicide and aggravated assault than do White females. Sometimes Simpson (1991) found that for certain types of personal crime victimization, African American female rates "for adults and juveniles are more similar to those for white males than those for white females" (p. 117). In addition, among juveniles, African American females were reported to be consistently more involved in assaultive crimes than were White females. Based on this and other research, Simpson concluded that gender alone does not account for variation in criminal violence.

Simpson also considered class and its effect on African American female criminality. By extrapolating on the well-established sociological truism that class often is related to violent crime, Simpson (1991) illustrated that the increasing marginalization of underclass African Americans was correlated with high levels of criminal violence (p. 118). When combined with changes in divorce laws, occupational segregation with low pay for women, and the rise of single-parent mothers living in poverty, this contributes further to understanding linkages between African American females and crime. Simpson offered additional clarity about class, race, and crime with her discussion of violence and the underclass, a term used to describe the bottom of the lower class. Whereas the lower class is disproportionately female and African American and, therefore, is relatively heterogeneous, the underclass is racially more homogeneous; it is primarily African American and young. "Its geographical terrain is center-city urban" (p. 119).

Whether caused by relative economic deprivation, absolute poverty, or some interaction of class with race and urbanism, violent crime rates are highest in underclass communities. These are urban communities that are disproportionately African American (Simpson, 1991, p. 119). Female-headed households in these communities are related to increases in juvenile and adult robbery offending for "both blacks and whites, but they have a greater effect on black homicide rates" (p. 119). Furthermore, the labor marginality of African American males has an impact on African American women and children. It can be argued, for example, that one consequence of African American male marginality is that African American females have extensive and shifting domestic networks composed of kin, non-kin, and pseudo-kin. Because the parents and guardians in these networks experience severe limitations on the social control they have over their children, African American females often are susceptible to criminal recruitment.

A similar line of reasoning, yet one that was not specifically focused on African American females, is found in Ogle, Maier-Katkin, and Bernard's (1995) theory on homicidal behavior among women. In their lead article in an issue of *Criminology,* these authors started with research findings indicating that the patterns of homicides by women are different from those by men. The differences between male and female homicides are quite striking. For example, homicide is overwhelmingly a male crime. According to Federal Bureau of Investigation statistics, in 1998 89.4% of the people arrested for homicide were males. But homicide is not exclusively a male crime. Women do commit homicide on occasion. As these data indicate, women were arrested for 10.6% of the homicides in 1998.

These differences suggest the need for a separate theoretical explanation of female homicidal behavior, one that recognizes and incorporates important structural, social, and cultural gender differences between males and females. In an effort to do this, the theoretical explanation offered by Ogle et al. (1995) reformulated three existing theories of criminal behavior and included an emphasis on situational stresses that women experience differently from males. For example, women tend to view themselves in the context of traditional sex roles and to be socially conservative. Women also perceive themselves as under "extreme life pressures that appear in many forms, especially depression" (p. 173). With these and other gender differences considered, Ogle et al. constructed a social psychological theory of homicidal behavior among females.

In general terms, Ogle et al. (1995) argued that men who kill do so out of a need to control a situation. Women who kill, on the other hand, tend to do so because they have lost control over themselves. For example, about 80% of the homicides by women involve killing intimates, especially in "long-term abusive relationships . . . and in pre- or postpartum periods" (pp. 173–175). Homicides by females generally occur in homes and often are spontaneous rather than planned.

Ogle et al. (1995) offered an empirically testable theory for various types of homicides by women. It focuses on structural, social, and cultural conditions, which generate strain for all women, which in turn produces negative affect. Women tend to internalize negative affect as guilt and hurt, unlike men who externalize it as anger directed at targets. For women, this results in a situation analogous to an over-controlled personality and results in overall low rates of deviance and crime. On occasion, however, extreme violence including homicide occurs, especially in long-term abusive relationships and pre- or postpartum environments. This theory is an important contribution to the feminist perspectives on crime because it specifically focuses on gender experiences unique to women.

Richie (1996) pursued a similar line of research on crime among some African American women. By focusing on the intersection of race, gender, class, and domestic violence, she argued that these women were essentially compelled into crime by their social circumstances. Her hypothesis was that their patterns of offending reflected economic marginalization, culturally constructed gendered roles for African American women, and their experiences with interpersonal violence.

A remarkable example of research along these lines is found in award-winning scholar Jody Miller's *Getting Played* (2008). This work employed a gendered, ecologically oriented theoretical framework for a comparative (girls and boys) examination of African American female youths' victimization in the slums of St. Louis and how this victimization is embedded in their everyday life. Miller relied extensively on paid interviews that provided a rich trove of personal accounts of the widespread occurrence of victimization—from mild forms of sexual harassment to gang rapes ("running trains"). She reports some of the most interesting and important findings on victimization to come out of research based on the intersection of race, class, gender, and—with her addition—on *place*. Although some of the findings tell us things we already know—for example, that victims and offenders often know each other—other results are more nuanced and complex.

Thus, if the incidents occurred in situational contexts of public space such as schools, parties, or often afterwards, the victims experience what Ruth Peterson (2008) calls an example "of the substantial isolation of poor African American girls, despite the fact that much of victimization occurs in social and public settings" (p. x). Witnesses who could have intervened, for example, did not because they thought it was a "private matter," that "she deserved it," or that they would experience retaliation. They feared being the target of comments made not only by friends and peers but also by adult men, including police officers (Miller, 2008, pp. 552–557). In fact, the girls had greater apprehension about sexual harassment from adult men than male peers.

Miller's research also examined the victims' and boys' own interpretations of the victimizations. In their own voices, she found that both girls and boys viewed the victimization as problems of individual character, not the result of the structural and situational context that they shared. Males used gender-based stereotypes such as "she was asking for it," "she wanted it," "her clothes," and "way she carried herself." Boys' behavior was also explained by stereotypes such as "boys will be boys" (Miller, 2008, pp. 32–66).

One of the consequences of such interpretations—a form of victim-blaming—is that victimized girls resort to various situational self-help strategies that actually increase the danger for them. Whereas changing clothing styles, walking with friends, avoiding being out at night, and other tactics do not attack the underlying structural cause of sexual victimization, they could—according to Miller—lead nonetheless to the continuation of victimization and degradation.

Miller's work is more than an important contribution to the understanding of victimization of African American girls in poor and disadvantaged neighborhoods. It also demonstrates that the dynamics of the harassment and violence against women are intrinsic to the gender stereotypes and gender inequality that are found in males' domination and control of public spaces. In other words, the social organization of public space in disadvantaged neighborhoods is highly gendered. Miller's final chapter contains policy recommendations on how to challenge street masculinities and bridge the gender divide. If implemented they would greatly reduce—perhaps even eliminate— the structural factors that cause girls to get played.

In another study, also based on data collected in St. Louis, Missouri, Mullins and Miller (2008) examined the temporal, situational, and interactional features of women's violent conflicts. Unlike much of the previous research that has focused on the immediacy of affronts that resulted in violence, Mullins and Miller's findings are different. They found that women's conflicts are produced by a long series of interactional sequences that are embedded in "broader macro- and meso-social contexts" (p. 58). Participants in violence took a longer view of the situation, sometimes trying to avoid conflict altogether or deferring it to a more opportune time. In other words, Mullins and Miller (2008) found that the evolution of violence was much more complex and nuanced than had previously been thought.

In an earlier work, Zhang, Chin, and Miller (2007) examined how organizational context and market demand shaped the extent and nature of how women were involved in Chinese transnational human smuggling. By using snowball sampling they were able to draw on interviews with 129 human smugglers, 106 of whom were men and 23 of whom were women. The research focused on the internal logic of an organized criminal enterprise and found that its strategies were gendered. According to Zhang et al. (2007, p. 715), human smuggling—where the recruiters can be men or women who are known as "snakeheads"—can be and often is viewed as an altruistic community service. This allows women to be involved in the illicit business while remaining within the normative bounds associated with being female caregivers. In fact, 53% of the women compared to 38% of the men indicated that helping others "was a facet of their motivation" (Zhang et al., 2007, p. 712). Additionally, because arranged marriages represented the safest way to smuggle humans, women were more likely to plan them than men.

The handling of documents, for example, was also gendered. The percentage of women smugglers who specialized in this activity was 21.7; for men, it was slightly more than half of this number at 11.3%. Men, on the other hand, dominated maritime operations and conducted business that at times involved corrupt governmental official (Zhang et al., 2007, p. 715). The gendered activities examined in this study of crime reflected the gendered patterns of inequalities in the larger social environment.

Today the interest in the intersection of race, class, and gender in criminology is still very strong. However, there has been a significant shift in the development of gender-centered analysis of crime. By 2006, Miller and Mullins (2006) could say that theories of gender were the major starting point in criminological analysis of women and crime. The same scholars at that time were of the opinion that this emphasis represented the best and most promising directions for the future of feminist criminology as "well as broader criminological thoughts" (Miller & Mullins, 2006, p. 218). Central to this promise then and now was the prospect of the development of theories that help to explain the "gender gap" and thus become more generalizable across gender.

Developments in this direction benefited from the insightful work of Daly and Chesney-Lind (1988) on what they consider the distinctive features of feminist theory. In the late 1980s, these included:

1. Gender is not a natural fact but a complex social, historical, and cultural product; it is related to, but not simply derived from, biological sex differences and reproductive capacities.

2. Gender and gender relations order social life and social institutions in fundamental ways.

3. Gender relations and constructs of masculinity and femininity are not symmetrical but are based on an organizing principle of men's superiority and social and political-economic dominance over women.

4. Systems of knowledge reflect men's views of the natural and social worlds; the production of knowledge is gendered.

5. Women should be at the center of intellectual inquiry, not peripheral, invisible, or appendages of men.

In a collective way, these points came to be recognized as key elements that distinguished feminist perspectives in criminology from conventional criminology or what has also been termed malestream criminology (see the previous chapter). According to Daly (2010) this list—or ones similar to it—were based on the received feminist wisdom of the 1970s that sex was a biological category from the "natural" world. Gender—it was thought then and for many theorists today—was a social category from the realm of the "cultural." This thinking was a major breakthrough because it let "gender take off" as a liberating force for understanding relations, differences, and identities. However, using the concepts of "sex" and "gender" was—despite its promise and success—undeniably an example of reducing complex matters to an either/or, dualistic, binary categorical format.

In the early 1990s, according to Daly (2010), this conventional wisdom began to change with "a seismic shift in feminist thought" as scholars began to rethink "the body." The new thinking focused on the relationship between sex and gender, and it focused in part on the idea that sex, rather than being a pre-social biological concept, was in fact socially and discursively constructed. This meant that body parts such as the genitals are given "a particular social significance in language and culture" that "are not the same across time and culture, but variable" (Daly, 2010, p. 231). One implication of this approach is that thinking about sex and gender dualistically can give way to new conceptualizations. These may recognize that the linguistic interplay that exists in the brain between the two may mean, in actual practice, that they are indistinguishable from one another. In other words, sex and gender may actually be "incorporated" or fused together in ways that make them indivisible except as linguistic constructs. Another implication is that new concepts needed to be developed to capture this idea. Concepts such as "embodied subjectivity," "physical corporeality" (Grosz, 1994, p. 22), and "embodied experience" (Lacey, 1997, p. 74) are helpful examples that might be integrated into feminists' thinking on crime (quoted in Daly, 2010, p. 231).

However intellectually provocative the idea of the incorporation of sex and gender into what is symbolically represented as "sex/gender," Daly observes that "to date this shift has had a minimal impact on the feminist work in criminology" (Daly, 2010, p. 231). She believes, however, that this will change. Until then, we will focus on how gender has been employed to advance our understanding of crime.

Masculinities and Crime: Doing Gender

Similar to other feminist scholars, including Daly and Chesney-Lind (1988), James Messerschmidt (1993) believed that traditional criminological theories provide an incomplete understanding of crime because they omit gender from their analysis. While sympathetic to patriarchal explanations, he also contended that they, too, are incomplete. Although gender is not ignored in the patriarchal perspective, in a sense, he argued, men are. There is a distinct tendency to place men in a single category where "women are good, men are bad, plain and simple. And it is this essential badness that leads to patriarchy and violence against women" (p. 43). But this tendency to see men in a unidimensional and stereotypical way ignores both how masculinity actually is linked to crime and, equally important, how various types of masculinity are related to different types of offending.

Messerschmidt (1993) began his analysis with the observation that males are socialized into a "hegemonic masculinity." This dominant cultural script means that males define or achieve their masculinity through "work in the paid-labor market, the subordination of women, hetero-sexism, and . . . driven and uncontrollable sexuality" (p. 82). It further involves "practices toward authority, control, competitive individualism, independence, aggressiveness, and capacity for violence" (p. 82). In their lives and in any situation, men must constantly "accomplish" or demonstrate their masculinity in ways that are consistent with this cultural script. They may achieve this goal in conventional ways such as through success in sports, school, and employment. But if this goal is blocked—if legitimate avenues to accomplishing masculinity are not available—then males must find other ways of showing their masculinity. This focus on goal blockage is similar to strain theory, and for both theories some adaptation to, or way around, this frustration must be made.

Messerschmidt indicated that crime is a central method—a critical resource—that males experiencing goal blockages use to "do gender" and announce their masculinity. In the face of emasculation, then, men can employ crime as a way of showing others that they "have guts," are fearless, or are "real men." Importantly, Messerschmidt argued that the extent to which masculinity is challenged and the responses to these challenges differ by race and class. Thus, different "masculinities"—tied to the structural locations—emerge and have varying impacts on the content of criminal behavior.

With regard to delinquency, for example, White middle-class boys are able to achieve masculinity through success in sports and in school, all the while knowing that professional work lies in their future. As a result, they accommodate to the emasculating experiences of school life. Once outside the classroom, however, they show their masculinity through nonviolent crimes (e.g., vandalism) or binge drinking. White working-class boys are less likely to achieve success in school, defining it instead as "sissy stuff." Accordingly, they manifest oppositional conduct in school such as pranks and other mischief. Outside the classroom they "do gender" through theft, fighting, or perhaps hate crimes. Finally, racial minority lower-class and working-class boys are likely to find school boring, unrelated to their future lives, and humiliating. In response, they evidence oppositional behavior that may involve physical violence. Outside the classroom, they take to the street, where they demonstrate masculinity through gang violence, robbery, and crimes in which victims are dominated and humiliated.

Messerschmidt (1993) also indicated that his perspective can account for a range of adult crimes, including the physical assault of women. In his view, wife beating is a "resource for affirming 'maleness'" (p. 149). He predicted that such battering will be more prevalent among men who are in economically precarious positions, whether in the working class or unemployed. In his view, these men "lack traditional resources for constructing their masculinity and, as a result, are more likely than are middle-class men to forge a particular type of masculinity that centers on ultimate control of the domestic setting through the use of violence" (p. 149). That is, the inability to prove masculinity in the public realm makes the demonstration of masculinity in the home all the more salient. In this sense, battering "serves as a suitable resource for simultaneously accomplishing gender and affirming patriarchal masculinity" (p. 150).

Today we conclude that Messerschmidt's 1993 work was important because it forced scholars to think more carefully about the features of maleness that may be implicated in crime causation and about how the intersection of race, class, and gender shape the gender-specific problems men face and how men respond to them. Graham and Wells (2003) and Mullins et al. (2004), for instance, found that "most men's interpersonal disputes with other men were grounded in their need to build and maintain gendered reputations" (Miller & Mullins, 2006, p. 238).

Additional advances in gendered perspectives are found in the empirical tests and theoretical work that has followed. Daly's (1998) work on *gendered pathways to lawbreaking, gendered crime,* and *gendered lives*—while itself a tour de force in feminist criminological thinking—has contributed significantly to recent and more complex advancements in feminist criminology that have gone beyond Messerschmidt's focus on masculinities (Miller & Mullins, 2006). We turn to these developments next.

Gendering Criminology

GENDERED PATHWAYS TO LAWBREAKING

From the early 1990s, feminist scholars have examined what is now referred to as "gendered pathways," an approach to explaining crime that is similar to life-course analysis (Miller & Mullins, 2006). Here, females' experiences are mapped to explore what led them to crime as well as desistence from it. Research found that young girls who have run away from home in response to neglect and abuse often move on the street into homelessness, unemployment, drug use, and survival sex (Chesney-Lind & Pasko, 2004; Gilfus, 1992; Miller & Mullins, 2006). An exclusive emphasis, however, on victimization as the key pathway to offending can minimize or overlook other indicators of gender inequality, including racial and economic marginality, school experiences, and drug and alcohol use.

GENDERED CRIME

Research on this aspect of gender and crime has renewed interest in "social situations that produce criminal events, as well as the individual decision-making and opportunity structures necessary for offending," an avenue of criminological inquiry that dates back to Sutherland (Miller & Mullins, 2006, p. 232). This emphasis is closely related to *gendered pathways to lawbreaking,* but it moves beyond individual motivation and the power of initial victimization and examines how "women navigate gender-stratified environments, and how they accommodate and adapt to gender inequality in their commission of crime" (Miller & Mullins, 2006, p. 233). *Gendered crime* analysis attempts to discover the contingencies within and across gender in order to "more precisely specify dynamic relationships between gender and crime" (p. 236).

By pursuing how situations influence men and women's crimes, research on this topic permits promising comparisons. For example, some women's crime is a response to opportunities for economic gain, recognition, and status enhancement as well as excitement and revenge. Prostitutes doing "sex work" for pimps may engage in "viccing"—a form of "rolling" a john—as a form of resistance to their victimization and vulnerability within sex markets cheapened by the drug economy (Maher, 1997). While at first glance "viccing" may look like an instrumental robbery, this "belies the reality that the motivations undergirding it are more complex and indeed, are intimately linked with women's collective sense of the devaluation of their bodies and work" (Maher & Curtis, 1992, p. 246).

Consider also that while women's relationships with men may explain why they are marginalized in male-dominated crimes such as organized burglaries, the end of such relationships also hampers or ends women's entry to other male-dominated burglary crews. Thus women's criminal opportunities were found to be restricted by situational changes (Mullins & Wright, 2003, cited in Miller & Mullins, 2006).

GENDERED LIVES

The concept of *gendered lives,* according to Daly (1998), emphasizes the "significant differences in the ways that women experience *society* compared with men" (p. 98, emphasis added). Miller and Mullins (2006) consider this aspect of feminist criminology as perhaps the most challenging because it requires systematic attention "to gender well beyond the analysis of crime" (p. 239). Compared to the attention given to studying the pathways to offending and the gendered nature of offending, less work has addressed gendered lives. Nonetheless, some important work has been published on this aspect of gendered crime. According to Miller and Mullins (2006), the works of Bottcher (2001) and Maher (1997) are important examples of this line of inquiry.

Bottcher's (2001) work focuses not on gender as individual action, but "instead on the gendering of social practices" (quoted in Miller & Mullins, 2006, p. 240). She identified three broad types of social practices: making friends and having fun, relating sexually and becoming parents, and surviving hardships and finding purpose. For example, Bottcher found that gender-segregated friendship groups placed high-risk males, compared with high-risk females, at greater risk of delinquent involvement. "Likewise, the meaning and rules guiding sexual relationships and childcare responsibilities had similar consequences" (Miller & Mullins, 2006, p. 240).

Miller and Mullins (2006, p. 240) consider Bottcher's approach to gendered lives to be important and notable because it emphasizes *practices* rather than individuals while at the same time it challenges the male-female gender dichotomy often found in studies of gender and crime. Furthermore, Bottcher demonstrates that gendered patterns of behavior are not universally applicable to all males or all females. Some male social practices encourage delinquent activity for either sex, while some female social practices appear to discourage delinquent behavior for both sexes (Bottcher, 2001, p. 904, quoted in Miller & Mullins, 2006, p. 240).

Another example of exemplary research on gendered lives is found in Lisa Maher's (1997) *Sexed Work.* It is a "consistent examination of the intersections of race, class and gender in shaping women's experiences and lives, and illustrates the strengths of feminist scholarship that moves beyond an exclusive emphasis on gender" (Miller & Mullins, 2006, p. 240). By blending feminist analysis with cultural reproduction theory she found, for instance, that contrary to some scholars who have argued that the drug trade opened new opportunities for women, gender inequality is institutionalized on the street. The stereotypes of women as unreliable and weak "limits women's participation in informal economic street networks" (Miller & Mullins, 2006, p. 241). More specifically, she found a rigid gender division of labor in the drug trade that was shaped along racial lines in which women were clearly disadvantaged compared to men. Excluded from the more lucrative aspects of the drug trade, Maher reported that women found sex work one of the few viable options for making money.

This finding and others challenge previous explanations and descriptions of women's participation in drug markets. It, for instance, contradicts the image of women crack users as desperate, pathological, and powerless individuals who will do anything for their next hit. Instead, Maher found that women are involved in any number of income-generating activities within the drug economy that follow occupational norms that govern their behavior. This finding supports the idea that women lawbreakers are less like dependent and passive victims and more like active, creative decision makers who often face contradictory choices. To think otherwise would be to deny the fact that crime in general and drug markets in particular involve a complex understanding of the relationship between structure and agency (Maher, 1997, p. 210, cited in Miller & Mullins 2006, p. 241).

The impact of "doing marriage" on desistance from crime is another way of thinking about gendered lives. Research conducted mostly in the United States, for instance, has found the effect of marriage on the reduction of criminal behavior for adults to be particularly robust across different samples (Bersani, Laub & Nieuwbeerta, 2009, p. 4). Similar findings are reported for the Netherlands. The results for this study indicated, however, that the effect of marriage on offending was significantly more favorable for men than for women. Thus, being "in the state of marriage" was "associated with a 36% decrease in the odds of a conviction for men and a 21% decrease in the odds of a conviction for women" (Bersani et al., 2009, p. 14).

Postmodernist Feminism and the Third Wave

Postmodernist feminism is one of the many feminist theoretical approaches to studying women and crime, although a relatively minor perspective compared to the larger body of literature on feminist perspectives on crime. As explained in the previous chapter, one of the most prominent features of postmodernist thought is an emphasis on deconstructing traditional explanations and categories of crime and offenders found in positivist science. Postmodernist feminism therefore seeks to deconstruct the "racial, class, and gender stratification that has resulted from modern Western civilization" (D'Unger, 2005, p. 563).

Similar to the emphasis placed on the media by cultural criminology, postmodernist feminism is also concerned with the constructed images of crime, including the images of women offenders. These images are more reflective of the ideas of the "privileged and powerful, whose interests they represent" instead of the truth (Chesney-Lind & Faith, 2001, p. 297; quoted in D'Unger, 2005, p. 563).

Closely related to postmodern feminism is a new direction of thinking about feminist issues. It is called "third wave feminism." It has yet to have much of an impact on feminist perspectives in criminology, but as Daly (2010) says, it is coming. Although it does not have an entirely different set of issues or solutions than concerned second wave perspectives and agendas, it does have distinguishing features. According to Snyder (2008), "What really differentiates the third wave from the second wave is the tactical approach it offers to some of the impasses that developed within feminist theory in the 1980s" (p. 175). One such impasse was the perceived rigidity of the second wave's ideology.

The second wave, for example, emphasized grand narratives of women's location in the social world that in effect tried to unify them. According to third wave feminism, one result was the creation of an identity for women that placed them in inflexible positions where they could not change sides. Females were placed against men, Blacks against Whites, and oppressed against oppressor. Third wave feminist replaces or rejects this effort by arguing that there is a wide array of discursive locations for women. This position attempts to move beyond conflict, and it endorses what has been called the "welcoming politics of coalition" (Snyder, 2008, p. 176). It emphasizes an inclusive and nonjudgmental approach that does not police or maintain the political boundaries that the second wave employed (see Table 10.1).

Table 10.1   Summary of Feminist Perspectives on Crime

Perspective	Description
Early Feminist Perspectives	
Liberal	Crime is due to gender socialization. As men and women become more equal in society, the amount and types of crime for men and women will become the more similar.
Marxist	Due to the capitalist mode of production, women assume subordinate roles in society. Their labor is exploited in the workplace and at home. The crimes they commit and their victimization reflect this subordinate position.
Radical	Men are born more aggressive and dominant. Men use crime, including rape, to control women.
Socialist	Patriarchy, which is based in capitalism, provides men with more opportunities to offend and enables their victimization of women.
Key Contemporary Developments	
Masculinities	Crime is a way in which men "do gender"—in which they show their masculinity.
Gendered lives	Pathways into crime and victimization are not general or gender-neutral. Rather, they are gender-specific, reflecting females' unique friendship groups, parenting roles, and hardships.
Intersection of race, class, gender, and place	The complexities of understanding victimization and criminalization are best understood by examining the interplay of each of these variables. Factors that contribute to victimization and criminalization often blur together.
Postmodern	Rejects criminological positivism and absolute truths. Attempts to deconstruct the class, race, and gender stratification that developed in the Western world.

Consequences of the Diversity of Feminist Perspectives

Although the debate over women's emancipation and crime has not been very fruitful empirically, greater attention has been given to women as victims and survivors of sexual and physical violence (Daly & Chesney-Lind, 1988). Aided by the women's movement, popular and important works (e.g., Brownmiller's [1975] famous *Against Our Will*), and numerous media reports of women as victims or survivors (e.g., the 1993–1994 case of Lorena Bobbitt severing her husband's penis after years of alleged physical and psychological brutalization, the 1994 double murder of O. J. Simpson's former wife Nicole and her friend Ronald Goldman), this topic—especially rape and intimate violence—has become central to the feminist perspective in conventional criminology and to left realism. It also has become central in the public consciousness ("Living in Terror," 1994; "Wife Tells Jury," 1994).

According to some observers, there are several reasons why the victimization of women as a topic has had such relative success compared to explaining female crime by the emancipation movement. First, victimization of women and girls can be tied to a number of feminist perspectives including Marxist, social, and radical feminisms. This expansion of the explanation of female victimization has permitted men's violence against women to be more easily linked to some of the more salient features of patriarchy, especially power. Victimization of women is one way in which patriarchal power can be defined and perpetuated. Second, this linkage between patriarchy and power moved "large numbers of grassroots feminists and some academic feminists to document the then-hidden forms of violence suffered almost exclusively by women" (Daly & Chesney-Lind, 1988, p. 513). Third, the whole milieu of the women's movement affected criminology, including nonfeminist criminologists, who had to "digest and deal with feminist scholarship" (p. 513). The increasingly documented record of men's violence against women had itself to be integrated into criminology; it could not be ignored. Finally, while these developments were taking place, more women and feminists were moving into criminology and other academic disciplines. Although this aspect of the feminist perspective in criminology has not been well charted, the impact of women on criminology has been debated.

One indication of this impact is found in the presidency of the American Society of Criminology (ASC). Founded in 1941, ASC has had 51 different presidents, only 5 of whom have been women, and all of these woman presidents have held office since 1989. The most recent female president of this organization, Julie Horney, was elected in 2005.

The Academy of Criminal Justice Sciences (ACJS), by comparison, has had 43 presidents since its first in 1963, and 5 of these have been women. These developments, especially the increased female membership in previously male-dominated professional organizations, no doubt were of great importance in developing a sense of urgency and affinity toward female victims (Daly & Chesney-Lind, 1988, p. 513).

Writing during the late 1970s, Rock (1977) observed in a review of Smart's (1976) *Women, Crime, and Criminology* that he doubted that "analytic losses" had been inflicted on criminological theory by not considering women. Today it is doubtful that this conclusion would be accepted as much more than sexist rhetoric. More recently, in a brief review of the impact of feminist work on criminology, Young (1992a) argued that by revealing the partiality and inaccuracy of conventional criminology regarding female crime, feminists sent a jolt "through the entire criminological enterprise" (p. 292). Slowly, he claimed, course syllabi were altered to include women and crime, new courses were designed, and conferences began to include sessions on women and criminal justice. Texts on deviance and criminology also were revised to reflect feminists' arguments including, in Britain, Downes and Rock's (1988) classic *Understanding Deviance*. In the United States, Sykes's (1978) *Criminology* contained only one feminist's interpretation of female property crime. Its second edition, however, contained an entire section titled "Feminist Criminology" (Sykes & Cullen, 1992). These changes and modifications were interpreted by one female observer as so influential as to have "revivified a discipline that had come to seem bogged down in its own internal arguments" (Young, 1992a, p. 290). The question remains, nonetheless, whether there is a "feminist criminology" today. The record appears to be mixed.

Addressing this question during the late 1980s, Gelsthorpe (1988) concluded that the term *feminist criminology* as used then created confusion. There are two difficult core elements that would identify feminist criminology. Certainly, according to Gelsthorpe, *a* feminist criminology cannot exist, just as there cannot be *a* feminist sociology, history, philosophy, or any other single feminist discipline. There are multiple feminist perspectives in criminology. Some feminists argue that men and women should be dealt with equally by the criminal justice system; others claim that they should be treated differently. Some argue that men and women differ in their capacity to commit crime; others disagree. Still others claim that feminist criminology should develop alternative interpretations of social reality; others argue that feminist criminology should focus on explaining the gross differences between men's and women's crime rates.

Although not dismissing the importance of these different emphases, Renzetti (1993) asked whether feminism has had enough of an impact to transform criminology/criminal justice education so that gender is a central organizing theme. Organizing her evaluation around curriculum, pedagogy, and campus climate for students and faculty, Renzetti concluded that feminist criminology/criminal justice education seems to remain at the margins of the "male-stream" (p. 219). In support of her conclusion, she pointed out that although women currently constitute more than 52% of the general undergraduate population, 40.4% of bachelor's degree recipients in criminology, and 68.4% of bachelor's degree recipients in sociology, they still are "largely invisible in our courses and textbooks" (p. 226). She stated that, at best, women are marginalized in special courses as special topics or in separate token chapters. They are not fully integrated into the criminology/criminal justice curriculum.

Campus culture also has contributed to the marginalization of gender issues in criminology/criminal justice. Drawing on the works of Goodstein (1992), Hall and Sandler (1985), and McDermott (1992), Renzetti (1993) reported that classroom interaction patterns among faculty and students, including sexist humor and language, "breeds other, less subtle forms of discrimination including sexual harassment" (p. 228). According to Stanko (1992), this contributes to women's marginalization because these experiences are reminders that they are "only women."

Feminist critiques of traditional pedagogy and research methodologies denounce the rigid separation of the "knower" from the "known." Greater emphasis is placed on qualitative methodologies employing interactive or participatory research strategies. As Daly and Chesney-Lind (1988) stated, "They are more interested in providing texture, social context, and case histories . . . , presenting accurate portraits of how . . . women become involved in crime" (p. 518). Feminist critiques object to insensitive quantification (Renzetti, 1993). More central to developing feminist criminologies is the emphasis on gender. This is not to de-emphasize the feminist perspectives' concern with stereotypical images of women and methodologies sympathetic to these concerns (Gelsthorpe, 1988), but gender is not only about women. As Renzetti (1993) explained, "The goal of feminism is not to push men out so as to bring women in, but rather to gender the study of crime" (p. 232). Theories that do not consider it not only are incomplete but also are misleading because gender carries great social, economic, and political significance (Gelsthorpe, 1988).

Additional consequences of the feminist perspectives on crime are found in a number of public social policies. Mandatory arrest for domestic violence, for example, is now widespread throughout the United States, although its long-term impact on domestic violence has been questioned. For example, one study found that domestic violence recidivism *increased* for arrested Black men but decreased for arrested White men (Sherman et al., 1992) Mandatory arrest policies implicitly assume that all batterers are the same. This assumption has been criticized because not all battering situations are the same, not all batterers are the same, and not all victims are the same (Hines, 2009, p. 126).

Mandatory arrest policies also create some unanticipated consequences for the criminal justice system. More arrests of batterers lead to more cases for prosecutors who often did not have their budgets and personnel increased, thus producing more strains for a system already heavily burdened with work (Hines, 2009, p. 126).

Furthermore, treatment for men who engage in domestic violence is based on a feminist model that assumes battering women is a problem that is created by the organization of a patriarchal society. Men, it is argued, use their power to batter women in order to control them. According to Hines (2009, p. 118), there is no evidence that such a treatment model works. Yet, such treatment models are usually ordered by courts and required by state law to be feminist-based. The recidivism rates of domestic violence for men who attend these programs are no different from those of male batterers who do not attend them. Although most batterers are men, female perpetrators of domestic violence do exist; treatment programs for them are scarce.

Changes in rape laws represent another example of the impact of the feminist perspectives on social policies. Traditionally, explanations of rape have focused on the sexual component of the offense with emphasis given to offenders' excessive sexual tensions or maladjustment and the manner in which the victims "asked for it." Since the early 1970s, the focus has changed from defining and prosecuting rape as a sexual act to treating it as an act of violence. Key to this shift in emphasis is recognizing that rape is an act of subjugation/domination reflecting cultural definitions of male roles. This interpretation emerged with the development of the women's movement and the feminist perspectives on crime and has been elaborated by several observers. Griffin

(1971), for example, claimed that rape is ingrained deeply in patriarchal societies and is used as a form of mass terrorism to deny women self-determination.

Others have given attention to what is now called "date rape," a topic unthinkable just a generation ago. The heart of the issue here is the question: When is sex considered sex, and when is sex considered rape? It developed in large measure as college women began to report being raped not by strangers but rather by men they knew well or casually (Fisher, Cullen, & Turner, 2001; Fisher, Daigle, & Cullen, 2010). As colleges and universities learned of this phenomenon, new reporting policies were created, and now the federal government and several states have laws requiring educational institutions to report crime statistics. Other responses include new college and university educational programs on acquaintance rape and consensual sexual relations (Celis, 1991; Karjane, Fisher, & Cullen, 2005).

Rape shield laws are another example of the impact of the feminist perspectives on criminal justice policy. These laws grew out of the need to protect a rape victim from being raped twice: once by the accused and then again symbolically by the accused's defense attorney probing the victim's past sexual behavior. Every state has adopted some type of rape shield law, but these laws vary considerably, especially regarding whether the victim previously had consensual sexual relationships with the defendant.

No one knows exactly how many women are raped annually, but it is clear that redefining rape as an act of violence, recognizing the existence of date and acquaintance rape, and developing rape shield laws represent dramatic changes in how rape is defined and prosecuted. But these changes are not the only evidence of the influence of the feminist perspectives on rape. More and more today, rape victims are finding justice through civil courts. Experts on rape say that the number of civil suits has grown from just a few in the 1970s "to a steady stream, as women have become less ashamed by rape and more aware of the legal options for fighting back" ("Many Rape Victims," 1991, p. A1).

Some Implications of Feminist Criminology for Corrections

Women comprise only 7% of the U.S. correctional population, but their number is rising even more rapidly than is true for men. In fact, by 2006 there were almost 200,000 women in prisons and jails in the United States, an approximate increase of 100% from 5 years before (Amnesty International, 2006). During the 30-year span between 1977 and 2007, however, the female prison population grew 832% (West & Sabol, 2008) and is now over 200,000. The male prison population grew 416% during the same time period (West & Sabol, 2009). The number of women on probation between 1997 and 2007 increased from 524,200 to 987,427, an 88% growth in one decade.

The so-called steel ceiling (often attributed to the so-called chivalry hypothesis) that used to divert women to correctional alternatives such as probation or other forms of community-based treatment, has been cracking for some time (Kruttschnit & Green, 1984), in part because women are committing a greater number of crimes of the sort that bring men to jail and prison, in part because of mandatory sentencing, in part because of the "war on drugs" that is sometimes called a "war on women," and perhaps in part because of a "vengeful equity" that has led the criminal justice system to respond to women's demand for equality with an equity that makes them pay for such demands (Chesney-Lind, 1998). At the same time, it is widely recognized that correctional systems were designed for men and that their deficiencies are magnified when dealing with women (Bell, 1998).

The human rights movement in general and feminist criminology in particular has called attention to several major problems facing correctional systems for women. Clearly, women bring different needs to prison. For example, approximately 80% of women inmates have dependent children, many of whom will have to be placed in foster care or even institutionalized themselves because there is no one else available to care for them. Equalitarian thought may insist that much the same is true for men, but it is still true than women are more family oriented than men, that they do most of the child caring, and that women inmates express more anxiety over their children than do male inmates (Allen, Latessa, Ponder, & Simonson, 2007).

Although the exact figure is unknown (with estimates of about 5%), a significant number of women enter prison pregnant. They need special diets, lighter work assignments, a less stressful environment, and other medical and environmental adjustments. Some female prisons have started programs, but these are rare. For example, Washington State has a nursery program, and California has initiated a Community Prisoner Mother Program in which inmate mothers live with their young children in seven small community-based facilities. Such programs are the exception, however, and their lack shows just how much the prison is still considered a "man's world." Meanwhile, recent reports still show that "prisons often shackle pregnant inmates in labor," a practice that is considered dangerous to both the woman and her child (Liptak, 2006).

Pregnancy, however, is only one example of the difference in the nature of medical problems encountered with women inmates. The National Institute of Corrections has found that gynecological services for women in prison are seriously inadequate (Amnesty International, 2006). A recent California study, for example, found that many women inmates had not had a Pap smear or mammogram for several years (Katayama, 2005), despite the fact that 5 years earlier, 32% of them reported gynecological and reproductive health problems (Stoller, 2000). This is partly because correctional medical services are designed for men and partly because medical facilities in most institutions for women are inadequate because of the small size of the prison population and the expense of special programming. The medical problems of women inmates constitute a serious problem because the cost of health care in state prisons for women is 60% more expensive than it is for men and still unsatisfactory (Katayama, 2005).

Even in women's prisons, the staff is still predominately male. Obviously, feminists concerned with patriarchy have special reason to be interested in the fact that women placed in prisons are under even more male control than those outside. Because they take away one's autonomy, prisons tend to foster a certain "dependency" in any case, and such institutionalization is unlikely to make women stronger. Furthermore, there is the continuing problem of sexual exploitation, sometimes in terms of return for favorable treatment, sometimes because of the psychological dominance inherent in the situation, and sometimes because of blatant rape (Human Rights Watch, 1998). Such reports are not uncommon, and to feminist criminologists and others they suggest the need for more women correctional officers (assuming that we are going to continue on our incarceration binge) and open policies that make it more difficult to retaliate against inmates who report sexual exploitation (Human Rights Watch, 1998).

Feminist criminologists and others have often commented on the abuse suffered by girls and women in a patriarchal society, and this is reflected in the problems women inmates bring to prison. A recent California study showed that some 57% of female inmates were physically or sexually abused prior to entering prison, compared with 16% of the men (Vesely, 2004). These women and others have special self-esteem issues along with a higher incidence of suicide attempts and drug abuse problems than do men (Allen et al., 2007). Programs such as boot camps may be especially hard on women offenders who have come from abusive relationships (MacKenzie et al., 1994).

On the other hand, the fact that prisons are still designed for men is reflected not only in the failure to face the different needs of women and their actual exploitation but also in the stereotypical programming often provided. In spite of the feminist movement, vocational programs still tend toward old favorites such as preparation for careers as a secretary, nursing assistant, or beautician. Because so many facilities for women are smaller, the rehabilitation programs, including vocational training opportunities, tend to be more restricted than in institutions for men. All this is compounded by the fact that because many states have only one major facility for females, it will often contain high, close, medium, and minimum classifications within it, meaning that women incarcerated because of minor drug dealing to support an addiction are mixed with violent offenders and that all the inmates must suffer from restrictions considered necessary to deal with the most problematic. Despite recognition of the need for "gender-responsive strategies" (Bloom, Owen, & Covington, 2003) that consider the special needs and problems of women inmates, all of these issues persist.

During the last decade, however, programs have emerged that address the gendered risks and needs of women offenders. One such program, "Moving On," was influenced by "three complementary approaches—relational theory, motivational interviewing, and cognitive-behavioral intervention" (Gehring, Van Voorhis, & Bell, 2010, p. 6). It has several principles and practical goals. They are treating women with respect and dignity; providing an environment that is supportive, empathic, accepting, collaborative, and challenging; assisting women to build a healthy and mutually supportive network; introducing an array of personal strategies, including decision making, problem solving, assertiveness skills, and emotional regulation; and assisting women with the challenges of reintegration (Gehring et al., 2010, p. 6; see also Bauman, Gehring, & Van Voorhis, 2009). A recent evaluation of the intervention—with 190 "Moving On" women probationers matched with 190 probationers not receiving the treatment—found promising results. The rearrest and conviction rates for "Moving On" probationers (but not rates of incarceration) were significantly lower than those for the comparison group (Gehring et al., 2010).

Conclusion

Feminist scholars have succeeded in "gendering criminology" in important ways. Where once it was permissible simply to ignore women offenders and victims or to attribute female criminality to sexuality and pathology, now it is clear that no theory will be complete—will be truly "general"—if it does not take into account the role of gender in both women's and men's crime. In the time ahead, an important challenge will be to determine how criminality is affected not only by gender differences but also by gender similarities. Although exceptions exist, feminist approaches have tended to unmask the unique experiences that shape women's criminal involvement. This perspective is understandable—and invaluable—given that these factors would have otherwise escaped consideration. Even so, there is a growing body of evidence that many risk factors for crime are similar for males and females, though they may express themselves in social relationships in different ways (see, e.g., Andrews & Bonta, 2003; Daigle, Cullen, & Wright, 2007; Moffitt, Caspi, Rutter, & Silva, 2001). Reconciling how gender differences and similarities converge to affect the development and specific manifestations of crime thus promises to be a theoretical and empirical puzzle worthy of careful consideration.

11

Crimes of the Powerful

Edwin H. Sutherland

1883–1950

University of Chicago and Indiana University

Author of White-Collar Crime Theory

Theories of White-Collar Crime

I n 1979, Jeffery Reiman published his influential book, *The Rich Get Richer and the Poor Get Prison.* This volume was not only an indictment of how street offenders are processed by the criminal justice system but also an exposé of the failure of the state to control the harms perpetrated by the powerful. In a chapter titled "A Crime by Any Other Name," Reiman illuminated how corporate practices endanger the health of workers and the public. Much like looking in a "carnival mirror," however, these acts are distorted and not treated as crimes. Rather, while corporations neglect workplace safety and place toxic chemicals into our environment—actions that injure, sicken, and kill thousands annually—the criminal justice system focuses its attention and condemnation exclusively on "typical" individual offenders. This inequality must be changed. It is time, urged Reiman, "*to let the crime fit the harm and the punishment fit the crime*" (1979, p. 195, emphasis in original).

Beyond the specific claims made, Reiman's book is important because it reflects a way of thinking that emerged in the 1970s. Crimes of the powerful—or what are known as "white-collar crimes"—had been largely ignored since the inception of criminology. In 1968, for example, the President's Commission on Law Enforcement and Administration of Justice devoted only 5 pages—in a volume of 814 pages—to a section on "'White-Collar Offenders' and Business Crime." The Commission concluded that "the public tends to be indifferent to business crime or even to sympathize with the offenders who have been caught" (1968, p. 158). But a few years later, it was as though scholars had awoken from a criminological hibernation and suddenly become aware of the crimes in the upperworld. Within a few short years, a spate of books appeared carrying titles such as *Crimes of the Powerful* (Pearce, 1976), *"Illegal But Not Criminal": Business Crime in America* (Conklin, 1977), *Corporate Crime* (Clinard & Yeager, 1980), *Corporate Crime in the Pharmaceutical Industry* (Braithwaite, 1984), *Wayward Capitalists* (Shapiro, 1984), and *The Criminal Elite* (Coleman, 1985). Scores of articles also were published during this period.

These works tended to embody three core themes. First, they claimed that a fundamental hypocrisy stained the American justice system. The ideal of equal justice before the law was a sham. Money and power allowed great crimes to be committed and, indeed, committed with impunity. Second, they claimed that the costs of the crimes of the powerful far outweigh the costs of the crimes of the poor. A single fraud could cost millions of dollars; exposing workers or the public to chemical toxins could silently sicken or kill thousands. Third, they claimed that the only way to stop these harms was to use criminal sanctions to bring law and order to the upperworld. In particular, send corporate executives—or corrupt politicians and shady physicians—to jail, and their shenanigans would soon stop.

When such a fresh way of seeing the world sweeps across criminology, it signals that a changed social context has caused numerous scholars to reshape how they interpret reality and what they believe merits study. As might be anticipated, this focus on the crimes of the powerful represented another rejection of mainstream criminology, which had been preoccupied with offenders located in city streets and not in corporate suites. Indeed, a key to the appeal of critical criminology at this time was a willingness to speak truth to power. In particular, it involved unmasking how structures of inequality in the United States were implicated in criminal conduct. In Chapters 7 through 10, we examined perspectives that questioned the existing status quo in America—whether this was racial, class, or gender inequality. In contrast to mainstream criminology, they argued that power was central to any understanding of the origins and efforts to label and control criminal behavior.

As seen in Chapters 7 through 10, the scholars who invented labeling, conflict, critical, and feminist theories in the 1970s and beyond were schooled in the sixties and early seventies. They witnessed extensive social turmoil—a context we described previously (e.g., assassinations, urban insurgencies, crackdowns on Vietnam War protests) (see, e.g., Brokaw, 2007; Collins, 2009; Gitlin, 1989; Patterson, 1996). These events triggered so-called new criminologies that sought to infuse the criminological enterprise with an appreciation for power and conflict. Those who turned their attention to white-collar crime at this time also were of this generation; they were impacted by the events of the day. Their particular concern was with how power placed the advantaged in the position to use unlawful means to pursue profit with no risk of going to prison.

There were, however, two special features of this era that were especially influential in pulling scholars into the study of white-collar crime specifically. First, the civil rights movement placed a great emphasis on the need to institute equal justice before the law.

In the South, the lynching of African Americans and the acquittal of Whites who murdered Blacks were especially symbolic of this injustice (Oshinsky, 1996). In 1968, the year in which Martin Luther King, Jr., and Robert Kennedy were shot to death, Richard Nixon assumed the presidency, his campaign fueled in part by the bold promise to restore "law and order" to America. This rhetoric was pregnant with a racially tinged appeal to White Americans—the so-called moral majority—that Nixon would use the criminal justice system to crack down on urban unrest and Black crime. This was the beginning of the "war on crime" that would legitimize a 40-year policy of mass incarceration.

But a fundamental hypocrisy, which was not lost on scholars, marked Nixon's trumpeting of the need for "law and order" in the nation's streets: Corruption was rampant at the highest levels of his administration. In a remarkable development, Vice President Spiro Agnew was forced to resign when it was revealed that he accepted kickbacks from contractors not only when governor of Maryland but also during his vice presidency. On October 10, 1973, he resigned his office, pleading no contest to charges of tax evasion (Patterson, 1996). Questions also arose about Nixon's own failure to pay taxes on improvements to his properties made by the government. Nixon responded: "I have never profited . . . from public service. . . . I have never obstructed justice. . . . I am not a crook" (quoted in Patterson, 1996, p. 776). Shortly thereafter, the Watergate scandal unfolded.

On June 17, 1972, five men were caught at 2:30 in the morning burglarizing the Democratic National Committee's headquarters, then located on the sixth floor of the Watergate Hotel and Office Building in Washington, D.C. They were attempting to replace a malfunctioning tap on a telephone that had been surreptitiously installed 3 weeks earlier. Investigations eventually traced the scandal to the president's staff, many of whom were subsequently convicted and imprisoned. These included John Mitchell, the nation's attorney general. Nixon was revealed to be deeply involved in covering up the break-in, including ordering the CIA to thwart the FBI's probe into the affair. In an event that stunned the nation, Nixon resigned the presidency on August 9, 1974. He was the first president to do so. Gerald Ford, who replaced Spiro Agnew when he stepped down, ascended to the presidency. In a controversial decision that seemed to place the former president above the law, Ford pardoned Nixon, ostensibly in an effort to heal the nation (Patterson, 1996; Watergate.Info, 2010).

Second, during this time, the consumer and environmental movements also were growing stronger. Activists—the most noteworthy being Ralph Nader—systematically documented the ways in which corporations rigged bids to inflate prices, defrauded consumers with false claims, sold products that were unsafe (e.g., automobiles like the Ford Pinto), and wantonly polluted the nation's air and water. Instances of business malfeasance were frequently seen on news programs such as *60 Minutes*. Americans now regularly watched reporters such as Mike Wallace elicit from executives denials of wrongdoing, only to turn around and show internal documents or hidden videotapes revealing untoward practices. In 1972, state attorneys general defined consumer fraud as a major concern (Benson & Cullen, 1998). And when Jimmy Carter was elected president, his attorney general, Griffin Bell, would comment in 1977 that white-collar crime would be his "number 1 priority" (Cullen, Link, & Polanzi, 1982). "Increasing numbers of Americans," observed Stephen Yoder (1979, p. 40) at this time, "have become aware that crime exists in the suites of many corporations just as it exists in the streets of their cities and suburbs."

Following World War II, the United States had become an economic powerhouse and a land filled with, in James Patterson's (1996) words, "great expectations." But as corruption and malfeasance were uncovered, these expectations were dashed. Confidence in corporations and in government plummeted. Indeed, mistrust of those in power grew so pervasive that commentators spoke of a "confidence gap" or "legitimacy crisis" (see, e.g., Lipset & Schneider, 1983). For example, in 1966, the confidence in those "running major companies" was 55%; 5 years later in 1971, it had dropped to just 27%. These and similar figures prompted Lipset and Schneider (1983) to conclude that "the period from 1965 to 1975 . . . was one of enormous growth in anti-business feelings" (p. 31). Such negative opinions remain in place today (Cullen, Hartman, & Jonson, 2009).

In this context, commentators came to talk about a "social movement against white-collar crime" (Katz, 1980; see also Cullen, Cavender, Maakestad, & Benson, 2006). Criminologists were part of this campaign. They mistrusted those who wielded influence—suspicious that they would use any means necessary to stay in office and to make profits. Similar to many other Americans, they were outraged by the Watergate scandal and repeated revelations of corporate wrongdoing. These cases demonstrated a crass willingness to abuse positions of trust and power to visit harms on the unsuspecting and the unprotected. Still worse, these upperworld lawbreakers often perpetrated their offenses behind the breastplate of righteousness—hypocritically preaching law and order while sneakily breaking the law with impunity. "The rich get richer and the poor get prison," Reiman's slogan, for many criminologists seemed to capture the essence of American justice. These inequities moved many scholars, in criminology and in other disciplines (e.g., law, business), to turn their attention to white-collar crime.

Scholars thus embarked on three lines of inquiry. First, to show the dimensions and costs of this criminality, some scholars wrote textbooks or compiled collections of readings conveying the outrageous misconduct of those wearing white collars (see, e.g., Ermann & Lundman, 1978; Friedrichs, 1996; Hills, 1987; Rosoff, Pontell, & Tillman, 2007). Most Americans are now aware of the enormous economic costs that can extend from even a single white-collar crime. Bernard Madoff's Ponzi scheme is alleged to have cost unsuspecting investors $65 billion. Over the past decades, one major scandal after another has occurred: insider trading on Wall Street, the savings and loan debacle, and the Enron fraud—to name but the most prominent. But what criminologists—and others—also sought to detail is the extensive *violence* perpetrated by corporations. Such entities not only unlawfully take money but also sicken, injure, and kill thousands of Americans annually. They have been documented to ignore safety standards in the workplace, to knowingly expose employees to lethal toxins (e.g., asbestos), to market unsafe products, and to dangerously pollute the air and water—so much so that whole communities have been devastated if not abandoned (see, e.g., Braithwaite, 1984; Brodeur, 1985; Brown, 1979; Cherniak, 1986; Frank, 1985; Mintz, 1985; Mokhiber, 1988; Nader, 1965; Peacock, 2003). It is estimated that the physical costs of white-collar crime rival or surpass those of street crime (Cullen et al., 2006). It is a form of criminality, scholars maintained, that should not be ignored.

Second, other scholars explored the difficult issue of how white-collar crime should be controlled (Benson & Cullen, 1998; Braithwaite, 1985; Cullen, Maakestad, & Cavender, 1987; Hochstedler, 1984; Simpson, 2002; Vaughan, 1983). The critical debate hinged on whether to use the criminal law to sanction upperworld offenders, especially when corporations were the lawbreakers. Historically, corporations that harmed people were either sued in civil court or regulated by government agencies. Scholars faulted these remedies for being ineffective (corporations continued to harm) and for being unjust (why should a burglar be sent to prison but not an executive who fixes prices?). Although the capacity of the criminal law to deter upperworld wrongdoing remains in question, most criminologists argued in favor of broadening its use and of imprisoning corporate leaders. It is instructive that today it is no longer uncommon for white-collar offenders, such as Bernard Madoff or the Enron executive Jeffrey Skilling, to receive lengthy prison terms (see Benson & Cullen, 1998; Cullen et al., 2006; Cullen et al., 2009).

Third, although less plentiful, scholars developed theories of white-collar crime. One approach was to use existing theories of crime to explain upperworld offending. If a perspective is a "general theory," it should explain all crime, regardless of the collar an offender is wearing. An alternative approach was to recognize that white-collar crime often occurs in a unique setting: as part of a legitimate occupation situated in a corporation or other organization. Crimes may be committed not only for personal gain—as in street offenses—but also to advance corporate profits and interests. Scholars argued that a "specific theory" is required to take into account these special circumstances.

The focus of this chapter is on *theories of white-collar crime*. Both specific theories and, where appropriate, general theories will be discussed. The chapter starts by exploring the pathbreaking work of Edwin H. Sutherland, the "father of white-collar crime." We then proceed to examine current thinking about white-collar crime in four sections: (1) organizational culture theory; (2) theories of organizational strain and opportunity; (3) the decision to offend, including neutralization theory and rational choice theory; and (4) state-corporate crime theory. We end by exploring the implications of theories for controlling white-collar crime.

The Discovery of White-Collar Crime: Edwin H. Sutherland

White-collar crime was "discovered" by Edwin H. Sutherland in the sense that he demanded that scholars pay attention to crimes committed in society's higher echelons. In this section, we thus start by discussing a historic address in which Sutherland used the concept of white-collar crime to criticize then-customary ways of viewing and explaining criminal behavior. We also explore the reasons why Sutherland had the insight and courage to be the scholar in his generation to focus on upperworld crime and its injurious consequences. Next, we explore why Sutherland's unique definition of white-collar crime proved controversial, and how the conceptual definition embraced by criminologists today has important theoretical implications. Finally, we consider Sutherland's use of differential association theory to explain white-collar crime. In so doing, he argued that competing perspectives were flawed because they attributed offending to poverty or traits said to cause poverty (e.g., feeblemindedness). By contrast, Sutherland claimed that differential association was a general theory because it could account for the crimes of rich and poor alike.

THE PHILADELPHIA ADDRESS

"The term 'white-collar crime,'" observes Gilbert Geis (2007),

> entered the English language on a cold, blustery winter evening in Philadelphia two days after Christmas in the year 1939, at which time the United States was suffering from the pangs of a wrenching decade-long economic depression, a period that poet W. H. Auden described as "a low, dishonest decade." (p. 1)

On Wednesday, December 27, Edwin Sutherland rose to deliver a presidential address to a joint meeting of the American Economic Association, whose president Jacob Viner had just concluded his presentation, and of the American Sociological Society. Sutherland, a faculty member at Indiana University, was president of the ASS. (The organization would change its name to the American Sociological Association so as to sport the less embarrassing acronym of ASA!) His address would be published shortly thereafter—February 1940—in the *American Sociological Review,* giving us an important record whose pages we will quote here.

In 1949, Sutherland would follow up his Philadelphia address with his classic book, *White Collar Crime* (oddly not using a hyphen to link White and Collar). Due to Dryden Press's fears that companies would file law suits for being called criminals, Sutherland deleted corporate names in this volume. The "uncut version" of this book, with the names restored, was issued in 1983 under the editorship of Gilbert Geis and Colin Goff; we cite this edition in this chapter (for accounts of Sutherland, see Gaylord & Galliher, 1988; Geis, 2007, 2010; Geis & Goff, 1983, 1986; Mutchnick, Martin, & Austin, 2009; Sheptycki, 2010; Snodgrass, 1972).

By the time Sutherland arrived in Philadelphia in 1939, he had been assiduously collecting news clippings on wayward conduct by professionals (doctors, lawyers), politicians, and those in the business world for more than a decade (Geis & Goff, 1983). An audience of economists and sociologists provided the perfect opportunity to share his findings on what he called "white-collar criminality." He began his Philadelphia address this way:

> This paper is concerned with crime in relation to business. The economists are well acquainted with business methods but not accustomed to consider them from the point of view of crime; many sociologists are well acquainted with crime but not accustomed to consider it as expressed in business. This paper is an attempt to integrate these two bodies of knowledge. (1940, p. 1)

Sutherland then proceeded to clarify his intent by noting that he was undertaking a "comparison of crime in the upper or white-collar class, composed of respectable or at least respected business and professional men, and crime in the lower classes composed of persons of low socioeconomic status" (p. 1). In short, he wished to explore lawlessness among those with high status who, in his day, wore "white collars."

Sutherland was quick to claim that his illumination of white-collar crime "was for the purpose of developing the theories of criminal behavior, not for the purpose of muckraking or of reforming anything except criminology" (p. 1). Writing in the first part of the 20th century, Muckrakers, such as Ida Tarbell, Upton Sinclair, Lincoln Steffens, and Charles Russell, authored exposés of political corruption and of the ways in which powerful captains in industry (the so-called robber barons) immorally preyed on workers and the public. They succeeded, according to Harvey Swados (1962, p. 9), in revealing "the underside of American capitalism." Upton Sinclair's (1906) *The Jungle,* which disclosed the unsanitary practices and deprave working conditions of the meat-packing industry in Chicago, is perhaps the most well-known example today of this brand of investigative reporting. These crusaders earned the disparaging name of "muckrakers" from President Teddy Roosevelt. Although a former supporter, he was angered by their attacks on political allies in the U.S. Senate and feared that the public might come to hate big business (Brady, 1989). In an April 14, 1906, speech at the Gridiron Club in Washington, D.C., Roosevelt criticized them for their negativity, doing so by referring to a passage from *Pilgrim's Progress:*

> In Bunyan's "Pilgrim Progress" you may recall the description of the Man with the Muck Rake, the man, who could look no way but downward with the muck rake in his hand; who was offered a celestial crown for his muck rake, but would neither look up nor regard the crown he was offered, but continued to rake to himself the filth of the floor. . . . Yet he also typifies the man who in this life consistently refuses to see aught that is lofty; and fixes his eyes with solemn intentness only on that which is vile and debasing. Now it is very necessary that we should not flinch from seeing what is vile and debasing. There is filth on the floor, and it must be scraped up with the muck rake; and there are times and places where this service is the most needed of all the services that can be performed. But the man who never does anything else, who never thinks or speaks or writes, save of his feats with the muck rake, speedily becomes, not a help but one of the most potent forces for evil. (Quoted in The Big Apple, 2010, p. 5; see also Brady, 1989; Swados, 1962)

In his Philadelphia speech, Sutherland's denial that he was engaged in muckraking was partially true (although, as we will see shortly, not totally forthcoming). He had criminological reasons for probing the wayward conduct of the rich and powerful. Unlike Shaw and McKay, who had focused mainly on immigrant youths in socially disorganized inner-city neighborhoods, Sutherland had long been fascinated with different kinds of offenders. His life history *The Professional Thief* (1937) is one notable example familiar to criminologists even today. But as Snodgrass (1972, p. 227) points out, Sutherland's "forays into business crime and professional stealing are only the most obvious and well-known examples. He also kept files and in some instances wrote on such esoteric subjects as lynching, bandits and outlaws, Indian-land frauds, circus grifting, kidnapping, smuggling and piracy" (p. 227). Sutherland's scholarly concern was whether a proposed theory of crime could account for diverse offenses. If not, then its claim for being a general theory was falsified. He believed that his theory, differential association theory, could explain varied crime types: They were all learned behavior (see Chapter 3).

Sutherland had a particular dislike for theories that explained crime by some sort of individual defect or pathology (Snodgrass, 1972; see also Laub & Sampson, 1991). White-collar crime, he felt, proved especially problematic for explanations based on crime as a lower-class phenomenon that linked offending to poverty or to "personal and social characteristics statistically associated with poverty, including feeblemindedness, psychopathic deviations, slum neighborhoods, and 'deteriorated' families" (1940, p. 1). Obviously, robber barons, shady physicians, and political bosses were not feebleminded and did not live in the zone in transition! And because they did not, then existing theories were hopelessly class-biased and wrong. As Sutherland (1940) poignantly asserted:

> The thesis of this paper is that the conception and explanations of crime which have just been described are misleading and incorrect, that crime is in fact not closely correlated with poverty or with psychopathic and sociopathic conditions associated with poverty, and that an adequate explanation of criminal behavior must proceed along quite different lines. The conventional explanations are invalid principally because they are derived from biased samples. The samples are biased in that they have not included vast areas of criminal behavior of persons not in the lower class. One of these neglected areas is the criminal behavior of business and professional men, which will be analyzed in this paper. (p. 2)

Other reasons existed for Sutherland to claim that his concerns were strictly criminological. If Sutherland had openly embraced a reformist muckraking position, he might have damaged, as president of the ASS, sociology's attempt to be seen as a real science and as a worthy discipline within academia. Further, professors in that day were vulnerable to political attack. For example, learning that Sutherland had challenged his criticism of parole boards, J. Edgar Hoover, the legendary director of the FBI, mandated that the agency would have "no contact" with Sutherland, including a refusal to send him FBI crime statistics (Geis, 2007). Earlier, E. A. Ross, a critic of the robber barons, was fired from Stanford University due to his "dangerous socialism" (Geis, 2007, p. 13).

Still, it is generally agreed that Sutherland was being disingenuous when he asserted in Philadelphia that his interest in exposing white-collar criminality was mere value-free science untainted by any dislike of those who exploited their high positions of trust to victimize (Geis, 2007; Geis & Goff, 1983; Snodgrass, 1972). Indeed, Sutherland's address manifested a tone of restrained anger. He chose language—"crooks" and "rackets"—intended to knock "respectable" lawbreakers off their pedestals by equating them with customary lower-class criminals.

Sutherland's (1940) disdain for upperworld offenders was apparent when he noted early in his Philadelphia address that the "'robber barons' of the last half of the nineteenth century were white-collar criminals, *as practically everyone now agrees*" (p. 2, emphasis added). "Present-day white-collar criminals," he continued, are merely "more suave and deceptive than the 'robber barons'" (p. 2). Their lawlessness was widespread. "White-collar criminality," Sutherland pointed out, "can be found in every occupation, as can be discovered readily in casual conversation with a representative of an occupation by asking him, 'What crooked practices are found in your occupation?'" (p. 2). In fact, said Sutherland, these schemes "are what Al Capone called 'the legitimate rackets'" (p. 3)—a disparaging comparison we imagine he made with some delight! He catalogued some of the illegal ventures that are "found in abundance in the business world" (p. 3):

> White-collar criminality in business is expressed most frequently in the form of misrepresentation in financial statements of corporations, manipulation in the stock exchange, commercial bribery, bribery of public officials directly or indirectly in order to secure favorable contracts and legislation, misrepresentation in advertising and salesmanship, embezzlement and misapplication of funds, short weights and measures and misgrading of commodities, tax frauds, misapplication of funds in receiverships and bankruptcies. (pp. 2–3)

Sutherland also sought to deflect any rejoinder that upperworld illegalities were mere peccadilloes that caused little damage. In a statement that future scholars would repeat in their own words many times (Cullen et al., 2006), he claimed that the "financial cost of white-collar crime is probably several times as great as the financial cost of all the crimes which are customarily regarded as the 'crime problem'" (pp. 4–5). A single offense, such as embezzlement by a bank official, or scheme, such as a financial fraud on investors, could procure hundreds of thousands, if not millions, of dollars. Among other examples, Sutherland made note of Swedish financier Ivar Kreuger, whose empire based on a pyramid or Ponzi scheme (giving 43% in returns to large investors) collapsed in the Great Depression, leading him to take his life in 1932 (Ivar Kreuger, 2010). "Public enemies numbered one to six secured $130,000 by burglary and robbery in 1938," Sutherland observed, "while the sum stolen by Krueger [*sic*] is estimated at $250,000,000, or nearly two thousand times as much" (p. 5).

There was, however, a more pernicious consequence when the advantaged offended. "The financial costs of white-collar crime, great as it is," said Sutherland, "it less important than the damage to social relations. White-collar crimes violate trust and therefore create mistrust, which lowers social morale and produces social disorganization on a large scale" (p. 5). By contrast, he asserted, "other crimes produce relatively little effect on social institutions or social organization" (p. 5). Subsequent scholars would call this the "social cost" of white-collar crime.

BECOMING THE FATHER OF WHITE-COLLAR CRIME

Why did Sutherland play such a prominent role in discovering "white-collar crime" in the sense of defining it and making it an object of criminological study? Why would he be the criminologist of his generation to earn the enduring legacy as the "father of white-collar crime"? Although speculative, scholars suspect that the answer to this question of "why Sutherland" lies in his early biography (Geis, 2007; Geis & Goff, 1983; Snodgrass, 1972). As Geis and Goff (1983, p. xx) note, the "emotional roots" of his views on white-collar criminality "lie deep in the midwestern soil of Sutherland's early home life."

On August 13, 1888, in Gibbon, Nebraska, "Edwin Hardin" was the third of seven children born into a family headed by an authoritarian, religious father. The father, George Sutherland, earned his divinity degree and would teach at and head colleges in Kansas and Nebraska, including Grand Island College, a conservative Baptist school from which Sutherland graduated in 1904. A yearbook photo shows Sutherland, who played fullback on the football team, "in uniform, looking suitably fierce and formidable" (Geis, 2007, p. 27). "The Sutherland parents, especially his father," as Snodgrass (1972, p. 221) notes, "were religious fundamentalists and followed all the austere and strict practices of the Baptist faith." Edwin, it is reported, rejected this fundamentalism and its acetic lifestyle, coming to enjoy activities such as smoking, bridge, golf, movies, and the like (Snodgrass, 1972). Still, his early experiences likely had two enduring influences on him that shaped his view of the upperworld.

First, as Geis (2007, p. 28) points out, "Sutherland undoubtedly absorbed in his youth the doctrine of the populist movement, which enjoyed particularly strong support in Nebraska." The state's Populist Party was founded in 1890 and would dominate Nebraska politics throughout most of Sutherland's formative years (Nebraskastudies.org, 2010). This movement sought to extend rights and protections to workers and farmers. It also was worried that powerful corporations would so consolidate their wealth and power as to threaten "the democratic control of industry" (Geis, 2007, p. 29). These populist sentiments were central to his view of corporate immorality. According to Snodgrass (1972):

> His theory of white collar crime was embedded in a generally hostile view of businessmen and business enterprises, particularly the modern, large-scale, monopolistic corporations. His dominant concern was with the concentration of industry, the power of which accrued as a result of the concentration, and the impact of these corporations on the economic system and the traditional social order. It was his view that the massive concentration of industry had occurred through criminal action, specifically the violation of anti-trust laws. (p. 269)

Second, by all accounts, Sutherland possessed unquestioned personal and professional integrity—traits that can be traced to his Baptist fundamentalist upbringing. This was a case, according to Snodgrass (1972, pp. 227–228), of Sutherland's father exerting "a primary differential association that most delinquents and criminals in Chicago never had." He carried with him into adulthood an "obsession with honesty," and "could not understand why all men did not reflect Cooley's altruistic primary ideals of love, loyalty, commitment, and honesty" (Snodgrass, 1972, p. 227). Sutherland had a particular repugnance for the hypocrisy of white-collar criminals who committed their offenses behind a veneer of respectability.

It appears that he preferred traditional criminals, including "Chic Conwell," the alias Sutherland (1937) gave to his professional thief, Broadway Jones. After working with Jones on *The Professional Thief,* Sutherland befriended him; they would correspond and visit in the upcoming years (Snodgrass, 1972). Indeed, in her critique of cultural deviance theory—the name she used for Sutherland's differential association approach—Kornhauser (1978) argued that the "strand of populist sentiment" in his work caused him to lose his "ethical neutrality" (p. 201). She claimed that Sutherland described the "slum boys who become delinquent" as "nice friendly lads, available for indoctrination in the delinquent subculture" but "the rich who violate antitrust laws, fraudulently advertise their products, and engage in many other wicked deeds" as an "abomination" (p. 201). She believed that "these sentiments . . . are a luxury affordable only by professors who, in the safety of their studies, are immune to the consequences of grimy-collar crime" (p. 201).

This final comment was a bit of a cheap shot that might be excused because Kornhauser was writing at a time when scholars were only beginning to demonstrate the devastating impacts of upperworld criminality. But Kornhauser was likely correct in concluding that Sutherland's sympathy extended only to poor, and not to rich, lawbreakers. Writing in the 1940s, Sutherland was fighting to puncture the pervasive class-biased stereotype of criminals as exclusively poor. He wished to scrape away the mask of respectability that allowed prominent community members to attend church on Sundays and engage in predatory business practices the rest of the week—with none of their neighbors the wiser. To accomplish this transformation in consciousness, he used the strategy of equating white-collar criminals with those typically seen as criminals. His point was that, in the end, white-collar crime, like street crime, is merely crime, and that its purveyors, like street criminals, are merely criminals. And once this equivalence is understood, then the special sin of upperworld offenders is revealed: They do not own up to being the criminals that they truly are.

Thus, in *The Professional Thief,* published 2 years prior to his Philadelphia address, Sutherland had noted that "many business and professional men engage in predatory activities that are logically the same as the activities of the professional thief" (1937, p. 207). In particular, white-collar and professional thievery both are organized and at times skilled illegal activities that, similar to any occupation, provide a livelihood. If there is a key difference, it is the extra level of dishonesty—the hypocrisy—that marks upperworld lawbreakers. White-collar offenders persist in denying their guilty mind—in denying that that they are in fact "criminal" (Benson, 1985; Conklin, 1977). "As a result," concluded Snodgrass (1972), "Sutherland seemed to have greater respect for the professional thief, and for the conventional offender, who openly acknowledged their criminality, than for the white collar criminal who deceived both the public and himself" (p. 269).

DEFINING WHITE-COLLAR CRIME

Who Is a White-Collar Offender? In the uncut edition of his 1949 book, *White Collar Crime,* Sutherland (1983) asserted that "white collar crime may be defined approximately as a crime committed by a person of respectability and high social status in the course of his occupation" (p. 7). Somewhat oddly—in the sense that the statement is made but left unexplained—he said that the "concept is not intended to be definitive" (p. 7). Rather, he merely wished to "call attention to crimes which are not ordinarily included within the scope of criminology" (p. 7). He made it clear, however, that white-collar crime covered only offenses that are a "part of occupational procedures." Other illegalities that high-status citizens engage in were thus excluded, such as "most cases of murder, intoxication, or adultery" (p. 7).

White-collar crime thus had two essential elements. The first was the *offender element:* The offender had to be of high status. The second was the *offense element:* The offense had to be occupationally based. The value of this conceptualization is that Sutherland sensitized scholars to a realm of lawlessness that heretofore had been largely ignored: crimes in which the rich and powerful use their occupational positions to accrue more wealth and power. But the Sutherland approach to defining white-collar crime—which we tend to favor—is open to criticism. Let us take, for example, the crime of embezzlement. At first blush, this might seem like a prototypical white-collar illegality. The situation is complicated, however, if we compare an embezzlement by a bank clerk making $20,000 a year to an embezzlement by a bank executive making $500,000 a year. Critics would contend that these are the same crimes: Both, after all, are embezzlements. Those in the Sutherland camp would counter that this argument misses the essential point of the concept of white-collar crime: The *offender element*—the high status—is what makes the construct meaningful. Thus, the executive's embezzlement differs qualitatively from the clerk's; it does not involve petty pilfering of a cashier drawer but rather swindling bundles of money under a cloak of trust and secrecy. To be direct, there is no solution to this debate. The important consideration is to know which definition of white-collar crime—with or without the offender element—a scholar is using.

Indeed, when empirical studies are examined carefully, an important discovery is made: A number of crimes that seem as though they qualify as white-collar—in the sense that they are not customary street crimes (e.g., robbery, burglary)—are in fact committed by lower-status offenders, some of whom are unemployed. Take, for example, an important study on white-collar criminal careers by Weisburd and Waring (2001). Defining white-collar crime as economic offenses that are achieved through "some combination of fraud, deception, or collusion," they examined involvement in these eight federal crimes: "antitrust offenses, securities fraud, mail and wire fraud, false claims and statements, credit and lending institution fraud, bank embezzlement, income tax fraud, and bribery" (2001, p. 12). Were these acts committed by the kind of high status offenders that Sutherland had in mind? Apparently not. Thus, two thirds of the offenders did not own or were not an officer in a business. By our calculations, nearly three fourths lacked a college degree, more than three fifths had more economic liabilities than assets, and about 45% were not employed steadily. Further, for those who were arrested for more than one offense, few (15.1%) committed only white-collar crimes. Indeed, Weisburd and Waring (2001) drew the following conclusion:

> *Many white-collar crimes do not require established occupational position or elite social status for their commission.* The skills needed for many of these crimes are minimal.

Lending and credit card institution fraud may be committed by anyone who fills out a loan form in a bank, while tax fraud may be committed by anyone who completes (or fails to complete) an Internal Revenue Service form. Mail frauds sometimes require little more than a phone or postage stamps. (p. 89, emphasis added)

Weisburd and Waring clearly are focusing on an important domain of crime: illegalities that may have grown more plentiful in America's postindustrial financial and service economy. But to call all these frauds and schemes "white collar" is to rob the term of the very meaning Sutherland gave to it. Alas, there is no criminological high court that decides how terms can be used. In the end, scholars are free to use the construct as they wish. It is thus a case of buyer—or, in this case, reader—beware.

Again, when reading any academic work, it is essential to know how the concept of white-collar crime is defined and what offenders are pulled under its umbrella. Most noteworthy, the theoretical implications of this definitional decision are potentially enormous. Recall that Sutherland focused on high status occupational offenders because he wished to show that bad traits, typically associated with poverty, did not cause all crime. But if a broad definition of white-collar is used—one that might include irregularly employed workers without college degrees—then theories touting bad individual traits might find empirical support as an explanation of white-collar crime. This very debate has erupted over whether Gottfredson and Hirschi's low self-control theory, which we covered in Chapter 6, is an adequate theory of white-collar crime. Using a broad definition that includes lower-status offenders, Gottfredson and Hirschi claim that their "general theory" is up to the task: As with other lawbreakers, those committing white-collar crimes have low self-control..

Critics point out, however, that this perspective offers an inadequate account for the kind of upperworld white-collar offending described by Sutherland (see also Simpson & Piquero, 2002). To rise into managerial positions where opportunities for these crimes exist, people have to display considerable self-control—not low self-control—first in education (getting good grades, completing college) and then on the job (showing up every day, performing well). It is unlikely that they could manifest the generality of deviance typical of those with low self-control—that is, engage in wayward acts analogous to crime (e.g., drug use)—and remain on an upward career trajectory. Further, because they are similar to other business practices, some white-collar crimes require self-control: the ability to plan and administer an illegal financial scheme, such as price-fixing, over time (see Friedrichs & Schwartz, 2008). It remains to be seen whether individual traits, such as self-control or perhaps risk-taking preferences, distinguish which white-collar officials do and do not break the law. The chief point here, however, is that how white-collar crime is defined and which kinds of people are allowed into the sample will stack the deck more in favor of some theories than others (see also Stadler, 2010).

What Is a White-Collar Crime? Crime is behavior that violates the criminal law—plain and simple. But as Sutherland understood, what counts as a *crime* for the rich and powerful is anything but plain and simple. Is it a crime, for example, for a corporation to knowingly market a product that, when used, results in a consumer's death?

In 1978, three teenage girls—two sisters and a visiting cousin—were killed due to a fiery crash of their Ford Pinto. The Pinto had a well-known history of fires following rear-end collisions (Dowie, 1977). The car's gas tank was placed only 6 inches from the back bumper, which, when impacted, would push forward and risk being punctured by bolts on the differential housing. Gas would leak, including into the passenger compartment, and an explosive fire would ensue. A local prosecutor in Elkhart, Indiana, Michael Cosentino, believed that Ford was responsible for the girls' deaths; their vehicle had burst into flames when struck from behind by a van. However, no homicide statute declaring corporate criminal liability for marketing a dangerous product existed. Instead, Cosentino innovatively used Indiana's reckless homicide statute to prosecute the company.

But Ford protested that this law was intended for individuals who had, for example, a finger to pull a trigger—not for a company selling a legal product. Ford also argued that as a federally regulated entity that did not violate any federal safety standards in manufacturing the Pinto, it was unconstitutional to then try to use state criminal law against it. Under the Supremacy Clause of the U.S. Constitution, said Ford, federal regulatory control "preempted" or precluded any criminal action by individual states. In the end, Cosentino experienced success and failure. He won the important battle to bring Ford to trial under this criminal statute, establishing that the sale of defective products can expose companies to criminal sanctions, including for violent crimes such as homicide. However, faced with difficult legal technicalities and a formidable opponent with deep pockets, he failed to earn Ford's conviction on the reckless homicide charges (Cullen et al., 2006).

In his Philadelphia address and subsequent *White Collar Crime*, Sutherland faced this very kind of issue. Even if a company or its officials injure others financially or physically, is it a crime? As noted, if controlled at all, many business harms traditionally were dealt with by two other legal systems: civil suits in which private citizens ask for damages and regulatory agencies in which government agents impose and enforce standards. Sutherland's response to this conundrum—misconduct that might be a crime but historically was not treated as such—was twofold. At issue, he understood, was the very boundary of criminology. What would count as the discipline's subject matter?

First, Sutherland noted that upperworld offenders avoid criminal penalties because they have the power to shape what laws are passed and to whom they are applied. In short, they make sure that their harms are not explicitly outlawed or, if so, are dealt with by the civil and regulatory systems—not prosecuted as crimes with the possibility of imprisonment. "Because of their social status," he observed, "they have a loud voice in determining what goes into the statutes and how the criminal law as it affects themselves is implemented and administered" (1940, p. 8). Second, the key standard for whether an act is a crime is not whether the person is formally convicted of a crime—as some of his critics contended (Tappan, 1947; see also Orland, 1980). To accept this approach, criminology would be beholden to the capacity of the rich and powerful to avoid criminal liability. Sutherland had an ingenious solution: "convictability rather actual conviction should be the criterion of criminality" (p. 6). Thus, regardless of what behaviors are prosecuted, it is the fact that an act is *potentially punishable* under the criminal law that makes it a crime.

Sutherland subsequently relied on this definition in *White Collar Crime,* a volume in which he studied the criminality of "the 70 largest manufacturing, mining, and mercantile corporations" over the lives of the companies, which was about 45 years (1983, p. 13). Recall that his 1949 book was reissued in 1983 with the names of corporations, excluded due fears of lawsuits, restored; this volume will be cited here. He examined the number of "decisions" finding wrongdoing against corporations rendered by civil courts, criminal courts, regulatory agencies, and in settled cases (for a complete explanation, see Sutherland, 1983, p. 15). Again, this measure of *corporate crime* can be criticized for using decisions in which criminal liability was not proven. Still, Sutherland's statistics surely grossly underestimated the extent of illegalities because many criminal acts never come to the attention of victims, the prosecutors, or regulatory officials.

Sutherland discovered that the 70 corporations had 980 decisions, meaning that they averaged 14 illegal acts apiece. Nearly all companies, 97.1%, had two or more adverse decisions. Citing these results, he concluded that the "criminality of the corporations, like that of professional thieves, is persistent: a large proportion of the offenders are recidivists" (1983, p. 227). "None of the official procedures used on businessmen for violating the law," he continued, "has been very effective in rehabilitating them or in deterring other businessmen from similar behavior" (p. 227). Even if his analysis were restricted to criminal convictions, noted Sutherland, it would reveal that 60% of the sample had been convicted, averaging four convictions each. "In many states," he poignantly asserted, "persons with four convictions are defined by statute to be 'habitual criminals'" (p. 23). Notably, subsequent studies have reached similar results (see, in particular, Clinard & Yeager, 1980).

Sutherland's choice of language was purposeful. He wished to debunk the myth that high social status or respectability insulated against criminal involvement. In his view, the labels applied to street offenders from city slums were equally appropriate for the persistent offenders who occupied corporate suites. They broke the law with regularity and thus earned the stigma of "recidivist" and "habitual offender." Snodgrass (1972) notes that Sutherland even titled a chapter in an early draft of *White Collar Crime* as "The Corporation as a Born Criminal" (p. 268). As quoted in Snodgrass (1972), Sutherland justified his selection of the term "Born Criminal"—made famous, of course, by Cesare Lombroso—in this way:

> The analysis which is presented below shows that approximately half of these corporations were criminal in origin and may, therefore, be called "born criminals," not in the Lombrosian sense of inheritance of criminality, but in the sense of overt behavior with intent to violate the law either in their conception and organization or in their behavior immediately after origin. (p. 268, fn. 67)

EXPLAINING WHITE-COLLAR CRIME

To reiterate, Sutherland had three purposes for inventing the concept, and writing about the reality of, white-collar crime. First, inspired by a concealed reformist or muckraking orientation, he wanted to bring the crimes of the rich and powerful within the scope of criminology. He wanted fellow scholars to focus on upperworld criminality—to show that these offenders were no better than more customary offenders and worthy of the same kind of stigma and punishment. Second, he wanted to debunk theories that pathologized offenders by focusing on individual traits supposedly associated with the poor. The very existence of white-collar offenders falsified these class-biased frameworks. It was difficult to argue that feeblemindedness was the key cause of crime when faced with a white-collar offender who attended Harvard University and was the chief of a major corporation. Third, Sutherland wanted to use this opportunity to trumpet his own approach to the study of crime, which favored developing a systematic *general theory* whose principles were sufficiently broad as to explain all forms of crime. This general perspective, of course, was his theory of differential association, which was discussed previously in Chapter 3 on the Chicago school of criminology. We now turn to Sutherland's explanation of white-collar crime.

"The people of the business world," Sutherland maintained, "are probably more criminalistic in this sense than are the people of the slums" (quoted in Snodgrass, 1972, p. 268). How can this be? After all, said Sutherland (1983), people in white-collar occupations come "from 'good homes' and 'good neighborhoods'" and do not have any "official records as juvenile delinquents" (p. 245). But when they enter the business or professional world, they are increasingly isolated from the conventional society of their upbringing. Old ways of thinking—old morals—are replaced by newly learned ways of doing things. That is, they experience differential association with white-collar workers who school them in the definitions of the situation and techniques that make illegal schemes possible. Indeed, newly on the job, "a young man with idealism and thoughtfulness for others is inducted into white-collar crime" (p. 245). He (they were virtually all males in Sutherland's day) either is ordered by managers or learns from coworkers to engage in unethical practices. He is taught an ideology that justifies breaking the law—"phrases such as 'We are not in business for our health,' 'Business is business,' and 'No business was ever built on the beatitudes'" (p. 245). Stated more theoretically:

> The hypothesis which is here suggested . . . is that white-collar criminality, just as other systematic criminality, is learned; that it is learned in direct or indirect association with those who already practice the behavior; and that those who learn this criminal behavior are segregated from frequent and intimate contacts with law-abiding behavior. Whether a person becomes a criminal or not is determined largely by the comparative frequency and intimacy of his contacts with the two types of behavior. This may be called the process of differential association. (Sutherland, 1940, pp. 10–11)

Recall that the Chicago school argued that criminal traditions, including those in the business world, emerge and are transmitted in those sectors of society that are *socially disorganized*. At these sectors, conventional institutions are weak (e.g., broken families) whereas organization for crime grows strong (e.g., juvenile gangs, adult rackets). Shaw and McKay argued that the zone in transition was a socially disorganized area in which criminal traditions were firmly entrenched and transmitted. Youngsters raised in these neighborhoods thus were likely to differentially associate with and learn criminal values and techniques. Importantly, Sutherland made the same argument about the upperworld. This sector of America, he asserted, also was socially disorganized: The forces organized against crime were weak, whereas the forces organized for crime were strong. The theoretical principle was the same for crimes of the rich and the poor; it was just the content that differed.

Thus, Sutherland identified at least four reasons why the forces aligned to fight white-collar crime are weak. First, due to their respectability and high social status in the community, the public—at least in Sutherland's day—do not think of businesspeople and professionals as "criminal." Accordingly, they do not rise up and call for the enforcement of the law against white-collar waywardness. Second, prevailing laissez-fair capitalist ideology, which Sutherland (1983) termed "anomie," provided a general justification for not intervening in business practices. As he noted, planning to regulate injurious conduct is condemned by business officials as "communistic" (p. 255). Third, business uses its influence to disrupt attempts to control it. This power is seen in its capacity to divert the criminal law from sanctioning untoward upperworld behavior. There also is "fraternization" between business officials and political elites, which undermines officials' motivation to stop illegal conduct (p. 256). Fourth, the potential victims of white-collar crime are, in comparison with their victimizers, weak. "Consumers, investors, and stockholders," observed Sutherland (1940), "are unorganized, lack technical knowledge, and cannot protect themselves" (p. 9). In fact, said Sutherland, "it is like stealing candy from a baby" (p. 9).

By contrast, white-collar crime is "organized crime." Sutherland (1983) argued that "corporate behavior is like the behavior of a mob" (p. 235). Beyond taking action to disorganize the public and potential enforcers (courts, regulatory agencies), business enterprises are organized for crime. Sutherland identified four important features of this criminal organization. First, companies develop, support, and transmit to new employees a criminal culture. This culture involves contempt for government and regulators, the valuing of illegal practices that neutralizes conscience and any feelings of shame, and rationalizations designed to redefine illegal acts as not truly criminal. Second, companies conspire both internally and with other companies to plan and carry out illegal acts (e.g., falsely advertise, fix prices). Third, the corporate veil of companies allows perpetrators of white-collar crime to remain anonymous and for criminal responsibility to be diffused across diverse officials in the organization. Assigning criminal liability to specific individuals thus is difficult, if not impossible. Finally, the rational nature of companies—after all, the purpose of business is to make profits—encourages the amoral selection of illegal practices that can be done secretly and that victimize the weak (i.e., consumers unaware of their victimization and who lack the resources to fight back). Business rationality is not used simply to foster "technological efficiency" but increasingly is aimed "at the manipulation of people by advertising, salesmanship, propaganda, and lobbies," leading companies to embrace "a truly Machiavellian ideology and policy." In the end, said Sutherland (1983), his analysis "justifies the conclusion that the violations of law by corporations are deliberate and organized crimes" (p. 239).

In sum, starting with his Philadelphia address, Sutherland had a defining influence on criminology. As Geis and Goff (1983) note, his introduction of the concept of white-collar crime "altered the study of crime throughout the world in fundamental ways by focusing attention upon a form of lawbreaking that had previously been ignored by criminological scholars" (p. ix). Scholars' interest in white-collar crime was modest, if not low, following World War II and until the turmoil of the 1960s and 1970s sensitized them to how social injustice and crime allowed the "rich to get richer the poor to get prison." As criminologists rediscovered Sutherland, they increasingly explored the causes and developed theories of white-collar crime. We turn next to the fruits of these scholars' labors.

Organizational Culture

An enduring legacy of Edwin Sutherland's approach to white-collar crime is to sensitize us to how the culture within a legitimate company can be criminogenic and transmitted to workers. Subsequent scholars have examined the sources and nature of this culture. A criminal culture consists of definitions that permit or encourage lawbreaking generally or in specific situations. In essence, criminal definitions tell workers either that it is "okay to ignore all these stupid laws" or "okay to break this one law" (e.g., okay to fix prices or to ignore this safety standard). They also might encourage negligent behavior by telling workers that certain risks—whether financial or health-related—are "not that bad" and should be ignored. In this section, we consider three theoretical perspectives that are in the Sutherland tradition.

UNETHICAL CULTURES

Along with Sutherland and a few other scholars (e.g., Gilbert Geis), Marshall Clinard played a defining role in shaping the study of white-collar crime. As Clinard (1952) noted, he was first alerted to this area by Sutherland when he was a graduate student at the University of Chicago. Although not a rigidly devout apostle, Clinard's perspective on upperworld criminality was heavily influenced by Sutherland. In particular, during his career, Clinard would focus on how differential association with varying ethical climates within corporations was a major cause of white-collar crime.

In his classic *The Black Market,* Clinard (1952) explored violations of regulations issued by the Office of Price Administration (OPA) during World War II. Price limits on rents and on scarce foods and consumer goods were imposed to prevent inflation (which would have made the war more costly). Rationing also was instituted to ensure that all citizens would have equitable access to products (e.g., meat, sugar, shoes, gasoline). With the nation's security hanging in the balance, it might have been expected that patriotism would result in universal compliance with these wartime regulations. This was not the case. Instead, Clinard reported that businesses widely and systematically violated these OPA rules. For example, to receive an allotment of beef whose prices were controlled, a wholesaler seeking illegal profits would require a store to take an under-weighted shipment, pay a kickback, or purchase an unregulated product (e.g., sausage) at an inflated price.

Clinard (1952) argued that OPA price violations reflected not "gangster and shady elements in business" (p. 293) but industry-wide practices; the barrel, not a few apples in the barrel, was rotten. Some of these fraudulent schemes were used prior to the war in other contexts; some were newly invented to circumvent the price controls imposed by the OPA. Given the "extensiveness of the black market and the kinds of violations," Clinard concluded that the illegality was due "primarily to subcultural transmission" (p. 299). Techniques for committing OPA illegalities were learned in conversation with other businesspeople. Definitions favorable to violating the laws were reinforced in the business community. Thus, government regulations were dismissed as "stupid," law violators did not lose status among other business executives, and a consensus existed that those in the industry would not "tattle" or snitch on one another (pp. 304–307). As Clinard (1952) asserted in language that Sutherland could have voiced:

> Most black market violations appear to have their origin in behavior learned in association from others, unethical and illegal practices conveyed in the trade as part of a definition of the situation and rationalizations to support these violations of law being similarly transmitted by this differential association. (pp. 298–299)

In 1980, with Peter Yeager, Clinard authored another classic study, *Corporate Crime*. Although not duplicating the methods used by Sutherland, this study replicated the thrust of *White Collar Crime*. For the years 1975 and 1976, Clinard and Yeager examined the illegality of 477 of the largest publicly owned manufacturing corporations in the United States. They recorded the number of criminal, civil, and administrative actions against these companies by 25 federal agencies. They realized that their measure did not take into account the large number of illegal acts that were not detected. Even so, they found that three fifths (60.1%) of the corporations had at least one action initiated against them and that the sample of firms averaged 2.7 federal cases of violation. Most stunningly, the analysis revealed that only 8% of the companies accounted for 52% of all violations, with these high-rate offenders averaging 23.5 infractions each (Clinard & Yeager, 1980, p. 116). We will return to this last finding shortly.

In *Corporate Crime*, Clinard and Yeager (1980) considered how corporate organization facilitated lawlessness. A key component of their perspective was the "culture of the corporation" (p. 58). Two factors made the internal culture a pivotal cause of illegal behavior. First, this culture is replete with "corporate defenses to law violations" (p. 68). These are beliefs or "rationalizations" that define why the laws regulating companies can be obeyed selectively. These include, for example, a belief that the free enterprise system makes any government regulation illegitimate. Another such defense is that an illegal act is permissible if it brings profits. And still another is that regulations are too complex to follow and that infractions are errors of omission. In a later section, we revisit these rationalizations or techniques of neutralization.

Second, corporations are well designed to "indoctrinate" their members into this culture. Employees occupy roles in which they are rewarded for embracing the "corporate mind" (p. 66). This structural location exposes workers to strong socialization pressures. They tend to be isolated from competing views of the world that might challenge their unethical beliefs. As Clinard and Yeager (1980) observed, "they tend to associate almost exclusively with persons who are pro-business, politically conservative, and generally opposed to government regulation" (p. 68). This isolation is enhanced by job transfers to new geographic locations and by overwork whereby the corporation becomes the dominant priority in a person's life. In this situation, "co-workers and higher-ups become 'significant others' in the individual's work and social life" (p. 63).

In a key insight, Clinard and Yeager noted that not all corporate cultures are the same; they vary in their support of unethical practices (see also Victor & Cullen, 1988). Recall that their empirical data support this conclusion. Some corporations—about 40%—had no violations in the 2-year window they examined. By contrast, an 8% minority of firms engaged regularly in illegal conduct. What distinguishes criminal from noncriminal corporations? Why do some companies have unethical cultures whereas others do not?

Clinard addressed this question with what might be referred to as a *managerial theory of corporate culture* in his 1983 book, *Corporate Ethics and Crime: The Role of Middle Management*. Based on interviews with 64 retired middle managers from Fortune 500 companies, Clinard (1983, p. 89) distilled a central conclusion: The culture or "ethical tone" of a company is heavily influenced by the orientation of top-level management, especially the chief executive officer (the CEO). Although corporate traditions exist, they are dynamic and will change if inconsistent with the views of high-level executives. CEOs promoted from within the firm due to professional or technical expertise are less likely to sponsor unethical practices. On the other hand, "financially oriented" CEOs—outsiders hired to produce profits—are likely to create a climate conducive to lawbreaking. The scandalous actions of Wall Street financial firms that led to the implosion of the worldwide economy in 2008 comprise but one example lending support to this proposition (see, e.g., Lewis, 2010).

OPPOSITIONAL CULTURES

"The theory of reintegrative shaming," observed Braithwaite (1989), "is unlike other theories of crime in the literature, with the notable exception of differential association, in that it does not exclude white collar crime from that which is to be explained" (p. 124). In Chapter 7, we reviewed shaming theory, noting Braithwaite's central thesis that reintegrative shaming lessened, whereas stigmatizing shaming increased, criminal behavior. Braithwaite, however, also is a major white-collar crime scholar, raised in criminology when the field was sensitized to upperworld criminality. He was prompted to conduct important studies, for example, of illegalities in the pharmaceutical and coal mining industries (Braithwaite, 1984, 1985). Not surprisingly, he understood the need for his theory to explain crimes of rich and poor.

As noted, an important source of white-collar crime is that cultures within organizations support it. According to Braithwaite, these cultures flourish for two reasons. The first reason, emphasized by many other scholars, is that illegal behaviors in the business world have been shamed only infrequently. When harmful acts are not morally rebuked—when moral standards are not defined and reinforced—criminal cultures can persist unimpeded. The second reason, unique to Braithwaite's perspective, is that efforts undertaken to shame and control untoward practices might have the unanticipated consequences of strengthening cultural values supportive of lawlessness. The core proposition is that *stigmatizing shaming fosters an oppositional subculture supportive of white-collar crime.*

Braithwaite (1985, 2002) has written extensively on the complexities of regulating corporate behavior. From firsthand observation, he is hardly naïve about the need to impose punitive sanctions, including criminal penalties, on recalcitrant companies and their executives. He has learned, however, that punitive sanctions that stigmatize frequently backfire among businesspeople already mistrustful of state regulation. When an adversarial approach is taken that seeks to stigmatize and humiliate, businesspeople are likely to respond with defiance (see also Sherman, 1993). They embrace the technique of neutralization of "condemning the condemners," seeing those imposing standards on them as unfair and illegitimate (Braithwaite, 1989, p. 127; see also Sykes & Matza, 1957). Stigmatized business officials, both within a company and across an industry, are also likely to bond with one another, thus creating solidarity and differential association supportive of illegal practices. Sharing oppositional views, they can form an "organized subculture of resistance that advocates contesting all enforcement actions" and "consistently challenging and litigating the legitimacy of the government to enforce the law" (p. 129).

Similar to responses to street crime, Braithwaite (2002) favors a more restorative approach that has a better chance to move firms to fix problems and reduce harms. Thus, reintegrative shaming would involve punishing "in a way that maintains dignity and mutual respect between enforcer and offender. Where possible, punishment should be executed without labeling people as irresponsible, untrustworthy outcasts, but instead inviting the offender to accept the justice of the punishment" (1989, pp. 131–132). Adopting the style of effective parenting, regulators should engage in warm but restrictive interventions. The preferred strategy is to secure compliance to stop illegal, harmful practices by appealing to offenders' moral side and *persuading them to stop the shamed behavior.* Persuasion is most effective when it is backed up by the threat of harsher penalties, including criminal sanctions (Braithwaite, 1985). According to Braithwaite, this approach has the potential to undermine the formation of oppositional cultures and to create a collective response in an industry to avoid fraudulent and injurious practices. Similar to Clinard, Braithwaite thus sees organizational criminal cultures as created and as potentially open to reform.

THE NORMALIZATION OF DEVIANCE

On January 28, 1986, the Space Shuttle *Challenger,* carrying a crew of seven, including New Hampshire teacher Christa McAuliffe, exploded 73 seconds after takeoff. The catastrophe was caused by the failure of an O-ring, a rubber-like seal on a joint on the Solid Rocket Boosters. In cold temperatures, the O-ring would potentially harden and not be able to prevent hot propellant gasses from igniting into a flame that could penetrate a tank containing liquid hydrogen and oxygen—all with lethal consequences. On the morning of the *Challenger* tragedy, the temperature reached only 36 degrees Fahrenheit, an unprecedented low for a launch from the John F. Kennedy Space Center in Cape Canaveral, Florida. Despite this knowledge of the O-ring dangers, NASA and Morton Thiokol, the manufacturer of the Solid Rocket Boosters, nonetheless chose to send *Challenger* into space (Vaughan, 1996).

In retrospect, observers—including a presidential commission investigating the incident—were appalled by the launch decision because it was "perfectly clear" that known, unacceptable risks were taken. Some commentators regard this decision as a crime (Kramer, 1992). How could anyone put seven lives in jeopardy when the O-ring faced the imminent prospect of lethal failure? In her remarkable account of the catastrophe, *The Challenger Launch Decision,* Diane Vaughan (1996) suggests two answers to this question: the standard theory of *amoral calculation* and her theory of *the normalization of deviance.*

When America's space program was initiated, it was administered as a scientific project in which cultural norms of technology and safety were preeminent. Well funded, there were few pressures for the program to prove its worth or to make money. By the 1980s, however, spacecraft hurtling into space had become commonplace, and NASA had to compete with other government agencies and priorities for funding. The space shuttle program was thus designed to employ reusable craft, such as *Challenger,* that could be launched frequently enough to be cost effective (i.e., by carrying private, military, and scientific payloads into space). This reconceptualization of missions— from space exploration to commercial enterprise—placed NASA under pressure to meet launch deadlines so as to remain politically viable (Kramer, 1992). According to Vaughan (1996), "Congress and the White House established goals and made resource decisions that transformed the R&D space agency into a quasi-competitive business operation, complete with repeating production cycles, deadlines, and cost and efficiency goals" (p. 389). The agency's technical, scientific culture thus increasingly was rivaled, if not supplanted, by a "culture of production" (p. 196).

In this context the standard criminological theory is that NASA and Morton Thiokol made the joint *amoral calculation* to launch *Challenger.* NASA could not afford further delays, whereas Morton Thiokol did not wish to disappoint its benefactor. In preflight deliberations, Morton Thiokol engineers expressed safety concerns but were overruled by upper-level management who approved the launch—presumably to please NASA officials who also were aware of the potential risks. According to Vaughan (1996), the "conventional explanation" argues that "production pressures caused managers to suppress information about O-ring hazards, knowingly violating safety regulations in order to stick to the launch schedule" (p. xii).

In her exhaustive study, however, Vaughan (1996) presents a more nuanced explanation. To be sure, the culture of production circumscribed those in the space program, penetrated the preexisting technical-professional culture, and created performance pressures. Within this contextual reality, work groups functioned to deal with a host of problematic technical issues in ways that would not unduly cause the shuttle program to fall off schedule. The O-ring problem was not new. But similar to other risks, the work group incrementally defined the growing hazard as manageable. Each time a difficulty with the O-rings emerged, it was analyzed and the associated risk was judged by the work group to be acceptable. According to Vaughan (1996), the record reveals "an incremental descent into poor judgment. It was typified by a pattern in which signals of potential danger—information that the booster joints were not operating as predicted—were repeatedly normalized by managers and engineers" (p. xiii).

Importantly, the "normalization of deviance"—the emergence of a worldview that neutralizes perceptions of danger—is a cultural set of beliefs and norms that guides decision making. Thus, when NASA and Morton Thiokol officials consulted in a conference call, their decision to launch *Challenger* did not suddenly disregard protocols and amorally calculate to risk the lives of seven crew members so as to avoid the inconvenience of another flight delay. Rather, the personnel followed work-group norms that led them to depreciate the risk of O-ring failure. "It was not amorally calculating managers violating rules that were responsible for the tragedy," concluded Vaughan (1996, p. 386). "It was conformity." According to Vaughan (1996):

> As a result, the decisions from 1977 through 1985 that analysts and the public defined as deviant after the *Challenger* tragedy were, to those in the work group making the technical decisions, normal within the cultural belief systems in which their actions were embedded. Continuing to recommend launch in the FRR [Flight Readiness Review] despite problems with the joint was not deviant; in their view, their conduct was culturally approved and conforming. (p. 236)

Vaughan (1996) sees her work as an investigation of the "sociology of mistake" (p. xiv). But it also might be envisioned as illuminating the sociology of corporate criminal negligence. There are, of course, many instances in which managers knowingly engage in fraudulent activity (e.g., price-fixing) and in marketing defective products. But what Vaughan has ingeniously uncovered is the process whereby managers and employees evolve ways of making decisions that unknowingly lead them, step by step over time, to deny that a hazard exists. This normalization of risk or deviance allows dangerous products to be marketed or not recalled promptly; it also can allow for risky financial practices that place investors in jeopardy of bankruptcy. The key point is that corporate cultures are complex and can facilitate illegal practices in diverse ways.

Organizational Strain and Opportunity

Businesses are goal-directed economic enterprises that seek to stay in business (avoid failure) and to increase market share and profits (meet rising expectations). Constant pressures to meet goals are often an everyday reality within many corporate environments. As Yeager and Simpson (2009) note, it is not surprising that "perhaps the most popular explanation for corporate offending is strain" (p. 357; see also Croall, 1992). Theorists have long recognized, however, that structurally induced strain can be adapted to in various ways. Using unlawful means requires access to criminogenic opportunities (Cloward, 1959; Cloward & Ohlin, 1960; Cullen, 1984). "Strain provides the motive for offending (the arousal of behavior)," Yeager and Simpson (2009, p. 337) observe, but corporate "offending will not occur absent opportunity and choice." Strain and opportunity thus are frequently seen as two intersecting factors that foster involvement in white-collar crime.

STRAIN AND ANOMIE

Traditionally, strain theory has been viewed as an explanation of lower-class crime and delinquency (see Chapter 4). To be more precise, it is a theory of individuals' transition to adulthood. Those thwarted in their quest for upward mobility—those denied a piece of the American dream—are most likely to offend. Because goal blockage is most pervasive in the lower echelons of society, poor youngsters are at risk of delinquency and, in turn, of adult offending.

As scholars realized, however, strain theory also has relevance for explaining the criminality of individuals who have made the transition successfully into adulthood but then enter a new social domain—corporate America—where expectations for goal achievement do not wane but intensify. As individuals, they are under performance pressures to meet goals or be denied mobility within the company. They may experience relative deprivation compared to other employees who are rising more rapidly up the corporate ladder (Passas, 2010). Pressures to deviate thus may be both acute and chronic.

Further, firms themselves exist in a competitive environment where they "are exposed to culturally approved goals" and "experience blocked opportunities" (Vaughan, 1997, p. 99). There is some evidence that financial difficulties are related to corporate violations (Shover & Scroggins, 2009; Yeager & Simpson, 2009). "Although the data are somewhat mixed," conclude Agnew, Piquero, and Cullen (2009),

> studies suggest that corporate crime is more common in for-profit companies, companies with relatively low profits, companies with declining profits, companies in depressed industries, and companies suffering from other types of financial problems (e.g., low sales relative to assets, small or negative differences between assets and liabilities, perceived threats from competitors). (p. 39)

This tendency, however, should not obscure that scandals frequently occur in firms, such as Enron and Goldman Sachs, where corporate earnings are skyrocketing. Executives, whose bonuses depend on stockholder returns, can exert inordinate pressure on underlings to produce ever-rising profits. At a broader level, this intense emphasis on economic success can generate not only strain but also—as Merton would predict—anomie. As may be recalled from Chapter 4, anomie is a condition where the norms regulating the use of legitimate methods to achieve goals are weakened or rendered ineffectual. In this circumstance, as Messner and Rosenfeld (2001) have noted, individuals are free to use the technically most expedient means to achieve desired ends. In the business world, this might involve defrauding clients or scrimping on costly worker safety conditions.

Notably, scholars have argued that anomie is rampant in many business organizations (Cohen, 1995; Passas, 2010; Vaughan, 1997; see also Waring, Weisburd, & Chayet, 1995). In this model, goal attainment becomes preeminent and there is a lack of concern for legitimate means. The ethical climate within the organization thus comes to emphasize, in Cohen's (1995) words, "instrumentalism," "individualism," "minimal interpersonal responsibility," an "emphasis on efficiency" and "cost control," and a "lack of concern for employees" (p. 196; see also Messner & Rosenfeld, 2001). Little attention is paid to business ethics and legal compliance. In this context, "criminal business practices" result (Cohen, 1995, p. 196).

CRIMINOGENIC OPPORTUNITIES

As will be explored more fully in Chapter 13, for a criminal event to occur, the opportunity to carry out the act must be present. This observation—no opportunity, no crime—might seem rather trite. But when taken seriously, it means that understanding crime requires studying in detail the nature of criminogenic opportunities. In this regard, Michael Benson and Sally Simpson (2009) have developed "an opportunity perspective" for the explanation of white-collar crime. Their insights illuminate the distinctive nature of upperworld opportunities.

Most generally, a criminal opportunity involves two components (again, see Chapter 13). First, there must be an *attractive target.* This might be a person to rob, property to steal, or someone's life savings to pilfer. Second, there must be an *absence of capable guardianship.* This might be a burly companion on a nightly walk, an alarm system on a house, or an accountant supervising one's funds. When there is an attractive target and no guardianship, a criminal opportunity exists.

With street crime, offenders must employ certain techniques to gain access to attractive targets and to avoid guardianship. Let us take burglary as an example. As Benson and Simpson (2009) note, burglars often have to use physical means to gain entry to a residence (e.g., kick in a door), have to be in a place where they have no legal right to be, have direct contact with a victim's living quarters, and commit their crime at a specific time and location. Burglary thus requires a certain physicality, nerve to enter a strange place, and risk of detection (someone could call the police or see and identify the offender).

By contrast, white-collar offenders typically have three very distinct properties. As Benson and Simpson (2009) explain, "(1) the offender has *legitimate access* to the location in which the crime is committed, (2) the offender is *spatially separated* from the victim, and (3) the offender's actions have a *superficial appearance of legitimacy*" (p. 80, emphasis in original). Let us take price-fixing as an example. An executive breaks down no doors but simply sits in an office or meets over lunch. When consumers pay inflated prices for the produce whose prices are rigged, the executive never sees these victims. And the scheme is undertaken behind the cloak of a respectable business firm.

The key point is that legitimate businesses—whether corporations or perhaps the office of a physician—provide access to vulnerable targets. Because the victimization occurs in the course of seemingly legitimate business activity, it is often undetected. In fact, in a complex society, the public is in a position where its members must trust that their money is not being swindled, that products are not being falsely advertised, that safety regulations are being followed, and that they are being billed for services actually provided. In many cases, however, citizens have little protection against the abuse of trust. As repeated scandals—small and large—show, white-collar officials have many ways to deceive the public. Behind closed doors, they can conspire to rig prices, manipulate financial statements, sell stocks soon to be worthless, set up Ponzi schemes, choose to ignore environmental standards, and so on. With the organization as a shield, they can conceal these illegal activities that often are difficult to distinguish from regular business activity (Benson & Simpson, 2009).

As Shover and Hochstetler (2006) have observed, the upperworld is filled with many temptations to profit by stepping outside the law. They call these temptations "lure," which they define as "arrangements or situations that turn heads. Like tinsel to a child, it draws attention" (p. 27). Lure becomes a criminal opportunity "in the absence of credible oversight" (p. 28). Their views thus capture the core components of opportunity: an attractive target and no guardianship. But an important reality is present in the concept of "lure": that the motivation to engage in a profitable white-collar crime might be produced by the presence of temptation. As has been said, in the case of white-collar crime, the opportunity may make the criminal.

In fact, the very nature of a capitalist system that expands, creates new industries, and offers innovative financial services may produce not only wealth and social improvement but a host of fresh and luring criminal opportunities (see, e.g., Calavita, Pontell, & Tillman, 1997; Jesilow, Pontell, & Geis, 1993). In his 1907 *Sin and Society*, sociologist E. A. Ross made this very point. "The sinful heart is ever the same," said Ross, "but the sin changes in quality as society develops" (p. 3). Trust in others is unavoidable in modern society. As Ross noted, "I let the meat trust butcher my pig, the oil trust mould my candles, the sugar trust boil my sorghum, the coal trust chop my wood, the barb wire company split my rails" (p. 3). The difficulty, however, was that such trust created criminal opportunities that businesspeople exploit. "The sinister opportunities presented in this webbed social life have been seized unhesitatingly, because such treasons have not yet become infamous" (pp. 6–7). According to Ross, the public was unaware of these new varieties of sins because they were committed at a long distance, anonymously, and behind the appearance of respectability. "The modern high-power dealer of woe," observed Ross, "wears immaculate linen, carries a silk hat and a lighted cigar, sins with a calm countenance and a serene soul, leagues or months from the evil he causes. Upon his gentlemanly presence the eventual blood and tears do not obtrude themselves" (pp. 10–11).

Deciding to Offend

Because many white-collar crimes are committed in organizational settings, scholars have—as we have seen—focused on how the cultural and structural contexts of companies are criminogenic. Other scholars, however, have focused less on this causal "background" and more on the "foreground" of white-collar crime. That is, when upperworld officials decide to offend, what are they thinking? Insights into the proximate origins of white-collar illegality tend to fall into two theoretical categories: techniques of neutralization theory and rational choice theory. These perspectives are reviewed below.

DENYING THE GUILTY MIND

Sutherland confronted the conundrum of how respectable people—many presumably raised in wholesome surroundings and churchgoers as adults—can nonetheless victimize others in the course of their white-collar occupations. How can they overcome their conventional respectability and, in Benson's (1985) phrasing, "deny their guilty minds"? In part, he noted that they learn "rationalizations" or criminal definitions that justify such offending in particular situations. To them, robbing someone on the street would be a "crime" and morally unthinkable, but robbing consumers by fixing prices is merely a "necessary business practice" that "everyone in the industry does." In *The Black Market,* Clinard (1952) echoed this insight and later called these definitions "corporate defenses to law violations" (Clinard & Yeager, 1980, p. 68). As might be recalled from Chapter 5, the more general statement of this perspective can be found in Sykes and Matza's theory of techniques of neutralization.

Donald Cressey (1950, 1953) applied Sutherland's views on "rationalizations" or definitions of the situation favorable to crime in his classic study of bank embezzlement (see also Jonson & Geis, 2010). Based on interviews with incarcerated embezzlers, Cressey argued that three factors intersected to enable respectable people to take "other people's money." These are sometimes called the "fraud triangle" (Jonson & Geis, 2010, p. 225). In essence, he developed an integrated theory that took into account motivation, opportunity, and decision making.

First, they had to face a "non-sharable problem" such as worries over unpaid gambling debts or excessive family expenses. Embezzlers, noted Cressey (1950), had a high incidence of "'wine, women and wagering'" (p. 743). Because they held respected positions in the community, money problems stemming from disrespected activities had to be kept secret. To ask for help risked tarnishing one's good reputation. Second, those having a non-sharable problem had to hold a position of financial trust and perceive that they could exploit the opportunity before them to pilfer funds. According to Cressey (1950), the embezzlers often told him that at a certain point, "'it occurred to me' or 'it dawned on me' that the entrusted funds could be used for such and such purpose" (p. 743). Third, embezzlers had to overcome the "contradictory ideas in regards to criminality on the one hand and in regard to integrity, honesty and morality on the other" (p. 743). Here, they had to invoke "verbalizations" or "rationalizations" that defined the violation of trust not as a crime but as something else—such as only "borrowing" money that would be paid back paid back at a later date. As Cressey (1950) observed:

It is because of an ability to hypothesize reactions which will not consistently and severely condemn his criminal behavior that the trusted person takes the role of what *we* have called the "trust violator." *He* often does not think of himself as playing that role, but instead thinks of himself as playing another role, such as that of a special kind of borrower or businessman. (p. 843, emphasis in original)

Shover and Hunter (in press) argue that the class status of white-collar offenders equips them with "cultural capital" that can be drawn on to deny a guilty mind. These offenders have "a level of verbal creativity and adroitness that enables them to argue self-interested interpretations of their fall from grace." They "deny criminal intent or that the acts were harmful," impute "malice and unworthy motives" to their accusers, assert that the "evidence that convicted them was false or misunderstood," and complain that they are "only 'technically guilty' or were hamstrung by circumstances that caused them to make mistakes that later would be labeled a 'crime'" (Shover & Hunter, in press). These "accounts" are efforts to deal with the stigma of a criminal conviction (Benson, 1985). But presumably they also are verbalizations that could be invoked prior to and during the commission of white-collar crimes. In this sense, the cultural capital of members of the upperworld supplies them with the verbal ability to neutralize guilt and to offend while maintaining outwardly, if not inwardly, the appearance of respectability.

The accounts or neutralizations used to justify white-collar crimes are not idiosyncratic—invented anew by each offender—but rather are patterned and repeated frequently (Shover & Hunter, in press). Verbalizations excusing the violation of financial trust voiced by Cressey's embezzlers in the 1950s could just as easily be heard today. This finding suggests that justifications for offending are deeply rooted in America's economic arrangements and corresponding cultural beliefs. As Benson and Simpson (2009) noted, the

world of business is imbued with a set of values and ideologies that can be used to define illegal behavior in favorable terms.... The availability of these norms and customs ... enable[s] potential white-collar offenders to interpret their criminal intentions and behavior in non-criminal terms. (p. 141)

For example, capitalism promotes a strong belief in free enterprise that can be used to define government regulation as illegitimate interference. Ignoring regulations thus may be justified—a guilty mind neutralized—by invoking the excuse that government controls were unfairly stifling the businessperson's right to make a profit.

Notably, the theory of techniques of neutralization has been used frequently in explanations of various white-collar crimes (Benson & Simpson, 2009; Shover & Hunter, in press). One example is found in Paul Jesilow, Henry Pontell, and Gilbert Geis's investigation of physician Medicaid fraud, *Prescription for Profit*. Jesilow et al. (1993) detail the structural conditions that created rampant opportunities for doctors to defraud the government. In essence, physicians were placed in positions of professional and financial trust. They were paid by the government to provide medical services to the needy. The Medicaid program, initiated in 1965, assumed that doctors would prescribe the appropriate services to patients and then invoice the government for the fee established by program guidelines. But with virtually no regulation or chance of detection, this fee-for-service arrangement created ubiquitous opportunities for fraud:

> Because doctors are reimbursed a fixed amount for each procedure, they can earn additional income by charging for a more expensive procedure than the one performed, double-billing for services, pingponging (sending patients back for unnecessary visits), family ganging (examining all members of the family on one visit), churning (mandating unnecessary visits), and prolonging treatment. (Jesilow et al., 1993, pp. 7–8)

Still, how could they succumb to this lure and engage in practices that were a gross violation of their professional ethics, not to mention the law? The answer was their ability to use techniques of neutralization. Jesilow et al. interviewed 42 physicians apprehended for Medicaid fraud. "To hear them tell it," Jesilow and colleagues reported, "they were innocent sacrificial lambs led to the slaughter because of perfidy, stupid laws, bureaucratic nonsense, and incompetent bookkeepers" (p. 148). All employed at least one of Sykes and Matza's (1957) techniques of neutralization (see Chapter 5). A common justification, "denial of a victim," was that they bilked Medicaid because they were not being paid what the service cost them to deliver. Another justification, "denial of injury," was that they were providing needed care that Medicaid would not permit. These beliefs were so pervasive that Jesilow et al. wondered if they comprised a "subculture of medical delinquency" (p. 175). Doctors could access these subcultural beliefs and employ them as "professional justification in lieu of defining their activities as deviant, illegal, or criminal" (p. 175).

WHITE-COLLAR CRIME AS A RATIONAL CHOICE

Business activity involves making profits—engaging in activities in which the benefits outweigh the costs. Rationality is rewarded with wealth; irrationality is punished, at times with bankruptcy. Thus, if any offenders' decisions should be guided by rational choice, it should be those in the business community. In fact, one explanation for why upperworld criminality flourishes is that the potential profits are high and, due to weak enforcement and the secrecy organizations provide, the costs are low. This perspective would also predict that managers who perceive offending to be profitable should be more likely to break the law. There is, however, a counterargument. To be sure, choices are made, but are they rational? Competing criminological theories argue that, beyond sheer profitability, many other factors shape the decision to offend. These might include social bonds, self-control, the strength of criminal versus conventional values, exposure to strain, and so on. The broader theory of rational choice is discussed in detail in Chapter 13. Here we are concerned with its applicability to white-collar crime.

In this regard, the most systematic perspective is Raymond Paternoster and Sally Simpson's (1993) *rational choice theory of corporate crime* (see also Paternoster &

Simpson, 1996; Simpson, Piquero, & Paternoster, 2002). They are interested in explaining the willingness of employees to commit crimes on behalf of the corporation. The key to this decision to offend is not the objective costs and benefits that might accrue but the *perceived utility* of the act. In this model, offenders use perceptions of what will occur to calculate whether a crime has utility (whether the benefits outweigh the costs).

Crass rational choice theories examine only the perceived certainty and severity of formal sanctions (e.g., penalties from regulatory agencies or criminal courts). As Paternoster and Simpson understand, however, two other kinds of costs also might shape the perceived utility of a corporate crime. One category is perceived informal sanctions. These might include costs to the company in terms of negative publicity or costs to the individual in terms of negative reactions from friends and family members. Another category is internally imposed sanctions, in particular the loss of self-respect that might be feared if laws were broken.

In deciding to offend, a corporate official thus must judge potential formal, informal, and self-imposed costs. But people do not arrive at a point of decision as empty vessels devoid of all morality. Variation in people's morality thus further shapes the criminal choice. If held strongly, moral beliefs can override perceived utility; some people will not do what they think is wrong. Other moral beliefs, however, are akin to situational ethics. They are "moral rules-in-use" that define "the acceptability of *particular* conduct within a *particular* context" (Paternoster & Simpson, 1993, p. 45, emphasis in original). These are definitions of the situation, or techniques of neutralization, that might justify illegal acts under some circumstances. Related to this point, the moral constraint of a regulation further depends on the "*perceived sense of the legitimacy of the rules and rule enforcers*" (p. 45, emphasis in original). Laws seen as unfair are less binding.

Beyond sanctions and controls, corporate managers must consider two other factors: the costs of complying with the law and benefits of not complying. This assessment then must be weighed against other legitimate options that might be available in the corporation or its environment for resolving a given problem. If lawful alternatives are closed off or seem too costly, the decision to offend gains in utility. Further, if officials have offended in the past, they may manifest behavioral stability. Even in the upperworld, it may be that in line with past "criminological research" on street criminals, "the best predictor of future offending is past offending" (Paternoster & Simpson, 1993, p. 47).

Paternoster and Simpson thus predict that corporate crime will be more likely when managers (1) perceive that formal and informal sanctions will be weak, (2) do not experience a loss of self-respect, (3) lack a strong morality or have internalized situational rules-in-use that justify the act, (4) view rules as unfair, (5) judge both the benefits of noncompliance and the costs of compliance as high, and (6) have broken the law in the past. Their tests of this model, mainly using vignettes to probe factors that affect managerial decisions, have produced supportive results (Paternoster & Simpson, 1996; Simpson et al., 2002; Smith, Simpson, & Huang, 2007). The question is the relative importance of "rationality"—assessments of costs and benefits—versus other factors. Morality appears to be the strongest predictor of the willingness to offend. Other individual traits, such as the desire for control, might shape perceptions of sanction effects and independently influence corporate crime decision making (Piquero, Exum, & Simpson, 2005). Most important, a strict rational choice model decontextualizes potential offenders. They occupy positions that are enmeshed in an ongoing corporate entity that has a history and habits, standard operating procedures, work group cultures, and a changing external environment. Perceptions of utility matter, but they are likely to be contingent on a host of contextual factors (Simpson et al., 2002).

State-Corporate Crime

At 8:00 a.m. on September 3, 1991, a fast-spreading fire broke out at Imperial Food Products, Inc., a chicken-processing plant in Hamlet, North Carolina. The fire erupted when a hydraulic line above a vat of grease burst, spraying flammable liquid into the 400-degree oil in the vat. The plant had no sprinklers, no windows, and few doors. Most egregiously, a metal fire exit door, later found to have dents in it from workers desperately seeking escape, had been padlocked shut on the orders of Emmett Roe, the company's owner. A truck also had been parked a few inches from the door, making it impassable. The owner's intent was to prevent workers who were pilfering chicken parts from sneaking out of the plant undetected. Of 90 workers, 24 died and 56 were injured. A snack-company employee servicing the vending machines also perished. Years later, another worker died from complications stemming from injuries on that day (Aulette & Michalowski, 1993; Cullen et al., 2006; Wright, Cullen, & Blankenship, 1995).

At first blush, identifying the white-collar criminal seems easy: Emmett Roe, due perhaps to rational choice and criminal values, decided to close off a fire exit to save a few chicken parts. Roe was duly indicted on 25 counts of involuntary manslaughter and, in a plea agreement, received a 19-year, 11-month sentence. He would be released after serving 4 and 1/2 years (Cullen et al., 2006). Justice was done—or was it? According to Ronald Kramer and Raymond Michalowski, this social construction of the deaths at the plant as an individual owner making a bad decision obscures the true nature of the offense (see also Aulette & Michalowski, 1992). In their view, this is an example of a *state-corporate crime* (Kramer, Michalowski, & Kauzlarich, 2003).

According to Kramer et al. (2003), a state-corporate crime "is defined as criminal acts that occur when one or more institutions of political governance pursue a goal in direct cooperation with one or more institutions of production and distribution" (p. 263). Essentially, Kramer and Michalowski are seeking to sensitize criminologists to the political economy of crime in the United States. The state and corporations are inextricably intertwined. Corporations are legal creations allowed to exist by the state; they also generate wealth and use their power to influence state policies. Although exceptions exist—such as when a Wall Street–banking scandal is disclosed—the state and corporate America share interests and ideology. Often in secretive ways, they do one another's bidding.

Kramer and Michalowski have distinguished two types of state-corporate crime. Both types involve the "coincidence of goal attainment, availability and perceived attractiveness of illegitimate means, and an absence of effective social control" (Kramer, 1992, p. 239; see also Dallier, 2011). First, *state-initiated corporate crime* "occurs when corporations, employed by the government, engage in organizational deviance at the direction of, or with the tacit approval of, the government" (Kramer et al., 2003, p. 271). Kramer's (1992) analysis of the Space Shuttle *Challenger* (discussed above) falls into this category (see also Vaughan, 1996). In his view, for a host of political and economic reasons, NASA was under intense pressure to attain the goal of regular, on-schedule space shuttle launches. Without Morton Thiokol's cooperation, however, the *Challenger* launch would have been cancelled. Wishing to please NASA, the company ignored internal engineering data on the potentially lethal risk of an O-ring malfunction in cold weather. Morton Thiokol managers approved the launch, and NASA proceeded to send *Challenger* into space. No external controls existed to ensure that this state-corporate decision was safe. Seven crew members perished as a result.

Second, *state-facilitated corporate crime* "occurs when government regulatory institutions fail to restrain deviant business activities, either because of direct collusion between business and government or because they adhere to shared goals whose attainment would be hampered by aggressive regulation" (Kramer et al., 2003, pp. 271–272). According to Kramer and Michalowski, the deadly fire at the Imperial Food Products plant qualifies as a state-facilitated corporate crime. To be sure, Emmett Roe was criminally negligent in padlocking the fire exit. But focusing on him misses the larger contextual reality that created this criminal opportunity and event. Roe was misguided, but he surely had no intent to visit harm on his employees. More telling, he would never have been allowed to chain the door shut if the state, in this case North Carolina, had properly enforced safety standards. To create a so-called favorable business climate—so as to attract companies—North Carolina embraced a policy of weak regulation. The plant building, which was more than 100 years old, had not been inspected by the state's Occupational Safety and Health Administration in 11 years (Cullen et al., 2006). More shocking, shortly before the fire, the North Carolina agency "had returned nearly a half million dollars in unspent OSHA money to the federal government" (Kramer et al., 2003, p. 277). In short, the state was equally, if not more, complicit than Emmett Roe in failing to prevent the fire and in placing corporate profits above worker safety and the potential harm to human life.

The unique contribution of Kramer and Michalowski's theory of state-corporate crime is that it forces consideration of the ways in which the state and corporations intersect to create criminal motivations and opportunities. As the most powerful actors in American society, their partnership in crime can produce inordinate harm with virtual impunity.

Consequences of White-Collar Crime Theory: Policy Implications

Perhaps the clearest lesson that can be drawn from existing theories is that reducing crime in the upperworld poses a daunting, if not insurmountable, challenge. As we have seen, unlike most street offenders, white-collar criminals occupy respected positions in society and often work within powerful companies. This corporate veil offers them secrecy and the ability to conceal their waywardness. In fact, their criminal acts frequently are intermingled with legitimate economic conduct—so much so that the two are indistinguishable to potential victims. In some firms, lawbreaking is supported by corporate customs (or habits). Deviant practices can become normalized so that participants fail to appreciate the recklessness of their decisions. Criminal cultures also prevail; these beliefs can be accessed to neutralize guilt or other moral restraints. The state at times conspires with corporations to ensure that regulations are weak or only weakly enforced. The use of criminal sanctions against executives historically has been limited. Much as Sutherland learned more than seven decades ago, when scholars peek into virtually any industry or profession, instances of unlawful behavior are easily detected.

Still, two approaches to combating white-collar crime are possible and have, in fact, been used to varying degrees. First, Sutherland—and E. A. Ross before him—warned that upperworld lawlessness flourished because society was disorganized. They were arguing that whereas white-collar offenders were organized for crime, American society was not organized against this form of illegality. The disorganization was due in part to public ignorance of the true effects of white-collar offenses and in part to the ability of powerful interests to use their influence to deflect efforts to control their conduct. However, the movement against white-collar crime that emerged in the 1970s has substantially altered this balance of power. Existing polling data are clear in showing that the public mistrusts executives, especially those from "big business," and is willing to punish them with the criminal law (Cullen et al., 2009). Further, the use of criminal sanctions against corporations and professionals has grown meaningfully in recent decades (Benson & Cullen, 1998; Cullen et al., 2006; Liederbach, Cullen, Sundt, & Geis, 2001). If nothing else, society is more positioned to uncover and fight revelations of white-collar crime when they emerge. State officials seeming to be too cozy with such offenders even risk being turned out of office. For this reason, officials are willing to send those found to have abused their trust—Bernard Madoff and Jeffrey Skilling are prime examples—to prison for much, if not all, of the remainder of their lives (see Cullen et al., 1987).

Second, there is a difference of opinion among scholars as to the extent to which criminal sanctions—especially those focused on corporate executives rather than simply their firms—can deter white-collar offending (Cullen et al., 1987; Simpson, 2002). Regardless, a general consensus exists that, similar to street crime, imposing harsh criminal sanctions is limited as a crime control strategy. Criminological theorists thus have urged that companies be persuaded or forced to engage in self-regulation (Braithwaite, 1985; Simpson, 2002). The focus should be on company managers taking steps both to change conditions that motivate offending and to close off criminal opportunities (Benson & Simpson, 2009). Such strategies might include management training in business ethics, the appointment of a compliance officer and staff to ensure that regulations are followed, opening lines of communication so that unacceptable risks or practices can be reported to upper-level executives, and the use of stricter accounting procedures to ensure that financial misconduct is not possible (Simpson, 2002).

Conclusion

Beginning with his 1939 Philadelphia address, Edwin Sutherland challenged criminology in two ways. First, he urged scholars to examine how conventional images of criminals were internalized not only by the public but also by the criminologists themselves. In Sutherland's view, criminology was fundamentally class biased. By studying offenders primarily in disorganized inner-city neighborhoods, scholars implicitly affirmed that crime was the exclusive province of the disadvantaged. In coining the concept of "white-collar crime," however, Sutherland dramatically revealed this vision of the crime problem to be false. Unfortunately, this message that crime existed at all levels of society largely fell on deaf ears until the events of the 1960s and early 1970s transformed criminologists' consciousness. A new generation of scholars, sensitized by prevailing events to social injustice, rediscovered Sutherland and revitalized the investigation of upperworld criminality. Since that time, repeated scandals—with enormous costs to American society—have regularly transpired and reinforced the need for criminologists to unravel the causes of, and expose the inordinate damage done by, the crimes of the powerful.

Second, having established the existence of white-collar crime, Sutherland upped the ante on what must be explained for a perspective to quality as a *general theory of crime.* As Braithwaite (1989) has noted, there is "nothing wrong with having a theory of a subset of crime, so long as a theory of rape, for example, is described as a theory of rape rather than as a general theory of crime" (p. 124). But scholars have tended to claim for their framework the status of a general theory even though they "adopt a class-biased conception of crime that excludes white collar offenses" (Braithwaite, 1989, p. 124). In Sutherland's aftermath, it is clear that any claim of theoretical generality will be scrutinized to see whether the perspective can provide an adequate explanation of crime in corporate suites and not just city streets. This task is particularly problematic for trait theories that attribute crime to pathological individual characteristics that are predicted to produce both crime and social failure (e.g., low self-control). The social Darwinism inherent in these perspectives cannot simultaneously account for how flawed people, who are predicted to fail at school and in the workplace, can rise into the upperworld where white-collar crimes take place.

In short, Sutherland—and the scholars who have followed in his footsteps—have cautioned criminologists not to take the concept of "crime" for granted. To be sure, investigating the crimes of the powerful is a daunting challenge because these offenders are often talented at concealing their criminality and do not welcome scholars snooping around and asking them to self-report on their waywardness.

Regardless, the components of a more systematic integrated theory of white-collar crime are emerging. As Yeager and Simpson (2009) note, it is generally agreed that white-collar offending is "caused by a confluence of motive, opportunity, choice, and constraint (lack of control)" (p. 338; see also Coleman, 1992; Friedrichs, 2010). Future research thus will profit from probing the specific content of these factors, their relative importance, and the ways in which they interact to foster upperworld crime.

12

James Q. Wilson

1931–

Pepperdine University and University of California, Los Angeles

Author of Conservative Theory

Bringing Punishment Back In

Conservative Criminology

As we saw in the earlier chapters, during the past two decades or so, many of the theoretical paths that had emerged during previous periods continued to develop but with a shift in emphasis in accordance with the context of the times. This chapter centers on the way in which the conservatism of the 1980s formed a context contributing to the revitalization of those perspectives that locate the sources of crime in the individual. In many ways, it was as if criminological theory had come full circle during the 1980s, with a return to the classical school depiction of crime as the result of individual actors' exercising rational choice or the positivistic portrayal of crime as the result of organic anomalies or psychological defects. These approaches almost invariably were set forth as rationales for the use of more punishment—especially by the state and the criminal justice system—as the solution to crime.

We have deliberately focused much of the contextual material on the United States of the 1980s and into the 1990s. Because it is difficult to analyze this period from a point too close to the time to allow for historical perspective, a more balanced analysis must await the work of future scholars. Our hope here is both to capture something of the flavor of the times and to encourage the reader to ponder the images portrayed in the mass media of the period as something more that "just news"—something amounting to the construction of a collective sense of social reality that, in turn, had its effect on criminological theorists.

As we have seen, the 1960s (and into the 1970s) was a time period in which existing assumptions about the social order were questioned and existing inequalities in race, class, and gender were laid bare. The state's raw use of power in ineffective attempts to suppress dissent over civil rights and the Vietnam War, among other things, served only to further undermine its legitimacy. This context was instrumental in nourishing the development and popularity of theories that linked crime to factors such as the denial of opportunity, state labeling and intervention, and inequality in power. The 1980s, epitomized by the election of Ronald Reagan, represented a culmination of the growing reaction to what many Americans—the so-called silent majority—saw as the excesses of the previous decades. It was a time when values of patriotism, religion, hard work, and individual responsibility for one's own fate were trumpeted. There was an ideological reaction against (1) the moral decadence and disdain for faith of the hedonistic secular culture and (2) the liberal welfare state whose policies were seen to make social dependents out of the poor and minorities.

This political sea change contributed to the emergence of theories that reflected conservative values. In general, there are two hallmarks of conservative theorizing. First, there is a denial that crime has any "root causes"—that criminal behavior is caused by structural arrangements in society, including inequality. Instead, *crime is attributed to individual choice*. This individual choice might be due to human nature, to rationality, or to moral defects from being raised in a permissive, immoral society. For conservatives, bad people choose crime and create a bad society, not vice versa. Second, in solving the crime problem, the focus is on placing more *restraints* or *controls* on individuals, usually in the form of greater discipline imposed in societal institutions such as the family, the schools, and the criminal justice system. There is a belief that the threat and use of harsh punishment—especially imprisonment—is the linchpin of effective social control.

We should note that unlike in the 1960s and early 1970s, most criminologists were not swept up by the times and moved to set forth conservative theories; only some were, as we will review in this chapter. Rather, if anything, criminology as a discipline was *oppositional* to the conservative political movement and to the punitive crime policies it set forth. Part of this opposition was ideological. A whole generation of criminologists who entered the field in the 1960s and 1970s were not about to simply surrender the view of crime they had come to embrace. But part of the resistance to conservative criminological ideas was that there was good reason to reject its main policy agenda: "getting tough" on crime through mass incarceration. For most criminologists, this approach was seen to ignore the complex root causes of crime, to be a convenient way to repress the poor and minorities, and to be largely ineffective in reducing criminal conduct.

Still, the persistence of conservative politics and thinking has likely weakened the hold that theories with strong roots in the 1960s have on criminologists. There is still a strong group of scholars who embrace critical and feminist theories. But there is also a wider ideological space now for other criminologists to pursue research outside these paradigms. These perspectives are less identifiable as liberal or conservative, though how they are applied can tilt them in one direction or the other. We will review these theories in Chapters 13 and 14. Now we return to our interest in conservative theories. We start with the context that has nourished their development.

Context: The United States of the 1980s and Early 1990s

To make our analysis more manageable, we concentrate on developments in the United States. As noted earlier, the turmoil of the 1960s was succeeded by a political backlash leading to the election of Richard Nixon to the presidency. Nixon was elected in part because of promises to get tough on crime and because he indicated that he would extricate the United States from the quagmire of the war in Vietnam by way of a "secret plan." He made "perfectly clear" his extreme distaste for the dress and drugs of the counterculture, and his election signaled the return to power of those who had little enthusiasm and tolerance for the civil rights, countercultural, and feminist movements of the time.

By the early 1970s, however, certain economic and political shifts that had been building for decades began to be felt. These underlying shifts had been very important to the sense of peace and prosperity of the 1950s and the radical developments of the 1960s, and they were to have major effects during the decades to come. The structural shifts were both economic and political. The economic dominance of the United States came into question during the 1970s, leading to a sense of something "gone wrong" by the end of the decade. Politically, the United States adapted with a shift to even more conservative political rhetoric during the 1980s that combined a cultural nostalgia for the "good old days" of the 1950s with spending policies that generated an enormous national debt.

THE ECONOMIC DECLINE OF THE UNITED STATES

By the close of World War II in 1945, the United States' share of world manufacturing was nearly 50%, a proportion so great that never before had it been attained by a single nation (and never since has it been). The United States came out of the war so strong economically that Henry Luce, one of the nation's leading publishers, proclaimed that it was the beginning of the "American century." Indeed, such a prediction was reasonable at the time. In 1945, for example, the United States owned two thirds of the world's gold reserves and all of its atomic bombs. It truly was the initial stage of the Americanization of the Old World. In view of the United States' vast economic power, it did little good for countries such as France to try to outlaw the importation and sale of items such as Coca-Cola (Kennedy, 1987, p. 29).

Given such immense resources, it is not surprising that during the late 1940s the United States continued to extend its military protection (some would say "capitalistic-militaristic imperialism") throughout the world. By 1970, the United States had more than 1 million soldiers in 30 countries and was a major supplier of military aid to nearly 100 nations around the world (Kennedy, 1987, p. 29). As the United States' military commitments began to increase after 1945, however, its positions in world manufacturing and world gross national production already were beginning to decline, slowly at first and then with increasing speed. Military expenditures under Presidents Eisenhower and Kennedy (prior to the military buildup in Vietnam) already were approximately 10% of the gross national product, but the economic burden was obscured by the fact that the United States' share of global production and wealth was about twice what it is now.

Something of the precarious nature of underlying economic conditions in the United States during the 1980s can be seen by recognizing that during these years under President Reagan, defense spending increased by at least 50% more than what was spent during the late 1970s under President Carter, Reagan's predecessor. Meanwhile, the national debt was growing at a remarkable speed. In 1980, the federal deficit was $59.6 billion; by 1985, it already was $202.8 billion and growing rapidly. Interest alone on the debt already was $52.5 billion by 1980, and by 1985 it had reached $129 billion (Kennedy, 1987, p. 33). By 1988, the United States had a $500 billion federal deficit, a far cry from when it owned two thirds of the world's gold reserves in 1945. Some observers termed this situation the "Argentining of America" (1987, p. 22). Many predicted that the United States' world economic position would continue to decline and that the 21st century would belong not to the United States but rather to countries in Asia, especially Japan. By the late 1980s, it was forecast that by 1995 private Japanese investors would own $1 trillion or 10% of all assets in the United States (p. 45).

Changes in other parts of the world also contributed to the United States' predicament. During this period, for example, some foreign countries increased their world production in areas where the United States once ranked supreme, such as in the production of textiles, iron and steel, shipbuilding, and basic chemicals. In addition, the United States faced stiff international competition in the production and distribution of robotics, aerospace technology, automobiles, machine tools, and computers (Kennedy, 1987, p. 29). And in food production, another area where the United States ranked supreme even into the 1970s, Third World countries and the European Economic Community began to export food, thereby competing with U.S. farm production and sales.

THE PERSISTENCE OF INEQUALITY IN THE UNITED STATES

During the 1980s, it became increasingly clear that the roots of inequality discussed by the strain theorists and conflict theorists ran even deeper in the United States than many had suspected. College attendance became more and more difficult for young people, some feminists began to rethink their position in the light of such slow progress toward gender equality, and racism reappeared in some especially ugly forms. Because of the underlying economic shifts outlined previously, the economic pie no longer was as large for the relative number of those hungry for a slice as it had seemed during the 1960s and early 1970s. Because the power of youths declined as the huge number of baby boomers who had fueled the protests of the 1960s moved toward middle age, and perhaps because many of the more energetic, educated, and articulate women and African Americans had been successful in obtaining their own slices of the pie (thereby becoming a part of the "system"), pressure for basic social change ebbed. By the end of the 1970s, Lasch (1978) already had cited the rise of a "culture of narcissism," and some were proclaiming the coming of the "Me Decade." More and more people seemed to focus on their own problems, even though basic social problems that had seemed to be on their way to at least partial solution began to worsen.

The race issue may be taken as an example. During the late 1960s, President Johnson appointed the Kerner Commission, charging it to investigate the causes of the racial riots of the time. The commission concluded that the riots were the result of White racism and cautioned that unless White attitudes and behavior changed, the United States would continue to move toward a system made up of two societies—one Black, one White—separate and unequal. Two decades later, it was evident that although some progress had been made, the impact had affected three different segments of the African American community quite differently ("20 Years After," 1988).[1] The African American middle class, a group that had relative economic success during the two decades prior to the 1980s, had achieved even more economic success, along with success in the political arena. But members of the African American working class during the late 1980s often experienced great difficulty in sustaining their economic position, being very vulnerable to the loss of jobs coming with downsizing and deindustrialization. Things were much worse, however, for a third group of African Americans who now seemed permanently "stuck" at the bottom.

This third group, the emerging "Black underclass," is composed of people who had succumbed to the economic and social strains inherent in slum living (Wilson, 1987). The mass media have described them as the "miserable human residue, mired in hardcore unemployment, violent crime, drug use, teenage pregnancy, and one of the world's worst human environments, [which] seems to be a partial perverse result of the very success of other blacks" ("20 Years After," 1988, p. A13). During the early 1970s, the African American median income was 57% that of Whites. In the years to follow, concerns about the seeming intractability of the Black-White income gap persisted.

Unfortunately, no reliable data exist on racial, religious, or ethnic violence for the entire United States prior to the 1990s ("Lack of Figures," 1987; "U.S. Had More," 1994). Only a few cities during the 1980s created special units for crimes motivated by hatred, but reports of racial strife and violence are on the increase. In mid-1987, the U.S. Department of Justice's Community Relations Service released information indicating that the number of racial incidents reported to the government had risen from 99 in 1980 to 276 in 1986 ("Lack of Figures," 1987, p. A13). In 1993, more than 7,000 hate crimes were reported to the Federal Bureau of Investigation (FBI) ("U.S. Had More," 1994). Corroborative information is present in investigations of arson and cross burning that have followed the move of minorities into mostly White neighborhoods. The Klanwatch Project of the Southern Poverty Law Project in Montgomery, Alabama, found that between 1985 and 1986 there were hundreds of acts of vandalism and other incidents directed at members of minority groups who had moved into mostly White areas ("Report Traces," 1987, p. A22).

Reports such as these were not restricted to the Deep South. In New York City alone, the number of reports of incidents of racial assaults in which African Americans and Whites were injured during the first 8 months of 1987 were the highest recorded by a special unit that had been investigating these acts for 7 years. There were 235 incidents in all of 1986, and 301 incidents were reported in the first 8 months of 1987. A similar trend was reported by the Mayor's Management Report, which stated that the number of bias incidents doubled between fiscal years 1986 and 1987 ("Reports of Bias," 1987, p. A19). Furthermore, the Anti-Defamation League of B'nai B'rith reported that anti-Semitic acts rose 12% in 1987. New York, the state with the largest Jewish population, led the nation in vandalism incidents, followed by California, New Jersey, and Florida ("Anti-Semitic Acts," 1987, p. A13).

Part of the mood that arose during the 1980s was symbolized by the Bernard Goetz case. Just 3 days before Christmas in 1984, Goetz, a middle-class White, shot four African American youths on a New York subway after one of them had asked for $5. In June 1987, a jury acquitted Goetz on the shooting charge. Throughout the proceedings, a debate raged over issues of racial tension, self-defense, and the perception that Goetz's acquittal would be interpreted as widening the circumstances justifying deadly force, at least of Whites against African Americans ("Goetz Case," 1987, p. A11; "Trial That Wouldn't End," 1987, pp. 20–21).

THE RHETORIC OF STABILITY

Those who do not wish to consider fundamental changes in the social order tend to locate the sources of social problems such as crime either in defective individuals or in failures of the institutions of socialization (Rubington & Weinberg, 1971). The increase in racial discrimination and violence outlined heretofore represented but one example, albeit a very important one, of tendencies to see certain *types of people* as the problem. As for the focus on the alleged *failures of our institutions of socialization,* it amounts to a belief that the "American way of life" is either exemplary or at least fundamentally sound but that the institutions of work, family, education, and religion are somehow failing to integrate potential troublemakers into the mainstream and are in need of reaffirmation and strengthening. If the first of these perspectives may be said to characterize President Nixon, then the second was well represented by the Democratic successor to the Nixon–Ford years, President Carter.

Carter's rise to the U.S. presidency is in itself an instructive example of the way in which economic and political changes affect public policy, including which criminological theories are implemented. Long before the underlying problems of the United States' economic decline were apparent to the mass media or the general public, important American financial interests were well aware of the situation and had responded with the creation in 1973 of the Trilateral Commission, a combined effort of powerful interests in America, Europe, and Japan. The one-time director of the commission, Zbigniew Brzezinski, summarized public disenchantment over the Watergate scandal, President Ford's subsequent pardon of Nixon, and a host of related signs of loss of faith in the system by writing that "the Democratic candidate in 1976 will have to emphasize work, the family, religion, and increasing patriotism to be elected" (Allen, 1977, p. 27). Partly as one aspect of the "southern strategy" to win back the electoral votes of the South, Carter was selected as a southern governor exemplifying these attributes. On Carter's election, Brzezinski became his national security adviser.

Carter was elected not only because of powerful supports from influential political kingmakers but also because he symbolized the mood of the times better than any other candidate. Indeed, it is important to remember that political figures in America are successful in large part because of their ability to mold themselves to the public mood so as to "become what the public want." Carter succeeded in reaffirming the work ethic to the extent that he often was called a workaholic, and he was well known for his familial affection for his wife, his daughter, and even his "bad boy" brother Billy. As a self-proclaimed born-again Christian with evangelicals in his immediate family, Carter represented some of the longing for a return to more traditional religious values and for moral leadership in the aftermath of the Watergate scandal. His presidency floundered, however, on his inability to reaffirm the power of patriotism with a victory over the hostage holders in Iran, and he in turn was succeeded by President Reagan, who became even more of a symbol of America during the 1980s than Carter had been in his time.

During the late 1970s, Carter spoke publicly of surveys showing a pervasive malaise in the United States. The public had lost much of its traditional faith in education as a force for socialization. Many people seemed to believe that the changes of the 1960s had destroyed the cohesion of the family and undermined its capacity to socialize and control American citizens. Religion seemed to offer personal solace to an increasing number of citizens, but perhaps it became less of a vehicle for community building and reform as more and more people turned to charismatic televangelists for spiritual direction.

By the end of the 1970s, the mood of the American public seemed to fit nicely with candidate Reagan's position that too many naysayers were spending too much time criticizing America and focusing on the dark side when what was needed was a patriotic public to "stand up for America." As indicated earlier, society as a whole seemed to be tired of social concerns and eager to turn its back on internal problems of economic decline, racism, poverty, environmental pollution, and all the rest—almost as if these problems did not exist. Although labeling theory received some lip service, it was not a time in which strain theories—much less conflict theories—were likely to gain much of a hearing.

The inauguration of Reagan as president suggested to some that the U.S. public was about to embrace the 1950s spirit once again. The theme of the Me Decade that had been proclaimed during the late 1970s began to take on a somewhat different tinge. Although the focus was much on the individual "looking out for number one," there was at the same time more and more condemnation of the permissive society. For example, the society as a whole seemed to view much more positively the somewhat cynical pursuit of individual material success in the "rat race," although it was growing much less tolerant of individualistic pursuits of pleasure in the form of sex and drugs. In fact, it became fashionable to call for surveillance over sexuality and for compulsory drug testing on the principle of restriction of individual rights for the protection of society.

The Embrace of Materialism. The America of the 1980s often seemed like a case of "life imitating art," if the mass media portrayals can be called art. The top-ranked television series in 1981 was the prime-time soap opera *Dallas,* featuring scenes of Texas high rollers led by the conniving Ewing oil baron "J. R." Later in the decade, the real-life Wall Street wheeler-dealer Ivan Boesky, who had summarized the thinking of many in his assertion that "greed is not a bad thing" ("The Eighties Are Over," 1988, p. 42), was named by Fortune as "Crook of the Year" (p. 44) after agreeing to relinquish illegal profits of $50 million and indemnify another $50 million and was later honored with a 3-year prison sentence. In 1984, *Lifestyles of the Rich and Famous* became a hit television show (p. 42), and in that same year *Newsweek* declared the real-life "Year of the Yuppie." In direct contrast to the yippies of the 1960s, the yuppies (young upwardly mobile urban professionals) represented a symbol of the times as a demographic cohort of self-centered materialists bent entirely on "making it," with no social conscience getting in the way.

The Reaffirmation of Traditional Sexual Preachments. By the early 1980s, the openly hedonistic sexuality that many associated with the rise of the 1960s' counterculture was declared a thing of the past in articles with titles such as "Sex in the 80s: The Revolution Is Over" (1984, pp. 74–78). A *Newsweek* cover story announced that the "Playboy Party Is Over," and *Time* referred to an era of "Sex Busters." Examples of the new mood abounded—congressional opposition to the Braille editions of *Playboy* that had been published monthly since 1970 under the National Library Service for the Blind and Physically Handicapped; a highly criticized government-sponsored study of how children were portrayed in *Playboy, Penthouse,* and *Hustler;* increased violence against homosexuals; and a hotly disputed Department of Justice report on pornography.

The debate over sexual values also reached into the nation's courts. For example, in *Bowers v. Hardwick* (1986), decided during the middle of the decade, the U.S. Supreme Court upheld (by the narrowest of margins in a 5–4 split) a Georgia law that made it a felony punishable by 20 years in prison for consenting adults to engage in oral or anal sexual relations. The case developed innocently enough when a police officer went to Hardwick's apartment to serve a warrant for carrying an open container of alcohol in a public place (part of a crackdown on drinking). The officer was ushered into the apartment by a guest and told that he could find Hardwick in his bedroom. There the officer discovered Hardwick engaged in oral sex with another man and made an arrest. In effect, the court ruled that individuals under Georgia law had no right to engage in such acts even as consenting adults in a private bedroom, taking the position that the state had a legitimate right to restrict and control such behavior as part of its responsibility for enforcing public morality even in private places.

In a matter also related to balancing privacy versus state interests, the 1980s witnessed the increasingly heated debate on the right of a woman to terminate a pregnancy through abortion. The courts were asked repeatedly to reconsider *Roe v. Wade* (1973), the Supreme Court case that had declared as unconstitutional laws prohibiting abortion, largely on the judgment that state interests were not compelling enough to justify interference in the personal privacy of women. As a difficult moral dilemma, the stance taken on abortion often cut across usual ideological lines (e.g., liberal Catholics vs. pro-choice). Even so, the growing pro-life movement was nourished by America's turn to the right as conservative politicians, religious leaders, and right-to-life groups played prominent roles in the crusade to overturn *Roe v. Wade*. They argued that the interest of the state in maintaining public morality, represented most fully in the moral choice to protect the lives of the unborn, was more than sufficient to allow state legislatures once again to restrict, if not criminalize, the practice of abortion.

Finally, the 1980s saw the emergence of AIDS as a growing threat. The mounting toll of AIDS initiated much discourse in the media and in political forums over existing sexual practices. For liberals, AIDS precipitated a call for more pragmatism in sexual relationships—the prudent selection of partners, the practice of "safe sex," and broader efforts to educate about the transmission of sexual diseases. For conservatives, the AIDS menace provided incontrovertible evidence of the dire consequences of the breakdown of sexual morality. Ultimately, conservatives warned, the problem of AIDS reflected a moral failing—the result of sex outside the boundaries of marriage—and could be eliminated only by a reaffirmation of traditional sexual values.

The "War on Drugs." Early in the Reagan administration, there were anti-drug campaigns, and in 1986 the federal government declared a new $250 million "war on drugs," a so-called war that was to extend the compulsory drug testing of employees in the world of business and athletics to federal employees ("Crack Down," 1986, pp. 12–13; "Trying to Say No," 1986). The announcement of the new testing policy came after a 5-month congressional study that had concluded that "urinalysis procedures were expensive, 'useless in most cases,' and often inaccurate" ("Drug Test," 1986, p. A15), but the policy was to be implemented despite the lack of any evidence that those to be tested had used illegal drugs or had committed any crimes at all. The federal government was described as being in a "frenzy" over drugs ("Fighting Narcotics," 1986, p. A25). Despite arguments that what employees do on their private time is their own business unless it affects job performance and that such testing replaces the due process assumption of innocence with an assumption of guilt until proven otherwise, a White House poll reported that the public was less concerned about the federal budget deficit and arms control than about drugs ("Crack Down," 1986).

The Department of Justice and the Supreme Court. Although it might not be true that the positions taken by units of federal government reflect the times exactly, they do provide clues. Thus, it is instructive to consider trends in both the Department of Justice and the Supreme Court during the 1980s. In both cases, there was increasing reluctance to focus on possible social conditions underlying crime and a tendency to demand harsher treatment of ordinary street criminals.

At the beginning of President Reagan's second term, his close friend Edwin Meese was appointed attorney general. In 1986, Meese declared that Supreme Court decisions should not necessarily be regarded as the supreme law of the land ("Edwin Meese," 1986, p. 9). He argued that Supreme Court decisions were binding only for those involved in specific cases ("Meese Says Court," 1986). A similar position was advocated by the chief of the Department of Justice's civil rights division, William Bradford Reynolds, who not only questioned the competence of a Supreme Court judge but also argued that efforts to achieve a radically egalitarian society would pose perhaps a major threat to individual liberty ("The President's Angry Apostle," 1986, p. 27). Reynolds was the leader of assaults on federal civil rights and school desegregation programs, on efforts to enhance equality in women's college athletics, and on efforts to halt discrimination in housing. According to Griffin Bell, attorney general under President Carter, "That's what he was hired to do" (p. 27).

In matters closely related to criminal justice, the leadership of the Department of Justice was equally conservative. Meese, for example, advocated overturning the *Miranda v. Arizona* (1966) decision, which required the police to inform criminal suspects of their legal rights, referring to the decision as "infamous" ("Meese Seen as Ready," 1987, p. A13). Chief Justice William H. Rehnquist sounded a similar refrain by calling for a limit on last-minute appeals filed by death row inmates. He argued that such appeals create "chaotic conditions . . . within a day or two before an execution" ("Death Row," 1988, p. A7). Unfortunately, the "chaotic conditions" that concerned Rehnquist resulted in part from the fact that the Supreme Court "made it clear that the constitutional right to assign counsel [did] not apply to litigation beyond the first appeal" ("The Gideon Case," 1988, p. A27). Meanwhile, Meese advocated execution for teenage killers ("Meese: Execute Teen-age Killers," 1985).

In addition to military spending and criticism of those at the bottom of the economic ladder, part of the politics of the 1980s in the United States revolved around a perceived need to reaffirm faith in the American way of life in the face of problems that seemed difficult to handle. On the 100th birthday of the Statue of Liberty in 1986, after 2 years of refurbishing at a cost of $39 million, the statue and all that it symbolizes was celebrated with the most expensive and extravagant party ever held in the world. It was attended by more than 5 million people. As testimony to the themes of liberty, freedom, and a new life—each central to the core of American values and symbolized by the Statue of Liberty—more than 25,000 immigrants took the citizenship oath on July 4, 1986, in 44 different locations across the nation. The reaffirmation celebration also was attended by foreign dignitaries, and no fewer than 35 foreign warships stood by in New York Harbor as a symbolic gesture supporting the American way of life. In keeping with the times, an expensive dance was held on Governors Island at a cost of $10,000 per couple ("Hail Liberty," 1986; "Sweet Land of Liberty," 1986).

THE LEGACY OF THE CONSERVATIVE POLITICAL AGENDA

To some extent, the conservative agenda was challenged by progressives during the 1990s, most notably with the election of President Clinton in 1992. Still, by 1994, Republicans—spouting a decisively conservative political agenda—had gained control of the Senate, the Congress, and a majority of state governorships. Many liberal policy initiatives were stifled (e.g., gay and lesbian rights in the military, national health care), and many conservative policies were implemented (e.g., welfare reform that set time limits on the period that women with children could stay on public assistance).

By all accounts, the U.S. economy became "bullish" as unemployment rates plummeted and inflation remained under control. "Dot.com" companies flourished, and young millionaires seemed almost commonplace. Still, this economic good news masked some persisting realities of economic inequality in America. Between 1973 and 1998, for example, the real hourly wages (i.e., wages adjusted for inflation) actually declined by 9%. During this same period, the percentage of total income in the United States owned by the richest 5% of American families increased from 14.8% to 20.7% (Wolff, 2001, p. 15). As Wolff (2001) noted, "As workers' wages have stagnated, economic inequality has worsened" (p. 15). Furthermore, unemployment figures are deceiving unless they are broken down by race and placed in a larger context. According to Currie (1998b), compared to the national average, unemployment is about twice as high for African American men, and this statistic does not include the huge number of these men who are incarcerated. As Currie observed, "In 1995, there were 762,000 black men officially counted as unemployed and another 511,000 in state or federal prison. Combining these numbers raises the jobless rate of black men . . . from just under 11 to almost 18 percent" (p. 33).

The continuing conservative influence throughout the 1990s can be seen in the steady rise in prison populations. During the presidential terms of Reagan (8 years) and George H. W. Bush (4 years), the number of people sent to local jails and to state and federal prisons were 448,000 and 343,000, respectively. During Clinton's 8-year tenure, this figure was 673,000. At the end of Clinton's presidency, the nation's incarceration rate stood at 476 per 100,000 citizens, compared to 332 for Bush and 247 for Reagan. For African Americans, the incarceration rate at the century's end was, 3,620 per 100,000 citizens, a rise from about 3,000 in 1992 (Gullo, 2001).

The conservatives' hold on American politics, though tenuous at times and in the so-called Blue or democratic-leaning states, continued into the first part of the 21st century. George W. Bush's narrow victory over Al Gore was followed up with election to a second presidential term—accompanied by Republican control of both houses of the U.S. Congress. Beyond the war in Iraq—largely justified by the neo-conservative philosophy to spread American values throughout the Middle East—conservatives pushed a right-wing policy agenda. This included the increased deregulation of corporations, free trade in the global economy that has undermined unions and afforded American workers few protections, cuts in social welfare programs (including to college student loans), cuts in taxes that disproportionately benefit the most affluent citizens, calls for a Constitutional amendment to ban gay marriage, support for faith-based initiatives funded by public monies, support by President Bush and Congressional leaders for the teaching of "intelligent design" as a legitimate alternative to evolution, proposals to spend huge amounts to patrol if not build a fence across the Mexican–American border, the appointment of very conservative judges to the U.S. Supreme Court, and attempts to expand executive power in the war on terrorism such as through unsupervised wiretapping of phone conversations. Meanwhile, "get tough on crime" rhetoric did not abate and policies to raise punishments were implemented with some regularity. By the end of George W. Bush's first term at the close of 2004, the nation's incarceration rate had climbed to the all-time high of 486 inmates per 100,000 U.S. residents (Harrison & Beck, 2005). As Garland (2001) suggests, the "culture of control" had become dominant.

Varieties of Conservative Theory

At the risk of simplifying complex arguments, five types of conservative theorizing emerged in the last two decades of the 20th century. First, efforts were made to revitalize the early positivist school's emphasis on ingrained individual differences, with a special emphasis on defects in human nature or in intelligence as the master predictor of criminal involvement (Herrnstein & Murray, 1994; Wilson & Herrnstein, 1985). Second, by revitalizing classical school principles, other scholars developed models that conceive of individuals as logical actors choosing crime when the benefit exceeds the cost. In short, crime occurs because, in our society, it "pays" (Reynolds, 1996). Third, others attempted to revitalize the psychological approach by suggesting that offenders persist in crime because they think differently rather than logically. They are held to have distinct "criminal minds" that make them pathological, if not psychopathic and beyond redemption (Samenow, 1984). Fourth, still others of a conservative bent went beyond an individualistic approach, but only to the point of linking crime to a distinctive type of social influence: the allegedly permissive culture—or "moral poverty"—that they trace to developments in the American society of the 1960s, such as those discussed in earlier chapters (Bennett, DiIulio, & Walters, 1996). Finally, there was the claim that public disorganization or incivility leads to crime not because of enduring poverty and other social ills but because the police tolerate it (Kelling & Coles, 1996; Wilson & Kelling, 1982). These perspectives are reviewed below.

Again, what makes these perspectives *conservative* is not simply the factors that the scholars argue underlie crime. For example, saying that offenders are impulsive, lower in intelligence, and likely to commit crimes when illegal behavior is greeted with tolerance rather than consequences does not make a perspective inherently conservative. In fact, these might be seen as individual and social risk factors for crime and be highlighted in a number of criminologists' writings. Rather, it is these theorists' consistent denial that capitalism or any form of economic inequality or concentrated disadvantage is implicated in America's crime rate—especially in inner cities—that defines them as conservative (Currie, 1985). In this paradigm, crime does not have "root causes" or, if it does, they are beyond any kind of government intervention that would redistribute wealth or extend safety nets to those in need (J. Q. Wilson, 1975). Instead, crime is seen as a choice—a choice by individuals who are impulsive, stupid, psychopathic "super-predators," calculating, raised in moral—not economic—poverty, and/or allowed to "break windows" without fear of consequences.

These perspectives also are defined by their policy recommendations. Thus, the identification of individual differences in offenders can lead to calls for early intervention programs and well-designed treatment programs (Andrews & Bonta, 2003; Farrington & Welsh, 2007). With few exceptions, however, these theorists typically take pains to use their theories to justify interventions aimed at controlling, not helping, the individuals at risk of committing crimes. There is an implicit, if not explicit, belief that offenders either are beyond change or are responsive only to sanctions that are painful. Their work thus attempts to "bring punishment back in" to criminology—to argue for its special utility. In particular, these scholars provide reasons for making formal control and, in particular, mass imprisonment the preferred policy for reducing crime. This embracing of "get tough" policies—even when their theories might recommend alternative policy interventions—is what earns them the status of conservative criminologists.

Crime and Human Nature: Wilson and Herrnstein

THE THEORY

A path-breaking, highly publicized example of the return of individualistic explanations of crime appeared in Wilson and Herrnstein's (1985) *Crime and Human Nature*. It was greeted by numerous reviews in well-known publications such as *Time, Newsweek, U.S. News & World Report, Vogue*, the *New York Times*, and the *Chronicle of Higher Education*. It also was treated as a news item by the Associated Press ("[Editorial]," 1985). The basic argument offered by Wilson, then a Harvard University political scientist, and Herrnstein, a Harvard psychologist, was based on a biosocial explanation of behavior that focused attention primarily on "constitutional factors" (Wilson & Herrnstein, 1985, chap. 3). Such factors, some of which are genetic, were treated by Wilson and Herrnstein as predisposing individuals to engage in criminal behavior. In their words, they wanted to explain "why some individuals are more likely than others to commit crime" and "why some persons commit serious crimes at a higher rate and others do not" (pp. 20–21). Their effort was directed toward explaining criminality rather than crime.

Although Wilson and Herrnstein (1985) did not give their theory a specific label or name, its biosocial focus was clear:

> The existence of biological predispositions means that circumstances that activate behavior in one person will not do so in another, that social forces cannot deter criminal behavior in 100 percent of the population, and that the distribution of crime within and across societies may, to some extent, reflect underlying distributions of constitutional factors. Crime cannot be understood without taking into account predispositions and their biological roots. (p. 103)

In certain respects, Wilson and Herrnstein's (1985) perspective hearkens back to the theories of Lombroso, Hooton, Sheldon, and the Gluecks (see Chapter 2) and to the positivist school of criminology. Although largely ignored by today's criminologists because of theoretical and methodological problems, Wilson and Herrnstein claimed that a distinctive body type exists that distinguishes criminals from noncriminals:

> Wherever it has been examined, criminals on the average differ in physique from the population at large. They tend to be more mesomorphic (muscular) and less ectomorphic (linear). . . . A corresponding argument is that the more muscular criminals are more likely to have biological parents who are themselves criminals. (p. 215)

On the importance of family and its contribution to crime, Wilson and Herrnstein (1985) stated that "bad families produce bad children" (p. 215). They buttressed this position by an extensive review of the work published on twins and adopted children (pp. 69–103).

The major argument here is that if there is a genetic connection to crime, it should show up when identical twins raised in the same environment are compared to fraternal twins also raised in the same environment. The findings suggest that if one identical twin has committed crimes, then the other twin is likely to have committed crimes as well. Research does not support this finding for fraternal twins.

Research on adopted children has focused on the criminality of the children compared to the criminality of their biological and adoptive parents. It is argued that children whose biological parents are criminals would engage in crime more than the adoptive parents. If genetics are connected to crime, then it would have an influence despite the environment. Some research findings suggest that the criminality of biological parents is more significant in explaining the criminality of children than is the criminality of adoptive parents.

Just how constitutional factors influence criminality is a point on which Wilson and Herrnstein expressed caution. They based their answer on what people consider rewarding and their ability to postpone gratification until sometime in the future. According to Wilson and Herrnstein, constitutional factors have an impact on the ability to consider future and immediate rewards and punishments. For example, aggressive and impulsive males with low intelligence are at a greater risk for committing crimes than are young males who have developed "the bite of conscience," which reflects higher cognitive and intellectual development.

Beyond the biological emphasis, Wilson and Herrnstein offered what Gibbs (1985) called an "operant-utilitarian" theory of criminality. In essence, Wilson and Herrnstein (1985) argued that "individual differences" rooted in biology are important to the extent that they influence subsequent social learning or how people interpret and are affected by rewards and punishments aimed at shaping their behavior. Still, from this point forward, they largely set forth a perspective suggesting that behavior is affected by the *consequences* that it evokes or that people think it will evoke. This part of their model is not dissimilar to other social learning theories. As Wilson and Herrnstein concluded:

> The larger the ratio of the rewards (material and nonmaterial) of noncrime to the rewards (material and nonmaterial) of crime, the weaker the tendency to commit crimes. The bite of conscience, the approval of peers, [and] the sense of inequity will increase or decrease the total value of crime; the opinions of family, friends, and employers are important benefits of noncrime, as is the desire to avoid the penalties that can be imposed by the criminal justice system. The strength of any reward declines with time, but people differ in the rate[s] at which they discount the future. The strength of a given reward is also affected by the total supply of reinforcers. (p. 61)

ASSESSING CRIME AND HUMAN NATURE

The evaluations of *Crime and Human Nature* did not uniformly heap praise on Wilson and Herrnstein (1985). The reasons were many, but they can be divided into two categories: conceptual/empirical and ideological. The conceptual/empirical criticisms focused on the authors' lack of "concern about the empirical applicability of their terms" (Gibbs, 1985, p. 383). They used concepts such as "ratio of rewards," "material and nonmaterial crime," "approval of peers," and "sense of inequity" (see Wilson & Herrnstein, 1985, chap. 2). But they failed to offer any numerical or operational expressions for the concepts. Consequently, there was a lack of clarity about what they were trying to communicate, thereby creating difficult obstacles to conducting research that would test their theory.

Another troublesome dimension of Wilson and Herrnstein's (1985) work was their claim that they were confining their attention to serious street crimes and predatory crimes such as murder, robbery, and burglary. Their failure to include offenses such as white-collar crimes under the label of predatory crimes and their failure to explain why burglary should be considered a street crime raised serious doubts about the generality and conceptual clarity of their theoretical arguments. As Gibbs (1985) stated, the relatively loose manner in which Wilson and Herrnstein used labels for crimes is "better suited for journalism than criminology" (p. 382).

An equally troublesome aspect of *Crime and Human Nature*, claimed some commentators, was that, although Wilson and Herrnstein (1985) gave the impression that they were objective in the selection and presentation of relevant literature, they might in fact have been selective in what they reviewed (Kamin, 1985). Although readers were given the impression that the authors' arguments were based on solid *science* and, therefore, should be *believed*, critics asserted that, in more than one instance, these arguments were based on shaky evidence. But readers were not given this information. For example, according to Kamin (1985), Wilson and Herrnstein's (1985) observation that young people growing up during the 1960s were less willing to delay gratification than were young people growing up 15 years earlier was based on a "single experimental study [taken] at face value" (p. 22). In addition, the authors failed to cite published research indicating that noncriminals were found to be more mesomorphic in body type than were delinquents (p. 24). Apparently, concluded the critics, Wilson and Herrnstein elected to emphasize resources that supported their arguments for constitutional factors related to crime rather than to address the literature that questioned or confounded this argument.

These criticisms, however serious, did not overshadow the ideological implications of some of Wilson and Herrnstein's arguments and the reasons why their arguments were given wide attention. On this point, Kamin (1985) stated,

> The Wilson and Herrnstein work ought not to be judged in isolation. Their selective use of poor data to support a muddled ideology of biological determinism is not unrepresentative of American social science in the sixth year of the Reagan presidency. The political climate of the times makes it easy to understand why social scientists now rush to locate the causes of social tensions in genes and in deep-rooted biological substrata. (p. 25)

Wilson and Herrnstein's work implied that certain biological predispositions found disproportionately among the poor and may be responsible for excessive criminal behavior. This message, particularly when read by those lacking criminological expertise and who do not pause to read the authors' caveats about their argument, suggests disquieting policy implications to some. We saw in Chapter 2 that early biological theories were used to justify repressive and physically intrusive policies. The historical record teaches that attempts to root crime in human nature exempt the social fabric from blame and lend credence to the idea that offenders are largely beyond reform and in need of punitive control.

But in fairness to Wilson and Herrnstein (1985), much criticism aimed at *Crime and Human Nature* occurred because many criminologists were themselves ideologically opposed to biological theorizing, again because of the repressive policies that, historically, such thinking had justified "in the name of science" (Gould, 1981). Wilson and Herrnstein were among the first criminologists to robustly try to "bring biology back into criminology," and in a sense, they paid a price for this. Today, as we will see in Chapter 14, many criminologists would not embrace their theory whole cloth, but they would agree that biologically based individual differences, such as impulsiveness and low intelligence, can affect social learning (and other criminogenic risk factors) and, therefore, play a role in criminal behavior.

But the reaction to Wilson and Herrnstein (1985) also was shaped by the conservative temperament that they manifested throughout *Crime and Human Nature,* especially in the book's final two chapters (see also its chap. 15). They provided pessimistic assessments of more progressive policy initiatives aimed at crime prevention such as correctional rehabilitation and more general social reforms (cf. Cullen & Gendreau, 2000; Currie, 1998b). They spent virtually no time in thinking innovatively about how governmental policies might be crafted to increase the utility of a conventional life course— what they called noncrime—over that of a criminal life course (Currie, 1985, 1998b).

Wilson and Herrnstein (1985) concluded mainly that the solution to crime, to the extent it is possible, lies in two domains: having parents punish and control children more effectively and having the government punish offenders more certainly, quickly, and perhaps harshly. They were too competent as scholars not to know that the evidence supporting deterrence is modest at best. Here, however, they embraced the largely untestable position that even if punishment does not deter all offenders, it does serve as an instrument of "moral education." In this scenario, punishment sets moral boundaries and teaches right from wrong. It creates in people a "subjective sense of wrongdoing . . . Punishment as moral education almost certainly reduces more crime than punishment as deterrence" (p. 495). This is plausible, but they could cite no evidence showing that criminal justice sanctions teach morality. They did not consider that such sanctions, or how they are applied, also may teach that the "system" is racially and class biased. And they did not consider that efforts to rehabilitate offenders may teach that we should value compassion and support, rather than vindictiveness, in society. Certainly, these messages may create moral values antithetical to crime.

Wilson and Herrnstein's taste for a punitive state is even more apparent in their justification of "personal responsibility." They recognized that theories that link crime not only to social context but also to ingrained biological differences are inconsistent with the idea that offenders exercise free will and should be punished for their misconduct. Offenders, after all, do not choose to be impulsive or less intelligent, and they do not choose to be born into families whose parents discipline them in ways that make them predisposed to crime. But Wilson and Herrnstein's conservative sensibilities prevented them from rejecting the logical dictates of their theory: If crime is caused, then offenders cannot be held "personally responsible" for their acts. Although the authors knew that this is where their theory leads, they feared that the failure to punish would erode society's moral message that we all should be responsible for our acts. Thus, they wrote:

> We know that crime, like all human behavior, has causes and that science has made progress—and will make more progress—in identifying them, but the very process by which we learn to avoid crime requires that the courts act as if crime were wholly the result of free choice. (p. 529)

At this point, conservative ideology trumps science and the underlying agenda of *Crime and Human Nature* is unmasked.

Crime and *The Bell Curve:* Herrnstein and Murray

By the late 1970s, the IQ theory that had fallen into disrepute decades earlier was being reasserted (Hirschi & Hindelang, 1977). Wilson and Herrnstein (1985) suggested that low intelligence was associated with inability to reason morally, reestablishing the notion that it represented not only cognitive but also moral backwardness. By the late 1980s, those accepting as facts the arguable assumptions that (1) intelligence is a single unitary faculty, (2) IQ scores are valid measures of this capacity, and (3) the capacity is essentially inherited rather than environmentally developed were using them to explain racial differences in crime and delinquency (Gordon, 1987). One might say, in somewhat oversimplified terms, that the early 1990s saw a restoration of elements of social Darwinism, with suggestions that African Americans, for example, were both cognitively and morally inferior by nature and that this inferiority explained much of the crime problem.

The thesis that intelligence is the most robust predictor of criminal behavior—indeed, of virtually all social behavior—received its fullest and most sophisticated statement in the best-selling book by Herrnstein and Murray (1994), *The Bell Curve: Intelligence and Class Structure in American Life*. These authors suggested that as the United States has moved into a postindustrial economy—a society that increasingly emphasizes knowledge, technical expertise, and merit—people's life chances and conduct have become increasingly determined by their cognitive resources. In the authors' view, the "cognitively disadvantaged" struggle to adjust in this new social context. They are more likely to fail at school, be unemployed, end up on welfare, produce illegitimate children, be irresponsible citizens, and (noteworthy for our purposes) commit criminal acts.

Herrnstein and Murray's (1994) work might have been dismissed as idle conservative speculation were it not for the fact that they ostensibly produced hard empirical data to back up their claims. To show that intelligence is the most potent predictor of human conduct, they analyzed data from the National Longitudinal Survey of Youth (NLSY). The NLSY contained both a measure of IQ and measures of criminal involvement. Herrnstein and Murray showed that even when controlling for socioeconomic status, those with lower intelligence scores have higher odds of committing self-reported criminal acts and of entering the criminal justice system. Thus, they argued for the "importance of the relationship of cognitive ability to crime" (p. 251). "Many people," observed Herrnstein and Murray, "tend to think of criminals as coming from the wrong side of the tracks. They are correct insofar as that is where people of low cognitive ability disproportionately live" (p. 251). In turn, this finding suggests that in dealing "with the crime problem, much of the attention now given to problems of poverty and unemployment should be shifted to another question altogether: coping with cognitive disadvantage" (p. 251).

But there are two problems with Herrnstein and Murray's (1994) argument. First, although research reveals that offenders generally have lower IQ scores than nonoffenders—at least in populations that exclude white-collar criminals—meta-analyses of existing studies report that intelligence is only a weak to moderate predictor of illegal conduct. That is, intelligence plays a role in crime causation, but it is just *a*—not *the*—determinant of criminal behavior (Cullen, Gendreau, Jarjoura, & Wright, 1997; see also Andrews & Bonta, 2006). Even in the NLSY data used by Herrnstein and Murray, a reanalysis of these data shows that intelligence explains less than 1% of the variation in self-reported crimes and only about 3% of the variation in how far a person "penetrates" into the criminal justice system (i.e., from nothing, to sentenced, to a correctional institution). Furthermore, when variables ignored by Herrnstein and Murray are included in a multivariate model, the relationship of IQ to self-reported crimes becomes statistically nonsignificant (Cullen et al., 1997). It is unclear why Herrnstein and Murray would conduct an empirical analysis that omitted known predictors of crime that could be measured by variables readily available in the NLSY.

Second, Herrnstein and Murray (1994) not only exaggerated the causal importance of IQ but also proposed solutions to crime that are based not on science but rather on pure conservative ideology. The finding that offenders have lower IQ scores than nonoffenders does not ineluctably justify punitive policies. For example, one might argue that this intelligence gap justifies intensive preschool programming aimed at stimulating children at risk for being cognitively disadvantaged or calls out for interventions aimed at assisting low-IQ youths as they confront the challenges posed by years of traditional schooling. But Herrnstein and Murray, rather than thinking along these lines, instead suggested that we fight the crime problem by creating a society governed by "simple rules" and certain punishments.

In Herrnstein and Murray's (1994) vision for a "virtuous" society, conduct rules would be clearly and boldly stated so that even the cognitively disadvantaged could understand them. "People of limited intelligence can lead moral lives in a society that is run on the basis of 'Thou shalt not steal,'" they contended, "but such people find it much harder to lead moral lives in a society that is run on the basis of 'Thou shalt not steal unless there is a really good reason to'" (p. 544). Of course, legal transgressions would have to be met with punishment, lest the message of right and wrong not be understood. And these punishments would have to "hurt" to be "meaningful" (p. 543). In short, the "policy prescription is that the criminal justice system should be made simpler. The meaning of criminal offenses used to be clear and objective, and so were the consequences. It is worth trying to make them so again" (p. 544).

This policy agenda would leave the typical criminologist incredulous (Cullen et al., 1997). The recommendations are based on pure speculation, and it is instructive that Herrnstein and Murray (1994) offered nary a citation to substantiate that the cognitively disadvantaged break the law because they are confused about what is or is not a crime and because they are insufficiently punished. Most noteworthy, this way of thinking about crime allowed Herrnstein and Murray to ignore the possibility that offenders, including those with low IQs, might best be diverted from crime through rehabilitation programs (Cullen et al., 1997). It also enabled them, as with other conservatives, to give the false impression that crime can be reduced mainly by manipulating laws and punishments, with no thought given to the possible untoward consequences of the socioeconomic inequalities that persist in American society.

The Criminal Mind

Although by the mid-1970s psychoanalytic theory had lost considerable standing and was to come under considerable attack by the early 1990s, partly because of new evidence suggesting that Freud had either misjudged or actually suppressed much of his critical case material and partly because of sharp feminist critiques, Yochelson and Samenow (1976) published a popular book, *The Criminal Personality,* arguing that crime is the result of pathological thought patterns constituting a "criminal mind." Offenders were described in the traditional language of psychopathy as manipulative, calculating, and largely immune to traditional efforts to treat them. Cognitive approaches to understanding offenders' thinking have merit, but simple constructs (e.g., a criminal mind) have not been demonstrated to exist across offenders. Importantly, Yochelson and Samenow's study was limited to a highly specialized group of offenders mainly from a hospital for the criminally insane; thus, the generalizability of their results is questionable. Still, by the 1980s, the notion that the criminal must be characterized by some pathological mentality, if only it could be identified, had gained more favor, with many different psychological characteristics suggested as the origin of crime.

Samenow (1984) subsequently extended his coauthored work with Yochelson in *Inside the Criminal Mind.* Samenow's perspective hinges on two central premises. First, "criminals think differently" (p. 23). They are egocentric, grandiose, impulsive, easy to anger, and insensitive to the pain of others. They externalize all blame, neutralize feelings of guilt (when they have them), and manipulate those around them for their self-gratification. Taken together, these thinking patterns comprise a "criminal mind"—a construct not dissimilar to Gottfredson and Hirschi's (1990) description of "low self-control."

Second, "how a person behaves is determined largely by how he thinks" (Samenow, 1984, p. 23). Appropriate thinking leads to conformity, but thinking rooted in a criminal mind produces chronic offending. In a way, this thesis overlaps substantially with the works of Sutherland and of Akers. The differential association and social learning perspectives argue that an important proximate cause of crime is the definition of the situation favorable to crime harbored by offenders. But Samenow disavowed such a connection with traditional criminology. Instead, he contended that "we are still a long way from solving the puzzle of causation" (p. 182); that is, we really do not know the origins of criminal thinking patterns. Criminal minds simply emerge seemingly at random, striking one child within a family but not the others. Oddly, Samenow rejected differential association theory, even though he eventually suggested that the solution to crime is to teach offenders how to change their minds from criminal to conventional ways of thinking.

Samenow (1984) was rigid in his assertion that criminals are rational and *choose* to commit crime. Unfortunately, he confounded the reality that offenders consciously make choices with the reality that those choices may be "bounded" or shaped by a range of factors. It is difficult, for example, to suggest that offenders are free-willed rational creatures when one also is arguing that they are controlled by criminal minds of unknown origin. This kind of confused logic led Samenow to propose that no sociological factors are involved in crime causation. Offenders are portrayed not as "victims" of their environment but rather as the knowing architects of their life situations. They are spanked, if not abused, by parents because they are intractably disobedient and unbearable; they associate with delinquent peers because they choose such "bad companions"; they fail at school because of their resistance to and boredom with learning; and they are unemployed because they prefer not to work. They *self-select* themselves into problematic life conditions; they are not produced by these conditions (see also Samenow, 1989).

To substantiate this position, Samenow (1984) systematically ignored criminological research showing that a range of social factors is causally related to criminal involvement. Samenow dismissed the possibility of *social causation* by citing anomalous cases in which social factors do not produce criminals as sociological theories would suggest. For example, disorganized inner-city neighborhoods are not responsible for crime because many of their residents do not turn to crime; moreover, crime is found in suburban areas. This reasoning is flawed, however, because it ignores the fact that exposure to certain social conditions increases the risk or probability of offending. Samenow's thinking is similar to those who would argue that smoking is unrelated to cancer because not all smokers are afflicted with cancer and not all nonsmokers are free from cancer. Yet with both crime and cancer, there are conditions that increase the risk of these outcomes and that almost certainly are implicated in their causation.

In the end, Samenow (1984) raised the important point that *how offenders think* is a worthy area for criminologists to investigate. Such thinking is likely an important *proximate* cause of crime—a factor that enters into why offenders in any given situation decide to break the law. Still, the explanation of crime does not end with what is inside the minds of criminals. It also must probe where such minds come from—that is, it must seek to uncover the more distal or *root causes* of crime.

Choosing to Be Criminal: Crime Pays

Rational choice theory represents an approach favored by many economists. Based on the concept of "expected utility," it proceeds on assumptions similar to Bentham's utilitarian assertion that individuals operate by rational decisions expected to maximize profits and minimize losses. That is, the central thesis is that people commit crime because it "pays"—because the benefits outweigh the costs. Gibbs (1975) argued that as sociologists became more interested in the deterrence issue, economists became interested in rational choice models of crime. There is little doubt that both shifts were products of the times.

In fact, over the past two decades, politicians frequently have trumpeted a crude rational choice theory in their public pronouncements on lawbreaking. They have argued that the criminal justice system is lenient in the punishments it hands out, another example of the permissiveness that infects societal institutions. This popular explanation of crime has helped to justify numerous "get tough" policies that have increased the harshness of the punishments given to offenders. After all, if people commit crime because it pays, then it follows logically that the solution to crime is to increase punishments sufficiently so that the costs of offending surpass the benefits. Furthermore, if laws are passed making punishments harsher and crime does not diminish, then this outcome is not seen as evidence that the underlying theory of crime is incorrect. Rather, the lack of deterrence is taken as proof that the costs of offending just have not been raised enough and that another round of "get tough" legislation is required (Clear, 1994; Currie, 1998b).

This kind of simplistic theorizing about crime is evident in the work of Morgan Reynolds, an economist who advocates expanding imprisonment (Reynolds, 1996) and whose views have been cited in major news outlets such as *Newsweek* (Reynolds, 2000) and in syndicated columns by conservative commentators (Ridenour, 1999). As Clear (1994) also pointed out, Reynolds's research and other studies like it have been "distributed among state elected officials [and] immediately translated into legislative initiatives and calls for tougher sentencing" (p. 99). In short, these ideas have had potentially important consequences. At the very least, they have helped to legitimate the kind of "get tough" policies that politicians have been inclined to propose.

"Most crimes are not irrational acts," asserted Reynolds (1996). "Instead they are committed by people who at least implicitly compare the expected benefits with the expected costs, including the costs of being caught and punished" (p. 7). He suggested that "the reason we have so much crime is that, for many people, the benefits outweigh the costs—making crime more attractive than other career options" (p. 7). And what about other conditions that criminologists—theoretically and/or empirically—have linked to an increased risk for criminal involvement? These hold little interest for Reynolds because they are largely outside his narrow model focusing on the "expected costs" of illegal conduct. In fact, he denied that most sociological factors are "root causes" of crime at all. "Despite much rhetoric to the contrary," Reynolds boldly claimed, "there is little evidence that such economic factors as poverty, a poor economy, low wages, or low income growth and high unemployment cause crime" (p. 16).

Reynolds (1996) supported his claims by presenting empirical evidence ostensibly showing that the crime rate in Texas varied from 1960 to 1995 according to the "expected cost" of offending—a factor he computed as the by-product of the probability of arrest times the probability of prosecution times the probability of conviction times the probability of imprisonment times the median sentence served for an offense (p. 8). He asserted that "Texas has shown in recent years that punishment deters crime and that when crime does not pay, criminals commit fewer crimes" (p. 23).

An immediate difficulty with Reynolds's (1996) work is that it is based on a simple bivariate analysis—a correlation of expected costs and crime rates across time. It does not consider the host of complex methodological factors that make such an analysis problematic (Spelman, 2000). Most noteworthy, the failure to control for other macro-level variables means that the statistical model used by Reynolds almost certainly is "misspecified." That is, because other causes of crime are not taken into account, his results probably are wrong. In this regard, Pratt and Cullen's (2005) meta-analysis of macro-level studies of crime shows that a number of sociological variables, such as unemployment and poverty, are stronger predictors of crime than are many measures of punishment (e.g., the ratio of people arrested).

There also is the larger question of whether criminal sanctions deter. There is a genuine dispute about whether increasing the certainty and severity of criminal punishment has a general deterrent effect and, if so, whether this effect is meaningful (cf. Nagin, 1998, Levitt, 2002, and Wright, 1994, to Currie, 1998b, and Lynch, 1999). With regard to specific deterrence, however, the evidence is unmistakable. Reviews and meta-analyses of existing studies reveal that longer sentences do not deter offenders more than shorter sentences, that imprisonment does not deter offenders more than community-based sanctions, and that monitoring and punishing offenders in the community (e.g., intensively supervising them, drug testing them) does not deter offenders more than traditional probation and parole (Cullen & Jonson, 2011; Cullen, Pratt, Micelli, & Moon, 2002; Cullen, Wright, & Applegate, 1996; Finckenauer, 1982; Gendreau, Goggin, Cullen, & Andrews, 2000; Gendreau, Goggin, & Fulton, 2000; MacKenzie, 2000, 2006; Nagin, Cullen, & Jonson, 2009). Notably, the failure to demonstrate that harsher punishment deters offenders from recidivating more effectively than more lenient punishment is strong evidence against a simple rational choice perspective.

More nuanced rational choice theories have developed that do not necessarily see formal state controls, such as imprisonment, as a solution to crime. We will explore these approaches in Chapter 13.

Crime and Moral Poverty

In addition to approaches stressing criminal minds and the choices that offenders make, some of the more conservative theorists have linked crime to a distinctive type of social influence: the permissive culture in U.S. society (Murray, 1984). Although minimizing the criminogenic effects of class and social inequality, they argue that families and schools are failing to discipline and punish effectively (see also Hirschi, 1983). Underlying these claims is the notion that unless humans' baser instincts are tamed, humans will embark on lives of crime.

The exemplar of this variety of conservative thinking is the book *Body Count*, penned by the rather noteworthy set of authors of William Bennett, John DiIulio, and Ronald Walters (1996). They contended that moral poverty is the key cause of crime, especially of the youthful "super-predators" who create havoc and mature into serious chronic offenders. These authors defined moral poverty as "the poverty of being without loving, capable, responsible adults who teach the young right from wrong" (p. 13). They elaborated on this basic definition in the following way:

> It is the poverty of being without parents, guardians, relatives, friends, teachers, coaches, clergy, and others who *habituate* (to use a good Aristotelian word) children to feel joy at others' joy; pain at others' pain; satisfaction when you do right; [and] remorse when you do wrong. It is the poverty of growing up in the virtual absence of people who teach these lessons by their own everyday example and who insist that you follow suit and behave accordingly. In the extreme, it is the poverty of growing up surrounded by deviant, delinquent, and criminal adults in a practically perfect criminogenic environment—that is, an environment that seems almost consciously designed to produce vicious, unrepentant predatory street criminals. (pp. 13–14, emphasis in original)

Thus, moral poverty is broadly conceptualized. It lacks a precise statement—a statement that would, for example, tell us the components of moral poverty and allow for a systematic test of its proposed relationship to crime. On another level, however, the idea of moral poverty is unremarkable. It would be difficult to find too many criminologists—including quite liberal scholars—who would dispute that crime is fostered when a child is raised by dysfunctional or deviant parents and has the misfortune of residing in a criminogenic social environment (Skolnick, 1997). The language of "moral poverty" may strike a different emotional or ideological chord, but the criminological message seems old hat.

Where Bennett et al. (1996) departed from traditional criminology, however, is in their analysis of the origins of moral poverty (Currie, 1998b; Skolnick, 1997). Here their conservative political leanings are manifested. Traditional criminologists might have linked poor parenting and criminogenic environments to *structural* conditions such as social disorganization, economic inequality, and a concentration of socioeconomic disadvantages. But Bennett et al. argued that "moral and not economic poverty is the real 'root cause' of the nation's drug and crime problem" (p. 193). That is, the problem is *cultural*, not structural.

Thus, since the 1960s, America has been in a "steep" moral "slide," Bennett et al. (1996) claimed, due to a growing "culture of permissiveness" (pp. 195, 199). The "influential rich and upper class" are blamed for trumpeting a lifestyle of "unfettered freedom mixed with moral relativism" that "eventually spread to the rest of society" (p. 198). Television and popular culture are culprits as well. And of course, the criminal justice system is taken to task for stressing rehabilitation rather than punishment, thereby apparently eroding offenders' sense of responsibility (see also Bennett, 1994).

Given this perspective, the solution to crime is to give children the moral guidance—the love and discipline—they need to grow into healthy adults, a prescription that few of us would dispute. But Bennett et al. (1996) also proposed a broader revolution in which a culture of permissiveness, with its tolerance for both small and major forms of deviance, is replaced by a culture of virtue. At the center of this culture is the ethos of personal responsibility and the notion that misconduct has negative consequences. In criminal justice, this means holding offenders responsible and punishing them.

They characterize corrections as being a "no-fault, revolving-door system" that allows too many violent and repeat offenders to walk the street, and they argue that there is too much "crime without punishment" (Bennett et al., 1996, pp. 83–84). In their view, prisons are not overused but underused, especially in inner cities where predatory criminals are particularly prevalent (see also DiIulio, 1994). The nation's daily prison and jail populations may have risen to more than 2 million inmates, but when "we spend less than a penny per tax dollar to restrain violent and repeat criminals, we are kidding ourselves about promoting public safety, doing justice, and sending an authoritative moral message about crime and disorder" (Bennett et al., 1996, p. 136). We need to rid ourselves of the idea that prison offenders can be treated; in fact, the whole "anti-incarceration outlook" was due to the fact people came to believe that "criminal behavior should be rehabilitated, not punished" (1996, p. 198).

Notably, Bennett et al.'s embrace of mass imprisonment and rejection of rehabilitation seem at odds with their theory of crime. If moral poverty creates offenders, then it seems that the solution would not be punishment but interventions aimed at changing antisocial values and personality traits (Andrews & Bonta, 2003; Cullen & Gendreau, 2000; see also Johnson, 2004). The victims of moral poverty, however, are not just "high-risk" offenders but rather "super-predators" (Bennett et al., 1996, p. 26; see also DiIulio, 1995). The difference is not a mere euphemism. "High risk" implies a likelihood, but not an inevitability, of reoffending; it implies that resources should be targeted to this population (Andrews & Bonta, 2003). But a "super-predator" is beyond redemption. It is as though, in the developmental process, a tipping point is passed—a line is crossed—after which a different species is created and reform is impossible. As Bennett et al. (1996, p. 26) warned, a "new generation of street criminals is upon us—the youngest, biggest, and baddest generation any society has ever known." Given this view, a policy of building and filling as many prisons as possible seems to make sense. Unfortunately, while there can be a reasonable debate over what the capacity of prisons should be to effect optimal crime control, the notion that offenders can be punished out of crime or are beyond rehabilitation lacks empirical credulity (Andrews & Bonta, 2003; Cullen & Gendreau, 2000).

On a broader level, Bennett et al. also reminded us of the importance of religion and the tendency over the past few decades to exorcise God from civil society. In their view, we need to "remember God" because "religion is the best and most reliable means to reinforce the good" (p. 208). It is noteworthy that President George W. Bush defined "faith-based" programs as an important policy initiative of his presidency (Dionne & Chen, 2001). He appointed John DiIulio, coauthor of *Body Count,* to administer the newly created Office of Faith-Based and Community Initiatives, which oversees these programs (Grossman, 2001). DiIulio, it should be noted, resigned not long into his term.

Broad cultural explanations of crime such as that proposed by moral poverty theory are difficult to assess empirically because measures of cultural change are in short supply (Pratt & Cullen, 2005). The problem is that a theory that is hard to measure cannot be tested and, most important, cannot be tested in comparison to competing explanations of crime—in this case, structural theories. Furthermore, although cultural factors undoubtedly are implicated in crime causation, Bennett et al.'s (1996) dismissal of structural and economic conditions—claiming, in essence, that they have no impact on criminality—is on shaky ground. As Currie (1998b) noted, "What's wrong in the conservative view is not the belief that the moral and cultural condition of American families and communities is important in understanding crime, but the denial that those conditions are themselves strongly affected by large social and economic forces" (p. 113).

Bennett et al.'s (1996) selective reading of the criminological evidence can be seen in their citation of the research of Robert Sampson and John Laub. They quote Sampson and Laub as confirming that family processes are strongly related to delinquency (Bennett et al., 1996, p. 60). This buttresses Bennett et al.'s claim that families, and the role they play in fostering moral poverty, are at the root of crime. What Bennett et al. failed to cite, however, is the scholarship by Sampson and Laub showing that the quality of family processes itself is affected by "urban poverty"—that is, that structurally based disadvantage is a salient source of family problems. As Sampson and Laub (1994) argued, "Poverty appears to inhibit the capacity of families to achieve informal social control, which in turn increases the likelihood of adolescent delinquency" (p. 538). In fact, contrary to the central message of Bennett et al., Sampson and Laub cautioned that "families do not exist in isolation (or just 'under the roof') but instead are systematically embedded in social-structural contexts" (p. 538). Ignoring these structural arrangements—that is, focusing only on moral poverty—leads to incomplete theoretical understandings and potentially misguided policy recommendations regarding how best to control criminal behavior.

Broken Windows:
The Tolerance of Public Disorganization

Why is street crime high in inner-city neighborhoods? As we have seen, extending from the Chicago school of criminology (see Chapter 3), a popular answer is that these communities suffer from social disorganization or, in more recent terms, from a lack of collective efficacy. In this perspective, social disorganization or weak collective efficacy was itself a by-product of "root causes," including the mass in-migration of ethnically and racially heterogeneous groups and the coalescence of conditions into "concentrated disadvantage." Accordingly, high crime rates were not due chiefly to the self-selection of deviant individuals into inner-city neighborhoods but to people's exposure to the structurally induced criminogenic conditions that prevailed in the areas in which they resided. Bad areas, not bad people, caused crime and delinquency.

James Q. Wilson and George Kelling (1982), however, proposed a different theory of community disorganization: the "failure to fix broken windows." Consider, they urged, a house in which a window is broken. If the pane of glass is replaced, it sends a message to passersby that the owner is present, cares about the property, and will not tolerate its disrepair. But if the window remains broken, this sends the very different message that this property has no guardian. The shattered glass now serves as an invitation to passersby to throw a rock and break another window. And if that is left untended, soon every window will be broken and the vandalism of the house will become complete.

Neighborhoods, observed Wilson and Kelling (1982), function in much the same way; the broken windows must be fixed. The spiral of decline in a community begins when public signs of social disorganization—sometimes called "incivilities"—are tolerated. "Disorderly people" are allowed to take over public spaces; these are not necessarily predatory criminals but rather "disreputable or obstreperous or unpredictable people; panhandlers, drunks, addicts, rowdy teenagers, prostitutes, loiterers, the mentally disturbed" (p. 30). The good people of the community are hassled for money, must endure catcalls, feel obliged to cross the street to avoid groups of uncertain characters congregating on the corners or blocking the sidewalk, and are confronted constantly with loud music, rambunctious behavior, and increasing litter. As their fear escalates, these reputable residents first retreat into their homes and, if possible, move to another neighborhood. The result is that the normal informal controls exercised in the neighborhood weaken. In turn, the area now becomes "vulnerable to criminal invasion" by more predatory offenders. When this occurs, the fate of the neighborhood as a crime-ridden community is sealed.

How can this process be halted, if not reversed? The obvious answer is to "fix the broken window" so that the whole neighborhood does not fall into social disrepair. The recipe for reform depends on understanding the causal chain hypothesized by Wilson and Kelling: disorder caused by disreputable people → breakdown in informal control → invasion of predatory criminals → high crimes rates in the neighborhood. To stop this spiral of decline, it thus is necessary to attack the initial factor in this causal chain. Thus, disreputable people are not to be allowed to take over public space and create disorder. Public disorganization must not be tolerated.

The challenge was to figure out who would fix the broken window of uncivil disreputable people. For Wilson and Kelling, this job would fall not on any social service agency but on the police. Their task would be to exert control over—including arresting—these deviants. In what has become known as "zero-tolerance" and "quality-of-life" policing (Eck & Maguire, 2000; Harcourt, 2001), officers would crack down on those loitering, bothering residents, sleeping in doorways, disturbing the peace, and soliciting customers for sex and drugs. Wilson and Kelling's theory is that this *formal control* would create conditions to allow *informal control* to flourish once again. With the broken window fixed, the reputable people would once again be able to come outside and control their neighborhood. Disorder would decline and the invitation to prospective criminals would be cancelled.

This theory draws its initial appeal from the cleverness of its title. "Broken windows" embodies a metaphor and story that resonate with most observers. After all, who hasn't seen an abandoned building with shattered windows and been tempted to toss a rock themselves? It also suggests that neighborhoods can be saved simply by returning police officers to the beat, where they will keep order by rousting those who do not belong and arresting those whose disturbances are more bothersome. There will be no need for a "Chicago Area Project" or similar social reforms that would cost money and return us to the welfare state. Further, there is some comfort in knowing that the world can be divided into two categories: the disreputable ("them") and the reputable ("us") (Harcourt, 2001).

But is it all this simple? The experience of New York City offers some supportive evidence. Informed by this model, the city's officers aggressively policed minor forms of deviance, from youths trying to wash windshields and then implicitly demanding money to jaywalking. Coincidentally, the city's rate of homicide and violent crime decreased dramatically. This natural experiment could seem to suggest that broken windows theory is correct: Control disorder and serious crime will fall (Kelling & Coles, 1996).

Commentators have questioned, however, whether New York's apparent success can be attributed to "zero-tolerance" policing and is, in fact, confirmation of broken windows theory. During the drop in crime, other policing strategies were employed that targeted serious crime directly (e.g., mapping crime and intervening in "hot spots"). Declining crime also can be traced to a confluence of other factors at this time, including economic prosperity and changes in the drug market away from crack cocaine and the violence it induced. More problematic, there is evidence that crime rates began to fall in New York prior to the police reform movement, and that cities that did not use broken windows enforcement tactics achieved comparable decreases in their lawlessness (Eck & Maguire, 2000; Harcourt, 2001). Furthermore, research indicates that the relationship between disorder and crime is complex and affected by a number of considerations, especially the economic status of the community (Harcourt, 2001; Taylor, 2001). It is clearly possible, for example, that serious crime produces social disorder, and not vice versa as broken windows theory posits. The relationship between disorder and serious crime also might be spurious, with both caused by an underlying neighborhood condition such as collective efficacy (Sampson & Raudenbush, 1999).

Wilson and Kelling's broken windows theory makes a contribution in alerting us to the complex interplay between formal and informal control and in combating the assumption, often found in leftist criminological writings, that policing strategies will have little impact on crime. If these enforcement interventions are effective, they will bring the largest benefit—within a relatively short time period—to inner-city neighborhoods that each day face the reality of disquietingly high rates of predatory crime (Kelling & Coles, 1996).

The theory's potential downside, however, is its implicit disinterest in why communities have people who are homeless, panhandling for money, alcoholic, mentally ill, loitering on street corners, and willing to commit predatory crimes. There is little consideration that these personal conditions were not created by social disorder but rather developed while growing up in families and communities affected by structural inequality and concentrated disadvantage. There is no concern for where disreputable people come from and, in the face of broken windows policing, where they go to. The challenge, it seems, is to transform them from being bothersome in the location where police are cracking down to being merely expendable individuals who go somewhere else (see also Harcourt, 2001). In the end, broken windows policing may well be implicated in community crime rates. But such policing must be placed within a larger political and social context if we are to arrive at an adequate theory of criminal behavior; and, equally important, the messages that broken windows theory sends about communities and people must be fully understood if the prudence of the policies they justify are to be accepted not only as effective but also as just (Harcourt, 2001).

Consequences of Conservative Theory: Policy Implications

At the core of conservative criminology has been the revitalization of individualistic theory. As noted previously, by looking inside people for the sources of crime, individualistic theories do not consider what is going on outside of people. There is a tendency to take the existing society as a given and to see crime as the inability of deficient individuals to adjust to that society. There is no consideration of how long-standing patterns of inequality in power and living conditions are implicated in these criminogenic "deficiencies." Thus, crime as a social problem is transformed into a problem of individual pathology; society is taken as good, and offenders are taken as bad. In this approach, criminology forfeits its "critical" potential; it risks masking, if not excusing, the inequities rooted in the social order.

It is instructive that, at least in the United States, the revitalization of individualistic theories has not resulted in an era of progressive practices in criminal justice. Although individualistic theories might not be (at least in most instances) inherently conservative, they often are used—or at times misused—to justify "get tough" policies. The events of recent years are instructive.

The revitalized conservative theories of the 1980s and early 1990s involved two policy agendas: incapacitation and deterrence. The major policy agenda of the new conservative theorizing centered around—and continues to center around—the incarceration of larger numbers of offenders for longer periods of time. Legislatures across America increased the harshness of criminal penalties, as seen most vividly in the abolition of parole release, in "three strikes" laws (mandating life imprisonment for three felony convictions), and in "truth in sentencing" laws (mandating that offenders serve at least 85% of assigned prison sentences). Not surprisingly, during the past four decades, the nation has experienced an "imprisonment binge" resulting in the United States having the highest rate of incarceration in the world (Irwin & Austin, 1994; World Prison Brief, 2009). In Travis Pratt's (2009) words, the nation has become "addicted to incarceration." Thus, on any given day, 1 in 100 adults in America is locked up. For African Americans, the figure is 1 in 11 (Pew Charitable Trusts, 2008).

To understand the growth in imprisonment, it is necessary to look to the past. For the 50 years prior to 1970, the nation's rate of incarceration had been relatively stable (standardized for population size examining the number of people in state and federal prisons per 100,000 Americans) (Blumstein & Cohen, 1973). Indeed, in 1970, the state and federal prison population in the United States was less than 200,000 (196,429). By the mid-1970s, however, conservative thinking about crime emerged and began to acquire credibility (J. Q. Wilson, 1975). The 1980s ushered in the conservative era and an explicit movement to get tough on crime. This policy agenda proved successful.

Today, the state and federal prison population exceeds 1.6 million—more than a sevenfold increase since 1970. When jail and other institutional populations are included, the daily total of offenders behind bars is about 2.4 million. It is estimated that the United States has an incarceration rate per 100,000 of approximately 760; Russia, the closest competitor of the United States, stands at 620 per 100,000. Our northern neighbor, Canada, has an incarceration rate of merely 116; it has remained relatively stable since the 1970s (Webster & Doob, 2007). The count for England and Wales is 154. Numerically, China is America's main competitor, but it houses about 750,000 fewer offenders. Put another way, although having only 5% of the world's population, the United States accounts for 25% of the 9 million people incarcerated worldwide (Cullen & Jonson, 2011; Pew Charitable Trusts, 2008).

Two aspects of conservative theorizing make incarceration seem like a prudent practice. First, in revitalizing classical theory, conservatives emphasize the need to ensure that crime does not pay and see lengthy prison terms as an effective means of increasing the cost of offending. Rational choice theory suggests that some will pursue crime so long as the benefits outweigh the costs. One *could* interpret this to imply that those who have "nothing to lose" will turn to crime, and thus that policies should be developed to ensure that citizens have enough of a "stake in conformity" that they are afraid to risk it by criminal activity (Sherman, 1993). Instead, the logic of rational choice is now taken to mean that lawbreakers will best be deterred from further offenses by long incarceration experiences and that those contemplating crime will be deterred from acting on their criminal impulses by examples of stricter incarceration of offenders. Second, in revitalizing positivist theory, conservatives are expressing their belief that a proportion of offenders, whether due to criminal minds or to criminal natures, are beyond reform and must be incapacitated behind thick walls and sturdy bars. As James Q. Wilson (1975) observed, "Wicked people exist. Nothing avails except to set them apart from innocent people" (p. 235).

There can be little doubt that such thinking has justified, if not actively encouraged, the ready use of prisons as a solution to crime. If conservatives gain comfort from rising inmate populations, then they have far less reason to be sanguine about the effects of this policy. As discussed earlier, research on the deterrent effects of imprisonment is equivocal at best. The general deterrent effects of prisons are open to question; evidence on specific deterrence of imprisonment is unfavorable to the conservative perspective. Getting tough does not seem to scare past or future offenders straight.

Similarly, although locking up offenders in large numbers reduces crime, its overall effect is modest and raises complex policy decisions. In computing the "incapacitation effect," scholars make a troubling assumption—that the alternative to incarceration is allowing offenders to "run free" on the streets. They do not assess, for example, the impact of being in prison versus being in a high-quality, community-based rehabilitation program. Thus, estimates of incapacitation effects construct a misleading, if not false, comparison. The real issue should be how much crime can be prevented through incapacitation as opposed to other forms of correctional intervention (Cullen & Jonson, 2011).

In any case, Visher (1987) calculated that a doubling of the prison population during the 1970s achieved only a small decrease (6%–9%) in crime rates for robbery and burglary. For the early 1980s, imprisonment was "only slightly more effective in averting crimes" (p. 519). More disturbing, with construction costs of $30,000 to $80,000 per cell and an annual cost in excess of $10,000 for housing an inmate, this modest reduction in crime was achieved only with a substantial drain on state treasuries nationwide (Camp & Camp, 1987). By the end of the 20th century, the average state and federal adult correctional agency budget was nearly $631 million, and total expenditures for all correctional agencies surpassed $32.7 billion (Camp & Camp, 1999). At this time, Spelman (2000) estimated that doubling the current prison population "would probably reduce the [FBI] Index Crime rate by somewhere between 20 and 40 percent" (p. 422). These projections are not inconsequential, but the cost of the policy intervention likely would exceed $20 billion. That amount of money, Spelman noted, "could provide day care for every family that cannot afford it, or a college education to every high school graduate, or a living wage job to every unemployed youth" (p. 420).

The financial burdens of prison during the fiscally tight times of the 1980s furnished ample motivation to search for alternative methods of social control. Still, in turning to community corrections, conservatives brought a distinctive look. These conservative times resulted in a redefinition of the meanings and importance of community, public, and privacy. A few years before the 1980s, it would have been highly offensive to the American public to turn a home into a prison—a bedroom into a cell. But by the mid-1980s, several states followed the lead of Florida and Kentucky in passing laws permitting house arrest and electronic monitoring (Ball, Huff, & Lilly, 1988; Ball & Lilly, 1985; Lilly & Ball, 1993). An idea rejected in the United States during the 19th century reappeared in the 1980s: charging offenders a daily fee for their supervision and keep. Although it had several drawbacks, by the late 1980s communities in California, Ohio, Michigan, and Maryland were charging inmates between $20 and $85 per day ("More Jails," 1987, p. A8).

But these high economic costs of incarceration and the search for reasonable alternatives were viewed by some observers as part of a larger trend. Early during the 1990s, it was reported that corrections in the United States and other parts of the world were part of a "corrections-commercial complex," not unlike the military–industrial complex that Eisenhower had warned of more than 50 years ago. The central point of this argument is that the costs of corrections cannot be explained by high crime and incarceration rates alone because corporations providing goods and services to corrections, as well as corrections officials and political interests, profit economically from "get tough" policies, including President Clinton's "three strikes" proposal for life sentences for repeat offenders. Observers of this development also contend that it is not only a U.S. phenomenon but also one that is transnational and that at times influences justice policies for profit in foreign countries (Lilly, 1992, 2006a, 2006b; Lilly & Deflem, 1993; Lilly & Knepper, 1992, 1993; Nellis & Lilly, 2010).

We should note that, beyond the imprisonment binge, public humiliation or public punishment reappeared during the 1980s. For example, in Portland, Oregon, a sex molester not only received the usual sentence of no alcohol, no drugs, counseling, orders to stay away from parks and school yards, and a jail term but also, on release from jail, was required to put a sign on his front door reading "Dangerous Sex Offender, No Children Allowed" ("Unusual Sentence," 1987, p. A10). A similar form of public humiliation was used in one Oklahoma city in which convicted drunk drivers were required to put bumper stickers on their cars reading "I am a convicted DWI, driving while intoxicated, DUI, driving under the influence. Report any erratic driving to the Midwest City police" ("Public Humiliation," 1986).

Even when offenders remained in the community, then, the more conservative policy approach of the 1980s and 1990s tended to pay less attention to the societal roots of crime than was the case during the 1960s and early 1970s. Operating on the assumption that social circumstances are either unimportant or unfixable, these policies tended to restrict their attention to finding inexpensive ways of incapacitating offenders (e.g., chemical castration, house arrest), to deterring crime by increasing its psychological costs (e.g., public humiliation), or to inculcating more acceptable values through advocacy of certain forms of traditional family values or the use of programs such as Drug Abuse Resistance Education (DARE), a drug abuse prevention program that consists of using police officers to go into schools for anti-drug discussions. As a passing note, evaluation research on the DARE program has reported that these programs have no effects on subsequent drug use by youths (Gottfredson, Wilson, & Najaka, 2002; Lynam et al., 1999; Rosenbaum, 2007).

Finally, as noted earlier, then-President George W. Bush initiated faith-based community programs that sought to combat a variety of inner-city ills, including crime (Grossman, 2001). Because the connection of religion to crime is complex (Evans, Cullen, Dunaway, & Burton, 1995), and because research showing the effectiveness of religious programming is in short supply (but see Johnson, 2004), it remains to be seen whether such interventions can reduce criminal involvement. Still, aside from their effectiveness, faith-based programs pose an interesting shift in thinking for those on the political Right: They reflect the view of "compassionate conservatism" (see Dionne & Chen, 2001). These programs are conservative because their focus on religion reflects the underlying belief that moral poverty, and not structurally induced poverty, is the root of crime. Insofar as structural issues are ignored, it is problematic whether services that are provided by faithful people and that induce offenders to become faithful will reduce crime. These programs are compassionate, however, because they portray offenders as having value and as capable of being reformed or "saved." To this extent, they reject the idea that offenders are inherently wicked and beyond redemption. Indeed, faith-based programs can result in offenders receiving services and support (Cullen, Sundt, & Wozniak, 2000). Whether this will occur when diverse community-based programs seek funding under the mantle of "faith" remains to be seen. Still, the idea of faith-based programs remains popular with the American public and may yet prove to be an enduring legacy of the conservative era (Cullen et al., 2007).

Conclusion

Our discussion of conservative theory reflects the central point of this book: Ideas about crime—or what we call theories—are a product of society that develop in a particular context and then have their consequences for social policy. Thus, the 1980s and early 1990s saw a return to ways of thinking about crime that, although packaged in different language, revitalized the old idea that the sources of lawlessness reside in individuals, not within the social fabric. The rekindling of this type of theorizing, we believe, was no coincidence. Like other theories before it, conservative theory drew its power and popularity from the prevailing social context. And as we have seen, conservative thinking has had its day in influencing the direction of criminal justice policy. Locking people up in unprecedented numbers makes sense only if one believes that prisons reduce lawlessness by deterring the calculating and by incapacitating the wicked. Unfortunately, there is little evidence that incarceration deters, and it appears that any incapacitation effect from imprisonment is achieved only at a substantial cost. Indeed, mass incarceration is extremely expensive, draining funds from hard-pressed taxpayers and limiting funds that could be spent on other forms of crime prevention and in areas such as education and health care. It is somewhat paradoxical that these theories that purport to be the epitome of rationalism tend to produce policy consequences of such dubious rationality.

As we move further into the 21st century, it remains to be seen whether conservative-oriented individualistic theories of crime will continue to enjoy favor. Although these theories have considerable public appeal—if for no other reason than their simplicity—research has yet to demonstrate that such explanations have greater practical value in reducing crime than does focusing attention and resources on the social environment. But neither is it clear that the near future will see a major shift away from these theories, the conservative explanations of crime, and the "get tough" policies of the past quarter century.

Some hope might be drawn from the historic 2008 presidential election of Barack Obama. It would seem that this election signaled the end of conservative dominance in American politics. It was not simply that the Democratic Party gained control of the White House and Congress but also that right-wing ideas on social and economic policies did not appear to work. One key symbolic event was the inability of the Bush administration to respond promptly and effectively as human tragedy unfolded in New Orleans in the aftermath of Hurricane Katrina. The collapse of banks and financial markets, which triggered a deep recession, further questioned conservative economic policies of minimal regulation of "free" markets and the widening inequality in America they produced. The need for government to control powerful interests and to provide safety nets for citizens thrown into financial ruin seemed important.

The weakening of the conservative worldview, however, does not mean that a new era of liberalism will arise. A seemingly intractable economic recession has stirred voter anger and created political uncertainty. Further, President Obama's own crime control policies are a mixed bag (e.g., he supports the death penalty and drug treatment). It is instructive that the proposed budget for the Department of Justice for fiscal year 2011 reduced spending for juvenile justice programs while allocating hundreds of millions to hire or retain police officers and to increase spending for federal prisons (Justice Policy Institute, 2010). If there is some reason to anticipate a halt to the 40-year-old mass imprisonment movement, it is that states faced with severe financial crisis no longer can afford the price tag of this policy. The Pew Center on the States (2010) reports that for "the first time in nearly 40 years, the number of state prisoners in the United States has declined" (p. 1). The decrease is slight—only 5,793 inmates or 0.4% of the total population. Still, in more than half the states (27), prison populations declined.

Beyond economics, a more fundamental shift in thinking might be taking place: the realization that America cannot imprison itself out of the crime problem. In a 2010 speech at the Pepperdine University School of Law, Supreme Court Justice Anthony Kennedy criticized three-strike laws and the excessively harsh prison sentences that are meted out (Williams, 2010). More telling, Governor Arnold Schwarzenegger (2010) proposed a constitutional amendment to the state constitution that would require California to spend more on higher education than on prisons. Shortly thereafter, the California legislature passed a parole reform law that sought to lower prison populations by more than 6,000 offenders. To reduce parole revocation for technical violation, the statute removed low-risk offenders from parole supervision and created drug and mental health courts for offenders to receive treatment rather than be reincarcerated. Among other changes, inmates completing rehabilitation programs (e.g., GED or vocational certificate) now can earn early release from prison (Archibold, 2010; California Department of Corrections and Rehabilitation, 2010; see also Petersilia, 2008). Whether these events are harbingers of more extensive policy changes remains to be seen, but, in the least, they signal a willingness among political leaders to question the wisdom of continued "get tough" policies.

Marcus Felson
1947–
Texas State University
Author of Routine Activity Theory

13

Choosing Crime in Everyday Life

Routine Activity and
Rational Choice Theories

Most criminological theories examine why it is that some individuals develop an orientation to commit crime—often called *criminality*—whereas others do not. Although they differ in many ways, these perspectives share the view that criminality is something that develops *over time*. Thus, they focus on the conditions that surround people as they are raised in disorganized communities, are ineffectively parented for years on end, spend their youth in schools that frustrate them or are unable to earn their commitment, associate with delinquents in a gang, or perhaps are incarcerated for a lengthy tenure. For these approaches, *crime*—the actual behavioral act of breaking the law—is implicitly assumed to be inevitable and thus not in need of any special explanation. Those individuals with strong criminality will take advantage of opportunities to offend and thus have more involvement in crime than those with low levels of criminality.

The theories examined in this chapter reverse this theoretical interest; they are concerned with crime and not with criminality. Their focus is not on what occurred in the distant past but rather with what is occurring *in the present situation*. For them, whether a criminal act will be undertaken presents a theoretical challenge and has important policy and crime-control implications. In particular, they reject the idea that offenders are like billiard balls, pushed and pulled into crime in a mechanical way. Instead, they assert that offenders are active, thinking participants in their criminal ventures. They made decisions; they make *choices*. And why they choose to commit a crime in one situation and not another is a challenging criminological question.

These perspectives are at times called "opportunity theories" because they contend that no crime can be committed unless the opportunity to complete the act is present (Wilcox, Land, & Hunt, 2003). This observation is such a truism that it seems almost superfluous to make. But once the concept of *opportunity* is not taken for granted but rather becomes the object of study, the importance of linking opportunity to crime becomes apparent. Thus, *the nature of opportunity* affects what, where, how, and against whom crimes are committed. In a given situation, it shapes what choices offenders make—whether, for example, they decide to burglarize one house and not another.

The focus on the proximate situation and opportunity—rather than on the distant past—also leads us to explore how offenders make the decision to break the law. One popular suggestion—often called *rational choice theory*—is that offenders are "rational" in the decisions they make. For example, they choose crimes that offer immediate gratification, that require little effort to complete, and that expose them to scant risk of detection and arrest. In this latter regard, although not quite the same, other criminologists study how the criminal choices people make are affected by their *perceptions* of whether they will be caught and punished. This is known as *perceptual deterrence theory*.

These perspectives also suggest that much, if not most, crime occurs in the context of the *everyday lives* that offenders and their victims lead. With some justification, scholars have linked crime to the "bad" conditions in society—things pathological and powerful enough to create criminality. By contrast, opportunity theorists tend to see crime as emerging from the routines that people—whether offenders or their victims—follow as they go about their daily lives. This idea was captured early on by Lawrence Cohen and Marcus Felson when they proposed their *routine activity theory*.

Finally, these approaches propose very practical methods for reducing crime. As we will see, they have little interest in attacking the "root causes" of crime that create criminality or in trying to rehabilitate those burdened with these inclinations. Instead, their concern is with *situational crime prevention*. Their thesis is that if opportunity is a necessary condition for crime to occur, then crime can be reduced by removing the opportunity to complete the act. This might seem like little more than common sense, but it is a policy that is theoretically based and with a measure of proven effectiveness.

We will revisit these themes of situation, opportunity, choice, and situational prevention as we learn more about the major perspectives in this paradigm. We start with the earliest and most well known: routine activity theory.

Routine Activity Theory: Opportunities and Crime

Again, traditionally, criminology has focused most of its attention on offenders and on what motivates them to commit crime (or, in the case of control theory, how offender characteristics stop them from acting on their motivations). It has long been pointed out that even if offenders desire to commit crimes, they cannot do so unless the *opportunity* to break the law is present (Cloward, 1959; Cullen, 1984). But this observation has been either ignored or treated as though it were "true but trite." The working assumption among criminologists has been that the extent to which individuals are involved in crime is determined by their criminal motivations or "criminality" and that crime rates across social locations are determined by the number of criminally motivated offenders in any given location. If opportunity plays a role in crime, then it is assumed to be minor.

Over the past two decades or so, however, another group of scholars has argued that the distinction between *criminality* and *crime* (or a *crime event*) is not trivial but rather consequential. Criminality—the motivation or predisposition to offend—may matter, but such willingness to break the law cannot be automatically translated into a concrete criminal act. Opportunity is a necessary condition for any specific crime to be committed. More broadly, these scholars argue that the distribution of opportunities and individuals' access to these opportunities shape, in important ways, why certain geographical areas have higher crime rates than other areas and why certain individuals are more involved in crime than other individuals.

As noted, perspectives illuminating the connection between opportunities and criminal events are called *opportunity theories;* they also often are grouped under the label of *environmental criminology* because they examine how features of the physical and social environment present or limit criminal opportunities (Bottoms, 1994). The most noted of these approaches, and the one we are concerned with here, is *routine activity theory*. Marcus Felson developed this perspective in conjunction with Lawrence Cohen in 1979 and has been its foremost champion since that time (see, in particular, Felson, 1998, 2002). This theory is described below in the section on the "chemistry of crime."

Felson, we should note, had no intention of being a criminologist. Starting his first academic position in 1972 at the University of Illinois, however, be became involved in a project trying to correlate social indicators (e.g., such as trends in the economy) with various outcomes. As a new professor, Felson was assigned crime, which no other scholar wished to study, as his dependent or outcome variable. At the time, crime rates were rising rapidly, but, somewhat perplexingly, the United States was becoming more affluent. Traditional criminological theories, which linked offending to disadvantage, thus did not seem to work. As a result, Felson traveled outside criminology for explanations. Writing in the third person, Felson recounts his intellectual pathway to routine activity theory:

> He then began to reflect backwards, searching for ideas. He re-read Amos Hawley's seminal 1950 book, *Human Ecology*, to help him think about crime in more tangible terms. He drew on ideas from his famous father's radiological work that had identified four densities of the human body, and wondered whether there were three or four elements whose physical convergence made crime likely. He put to use his own Ph.D. on the stratification of consumer behavior, wondering whether some consumer goods were more likely to be stolen. Professor David Bordua told Felson about Patrick Colquhoun and other writings on crime opportunity. This produced the routine activity approach, the equations and the theory connected to it. (Clarke & Felson, 2011, p. 253)

The focus on opportunity suggests a *pragmatic* approach to preventing crime: Decrease opportunities for offending, and crime will be reduced. As we will see, the advice to reduce crime opportunities often leads to a focus on aspects of the environment that are most easily manipulated, such as whether a house has a burglar alarm and whether a store minimizes the amount of money in its cash registers. Although not without merit, the tendency of this perspective to focus on the pragmatic and to avoid discussing issues of inequality and power—and how they structure criminal opportunities (Maume, 1989)—is an implicit ideological decision. Notably, Felson (1998) criticized conventional criminology for its politicization: "Right-wing, left-wing, or whatever your agenda, if there is something you oppose, blame that for crime; if there is something you favor, link that to crime prevention. If there is some group you despise, blame them and protect others" (p. 20). Felson had a point—few criminologists fully escape their ideological preferences—but he and other environmental criminologists often are not reflective of their own implicit politics.

In fact, Felson has a preference for linking crime to the mundane or "everyday" features of society—an issue we will return to later. Felson (1998) used the term "pestilence fallacy" to describe the tendency of criminologists to treat crime "as one of many evils that comes from other evils in society" (p. 19). But his a priori assumption that crime is largely unrelated to the "evils of society"—to social problems—limits what he will systematically explore as related to crime causation. Similarly, he used the term "the not-me fallacy" to describe the supposedly mistaken assumption that "most individuals would like to think that they are *fundamentally* different from serious offenders in their willingness to commit crimes. . . . Everybody could do at least some crime at some time" (pp. 10–11, emphasis in original). Felson's preference to see criminality as largely evenly distributed across society allowed him to ignore those social conditions—including factors related to socioeconomic inequality—that might create stronger criminal motivations in some people than in others.

This pragmatic focus and tendency to avoid structural and political issues make this theory attractive in the current social context. After the strong tilt of social and crime policies to the right during the 1980s and early 1990s, the nation might be considered to be in a "post-conservative" period. Many conservative themes remain vital in American politics, but so do progressive ideas (Dionne, 1996). Although the base members of the political parties might take rigid ideological positions and fight one another vigorously, the nation has moved into something of an ideological standoff or gridlock. Most citizens hover around the center of the political spectrum. Although not rejecting social welfare and free market ideas, Americans have become suspicious of liberal efforts to build a large welfare state and of conservative efforts to create a state that simultaneously imposes right-wing morals on others and gives corporations and financial institutions unfettered discretion to do what they wish. In this context, ideology often is eschewed in favor of doing "what works" or "what seems to make sense." Indeed, within the criminal justice system, there is increasing emphasis on employing evidence-based or "what works" approaches that can be shown to reduce crime (see, e.g., Cullen & Gendreau, 2001; MacKenzie, 2006; Sherman, Farrington, Welsh, & MacKenzie, 2002; Welsh & Farrington, 2006, 2009).

The attractiveness of opportunity theories of crime is that they avoid larger discussions of whether the United States is excessively unequal or excessively morally permissive and argue that crime can be prevented meaningfully without a major cultural or social revolution. Instead, by changing a few locks and installing a few alarms—or similar modest interventions—we can make ourselves safer. These policy recommendations are, in fact, potentially important; criminal opportunities do matter. Decreasing them can save lives. Further, these theories argue against "get tough" policies, instead showing the offenders' choices are shaped not so much by uncertain threats of imprisonment but rather by the obstacles to criminal gratification found in the immediate situation. Even so, the risk in trumpeting pragmatic policies is that it suggests that crime is not influenced by the "other evils in society." In the end, fundamental "social evils" or "root causes" might be difficult to change, but this does not alter the reality that such causes exist or obviate the need to counteract their untoward effects on individuals (see, e.g., Currie, 1985, 2009).

THE CHEMISTRY FOR CRIME: OFFENDERS, TARGETS, AND GUARDIANS

Just as a chemical reaction cannot occur without all of the necessary ingredients mixed together, Felson (1998) suggested that the "chemistry for crime" (p. 52) requires all of the necessary ingredients. Again, environmental criminologists argue that a criminal event involves not just a person willing to offend but also the opportunity to act on these motives. Felson's special contribution is that he helped to demarcate the key elements of opportunity.

In his classic article with Lawrence Cohen, Felson noted that "each successfully completed violation minimally requires an offender with both criminal inclinations and the ability to carry out those inclinations" (Cohen & Felson, 1979, p. 590). This is another way of saying that a criminal event requires a "motivated offender" who has the opportunity to act on those motivations. This opportunity or "ability to carry out" a crime, in turn, involves two elements.

First, there must be "a person or object providing a *suitable target* for the offender." Second, there must be an "*absence of guardians* capable of preventing violations" (p. 590, emphasis in original). Cohen and Felson chose the term "suitable target" rather than "victim" because they meant to include not only people but also property. They chose the term "capable guardians" rather than "police" because they meant to include not only law enforcement personnel but also all means by which a target might be guarded. Most often, such guardianship is provided informally by family members, friends, neighbors, or other members of the public. But guardianship also can be furnished by other means, such as dogs and security cameras.

Cohen and Felson (1979) limited their analysis to "predatory crime" or crime involving offender-target contact. For these transgressions to take place, the three essential ingredients—motivated offenders, suitable targets, and an absence of guardianship—must *converge in time and space* (see Table 13.1). In a truly innovative insight, the authors suggested that the major determinant of this convergence was the *routine activity* of people in society (and thus the naming of their perspective as "routine activity theory"). The term *routine* carried two meanings. Most important, it was a technical term that referred to the "everyday activities" that people in society followed—when and where they worked, attended school, recreated, and stayed home. More implicit, the term *routine* meant to imply the mundane in life, not the special or abnormal. Cohen and Felson were calling attention to the fact that the amount of crime was influenced not by the pathological features of society but rather by its normal organization. In Felson's (1998) words, this approach "emphasizes how illegal activities feed on routine legal activities" (p. 73). As Bottoms (1994) put it, "Routine activities theory in effect embeds the concept of opportunity within the routine parameters of the day-to-day lives of ordinary people" (p. 605).

Table 13.1 Summary of Cohen and Felson's Routine Activity Theory

Elements of a Crime	Definition	Key Criminological Point
Motivated offender	Person who has the propensity or inclination to offend.	Most traditional theories of crime explain why some people are motivated to offend. They do not examine the elements of criminal opportunity, which involve a target and guardian.
Suitable target	An object—person or property—that the offender would like to take or control.	A crime cannot take place without a suitable victim. Suitable can mean attractive in the sense that the object can provide reward (e.g., money) or be transported (e.g., a computer versus a refrigerator).
Absence of a capable guardian	Guardians can be friends or family, security personnel, or dogs. A person can be a guardian of his or her own person or property.	The absence of guardians allows crime. The presence of guardians prevents crime.

Cohen and Felson were particularly interested in explaining changes in *crime rates* over time. In this sense, their perspective initially was stated as a *macro-level* theory of crime. Jumps in crime rates typically had been attributed to social problems in America that enlarged the pool of motivated offenders. By contrast, Cohen and Felson (1979) showed that "substantial increases in the opportunity to carry out predatory violations have undermined society's mechanisms for social control" (p. 605) and heightened lawlessness independent of the characteristics of offenders. They maintained that the "convergence in time and space of suitable targets and the absence of capable guardians can lead to large increases in crime rates without any increase or change in the structural conditions that motivate individuals to engage in crime" (p. 604). As Felson (1998) reminded us, "Offenders are but one element in a crime, and perhaps not even the most important" (p. 73).

Thus, Cohen and Felson (1979) argued that a key reason for rises in predatory offenses was that "since World War II, the United States had experienced a major shift of routine activities" away from the home (p. 593). Because homes were now increasingly left unattended during the day, they had become candidates for burglary because the attractive targets within them no longer were as vigilantly guarded. Similarly, as people spent more time at large in society—going to and from work, school, and leisure activities—they were more likely to come into contact with motivated offenders in circumstances where guardianship was lacking. Thus, the possibility for robbery and assault was increased. Furthermore, Cohen and Felson linked property crimes not to economic deprivation but rather to the production of goods that were "expensive," "durable," and "portable"—and that had become more common in an increasingly affluent society. Thus, electronic goods and automobiles were stolen more often than food and refrigerators because the former were more suitable targets; they could be moved, be used again, and bring payoffs. Contrary to much criminological thinking, prosperity could bring about higher, rather than lower, crime rates by expanding the number of attractive targets available to motivated offenders.

The focus on the possible crime-inducing effects of prosperity is illuminating, but it also shows the tendency of routine activity theory to ignore the potential role of poverty and inequality in generating crime opportunities—not to mention in generating criminal motivations. To be fair, Felson (1998) suggested that "poverty areas" may increase "temptations" and decrease "controls." For example, poor people are more likely to live next to "shopping malls, commercial strips, warehouses, loading docks, parking lots and structures, train yards, factories, bars and taverns, medical facilities," and so on (p. 35). These places provide opportunities to offend; that is, there is "more to steal and fewer people at night to watch things, as well as stragglers who are easy to attack" (p. 35). Similarly, in poor areas, there is a larger market for secondhand goods, inducing the establishment of stores that may encourage theft by their willingness to fence stolen property. Such areas also may contain more vulnerable populations, such as immigrants, who are easier to victimize. Still, Felson failed to develop systematically within his theory how the political economy shapes illegal opportunities and shapes the social distribution of crime. He might have considered, for example, the innovative work of Maume (1989), who presented macro-level data showing that the effects of inequality on rates of rape are mediated by routine activities (see also Cao & Maume, 1993). "One of the hidden costs of inequality," Maume (1989) concluded, "is that some people are constrained to live risk-prone lifestyles; these people are burdened most heavily by the problem of crime" (p. 524).

Numerous tests of routine activity theory using macro-level data have been conducted. Pratt and Cullen's (2005) meta-analysis showed that most of the tests have measured the guardianship component of the theory. Even so, the results generally support the conclusion that routine activities influence rates of crime across ecological units. We should note, however, that routine activity theory has been used on the *micro level* to explain the behavior of *individuals*. Thus, research reveals that when youths are engaged in structured conventional activities or routines, they are less likely to offend. This insight is consistent with Hirschi's (1969) ideas on the social bond of involvement (see Chapter 6). Wayne Osgood and his colleagues (Osgood, Wilson, Bachman, O'Malley, & Johnston, 1996), however, noted that Hirschi's work did not specify what uninvolved adolescents do with their time. For Osgood et al., the key routine activity that fosters crime is time spent in unstructured socializing with peers, especially without adult authority figures present to supervise them. This kind of routine activity exposes youths to situations (e.g., riding around or partying with friends) that are likely to offer the lure and opportunity for crime and other deviant activities (Osgood, 2011).

There also is a growing body of scholarship showing how routine activities can affect who in society is most likely to be *victimized* (see, e.g., Fisher et al., 2010; Fisher, Sloan, Cullen, & Lu, 1998). A perspective similar to routine activity theory had been independently developed by other researchers (e.g., Garofalo, 1987). Based on victimization surveys, they noted that some individuals were more likely to be victimized than others. They explained this differential victimization by a "lifestyle model" in which those whose lifestyles or routine activities are riskier—exposing them to potential offenders—are more likely to experience a higher level of personal victimization (Garofalo, 1987). For example, people who go to bars, drink, stay out late, and walk home alone are more likely to be victimized than people who spend their evenings at home with family members. Studies tend to support a lifestyle or routine activity explanation of victimization (Fisher et al., 2010). However, the results are at times inconsistent and often based on incomplete research designs (e.g., failure to control for the effects of community context) (Meier & Miethe, 1993). Continued systematic research thus is needed (Kubrin et al., 2009).

VIEW OF OFFENDERS

Although largely disinterested in why people are motivated to offend, routine activity theory is most compatible with rational choice theory (to be discussed below) and Gottfredson and Hirschi's (1990) theory of low self-control (see Chapter 6) (Eck, 1995). Clarke and Felson (1993), for example, suggested that "the routine activity and rational choice approaches, though differing in scope and purpose, are compatible and, indeed, mutually supportive" (p. 1). In any case, the link between these seemingly dissimilar perspectives is the notion of the *effort* it takes to commit a crime or, conversely, the ease with which a crime can be accomplished. Felson (1998) argued that crime is less likely to occur when it is made less attractive. Because offenders are guided by the lure of "quick pleasure" and the avoidance of "imminent pain" (p. 23), anything that makes crime harder to commit also makes it less likely to occur. In fact, Felson noted that attempts should be made to "assist in self-control" by not making opportunities to crime too tempting to those deficient in self-control (p. 48).

From a rational choice perspective, these limitations placed on opportunity are "costs" that reduce the "expected utility" of crime. And from a self-control perspective, making crime harder to commit makes it less immediately gratifying. Although low self-control undermines the ability to resist criminal opportunities, the lack of such steadfastness also makes it less likely that an offender will engage in a crime that requires diligence or overcoming barriers. In the end, these perspectives share with routine activity theory the view that, in any situation where a crime event could transpire, the decision to offend will be influenced by the ease or difficulty with which the offender's search for gratification can be satisfied.

Again, routine activity theory does not rest inherently on a particular view of what motivates people to become offenders; it argues only that for crime to occur, there must be motivated offenders. Conceivably, then, this perspective could explore its compatibility with social learning theory, strain theory, feminist theory, and any other motivational theories of crime. Thus far, however, these potential linkages have not been explored systematically. Rather, implicitly or explicitly, routine activity theory is based on a view of offenders as gratification seekers who wish to "gain quick pleasure and avoid imminent pain" (Felson, 1998, p. 23). As we will see shortly, this working assumption about offenders is a chief reason why routine activity theory, similar to rational choice theory, leads to the policy recommendation to fight crime through "situational crime prevention" measures.

Finally, although routine activity theory was not developed to focus in detail on offenders, scholars within "environmental criminology" have taken up this task (see, e.g., Bottoms, 1994; Brantingham & Brantingham, 1993; Eck & Weisburd, 1995). Because crime events involve the interaction of offenders and targets in time and space, these scholars argue that it is necessary to study not just the routines of potential victims but also the routines of potential offenders and how they select their targets to victimize. This is often called *offender search theory*.

Offenders do not wander randomly looking for crime opportunities; rather, they engage in patterned behaviors, typically traveling to certain areas but not others and traveling only so far from home. They develop "cognitive maps" of their environment and tend to commit crimes in places that are familiar to them (Bottoms, 1994). They also evolve "mental templates"—holistic conceptualizations that are based on experience and routines—that are used to "predefine the characteristics of a suitable target or suitable place" and to "identify what a 'great' chance [for crime] is or what a 'good' opportunity would be or how to search for chances and opportunities" (Brantingham & Brantingham, 1993, pp. 268–269; see also Meier & Miethe, 1993). In short, offenders play an active role in producing criminal opportunities. Where they are willing to travel and how they interpret their social environment when they get to their destination help to determine which targets they come into contact with and which targets they see as attractive and capable of being victimized. Ultimately, the distribution of offenses across time and space will be a by-product of this intersection between the routine activities of both victims and offenders (Brantingham & Brantingham, 1993).

POLICY IMPLICATIONS: REDUCING OPPORTUNITIES FOR CRIME

A core theme of this book has been that a given set of theoretical ideas logically suggests—or makes "sensible"—a limited range of policies for confronting crime.

Routine activity theory is no exception to this theme. Because of its general disinterest in offenders, except insofar as offenders are portrayed as gratification seekers, this theory forgoes any thoughts of how steps might be taken to change the criminality or motivations of lawbreakers. Accordingly, there is no thought given to rehabilitation, early intervention programs, or social reforms that might attack the root causes of crime. Instead, for Felson (1998) and others in this theoretical camp, the key to stopping crime is to prevent the intersection in time and space of offenders and of targets that lack guardianship. This is another way of saying that because the fundamental cause of crime events is *opportunity,* the chief way of preventing crime is to *reduce such criminal opportunities.* On a broad level, this means making targets less attractive and supplying targets with capable guardianship.

The idea that opportunities are key to fighting crime is not fully new or linked only to contemporary criminological theories. According to the National Advisory Commission on Criminal Justice Standards and Goals (1975), "Of all the things a citizen or a community can do to reduce crime, the most immediate and most direct approach is to eliminate obvious opportunities for criminals." To "make crime . . . more difficult," the commission recommended "locked cars, well-lighted streets, alarm systems, and properly designed and secure housing" (p. 146). Some years earlier, Jane Jacobs (1961) noted that city streets were safe "precisely where people are using and most enjoying [them] voluntarily and are least conscious, normally, that they are policing" (p. 36). But when cities are designed so that public places are deserted or not crowded, guardianship decreases and opportunities emerge for "barbarism to take over many city streets" (p. 30). Similarly, Oscar Newman (1972) assessed how the architectural design of residential environments, such as public housing, might reduce criminal opportunities by creating "defensible space." It was possible, Newman asserted, to foster

> extremely potent territorial attitudes and policing measures, which act as strong deterrents to potential criminals . . . [by] grouping dwelling units to reinforce associations of mutual benefit; by delineating paths of movement; by defining areas of activity for particular users through their juxtaposition with internal living areas; and by providing natural opportunities for visual surveillance. (p. 4)

And we would be remiss if we did not mention C. Ray Jeffery (1977), who championed the broader notion of "crime prevention through environmental design."

Routine activity theory and, more generally, environmental criminology have built on these insights to argue for a systematic approach to blocking criminal opportunities (Wikstrom, 1995). Again, these perspectives share with rational choice theory the notion that "making crime difficult" also makes it less likely to occur. Therefore, they suggest that efforts should be made to "design out crime" (Felson, 1998, p. 58) or to introduce "situational crime prevention" (Clarke, 1992). In Clarke's (1992) words:

> Situational crime prevention comprises opportunity-reducing measures that are (1) directed at highly specific forms of crime (2) that involve the management, design, or manipulation of the immediate environment in as systematic and permanent [a] way as

possible (3) so as to increase the effort and risks of crime and reduce the rewards as perceived by a wide range of offenders. (p. 4)

Based on this definition, Clarke (1992, p. 13) provided a useful way of understanding how crime opportunities can be blocked or made less attractive (see also Bennett, 1998; Clarke, 1995). First, there are strategies that seek to *increase the effort* needed to commit a crime. These might include using more effective physical barriers to crime, such as a lock on a vehicle's steering wheel, a higher fence around one's property, and a buzzer to gain entry into a building. Second, there are strategies to *increase the risks* of attempting to commit a crime. These might include a host of ways of increasing the chances of detection, such as installing burglar alarms, using merchandise tags and detectors at store exits, hiring security guards, and increasing the ease with which neighbors can see one's house (e.g., pruning bushes, improving lighting). Third, there are strategies aimed at *reducing the rewards* of crime. These might include limiting the cash kept in a store, marking property with an owner's identification to make its sale more difficult, and having removable (as opposed to permanent) CD players to make breaking into automobiles less tempting.

We should note that Clarke came upon the idea of situational crime prevention as the result of his involvement in two studies that he thought would yield different results. Trained as a clinical psychologist, Clarke took a position as a researcher for the United Kingdom's juvenile training schools. One experimental study he conducted with Derek Cornish showed that, surprisingly, the reconviction rates for youths in a therapeutic community did not differ from those in the control condition (a regular house in the training facility). This finding taught him that changing criminal motivation was a daunting challenge. In a previous study, he found that a range of psychological tests and demographic variables were, also surprisingly, unable to identify which youths did and did not abscond from the training schools. Although he was unable to predict which individuals ran away, he discovered that, across 88 schools, rates of absconding nonetheless varied widely. This finding taught Clarke that "school environments were much stronger determinants of absconding than the dispositions of the residents and also suggested that changes made to the regimes and environments could substantially reduce absconding rates" (Clarke & Felson, 2011, p. 251). Together, then, these studies persuaded Clarke that changing offenders—removing the motivation to offend—was quite difficult. Instead, the best bet to reduce crime was to tinker with environments so that the choice of crime would be knifed off or, in the least, discouraged. This line of inquiry eventually prompted Clarke (1980) to write a classic essay called "'Situational' Crime Prevention."

Similar to Clarke, Felson (1998) suggested that blocking crime opportunities can be heightened through three strategies. First are *natural strategies,* where space is designed in such a way that people are channeled "to go where they will do no harm or receive no harm" (p. 150). For example, signs and access to only certain doors might "naturally" lead people to enter a building only through a door where surveillance is high (i.e., many people are around). Second are *organized strategies,* where security guards are hired for the express purpose of making crime difficult. Third are *mechanical strategies,* where "alarms, cameras, and other hardware are employed to control access and provide surveillance" (p. 150).

In a like vein, Eck (2003) has expanded the concept of guardianship to that of "controllers." His concern is with studying, in Felson's (1995) words, "those who discourage crime." Consistent with routine activity theory, he sees crime as the intersection of *offenders* and *targets* in a particular *place.* He uses a "crime triangle" to visually represent this convergence and how controllers can limit opportunity and thus crime (see Figure 13.1). Hence, "each of these three elements . . . has a potential 'controller'—a person (or people) whose role it is to protect them. If a controller is present, then the opportunity for crime either is diminished or vanishes" (Cullen, Eck, & Lowenkamp, 2002, p. 32).

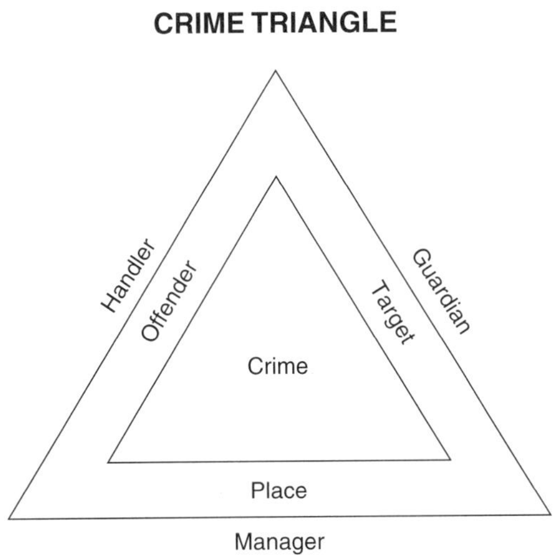

CRIME TRIANGLE

Figure 13.1   Eck's Routine Activity Crime Triangle

SOURCE: Adapted from Eck, J. E. (2003). Police Problems: The Complexity of Problem Theory, Research and Evaluation. In J. Knutsson (Ed.), *Problem-Oriented Policing: From Innovation to Mainstream* (Vol. 15, pp. 67–102). Monsey, NY: Criminal Justice Press, page 89.

In his terminology, offenders have "handlers," usually people they know such as spouses, family members, neighbors, or clergy. Because of their mutual bonds, offenders would not wish to break the law right in front of these individuals. As a result, the presence of a handler often precludes a crime from occurring. Targets (or victims) have "guardians." It might be that a person is a guardian over his or her own property; or, when out late at night, it might be that friends in a group will provide mutual protection, thus discouraging a potential offender's attempt at a predatory crime. Finally, "managers" are those who are responsible for the proper functioning of a particular place. This might include a store clerk, a custodian, a teacher in a classroom, a doorman, or a receptionist. Even if preventing crime is not on their minds, the mere presence of place managers can discourage an offender from considering this spot available to victimize.

As we have just seen, advocates of opportunity reduction usually emphasize the importance of strategies that do not involve the use of police. In large part, this is because crime prevention is most effective when it is a permanent and/or natural feature of a place location. By contrast, police surveillance often requires planning and typically is episodic, with officers discouraging crime mainly when they are physically present in a location. Even so, one type of law enforcement is potentially consistent with routine activity theory and situational crime prevention: "problem-oriented policing" (Eck & Spelman, 1987; Goldstein, 1979; Sherman, 1997). From this perspective, police would define a problem, such as juvenile violence in a neighborhood or drug selling in a particular building, and then devise strategies to make these specific offenses more difficult to commit. This approach might involve training officers to use legal means to search juveniles and seize their guns (Sherman, 1997). These tactics might discourage youths from carrying firearms in public spaces, thereby diminishing the occurrence and lethality of violent acts. Or it might involve officers' meeting with apartment building owners to encourage them to manage their properties more safely (e.g., evicting drug dealers) or risk forfeiting their buildings through civil litigation (Eck, 1998). Efforts also might be made to map crime patterns to discover the likely residences of, and routes to crime locations taken by, offenders— all in hopes of focusing police investigations on certain locations and increasing the chances of apprehending repeat criminals (Rossmo, 1995). Places where crimes recur repeatedly are sometimes called "hot spots" (Sherman, Gartin, & Buerger, 1989).

Notably, research studies (although not always of high quality) generally suggest that efforts to design out crime or to effect situational crime prevention achieve reductions in crime (Clarke, 1992; Eck, 1997; Felson, 2002; Felson & Boba, 2010; Guerette, 2009; Welsh & Farrington, 2009). Because these findings are consistent with routine activity theory (and rational choice theory), they might be taken as evidence that these perspectives have merit.

In any case, there is a serious challenge to the optimistic conclusion that blocking opportunities for crime "works"; few evaluations of crime prevention programs systematically take into account what is known as the "displacement effect." Simply put, displacement is the possibility that when crime is made more difficult in one location, offenders will move on and commit their crimes in another location. Crime is not reduced but rather repositioned to another target or place. Thus, if one house has an alarm, then offenders will break into the house next door; or if the police are "cracking down" on drug pushing on one street, then offenders will sell their drugs a few streets away. Importantly, although displacement cannot be discounted, it is unlikely that crimes blocked in one place are displaced 100% to another place (Barr & Pease, 1990; Clarke, 1992, 1995; Eck, 1997; Eck & Weisburd, 1995; Felson, 1998). Felson (1998) called this critique of crime prevention the "displacement illusion" (p. 140). Similarly, Eck and Weisburd (1995) concluded that "there is a growing body of evidence that suggests that displacement is seldom total and often inconsequential or absent" (p. 20; see also Guerette, 2009).

There are at least three reasons why displacement is limited. First, when a crime is thwarted, finding a new opportunity to offend often takes time and effort. To the extent that searching for new targets is "costly" or requires self-control, offenders may decide that committing a crime is not "worth the effort" (Gottfredson & Hirschi, 1990, p. 256). As Clarke (1992) observed, displacement will take place only when "alternative crimes present similar rewards without unduly greater costs in terms of risks and effort" (p. 22). Second, a particular type of crime may have "choice structuring properties" (p. 24). For example, burglary offers an offender a means of securing funds without having to confront a victim, have access to a weapon, and threaten—or actually use—violence (Wright & Decker, 1994). If a burglary is prevented, however, the offender might not be willing or able to engage in another crime, such as robbery. Third, to the extent that crime opportunity flows from the routine activity of offenders, a disruption in a community's criminal opportunity structure may diminish offending until new routines and cognitive maps are developed.

Rational Choice Theory

As we saw in Chapter 12, some theorists embrace a simple view of why people choose crime: It pays. This calculation is reduced to a simple balancing of how much can be gained from crime (usually money) and how much can be lost in terms of being caught and punished by the state. Leniency is thus seen as the chief cause of crime, and more certain and lengthier prison terms are seen as the chief solution to the lawlessness in society. However, other scholars—Derek Cornish and Ronald Clarke most prominent among them—have attempted to construct a more sophisticated approach to the decision to offend, which they have termed *rational choice theory*. This perspective, as alluded to above, leads to a preference to control crime not through state criminal sanctions but through more informal situational crime prevention.

RATIONAL CHOICE AND CRIME

In their model, Cornish and Clarke (1986) understood that people are not "empty vessels" when they approach a situation in which a crime might be committed (see also Clarke & Cornish, 2001). They bring with them "background factors" that include many of the influences articulated by other theories of crime, such as temperament, intelligence, cognitive style, family upbringing, class origin, neighborhood context, and gender. These factors create "criminal motivations—deep-rooted inclinations or dispositions to commit crime" (Clarke & Cornish, 2001, p. 33). Such motivations are important in giving people a "taste" for crime or in increasing the probability that crime will be "subjectively available" to people—something they would consider doing (Wright & Decker, 1997).

From a rational choice perspective, the problem with traditional theories is that they stop their analysis of crime causation at this point. But in the end, crime is not simply due to underlying motivations or predispositions; it also involves a concrete *choice*—or, in fact, a *sequence of choices*—that must be made if these motivations are to result in an actual criminal act. In a sense, criminologists have assumed that if a person is criminally motivated, then the individual will commit more crimes than if he or she is not so motivated. This assumption may be generally accurate, but it leaves open how those motivated to offend decide to break the law in any given situation. The rational choice perspective urges us to take seriously how offenders *think* so as to predict when criminal events will occur. Therefore, central issues include the decision to commit a particular type of crime, how an area in which to commit the crime is selected, how "targets" or victims are selected, how offenders decide to take steps to avoid detection, and how offenders decide to recidivate. As Clarke and Cornish (2001) pointed out, traditional theories are more interested in the "wider social and political contexts that mold beliefs and structure choice. Consequently, they have taken little interest in the details of criminal decision making." By contrast, for the rational choice perspective, "it is these details that must be understood" (p. 32).

A key assumption of rational choice theory is that the decisions that offenders make are "purposive." That is, they are "deliberate acts, committed with the intention of benefiting the offender" (Clarke & Cornish, 2001, p. 24). Offenders are not perfectly rational; rather, their rationality is said to be "bounded." This means, in effect, that offenders make choices that might be based on limited information, made under pressure, insufficiently planned, and/or attentive only to the immediate risks of apprehension rather than to the long-term consequences of their actions. On reflection, such choices might be judged as ill advised, if not foolish. Still, within this context, offenders are not acting irrationally but rather are endeavoring to satisfy needs (e.g., money, status, revenge) and to avoid negative consequences (e.g., arrest, being shot by an intended victim). As Clarke and Cornish (2001) pointed out, offenders "are generally doing the best they can within the limits of time, resources, and information available to them. This is why we characterize their decision making as rational, albeit in a limited way" (p. 25).

The theoretical focus of rational choice theory has important, but limiting, policy implications on how best to reduce crime. Because of their interest mainly in the situational dynamics that affect the decisions offenders make, rational choice scholars lump many of the strongest predictors of crime under the catch-all heading of "background factors." These factors hold little theoretical interest for these scholars, who do not think seriously about how interventions might be designed to change these "root causes" and reverse the effects they have on offenders. Not surprisingly, rational choice theorists ignore the evidence showing the effectiveness of correctional rehabilitation programs and accept the incorrect notion that treatment programs have been a "catastrophic failure" (Cullen, Pratt, Micelli, & Moon, 2002). Instead, they are predisposed to believe that "every act of crime involves some choice by the offender. He or she can be held responsible for that choice and can be legitimately punished" (Clarke & Cornish, 2001, p. 34). At this point, they slip into conservative ideology. One might just as easily argue, for example, that every act of crime involves *only* some choice. Because crime acts are at least partially bounded or caused by background factors, one must question the assumption of free will and the legitimacy of punishing offenders.

Rational choice theory, however, suggests a more promising approach to reducing crime: *situational crime prevention* (Clarke, 1992). By studying how offenders make decisions to, say, commit burglaries, steps may be taken to reduce such *opportunities* for these offenses to occur. In this approach, crime is prevented not by changing offenders but rather by changing aspects of the *situations* in which offenses typically occur (e.g., installing burglar alarms, obtaining dogs). The focus is on making crime more difficult to commit or less profitable so that it becomes a less attractive choice. Recall that we considered this initiative earlier in the chapter when we discussed routine activity theory.

Finally, rational choice theory represents an important advance in criminological thought because it shows the need to study offenders not as "empty vessels" propelled to commit crime by background factors but rather as conscious decision makers who weigh options and act with a purpose. The danger in rational choice theory, however, is that offenders will be treated as though they were *only decision makers*. When this occurs, the context that affects why they come to the point of breaking the law is ignored, and commentators begin to recommend harsh criminal justice policies that ignore the social context and focus only on making crime a costly decision.

ARE OFFENDERS' CHOICES RATIONAL?

The complexity involved in understanding the choices made by offenders was illuminated poignantly in the ethnographic studies of burglars and robbers in inner-city St. Louis, Missouri, by Wright and Decker (1994, 1997; see also Shover, 1996; Tunnell, 1992). Echoing the observations of rational choice theorists, Wright and Decker showed that offenders make a series of choices about whether to offend, which targets to victimize (e.g., houses, individuals), how to complete the crimes effectively, and how to avoid detection. These choices, moreover, are in response to pressing needs, typically for money. Thus, crime is a solution to the problem of the desire for quick cash. Offenders choose to burglarize and rob to get money to achieve desired goals. By all accounts, they are eminently rational.

Appearances of rationality, however, are potentially misleading. As Wright and Decker showed, the very "desperation" that offenders have for money is a by-product of their participation in a "street culture" that values conspicuous consumption of consumer goods and constant partying, an activity that often leads to drug abuse, if not addiction. Imbued with these cultural imperatives, offenders follow life paths in which they enter into a cycle of spending all of their cash on goods and drugs—on immediate gratification—and then face the pressing need to secure more funds through the only means available to them—crime. Although offenders "made conscious choices," observed Wright and Decker (1994), "their offending did not appear to be an independent, freely chosen event so much as it was part of a general flow of action emanating from and shaped by their involvement in street culture" (p. 205). Their exposure to this "code of the street" (Anderson, 1999), moreover, is itself determined by the misfortune of being raised in a socially disorganized area where legitimate work opportunities have largely vanished (Sampson & Wilson, 1995; Wilson, 1987, 1996). It is this larger cultural and structural context that explains why the commission of burglary and robbery is disproportionately a "rational choice" for residents in inner-city communities as opposed to those in middle-class communities.

These observations raise the broader question of whether one is justified in calling the decision making of offenders—and in calling this theoretical perspective—*rational* choice. Scholars in this paradigm are likely to argue that the emphasis on how offenders make a series of calculations as they go about their crimes corrects the neglect of decision making by traditional criminology. They also have a point when they claim that offenders' conduct is purposive, that how offenders think influences when crime events take place, and that in any crime situation offenders "seek to gain quick pleasure and avoid imminent pain" (Felson, 1998, p. 23). From a larger perspective, however, one might question how rational the choice of persistent offending is when this lifestyle often "knifes off" conventional opportunities and rewards and leads to a life characterized by the risks of violence and injury, drug abuse and addiction, and arrest and imprisonment. If one's relative "chose" this path in life, then we might question whether this person would be seen as the most rational member of the family.

The real qualms about the use of the term *rational* are scientific: Criminologists in this perspective fail to provide any clear criteria that could be used to assess whether or not choices made by offenders are rational. Indeed, the mere fact that offenders make choices frequently is the only evidence used to assert that they make rational choices. Because no standards are presented to judge rationality, no "choices" can be shown to be nonrational; in essence, the thesis that offenders make rational choices becomes an assumption that is difficult, if not impossible, to falsify. Furthermore, even when scholars admit that choices are "bounded" or structured by social and cultural factors—not to mention by individual traits—they cling to the fiction that the decisions are best portrayed as rational. But it is difficult to see what is gained by stressing the rational component of criminal acts when the calculations offenders make are so substantially shaped or determined by a range of other factors. If crime is an appealing or gratifying choice because offenders hold antisocial values, wish to relieve strain, are impulsive, and so on, then characterizing this choice as rational may distort more than it illuminates.

Indeed, the larger danger is that the emphasis on rationality will become an ideological commitment rather than a theoretical one. And as we have noted, such ideology leads to the advocacy of certain policy interventions—criminal justice sanctions, situational crime prevention, and so on—and to a disinterest in, if not skepticism about, other policy interventions that would attack the conditions that place individuals in situations where crime is, among all their choices, the most rational one available.

Finally, it should be realized that simple rational choice theory—sometimes called "neoclassical economics"—is under reconsideration more generally from "behavioral economics," which merges the insights of economics with those of social psychology. Thaler and Sunstein (2008) call this the "science of choice" (p. 7). This perspective argues that choices involve not merely incentives, as economists contend, but also social psychological processes. Economists make the mistake of assuming that "each of us thinks and chooses unfailingly well" (p. 6). Of course, incentives matter; raise the price on gasoline, candy, or cigarettes and consumption goes down. Yet, as Thaler and Sunstein point out, economists make the mistake of assuming that "homo economicus [economic man] can think like Albert Einstein, store as much memory as IBM's Big Blue, and exercise the willpower of Mahatma Gandhi" (p. 6). Humans, however, are more fallible than this. "Real people," they note, "have trouble with long division if they don't have a calculator, sometimes forget their spouse's birthday, and have a hangover on New Year's Day. They are not homo economicus; they are homo sapiens" (pp. 6–7).

The issue is even more complex than this. Social psychological research has long demonstrated that people's decisions are systematically biased by the methods or shortcuts they employ when making choices (see, e.g., Tversky & Kahneman, 1974, 1981). When assessing risks in uncertain situations, humans' brains are not designed to weigh all information in a coolly rational fashion. To do so would take calm emotions, the willingness to devote considerable effort to carefully collect and dissect information, and good analytical skills. But humans do not always remain calm, do not always have the time or disposition to spend a lengthy time pouring over the data, and do not always analyze well. Rather than be strictly "rational"—rather than weigh all costs and benefits—they thus employ rules of thumb or decision strategies that social psychologists call "heuristics." These heuristics allow them to make decisions quickly and that are psychologically comforting; such decisions require less effort and "feel right." From a purely objective standpoint, however, these choices would be seen as irrational—as misjudging risks (costs) and benefits.

Take, for example, the use of the "availability" heuristic, "situations in which people assess the frequency of a class or the probability of an event by the ease with which instances or occurrences can be brought to mind" (Tversky & Kahneman, 1974, p. 1127). Thus, in deciding whether to fly or drive to a destination, knowledge of a recent plane crash might lead people to conclude that driving is safer than boarding a plane; knowledge of a recent car crash might lead to the opposite choice. Certainly, no data are collected, and these travelers have no way of knowing whether their assessment of risk is accurate. Due to "status quo bias," people may decide to purchase a vehicle they are now driving—to stick with what they are currently doing—rather than to purchase an alternative vehicle that is less expensive and of higher quality. Many people also overestimate their abilities due to an "optimism bias." As Thaler and Sunstein (2008) observe, "ninety percent of all drivers think they are above average behind the wheel, even if they don't live in Lake Wobegon" (p. 32).

Behavioral economics holds important implications for understanding the decision to commit crimes (Nagin & Paternoster, 2010). Such choices—which involve assessment of risks (punishment) in an uncertain environment—undoubtedly are affected by the same heuristics that bias human decision making generally. Offenders may assume that they will not be caught because they are among the top 10% of all criminals. They may assume that, since lightning never strikes twice in the same place, it is safe to break the law because their recent arrest makes it unlikely that they will be caught again (see the discussion of the "resetting effect" below). Or they may assume that because they had a "big score" (secured a lot of money) in their last street robbery, they might obtain the same amount if they repeated the robbery in the same place. Again, these kinds of biases are likely a core reason why harsh penalties that seek to increase the costs of crime are of limited effectiveness. "Get tough" laws are predicated on the assumption that offenders think carefully about the potential sting of criminal penalties. But as behavioral economics teaches us, decisions in risky, uncertain situations are often controlled by heuristics that humans—including offenders—are able to resurrect in the moment. It is not that incentives do not matter at all; it is just that they are but one component, and at times a minor component, influencing the choices criminals make.

The key point is that behavioral economics shows us that all decision making is shaped by complex processes that cannot be reduced to the neoclassical view that judgments about utility—costs and benefits—rule human choice (Thaler & Sunstein, 2008). Much more than utility is involved, an insight that perceptual deterrence theorists have increasingly understood.

Perceptual Deterrence Theory

Another line of inquiry that focuses on the decision to commit a crime is *perceptual deterrence theory*. Following in the classical school tradition, this approach proposes that individuals refrain from breaking the law when the costs outweigh the benefits. However, the main proposition is that the decision to offend depends on the *perceptions* of costs and benefits and not on actual or *objective* risks of being sanctioned or gaining rewards.

Although there are similarities, this perspective differs from rational choice theory in three ways. First, the perceptual deterrence perspective does not assume rationality. It is predicated on the view that perceptions lead to behavior. However, although perceptions may be rational, they also may be based on wild misperceptions of reality. Second, perceptual deterrence theory—as the name "deterrence" suggests—has traditionally focused more on perceptions of legal punishments. The approach has largely neglected the situational aspects of offender decision making, which are central to rational choice theory. Third and related, the policy proposals from perceptual deterrence theory are often unstated or unclear. Because the link between actual levels and perceptions of punishment is uncertain, it is problematic whether simply "getting tougher" on crime will translate into reductions in the decision to offend. Campaigns to raise perceptions of punishment face daunting challenges, not the least of which is reaching the intended audience (high-rate offenders) and ensuring that the effects of the publicity do not decay over time. In any case, whereas rational choice theory focuses on crime situations and on how they may be structured to discourage offending, perceptual deterrence theory lacks a situational focus and thus is largely silent on the efficacy of informal (as opposed to legal) crime prevention strategies.

Perceptual deterrence theory arose in large part because of the weaknesses in traditional deterrence theory. Initially, scholars explored the impact of objective levels of punishment either across macro-level units (e.g., whether states who punish more have lower crime rates) or across individuals (e.g., whether people who are punished more commit less crime). As noted previously (see Chapters 7 and 12), these studies have produced mixed results that suggest that legal sanctions have modest deterrent effects and that, under certain conditions, may increase criminal involvement (see also Akers & Sellers, 2004). One response was to wonder whether the lack of deterrence might be because people's awareness of their risks of being caught and sanctioned did not correspond with the objective risks of punishment. It made sense that what should matter more is what individuals *thought were their chances of getting caught* (the "certainty" of punishment) and what they *thought would happen to them after they were caught* (the "severity" of punishment). In short, the theory was that perceptions of reality—in this case punishment—and not actual or objective reality are what shape behavior.

This approach also had another major advantage: It seemed easy to test empirically. Whereas objective punishments exist at a distance from individuals, perceptions are inside and carried by people as they pursue their daily lives. The measurement task thus was straightforward: develop questions that ask people to consider what would happen if they were to commit a crime. What did they think would be their chances of being punished? If caught, how severe did they think the punishment would be? In this way, it would be possible to assess each individual's perception of the certainty and severity of punishment. Then, on the same survey, each person would be asked to self-report his or her level of involvement in crime or delinquency (high school samples were usually used). The final task was to correlate these two measures—to see if people who thought that punishment was certain and/or severe were less likely to commit criminal acts. If so, then perceptions of punishment could be said to have a deterrent effect.

The early studies tended to show that there is a deterrent effect, especially for certainty of punishment: If the respondents thought they had a high chance of being caught and punished (regardless of how severe the resulting punishment might be), they were less likely to report breaking the law. The limitations of this research, however, soon became apparent. In particular, these studies were mostly bivariate (they measured only two variables—punishment and criminal participation) and were cross-sectional (conducted at one point in time). Subsequent studies attempted to rectify these shortcomings. This research used designs that were multivariate (measures of other known causes of crime, such as antisocial values or social bonds, were included) and were longitudinal (the respondents were studied at two or more points in time to determine if perceptions predicted future behavior). In these more methodologically rigorous investigations, the effects of the perceptual deterrence factors—including certainty of punishment—were substantially diminished (Paternoster, 1987).

More recently, Pratt, Cullen, Blevins, Daigle, and Madensen (2006) undertook a meta-analysis of the existing perceptual deterrence research. They analyzed 40 studies that reported 200 effect size estimates of the relationship between measures of deterrence and measures of crime/deviance. Across all the analyses, they found that the deterrence effect ranged from "modest to negligible" (i.e., from approximately .20 to zero; 2006, p. 383). There was some evidence that perceived certainty of punishment had a consistent deterrent effect on participation in white-collar crime, such as failing to comply with regulatory standards, not reporting taxes, and financial fraud. However, the most important finding involved what transpired when Pratt et al. (2006)

examined studies in which stronger research designs were employed—especially those that introduced controls for known predictors of crime such as delinquent peer influence, antisocial attitudes, and/or low self-control. In these studies, the impact of the measures of perceived legal sanctions were "substantially reduced—often to zero" (p. 384).

These findings suggest two conclusions. First, perceptual deterrence due to legal sanctions is likely a weak cause of crime whose effects are dwarfed by a range of other factors (such as those identified by other theories of crime). Second, for perceptual deterrence theory to contribute meaningfully to our understanding of crime, it must develop a richer perspective of how deterrence is specified by the nature of costs and benefits, by individual differences, and by the complex ways in which perceptions are formed and influence behavior. Fortunately, scholars have made important strides in this direction.

First, Williams and Hawkins (1986) observed that beyond the cost of being arrested and punished by the state in some way (e.g., probation, prison sentence)—which are referred to as "legal costs"—state sanctions can have detrimental consequences on other facets of a person's life. They identified three general categories of what scholars refer to as "non-legal costs": the "stigma of arrest (social degradation and loss of respect due to being caught)"; "commitment costs (cost of arrest for future goals such as employment and education)"; and "attachment costs (loss of friends due to being caught)" (1986, p. 562). Similarly, Grasmick and Bursik (1990, p. 838) noted that in addition to "state-imposed costs" or legal sanctions, there were "socially-imposed costs" (embarrassment or loss of respect) and "self-imposed costs" (feelings of shame or guilt). A complete deterrence model thus would include people's perceptions of both the legal and non-legal costs of being detected committing a crime. There is evidence that perceptions of non-legal costs are inversely related to criminal involvement (Pratt et al., 2006).

Second, perceptual deterrence theory has focused disproportionately on the "cost" side of the cost-benefit calculus. In most studies, the potential benefits of crime are unmeasured or are viewed as limited to how much money a crime might bring. Research indicates, however, that committing a crime can have a range of positive returns, including—among other benefits—fun, excitement, feelings of power, taking revenge, defending one's honor, achieving respect, and increasing one's status in a peer group (see, e.g., Bordua, 1961; Katz, 1988; Matsueda, Kreager, & Huizinga, 2006). Although the studies are limited, there is evidence that the perceived benefits of crime may be positively related to criminal involvement.

Third, perceptual deterrence theory needs to explore more systematically how perceptions of punishment are shaped by individual differences among potential offenders. In particular, there is evidence consistent with the conclusion that people who are low in self-control, present-oriented, and/or impulsive see the rewards of crime as more attractive (given their preference for immediate gratification) and see the costs of crime as less important (given that any forthcoming punishment would be delayed and not immediate). That is, the perceived utility and consequences of crime are not judged equally by all potential offenders but rather are filtered through people's personalities (see, e.g., Nagin & Paternoster, 1993; Nagin & Pogarsky, 2001, 2003).

Fourth, perceptual deterrence theory must take into account the complex ways in which perceptions are formed and influence behavior. For example, Pogarsky and Piquero (2003) have explored the counterintuitive finding that when people are punished for an offense, it tends to *lower* their perceptions of the certainty of punishment and to *increase* their offending. Traditional deterrence theory, of course, would assert that punishment teaches that crime does not pay, thus heightening individuals' perceptions that offending will be sanctioned. In contrast, Pogarsky and Piquero suggest that perceptions of certainty might decrease due a phenomenon they term "resetting." For comparative purposes, they cite the "gambler's fallacy" in which gamblers who lose several times in a row believe that they are "bound" to win the next hand or roll of the dice (this is a fallacy because each hand or roll of the dice is independent of all others and the probability of any particular outcome occurring is the same every time). In a similar way, offenders who are caught and punished may believe that, having already experienced a streak of bad luck, they are due for good fortune the next time the chance to break the law arises. As a result, they "reset" their perceptions of the certainty of punishment at a lower level because they surmise that "they would have to be exceedingly unlucky to be apprehended again" (Pogarsky & Piquero, 2003, p. 95).

Warr and Stafford (1993) also have elaborated on the nature of deterrence. They observe that the perceptions of the risk of being punished are shaped not only by the experience of being punished but also by the experience of *avoiding punishment*—that is, of not getting caught. Warr and Stafford further note that punishment and avoidance of punishment can be "direct" (experienced personally) or "indirect" (experienced vicariously by seeing whether others one witnesses committing crimes are punished or avoid punishment). When these dimensions are cross-tabulated, the result is four potential sources of perceptions: direct punishment, indirect punishment, direct avoidance of punishment, and indirect avoidance of punishment. The failure to take into account these multiple sources could lead to an incomplete understanding of why people believe their chances of being punished are high or low.

As a final example, other scholars have investigated how perceptions of punishment are part of a learning process that occurs over time (see, e.g., Pogarsky, Kim, & Paternoster, 2005). In this regard, Matsueda et al. (2006) propose a "Bayesian learning process" (based on Bayes's probability theorem) in which current perceptions of the risk of punishment are based both on past risk perceptions and on newly acquired information. Perceptions thus are not static but dynamic, always being updated as individuals' experiences with punishment change. These experiences include both whether they have been punished or avoided punishment, and the extent to which they have delinquent peers (which presumably lowers risk perceptions). The decision-making model is further complicated by three additional factors. These include "opportunity costs" for committing a crime (whether it endangers school or employment), the "psychic returns" from crime such as excitement or being "cool," and perceived opportunities "to get away with crime."

Taken together, these various lines of inquiry suggest that a rudimentary perceptual deterrence theory that only seeks to link perceptions of certainty and severity of punishment to crime will have limited explanatory power. The challenge is to explore the diverse sources of these perceptions, the process by which perceptions are formed and reformed, the cognitive processes by which people with different personal traits assess the costs and benefits of particular criminal acts, and the way in which structural and situational conditions shape both risk perceptions and the opportunity to act on these perceptions.

Conclusion

Whereas the vast majority of criminological theories focus on the "background" of crime, the theories of offender choice and opportunity argue for the causal importance of the "foreground" of crime (see Katz, 1988). Thus, they generally do not consider how criminal motivations or "criminality" develops over time from the interaction of individual traits, socialization, and structural location. Instead, their concern is largely with the present and with the crime-inducing factors that are proximate or contemporaneous with the criminal act that is about to occur—or, in some cases, not occur. This theoretical decision to ignore the "background" of crime means that much about crime causation is omitted. But it also means that special insights are made. We will close this chapter by reiterating three of them.

First, theories of choice and opportunity remind us that offenders are active and not passive participants in the decision to break the law. In Laub and Sampson's (2003) terms, they have "human agency." They may arrive at a situation prepared to commit a crime because of "background" factors, but a crime event is not foreordained; they will make decisions that will lead them either to offend or not offend. Second, once in these situations, what determines whether potential offenders will commit a crime is an important criminological question. The theories in this chapter alert us to the need to consider how the decision to offend is affected by perceptions of costs and benefits and by situational factors such as the attractiveness of targets and the presence of controllers. Third and perhaps most important, the theories have important policy implications. In particular, they identify how crime prevention might best be achieved not by the threat of legal sanctions but by efforts to make any given place less conducive to crime—that is, by reducing opportunities to offend.

Charles Darwin
1809–1882
Author of Evolutionary Theory

14

The Search for the "Criminal Man" Revisited

Biosocial Theories

As suggested in Chapter 2, the earlier biological theories of Lombroso and others had achieved their prominence not only because of the social and political climates but also because biology, especially in the form of Darwinism, had become the most fashionable science of the late 19th century. The social Darwinism of the early 20th century made it plausible to trace criminality to some form of biological inferiority or "unfitness." But despite the efforts of theorists such as Hooton (1939) and Sheldon (1949), by the 1960s biology had lost much of its influence in criminological theory. Some of this was due to the realization that social Darwinism had been used as an excuse to "blame the victim." The earlier luster of biological criminology was also tarnished by exposure of the uses of such theory under the Nazis, where millions of Jews, Slavs, gypsies, and others were branded "inferiors" and killed by the state (Bruinius, 2006; Cornwell, 2003). But to some extent, the biological sciences also were eclipsed by the scientific prominence of natural sciences such as physics (with its thermonuclear weapons and highly publicized efforts to travel to the moon) and by the influence of the rapidly growing social and behavioral sciences that grew enormously during the 1960s.

By the mid-1970s, however, interest in newer biological approaches again was in the wind. E. O. Wilson (1975) published his very influential book, *Sociobiology*, which proposed to interpret all of the new discoveries of the social and behavioral sciences in essentially biological terms. During these years, biology was being revitalized by new discoveries in the laboratory, especially in the subfield of genetics as exemplified by the explosion of work on recombinant DNA. Dawkins (1976) made headlines with his book asserting that genes were "selfish." He opened the path to a return to varieties of Darwinism, proposing that virtually all of human behavior could be understood as biologically determined activities of individuals directed toward reproduction so that their "selfish" genes would be passed on to future generations.

In the late 1970s, Lee Ellis (1977) published in *The American Sociologist* an article titled, "The Decline and Fall of Sociology, 1977–2000." Ellis pointed to several problems facing that discipline and hearkened to scientific approaches that promised greater results in the future. Having grown up on a farm in Kansas, he had experienced what he later called a "barnyard education" that led to a lifelong interest in sex differences among animals, including the human animal (Ellis, 2003b). Ellis offers an excellent example of the criminologist who was becoming increasingly disillusioned with the social sciences and intrigued by the possibilities of a biological or biosocial approach. Unlike the sociologists of the 1920s who were struck by the "social disorganization" of Chicago or the conflict theorists of the 1960s who were so impressed by the social struggles all around them that they began to see conflict as the core of social life, a few criminologists such as Ellis—starting their careers at a time when the social sciences were coming under attack and the biological sciences were making some stunning advances—were placing their hope for criminological theory in a biological or biosocial approach that might overcome the defects of early biological criminology.

In the year following Ellis's piece, one of the authors of this book published an article in *The American Sociologist* arguing for a "general systems theory" approach to social theory that would allow it to build upon biological foundations without falling into reductionism (Ball, 1978b). Such possibilities are even more relevant to criminology, which has always stressed its *interdisciplinary* nature. Unfortunately, the fact that most criminologists are pigeon-holed by academic discipline has meant that the field is usually "interdisciplinary" only in the sense of allowing representatives of different disciplines to contribute to its professional meetings and publish in its journals. One promising development may be seen in Anthony Walsh's (2009) recent *Biology and Criminology*, which examines nearly all of the theoretical perspectives that we have covered earlier in the light of biosocial criminological theory. Meanwhile, Ellis, Beaver, and Wright (2009) have recently developed a *Handbook of Crime Correlates* that draws together many of the correlates of crime from different fields, including the biosocial.

Another of the authors of this book has recently suggested that biosocial criminology is a "broader and more powerful" paradigm than sociological criminology and is likely to be the major paradigm of the 21st century (Cullen, 2009). We are by no means committed ideologists. This entire volume has been dedicated to showing that theoretical perspectives are products of particular social contexts and carry profound policy implications. The sociological and psychological theories that dominated much of the 20th century are obviously laden with ideological bias, sometimes to the extreme. But it would be the height of arrogance for the rising biosocial criminology to insist that it is free from all value judgments. Indeed, these are obvious in the sort of "crime" it regards as important and in its essentially positivistic definition of "science." We can only hope for genuinely interdisciplinary work informed by humility rather than hubris. The various conflict theories that we have examined—whether Marxist, feminist, "critical," left realist, or other approaches—are surely ideologically biased. But they have alerted us to the role of social power in determining which criminological theories thrive, and they have made clear the appeal of biosocial theory to the powerful, who are much more willing to support and reward criminologists who support what used to be called the "Establishment" than those who critique it. We must not throw out the baby with the bath water.

Meanwhile, more major "breakthroughs" were being hailed in biology, lending additional momentum to the few biologically oriented criminologists. With the successful cloning of a sheep, the famous "Dolly," a number of publications predicted that cloning of humans was soon to arrive. Indeed, some writers asserted that this already had been accomplished but was being hidden from the public. The media frenzy surrounding biological research soon came to resemble the heyday of the astronauts during the 1960s.

In a 1985 article hearkening back to the science fiction films of the 1950s, a major newspaper reported that the brains of executed prisoners were being studied at the University of Florida ("Officials to Probe," 1985). A study at the New York University Medical School soon after reported that convicted murderers had suffered head injuries at one time ("A Study of 15," 1986). In 1985, two medical doctors reported at the Fourth World Congress of Biological Psychiatry that 90% of all excessively violent people had brain defects ("Brain Defects," 1985, pp. A17, A19).

Although the claims of most of this type of research were presented in rather general terms such as "brain damage" and "neurological defects," reports identifying a *specific* defect began to appear in 1987. One report that made the front page of the *New York Times,* for example, linked manic depression to a "specific genetic defect" ("Defective Gene," 1987, p. A1). As some noted at the time, the medical claim that a specific genetic structure can explain a specific form of behavior is only a short step from the creation of policies that would require "special treatment" for people with such an identifiable genetic structure. By the late 1990s, the cover of *U.S. News & World Report* prominently posed the question "Born Bad?" (1997), calling attention to inside coverage on the biological sources of crime.

In 1988, the National Research Council, the operating arm of the National Academy of Sciences that Congress chartered to advise the federal government, called for a 15-year study effort "to diagram the possible 100,000 human genes" ("Scientists Urge," 1988, pp. A1, A13). At an estimated cost of $3 billion, it would "set the stage for constructing the ultimate physical map—the complete DNA sequence of the human genome" (p. A13). By the late 1980s, DNA "fingerprinting" already had been used to obtain convictions for rape and murder in both the United States and England, and there was increasing talk of identifying the DNA structure of convicted criminals as a basis for screening the public for potential offenders with similar DNA structures.

By 2005, the Human Genome Project was complete, and leading biologists hailed a "new era" that would be a matter of "filling in the blanks" in research. The impact on criminology was enormous. Forensics courses proliferated in criminal justice programs at colleges and universities across the United States. Graduate forensics programs were widely developed. Funds became available through the Department of Defense for research and development of "non-lethal" weapons that could be used to immobilize enemy troops (or civilians who seemed to pose a threat), often using biological agents or technological gadgets that produced immobilizing effects upon the human body.

The public was entranced by these developments. The most popular television show of the new century, *CSI: Crime Scene Investigation,* focused on crime scene investigation employing all sorts of biological and technological techniques and became so influential that it produced two spin-off shows, *CSI: Miami* and *CSI: NY.* Television commercials were full of ads touting medications that would solve problems from obesity to impotence. Medications such as Ritalin had become so popular as a means of assisting children with ADHD (Fumento, 2003) that some school districts were refusing to accept certain children unless they agreed to such medication. Public policy seemed clearly ready to accept the findings of biological researchers and employ such interventions as the findings suggested. This in turn has encouraged more biological research, so that the field is gaining impetus rapidly. Meanwhile, the FBI Combined DNA Index System (CODIS) now complements its Automatic Fingerprint Identification System (AFIS) as a means of matching offender characteristics and aiding in identification and apprehension of offenders.

At the same time, biological and biosocial criminologists would stress that in many ways the context of the times was making it very difficult for them. Criminology had been dominated for decades by those who traced crime to social circumstances, and surveys conducted at both the beginning and the end of the 1990s indicated that about 85% of American criminologists remained strict "environmentalists" (Ellis & Hoffman, 1990; Ellis & Walsh, 1999). Those advocating a biological approach to crime or criminality sometimes describe serious difficulties faced due to academic misunderstandings of their work by criminologists with little knowledge of biology and to career obstacles placed in their paths by more traditional criminologists of all kinds (Ellis, 2003b). In a society so easily inflamed over racial and gender issues, some consider it too provocative to even suggest that racial or gender differences might play a much bigger part in crime than we have been willing to admit. One surely risks being charged with racism and/or sexism in an atmosphere of political correctness, even if one's work is purely scientific. Too many wonder about motives.

Still, even with the resistance of many sociologically oriented criminologists, it is clear that the current resurgence in biological theorizing will persist and grow in the foreseeable future. Thus, the search for the "criminal man"—the search for a biologically distinctive offender—will continue. At issue, however, is whether this search will contribute to the view that criminals are a distinctive, dangerous class of people who are inherently wicked and beyond redemption. As we will see, most current theories are more nuanced than this, rejecting the idea that biology translates into a predestined fate and instead suggesting that genetic traits interact with the environment to shape human behavior. For such biological theorists, *everything* that is not specifically genetic is "environmental," including the womb. Thus, even the few who argue that some may be "born criminal" do not insist that this is the result of inherited criminal tendencies. These approaches are typically called "biosocial theories."

Although there is a great deal of overlap in their theoretical concerns, these biosocial approaches may be categorized in terms of *evolutionary psychology*, *neuroscience*, and *genetics* (Walsh & Beaver, 2009). We consider each of these categories of theorizing. In addition, we examine theoretically relevant research on *biological risk and protective factors* and on *environmental toxins*. Finally, we end with a discussion of the potential policy advances and dangers of the rising popularity of biological theorizing.

Evolutionary Psychology: Darwin Revisited

All of these research projects have led to several new efforts to formulate theories based on evolutionary dictates (Ellis & Walsh, 2000). The field is sometimes called evolutionary psychology. The differences between males and females in tendencies toward law violation, for example, is traced to contrasting natural selection processes among primates that tend to lead to survival and reproduction of more aggressive males and more social females (Dunbar, 2007). Some of the more prominent theories include cheater theory, r/K theory, conditional adaptation theory, alternative adaptation theory, and evolutionary expropriative theory (Ellis & Walsh, 2000; Fishbein, 2001; Walsh, 2009).

THEORETICAL DIVERSITY

Cheater theory argues that some males have evolved "alternative reproductive strategies" that unconsciously ensure that their genes are passed on to succeeding generations (Mealey, 1995). Whereas normal "dads" find reproductive opportunities by fulfilling female desires for a mate who can support offspring, the cheating "cads" use force or deception to impregnate females. The theory equates persistent criminals with "cad" types. According to this application of Darwinian theory, some males may develop "cad" behavior as a result of environmental experiences while others inherit the tendency, but the reproductive style of these biosocial "cheaters" is also reflected in their disregard for the law.

The *r/K theory* (sometimes called "differential K theory") stresses that organisms vary in their approaches to reproduction, with rapidly producing organisms following an "r strategy" emphasizing production of large numbers of offspring without spending much time caring for each one, while other organisms follow a "K strategy" involving slower reproduction with careful devotion to each offspring (Ellis, 1987, 1989; Rushton, 1990, 1995). According to r/K theory, criminal behavior should be more associated with the "r strategy." Markers for this would include low birth weight, large numbers of siblings, and earlier involvement in sexual activity.

Conditional adaptation theory maintains that antisocial behavior is part of an overall adaptive response to an unstable or hostile environment (Belsky, 1980). The theory argues that in an adaptive response designed to ensure reproduction, children who live in such environments will enter puberty early and engage in early sexual activity. Such an adaptation pattern is said to be associated with antisocial behavior. Interestingly enough, this is one example of a criminological theory based on a Darwinian perspective that leads to policy implications that would focus on environmental change rather than, for example, on intervention following a medical model. It suggests that development of a stable, caring environment would reduce, if not eliminate, these antisocial adaptations.

Whereas conditional adaptation theory assumes that most humans have essentially the same genetic potential for antisocial behavior, *alternative adaptation theory* assumes that some people inherit a greater tendency to engage in antisocial behavior (Rowe, 1983, 1986). These are said to be individuals who are driven more by mating urges than by parenting urges. According to alternative adaptation theory, they should be individuals of lower intelligence who are aggressive sensation-seeking types with a strong sex drive.

Like conditional adaptation theory, *evolutionary expropriative theory* assumes that all humans have an equal genetic potential for antisocial behavior (Cohen & Machalek, 1988; Vila, 1994, 1997; Vila & Cohen, 1993). The theory asserts that humans are genetically driven not only to seek mates but also to acquire resources in order to ensure reproduction. Whereas some people do this through productive strategies entailing creation and development of such resources, others expropriate resources by victimizing others. According to the theory, such expropriative behavior should increase when available resources are inadequate or threatened.

ASSESSMENT

As can be seen from this cursory review, even these evolutionary theories are rarely as purely biological as the media make them appear. They are generally "biosocial" rather than biological, although they tend to emphasize nature much more than nurture in their arguments. Thus, they are often called evolutionary-ecological theories, sometimes actually stressing the impact of the environmental (ecological) forces over the genetic factors.

Some have criticized evolutionary theories based on the notion of "survival of the fittest" for being tautological—true by definition. Actually, it is stating no more than the obvious, for we are defining the "fittest" as those who have survived. What most people do not realize, however, is that Darwin's *unit of analysis* was the *species*, not the individual member of the species. There are many traits that may facilitate individual success at the peril of the species as a whole. Indeed, many would explain problems such as global warming this way. Aggressive exploitation of the environment may make an individual entrepreneur much more successful than the workers who occasionally die in the factories, one particular corporation more successful than those it drives out of business, and one nation more successful than those without such natural resources, but such "success" may endanger *homo sapiens* in general, not to mention many other species.

Furthermore, Darwin was really concerned with the *process* by which all this occurs, which he termed *natural selection*. There is a tendency for some evolutionary psychologists to write as if a trait that has survived must be *ipso facto* a desirable one, simply because it contributed to survival. But the principle of natural selection is focused on the past, not the present or the future. Traits that contributed to survival in a hunting/gathering environment may be counterproductive today. On the other hand, many evolutionary psychologists study the presumed evolution of "positive" traits such as empathy and altruism, which also contribute to survival, not only of the individual but the species as a whole. Although biosocial criminologists explain their focus upon negative traits as "natural" to anyone searching for the sources of law violation, this emphasis might be balanced by greater attention to positive traits or at least the positive aspects of so-called negative traits, for at least three important reasons. First, it may be that characteristics such as a powerful sex drive, early involvement in sexual activity, and even a predilection toward mating rather than parenting, may be positive as well as negative. Thus, for example, Felson and Haynie (2002) found that boys entering puberty early were more prone to delinquency and other antisocial acts but that they were also more autonomous, better adjusted psychologically, and had more friends. Second, it is likely that any person who is presumably predisposed to law violation because of some factor such as a "cad" mentality or an "r strategy" will also be characterized *to some extent* by "protective" factors such as empathy and altruism, so we are always dealing with a combination of factors at the individual level, and (as pointed out Chapter 5 with respect to containment theory) these other "self-factors" may operate to "insulate" them from more antisocial behavior. The third reason for greater attention to positive factors lies in the evidence (seldom appreciated by biosocial theorists) that law violation is also associated with such attributes as energy, courage, and loyalty, something always understood by gang members (Cloward & Ohlin, 1960; Katz, 1988).

As Rowe (2002) points out, evolutionary theories are very hard to prove or disprove. Even if we could establish that humans are genetically driven to expropriate (i.e., steal) resources from one another, which would be a feat in itself, how could we prove that this was in fact the result of natural selection in terms of "survival of the fittest"? It might very well be that early humans were much more "expropriative" and that evolution is actually operating to *reduce* such tendencies. This is a more important consideration than it may seem, because evolutionary theories carry a value judgment to the effect that the behaviors they cite are "useful," "valuable," "effective," and "desirable" in terms of human survival.

Neuroscience: Neurological and Biochemical Theories

The second major approach of biosocial theory comes through neuroscience, which has made impressive advances the past two decades. Here biosocial theories tend to acknowledge the importance of learning, but they emphasize the extent to which the learning and conditioning of behavior occur differently for different individuals because of their neurological or biochemical variations and the way in which the environment (including the womb) affects these neurological and biochemical variations, which are called *polymorphisms* (Fishbein, 1990; Moffitt, 1983; Zuckerman, 1983). Thus, for example, psychopaths are said to be impulsive and unemotional because their autonomic nervous systems (ANSs) are more difficult to condition due to low arousal thresholds, while their low levels of fear and anxiety also seems to be associated with a chemical imbalance between the behavioral activation system (BAS), which is very sensitive to dopamine, a neurotransmitter that facilitates goal-directed behavior, and the behavioral inhibition system (BIS), which is sensitive to serotonin, a neurotransmitter that helps us to manage risk. According to some biosocial theory, this tends to make them more determined to achieve their goals while less able to manage the risks associated with that, persisting in trying to get what they want even when the danger of punishment is high. Some such theories suggest that this behavior may be associated with a dysfunctional amygdala, a neurological component central to the limbic system of the brain, which is also reflected in low startle responses and a limited capacity for empathy (Bartol, 1995).

Like the earlier biological theories, the newer biosocial theories have focused considerable attention on family studies aimed at locating genetic factors by examining behavioral similarities among members of the same family, although today they stress behavioral characteristics such as hyperactivity or attention deficit hyperactivity disorder (ADHD). Twin studies have been very popular, despite their methodological deficiencies. Fishbein (1990) has noted a number of biochemical differences between control groups and individuals with psychopathy, antisocial personality, violent behavior, or conduct disorder, pointing especially to levels of certain hormones, neurotransmitters, peptides, toxins, and metabolic processes as well as psychophysiological correlates, such as electroencephalogram (EEG) differences, cardiovascular differences, and electrodermal variations in the case of psychopathy.

MEDNICK'S BIOSOCIAL THEORY

The 1980s saw publication of a series of works in criminal biology, with the biosocial theory of Mednick and his associates, which had begun to appear during the late 1970s (Mednick, 1977; Mednick & Christiansen, 1977; Mednick & Shoham, 1979), attaining special prominence (Mednick, Gabrielli, & Hutchings, 1984; Mednick, Moffitt, & Stack, 1987; Mednick, Pollack, Volavka, & Gabrielli, 1982; Mednick, Volavka, Gabrielli, & Itil, 1981). Eysenck's (1964) earlier work, which associated low cortical arousal with extroverted personality, lower "conditionability," and a consequently weaker moral conscience, had received some attention during the mid-1960s, but the social context was more accepting of biosocial approaches by the time Mednick's (1977) similar theory appeared during the late 1970s. Mednick proposed that certain high-risk individuals have inherited an autonomic nervous system (ANS) that is less sensitive to environmental stimuli and that such slow arousal makes it less likely that they will develop the responses necessary to inhibit antisocial behavior. He focused on the ANS in an effort to discover why the individuals representing approximately 1% of the population who seem to be responsible for more than half of the crimes reported appear either unable or unwilling to learn from their continued mistakes, stressing that conformity is learned and that the key involves learning of the fear response that becomes associated with consideration of antisocial behavior. Mednick's theory stresses that an individual with a normal ANS will experience a reduction of fear immediately on inhibiting antisocial activity and that, because this fear reduction is a powerful reinforcement, that person will learn to inhibit such activity. He maintains that if the ANS reduces the fear too slowly or ineffectively to produce the necessary reinforcement, normal inhibition may fail to develop.

Theories suggesting faulty chromosomes (e.g., the XYY syndrome) that had produced much confusion and little research support during the 1960s and 1970s were supplanted during the 1980s by more careful research leading to theories arguing for disorders of the central nervous system (CNS) such as the *cortical immaturity hypothesis* (Mednick et al., 1981). By the late 1980s, this work had crystallized with the publication of an influential volume (Mednick, Moffitt, & Stack, 1987) including an elaborate study of 14,427 Danish adoptees (Mednick, Gabrielli, & Hutchings, 1987). Newer theory centered on the CNS by suggesting variations of a *hypoarousal hypothesis* (Volavka, 1987).

"Hypoarousal" is measured by low pulse rate, low blood pressure, and reduced skin conductance of electricity, some of the same variables measured by the polygraph. Such arousal is controlled by the ANS. With the polygraph we are searching for uncontrollable indications of lying through an involuntary increase in pulse rate, slight sweating that increases skin conductivity, and so on. Subjects characterized by hypoarousal show much lower responses to environmental stimuli than "normal" subjects. *Fearlessness theory* suggests that these low levels of arousal are markers for low levels of fear and that such relatively fearless types are simply less likely to avoid situations that bring "trouble with the law." *Stimulation theory,* on the other hand, argues that such low arousal represents an unpleasant psychological state resulting in such types actually going "looking for trouble" as a means of getting enough sensory stimulation to avoid boredom and the "blahs." Some have associated both tendencies with low levels of dopamine. By the beginning of the 1990s, Gottfredson and Hirschi (1990) already had identified several major defects in the Mednick et al. research. Still, the advocates of biosocial theory continued to stress the biological forces at work in criminality (Brennan, Mednick, & Volavka, 1995). Among 48 biological, psychological, and sociological predictors of violence studied in the late 1990s, Farrington (1997) found low resting heart rate to be the strongest and most consistent predictor of crime.

BRAIN DEVELOPMENT AND CRIME

As often happens in science, some of the most dramatic developments in biologically oriented research have come about because of the development not of new theoretical orientations but rather of new observation techniques. These new techniques include computed tomography (CT), magnetic resonance imaging (MRI), functional magnetic resonance imaging (fMRI), positron emission tomography (PET), and single photon emission tomography (SPECT), all of which represent significant breakthroughs in brain imaging (Raine, 1993). Some of these technologies allow for close examination of brain structure, searching for structural anomalies that could never have been detected before, and others permit observation of the actual brain functioning as it processes various stimuli. Biosocial criminologists have concentrated on the possibility that structural abnormalities of the brain, especially of the frontal or temporal lobes, may be associated with brain dysfunctions that are, in turn, associated with criminal activity. Raine (1993) even suggested that current research leads to the hypothesis that violence may be associated with frontal lobe dysfunction, whereas sex offenses may be associated with temporal lobe dysfunction.

The frontal lobes are associated with the executive cognitive functions (EFCs) of the brain that are implicated in planning, inhibition, and the ability to learn from experience (Comings, 2003). They have a great deal to do with ability to interpret emotional cues and regulate responses. Thus, the *frontal lobe dysfunction hypothesis* has received increased support in recent years, and biosocial criminologists such as Yang, Glenn, and Raine (2008) have drawn attention to its implications for the law.

The two major divisions of the frontal lobes are the dorsolateral and orbitofrontal lobes. Deficits in the former seem to be associated with distractibility and poor executive functioning in general and may be particularly associated with conduct disorder (CD). The latter are the site of key connections between the brain and the ANS, and evidence suggests that both conduct disorder (CD) and antisocial personality disorder (ASPD) may result from orbitofrontal deficits. Such deficits are diagnosed in several ways, including not only brain lesions but also by low glucose metabolism in these areas. The latter indicates relative inactivity and suggests a deficiency of neurons necessary to adequate functioning. Sources of such deficits include inherited deficiencies, environmental insults, and brain injury. Some research has suggested that subjects with behavioral inhibition dysfunction show less activity in the left dorsolateral prefrontal cortex, posterior cingulated gyrus, and bilateral temporal-parietal regions than children who are better able to control their behavior (Rubia et al., 2008). However, the relationship between the dorsolateral lobes and the orbitofrontal lobes is proving to be quite complex, as we will see below.

In a study focused specifically upon a subgroup of "intractable" offenders, Fishbein (2003) has offered a theory that traces much of the problem to an impaired hypothalamic-pituitary-adrenal axis (HPA). This is a circuit operating from the brain to the adrenal glands, and she believes such impairment to have an especially negative effect upon the executive cognitive functions (EFCs) that (among other things) regulate emotional responses such as impulsive aggression. The biosocial thrust of the theory is clear. Fishbein maintains that HPA impairment is associated with environmental stress. She argues that in response to chronic or severe stress, increased cortisol levels cause the structures comprising the HPA to shrink, leading to memory and cognitive decline, along with depressive and other affective disorders. An exhausted HPA is said to deplete cortisol, with results including inability to regulate emotion satisfactorily. According to Fishbein, the process may occur consequent to childhood stress that prevents development of the HPA in the first place or later stress that damages it.

There is a long history of tracing various forms of deviant behavior to the possibility of head injury, and this notion has been revived because of the recent research linking brain dysfunctions with "criminality." Damasio's (1994) *somatic marker hypothesis* maintains that decision making entails both cognitive and emotional processing to evaluate the reward value of various behavioral choices, with "somatic markers" being formed in the brain as a consequence of various action-outcome sequences experienced in the past. The ventromedial prefrontal cortex (VMPFC) is the key region for these markers, which tend to provide a template by which to evaluate future actions. Patients with damage to the VMPFC frequently suffer from behavioral disinhibition, social dysfunction, poor decision making, and lack of insight into their behavioral problems (Yang et al., 2008). As some have pointed out, of course, such a theory cannot account for calculated, instrumental aggression (Blair, Mitchell, & Blair, 2005).

BIOCHEMICAL THEORIES

Recent biochemical theories have focused on sex hormones and neurotransmitters. For males, *sex hormone theory* has been concentrated on possible connections between testosterone and aggression. Part of the problem here is that testosterone may be as much a product of aggression as a cause, so the causal relationship is difficult to untangle. Nevertheless, the testosterone factor has been reemphasized in recent years, especially by Ellis (2003a, 2005), who has used it to develop an *evolutionary neuroandrogenic theory* of criminality. Because the theoretical emphasis is upon the biochemical effect of testosterone, we have elected to examine this theory here rather than in the section devoted to those theories stressing the evolutionary aspects of criminality.

Ellis stresses that the biochemical effect of testosterone tends to lower neurological sensitivity to environmental stimuli, which is conducive to "acting out" and problems with emotional control, and that testosterone also tends to impair higher thought by producing a shift in the functioning of brain hemispheres. The evolutionary aspect of the theory suggests that the "competitive-victimizing" behaviors associated with testosterone have emerged through natural selection, the key process in the Darwinian theory of evolution. One need not accept the second component of the theory to accept the notion that testosterone is associated with law violation.

Biosocial theorists who favor a testosterone-based theory of criminality use it to explain why the male crime rate is higher than that for females in every society for which we have acceptable data. As for biochemical factors affecting females in particular, various theories have suggested hormonal shifts occurring several days before menstruation and leading to a premenstrual syndrome (PMS) characterized by seriously distorted judgment and tendencies toward violence, along with postpartum depression syndrome occurring as a result of hormonal imbalances following pregnancy and childbirth, which has been used as a defense in infanticide cases.

Neurotransmitters are chemicals mediating signals between brain neurons. Of these, serotonin, dopamine, and norepinephrine have received the most attention. Animal studies have shown that serotonin may operate to inhibit aggression, and there is evidence associating low levels of serotonin with both habitual violence and suicide. Animal studies also have shown some association between (low levels of) serotonin, (low levels of) dopamine, and (high levels of) norepinephrine and aggressive behavior under certain circumstances. The problems here, however, are extremely complex. Just as high levels of testosterone, for example, may be as much the product of aggression as the cause of aggression, so too may low levels of serotonin be as much the product of environmental stress leading to chronic depression as the cause of phenomena such as suicidal behavior. The association between biochemical factors and antisocial behavior is perhaps the clearest example of the "chicken or egg" problem facing criminological theory. Which causes which? Are both interrelated in some complex feedback process? If the associations eventually are established, then should policy focus on changing the biochemistry of high-risk individuals (e.g., with pills or injections designed to produce the "proper" level of serotonin or dopamine) or on changing the environment that produced these unusual physiological consequences in certain individuals? Such decisions are, of course, political ones, but they depend in part on whether criminological theory tends to focus on "the chicken" or "the egg." It is not surprising that biologists wish to focus on one and sociologists wish to focus on the other.

These differences are even expressed in the way terms are defined. In Chapter 8, for example, we saw that the sociological conflict theorist, Austin Turk, distinguished between "crime" and "criminality," defining the latter as the result of the *criminalization process*, which he then analyzed in terms of power relationships. The biosocial theorist, Walsh (2009), on the other hand defines "criminality" as the *predisposition* to engage in law violation. Thus, Turk can explain criminality as a consequence of power struggles over which behavior should be regarded as crime, Walsh can explain it as a biological predisposition to engage in antisocial behavior, and both can be right. It will always be difficult for theorists to talk across disciplines when they choose to define their terms differently.

Of the various environmental factors influencing physiology, biosocial theorists have focused on diet, with special attention given to protein intake, refined carbohydrates, food allergies, vitamin deficiencies, exposure to lead or cadmium, and certain substances such as monosodium glutamate (MSG), caffeine, and chemicals found in chocolate (Curran & Renzetti, 1994). Serotonin, for example, is produced by an amino acid found in high-protein foods. Thus, it has been suggested that aggression due to low serotonin levels may be corrected through a high-protein diet. Diets high in refined carbohydrates, such as those found in sugar and most junk foods, produce large amounts of glucose in the bloodstream, and this may trigger excessive insulin release and a dopamine increase leading to impaired behavioral control. As early as the end of the 1970s, this carbohydrate theory was used to establish diminished capacity in the famous "Twinkie defense" involving former San Francisco Supervisor Dan White, who confessed to killing Mayor George Moscone and Supervisor Harvey Milk (Curran & Renzetti, 1994). White was convicted of manslaughter rather than of the more serious charge of murder.

In any case, polymorphisms for genes regulating dopamine and serotonin have shown significant relationships to "criminality" and intervening variables associated with it. Certain variants in the dopamine genes DRD2 and DAT1 have been linked to greater involvement in serious and violent crime, while particular polymorphisms in MAOA, DRD2, and DATI have been associated with serious and violent delinquency in young males and specific variants of DRD2 have been tied to greater likelihood of delinquency victimization (Guo, Roettger, & Shih, 2007; Regoli, Hewitt, & DeLisi, 2010). Polymorphisms of DRD4 have been linked to ADHD (Shaw et al., 2007). Many such relationships have been found, although sometimes they seem to operate only in the presence of other variables, such as high-risk family environments. Because there are more than 25,000 genes in the human genome, the possibilities for research are endless.

One of the most perplexing problems for criminology has always been the need to account for the remarkably sharp rise in law violation during adolescence followed by a dramatic decline in the late teens and early twenties, at least if we focus upon the "street crime" that seems to be the exclusive preoccupation of biosocial criminologists. In the 19th and early 20th century, the notion of this developmental period as one of *sturm und drang* characterized by impulsivity, rapid mood swings, and a generally negative attitude toward authority was generally accepted as a fact. Toward the mid-20th century, however, writers such as Margaret Mead (1928) and others gained considerable popularity by arguing that their cross-cultural data clearly showed adolescence to be a fairly tranquil period in preliterate societies (Ball, in press). This interpretation remained influential for decades until (1) delinquency rates soared, (2) some of the data were challenged, and (3) biosocial investigations began to reveal some of the neurological and biochemical factors behind the "stormy" nature of adolescence that had long been depicted in art and literature.

As mentioned earlier, biosocial theories such as Ellis's evolutionary neuroandrogenic theory points to the dramatic biochemical increase in testosterone during male adolescence as a major factor in increased antisocial behavior. However, MRI studies also show that juveniles have more dominant nucleus accumbens activity compared to the amount of activity in the prefrontal cortex (PFC) relative to younger children and adults. The nucleus accumbens is a neurological structure associated with the push toward immediate gratification and "Pavlovian reward learning" (Day & Carelli, 2007), while the PFC tends to operate as an impulse inhibitor associated with the "executive functions" that manage behavior more systematically. There is increasing evidence that the accumbens tends to develop earlier than the orbitofrontal cortex and that this might be associated with elevated risk taking in adolescents (Galvan et al., 2006).

Perhaps the powerful effects of testosterone, which Ellis attributes to the natural selection process that gave males with high testosterone levels significant mating advantages, may be combined with powerful nucleus accumbens activity and low PFC activity in such a way as to increase the odds of antisocial behavior even more. This might explain in part why testosterone seems to be more clearly associated with *dominance* than *aggression*, for it might be that adolescent males whose testosterone is very high might still tend to channel it into dominance activity such as sports or successful domination of a teenage group's activities because of a more favorable balance between nucleus accumbens activity and PFC activity (Ball, in press). This is a researchable question.

Some of this work is cited in our discussion of "life-course" theories in Chapter 15. Among the best-known is Moffitt's (1993) life-course persistent/adolescence-limited theory, which makes the important distinction between two groups of adolescents—a small percentage that shows antisocial behavior in childhood extending through adolescence into adulthood and a very large group including most adolescents that shows little such behavior in childhood only to experience a surge of antisocial behavior during adolescence and then "age-out" and cease their antisocial activity as they enter adulthood. The first group is said to be characterized by factors such as low IQ, hyperactivity, inattentiveness, negative emotionality, and low impulse control, while the second found antisocial behavior during adolescence functional and adaptive given the excitement it afforded, the autonomy it asserted, and the friends it drew. It was the neurological and biochemical problems of the first group, *interacting with the environment*, that led to their childhood misbehavior and reduced the chances that they would come into conformity in adulthood. The antisocial behavior of the second group amounted to what adults often summarized as "going through a phase."

Genetics

The genetic approach may be subdivided into behavior genetics, molecular genetics, and epigenetics (Walsh & Beaver, 2009). Each of these seeks to explore relationships between *genotypes* (genetic endowments) and *phenotypes* (observable traits and behaviors). Examination of genotypes is a question of where to look for the "independent variable" that is associated with the traits and behaviors in which biosocial criminology is interested. With respect to the phenotypes, the old Lombrosian dream was to tie particular genotypes such as head shape to particular forms of crime such as robbery. That dream is long gone. Biosocial criminologists now search for *intervening variables* that seem to *link* certain genotypes to *general* categories of law violation such as violent crime, which they define as direct physical violence against another rather than as indirect violence through, for example, adulterating and mislabeling baby food or deliberately exposing employees to asbestos that has a high probability of causing cancer some decades later. This is the more conservative definition of "crime." That is why the theories now seldom focus upon explaining "crime" but rather emphasize the explanation of intervening variables such as "impulsivity" or "fearlessness."

Furthermore, the effects of these intervening variables are often more complicated than they first appear. Thus, IQ is clearly related to delinquency, but the nature of the relationship is not so obvious. The old belief that crime was simply the result of stupidity represented a tempting but probably unwarranted leap beyond the data. Intelligence as measured by IQ tests may not predict delinquency directly, but it is tied to poor school performance, and school failure is perhaps the most effective predictor of delinquency, even for very bright but alienated students. To take one simple example of the way some of these chains of intervening variables may operate, we might consider the pattern by which (1) low IQ is associated with low grades in school, (2) such low grades are associated with disliking the school environment (and perhaps to being pushed away by teachers), (3) alienation from the school environment is associated with rejection of authority (and dropping out), (4) rejection of authority and dropping out are associated with loss of the social bonds contributing to social control, and (5) these plus other factors are associated with increasing likelihood of delinquency. It is worth exploring these "chains" because we may wish to intervene at any point to reduce the probability of delinquency. We could give extra help to those with low IQ scores to break the link to school failure. We could intervene to reduce the odds that low grades would lead to extreme alienation and encourage teachers not to push away the problem students. We could intervene to reduce the chances of dropping out as a consequence of the earlier issues. We could develop programs to help keep dropouts out of trouble. And it is obvious that such interventions are not mutually exclusive; we could intervene at all links of the "chain" simultaneously. In short, the evidence that IQ is highly heritable and is associated with delinquency does not mean that delinquency is a necessary outcome of a low IQ score.

BEHAVIOR GENETICS

Behavior genetics tries to tease out the "heritability" component from the environmental contribution to various traits and behaviors, such as IQ or "violence." Biosocial theorists point to studies showing significant heritability coefficients (.20–.82) for traits such as fearlessness, aggressiveness, sensation seeking, impulsivity, and low IQ (Walsh & Beaver, 2009). Behavior genetics began with the old observation that antisocial behavior is more common among males than females in virtually all societies. Males also tend to exhibit less self-control in general and are much more aggressive in the nature of their criminal activity than are females engaged in crime. Girls show fear much earlier than do boys and tend to be more fearful as adults, even when research controls eliminate the effects of socialization. These and other differences have long suggested that the genetic differences between males and females might be associated with crime. The behavior genetics approach is also behind those biological and biosocial theories tracing intervening variables that increase the likelihood of criminal activity to obvious genetic differences such as those associated with skin color. We must keep in mind, however, that what an individual inherits is not a specific behavior but rather a tendency to respond to certain environmental forces in terms of general predispositions (Fishbein, 1990).

MOLECULAR GENETICS

Molecular genetics focuses upon analysis of the detailed *processes* by which genetics has its effects upon traits and behaviors, with special attention to the deep molecular structure of substances such as deoxyribonucleic acid (DNA). This particular approach received an enormous impetus with completion of the Human Genome Project, which has provided the DNA "map" so necessary to progress in the field and should be expected to progress in the field. We may expect a significant increase in molecular genetics research attempting to explore any connections between DNA characteristics and intervening variables that might lead to law violation.

It is easy to understand the enthusiasm for DNA research. Nearly 1,200 diseases, including cystic fibrosis, sickle-cell anemia, and Huntington's disease, are caused by a single gene (Beaver, 2009). Still, there is nearly unanimous agreement that there is no such thing as a "gene" for crime, if by that is meant a so-called OGOD (One Gene–One Disorder). Human behavior is too complex for that to be the case. Instead, the more genetically oriented theories point out that whether a genetic *predisposition* toward criminal activity is *encouraged* or *discouraged* depends upon the *environment*. Single-parent families, for example, may fail to control certain genetic predispositions that would be controlled with greater supervision.

The relationship between gene and environment is referred to as the "gene × environment correlation" (rGE), and the three main types of rGEs are termed passive, and active, and evocative (Moffitt, 2005; Rutter, 2007). Passive rGEs arise because in a way children inherit *both* their genes *and* their environment from their parents, and there is already an existing correlation between the two, the environments having been created or sought out to some extent by the parents themselves. Passive rGEs are simply given to the child as a package (Rutter, 2006).

Active rGEs, on the other hand, reflect the tendency for people to seek out environments to which they are predisposed, environments that may amplify their predispositions. Sutherland's differential association theory stressed that the likelihood of people becoming criminals is in large part a matter of the attitudes and techniques they pick up from people with whom they associate most frequently and intensely for long periods of time, but many students ask why the criminals were attracted to these associates in the first place. Recent research has suggested that such an rGE relationship is involved in delinquent peer group formation (Beaver, Wright, & DeLisi, 2008). Perhaps gang members and potential white-collar criminals are attracted to associates who simply reinforce their predispositions. This is a wonderful example of the feedback processes possibly at work in the rGE patterns, where one side of the equation reinforces the other in a "deviance amplification spiral."

Evocative rGEs are a reflection of the fact that different people also evoke different responses from their environments even when the latter are identical. In some ways, people make their own environments to a certain extent (Scarr & McCartney, 1983). Gender research shows that boys and girls are treated so differently in the family or school that they might as well be in different places. Teachers respond in one way to the noisy child and another way to the quiet child. Gangs may be eager to enlist the services of the muscular mesomorph while rejecting the skinny ectomorph even if he pleads for initiation. Did the boy choose the gang, or did the gang choose the boy (Maddan, Walker, & Miller, 2008)? Girls who reach puberty early are more likely to "get into trouble," but this may be in part a result of the boys who are attracted to them. Did the girl's early onset of puberty operate as an active rGE, or are we witnessing the operation of an evocative rGE by which the environment responds to her very differently than to other girls of her age and level of social maturity?

EPIGENETICS

The third genetic approach, epigenetics, is based on the startling new discovery that environmental factors tend to alter gene functioning without affecting the molecular structure of DNA at all by activating or deactivating particular aspects. Epigenetics is scarcely more than a decade old. It has actually revived interest in the long discredited Lamarkian notion suggesting the inheritance of characteristics *acquired* by parents that were never part of the gene pool (Bird, 2007). "Epigenetics" means "in addition" to the genes, and it really refers to any process that alters the functioning of the genes without changing the sequence or basic structure of the inherited DNA.

Genes contain a wide variety of "polymorphisms" or possible variations, depending upon environmental stimuli. While fields such as molecular genetics focus upon the "hardware" of the genes, epigenetics studies processes by which the "software" is "programmed" to "turn on" some genes, "turn off" others, and "modulate" still more. Epigenetics is demonstrating that while genes contain the coded information for the structuring and functioning of the organism, they need more specific "software" instruction themselves. These are found in an array of chemical "switches" called the epigenome, which operates to adapt the organism to the particular environment in which it will find itself.

Everyone knows, for example, that what a pregnant woman eats and drinks will affect the fetus (Manning & Hoyme, 2007). This is not because these substances alter the basic genetics, which are passed on by a combination of the DNA of both parents, but because the *expression* of these inherited predispositions is altered through an "imprinting" process. Many of the phenotypic differences between genetically identical organisms (e.g., identical twins) may be the result of these epigenetic effects of the environment upon the genes, and some effects may be "passed on" to offspring.

Epigenetics has shown that the diet of the pregnant woman may actually affect her grandchildren and great grandchildren because some epigenetic effects seem to be carried through several generations. Epigenetic research has now shown that factors such as poverty, parenting, and environmental toxins also have a significant effect by altering the expression or "behavior" of the inherited genes over several generations. Evidence is emerging, for example, that some of the recent increase in childhood autism may be a result of "gene silencing by epigenetic mechanisms" through which environmental experiences "turn off" or modify certain genes associated with normal social interaction (Lopez-Rangel & Lewis, 2006). Research has shown as well that child abuse may lower the expression of glucocorticoid receptors, which affects the instructions provided in the DNA inherited from the parents and alters the hypothalamic-pituitary-adrenal (HPA) function mentioned above, a consequence associated with antisocial behavior. The genetic inheritance is not altered, but the epigenetic effect of the abuse seems to shift its expression by "adjusting" the imprinting on the epigenome. Thus, it is possible that "antisocial behavior" is passed down several generations not by genetics as usually defined but by a cycle of abuse that is imprinting the epigenome of each successive generation.

If this proves to be the case, it suggests that efforts to interrupt the cycle of violence may represent a better preventive strategy than policies such as eugenics aimed at "sterilization" of parents to prevent their reproduction, although even a child removed from such parents at birth may be prone to abuse his or her children because of the abuse experienced by one or more of the biological parents. As Anthony Walsh (2009) puts it, "To analogize the relationship between the genome, polymorphisms, and epigenetics; if the genome is an orchestra and polymorphisms represent the musical variety it can produce, epigenetics represents the conductor governing the dynamics of the performance" (pp. 50–51), adding that, "Epigenetics contains some suggestive lines of evidence that may open up whole new vistas for criminologists" (p. 51).

Biosocial Risk and Protective Factors

RISK FACTORS

The more sophisticated biosocial approaches rarely make the simple argument that some individuals are inherently criminal. Instead, they trace antisocial behavior to many biological *risk factors* that increase the odds of delinquency and criminal behavior, especially if combined with any negative environmental conditions (Fishbein, 1990, 1997). One example of this process is cited in the alleged link between low IQ or learning disabilities and criminal behavior mentioned earlier. Although neither low IQ nor learning disability is considered inherently criminogenic, the model suggests that in the absence of appropriate intervention a child will become frustrated with mainstream activities such as school, suffer drastically lowered self-esteem in general, and begin to interact with others experiencing similar frustrations as they move in the direction of delinquency and crime (Fishbein, 1990).

Intelligence is a highly complex concept that refers in general to the "adaptability" of a subject. Molecular biologists do not expect to discover any specific gene for intelligence or, indeed, for any specific form of adaptability. IQ is often defined operationally as "that which is measured by an IQ test," and it reflects an effort to measure that sort of narrow intelligence associated with what used to be called "book learning." According to Walsh (2003b), the "heritability coefficient" (i.e., the degree to which variation in a phenotypical trait such as intelligence is genetically influenced) for IQ is approximately .60, which is quite strong, because the heritability coefficient, unlike the correlation coefficient, is designed to measure the amount of variance accounted for without being mathematically squared. Thus, the .60 heritability coefficient indicates that 60% of IQ is inherited.

One of the more interesting variations on the basic IQ theme has to do with the notion of "intellectual imbalance" (Walsh, 2003b). Sociopaths, for example, tend to show much higher scores on the *performance* portion of IQ tests than they do on the *verbal* part of the test. PET scans show that mental work required to handle the performance portion is centered in the right hemisphere and that verbal work is centered in the left hemisphere. There is some evidence that this type of imbalance is the result of child abuse/neglect and that children who spend a great deal of time in a constant, low-level state of fear tend to develop a focus on the visual-spatial (nonverbal) cues of danger that are processed by the right hemisphere, resulting in a state of "frozen watchfulness," to the detriment of the verbal abilities processed by the left hemisphere (Walsh, 2003b).

According to biosocial theory, two additional risk factors that may predispose one to delinquency/crime are attention deficit hyperactivity disorder (ADHD) and conduct disorder (CD). The "attention deficit" in ADHD involves chronic inattention while the "hyperactivity" aspect is self-explanatory, with both aspects aggravated by "impulsivity" in ADHD. There is disagreement among the biosocial theorists as to the levels of ADHD, with theorists such as Rowe (2002) putting the figure at 3% to 5% of boys and Comings (2003) putting it at 8% to 12%, but ADHD is probably the most common psychological disorder among boys. Some highly publicized recent research seems to have succeeded in tracing ADHD to a particular seven-repeat allele of the dopamine D4 gene (DRD4) (Shaw et al., 2007). Recent research also has shown a gene × gene interaction between DRD2 and DRD4 in the etiology of CD and antisocial behavior (Beaver et al., 2007).

Some biosocial theorists have suggested that ADHD is a significant factor in adolescent drug use and in adult crime. In fact, there is now substantial evidence, including meta-analytic research, showing that ADHD is a significant risk factor in a range of antisocial behaviors (Barkley et al., 2002; Barkley, Fischer, Smallish, & Fletcher, 2004; Pratt, Cullen, Blevins, Daigle, & Unnever, 2002). However, Comings (2003) has questioned this relationship, pointing out that ADHD is often accompanied by conduct disorder (CD)—something that is known as "comorbidity." Trying to untangle the etiology here, he cites evidence that it is the presence of the comorbid CD rather than the ADHD itself that produces adolescent substance abuse and the adult crime. That is, there are many children who are being diagnosed with ADHD because of their relative inattention, impulsivity, and hyperactivity, but these children may actually be no more "at risk" for adolescent substance abuse and adult criminality than control groups. According to Comings, only those diagnosed with ADHD *who are also diagnosed with the comorbid CD* are more likely to engage in these behaviors than "normal" subjects. It may be the CD, which is characterized by aggression toward people and animals, destruction of property, deceitfulness or theft, and serious violation of rules, that is the problem, while relative inattention, impulsivity, and hyperactivity in themselves do not contribute much to either adolescent substance abuse or adult criminality. Although not all scholars would agree with Comings's perspective (see, e.g., Barkley et al., 2002), if this turns out to be the case, it would be an example of success in teasing out the criminogenic effects of co-occurring risk factors.

If Comings is correct, this finding will also demonstrate a clearer relationship between diagnostic categories used to label behavioral disorders and deficits in specific areas of the brain, something that has intrigued biological theorists from the earliest days. He maintains that (1) both ADHD and CD can be traced to pathologies of the frontal lobes, but that (2) while dorsolateral deficits of the frontal lobes render one susceptible to ADHD, the comorbid appearance of CD is likely only with deficits also occurring in the orbitofrontal area of the frontal lobes. According to his analysis, further research should produce a genetic test involving the examination of only 10 to 20 genes that would predict that subgroup of children with CD who are at the *highest risk* for antisocial behavior.

As biosocial research continues, additional genetic polymorphisms are being linked with ADHD and CD, illustrating once again that things are not as simple as they sometimes seem to the media. For, example, the low levels of prefrontal cortical functioning linked to such behavioral disorders have been linked to a variant of the COMT gene (Thapar et al., 2005). Meanwhile, a variant of the 5HTT serotonin transporter gene has been tied to substance abuse, school problems, and aggression in adolescents (Gerra et al., 2005) and extreme forms of violence in Chinese males (Liao, Hong, Shih, & Tsai, 2004).

Biosocial criminology is characterized by its own terminology, and some very common terms are used in different ways by those working in the biological tradition than they are used by those trained in the social sciences. Even the basic term, *environment,* is used differently, with the biological and biosocial theorists including, for example, the uterus (in the case of the fetus). When biosocial theorists refer to "risk factors," they are sometimes distinguishing between causal factors of considerable power and those that merely increase the odds of antisocial behavior somewhat. Among the latter are functional impairment of the left angular gyrus (an area lying within the parietal lobe of the brain) and the corpus callosum (which connects the two brain hemispheres) in murderers (Raine, 2002).

Some of the most appealing work in biosocial theory has come from the search to identify certain genotypes with substance abuse such as alcoholism. This is probably not surprising, because (1) alcoholism is somewhat more *specific* in nature than some very general (and rather vague) phenotypes such as "antisocial behavior" and because (2) the factors involved in the independent variable (i.e., alcoholism) are presumably at least in part as biological in nature as are the biological predictors (e.g., dopamine deficits) used in the theoretical approach. As we have noted earlier, variables such as "antisocial behavior" tend to be social constructs, and they vary enormously from time to time and place to place, so that all sorts of contrasting behaviors have been labeled as "antisocial" by people with different notions of what is prosocial. While this is also true to some extent of alcoholism, there is much more cross-cultural consensus as to the point where putting "too much" of a particular chemical substance into the body becomes a problem. While we noted that there is a problem of logic involved in using a *biologically defined* independent variable to predict a *socially defined* dependent variable, in the case of substance abuse the major defining characteristics of both sets of variables may well be biological (although this remains to be determined).

However, we should call attention to several important points here. First, alcoholism is not a crime. For more than 200 years, our criminal law has been based to a large extent upon the principle that *we do not criminalize traits* but only overt, harmful *behaviors.* It is the destructive behavior associated with drunkenness that tends to be criminalized. Ironically, the more biosocial researchers have been able to show that alcoholism is associated with genetic predispositions, the more we have defined it as an "illness" rather than a "crime." It is also worth remembering that homosexual behavior among consenting adults in private was still a crime until recently when both public opinion and increasing evidence suggesting that homosexuality may be associated with certain biological sources came together to virtually legalize the behavior.

PROTECTIVE FACTORS

As Tibbetts (2003) points out, some of the most interesting theories here are those that attempt to integrate the biological approach with some of the theories described earlier in this volume, including the social control and self-control theories, which happen to be the two most favored theories among criminologists (Ellis & Walsh, 1999). As we saw earlier, control theories emphasize *protective* factors that operate to counteract risk factors. One important factor in control theories, for example, is *empathy,* which provides an ability to appreciate others' points of view and produces a sense of identification with them. The heritability coefficient for empathy is approximately .68 (Tibbetts, 2003), even higher than that for IQ. Empathy is one example of a protective factor that, rather than increasing risk of criminality, serves to insulate the subject to some extent from the risk factors themselves. Biological factors can apparently work in both directions. Raine (2002) has pointed out that until recently there had been no research (and apparently little theoretical interest) in biological factors that might serve to *protect* against antisocial behavior, but this is slowly changing. The search for such protective factors may yet turn out to be an area of significant growth in biosocial criminology.

One of the most obvious possibilities in the search for protective factors of a biological nature would simply involve research into the opposite end of the spectrum of risk factors. If orbitofrontal lobe deficits are held to produce antisocial behavior by *reducing the activity of the ANS,* which helps control emotion, is it possible that *high ANS activity* might protect one against involvement in such antisocial behavior? Raine (2002) reports evidence that this is indeed the case, citing a 14-year prospective study showing that "desistors" (boys who were antisocial during adolescence but did not go on to criminal behavior as adults) exhibited greater ANS activity (with a concomitantly higher arousal level as indicated by factors such as heart rate) than did those who continued their antisocial behavior into adult criminality.

According to Rowe (2002), "kin altruism" is also a protective factor in general, with some statistics showing that the rate of fatal child abuse against a stepchild by a stepparent runs 40 to 100 times greater than that against a biological child by a biological parent (p. 57). This is interpreted as suggesting that biological kin have a greater affinity for one another that serves to reduce the violence that might otherwise be higher. Of course, as Rowe points out, it also fits the "evil stepparent" stereotype of folklore.

Rowe notes that kin altruism may also operate in the opposite direction, as a risk factor, if the kin are involved in criminal activity. Here it would produce co-offending. Of course, because according to this theory, kin altruism should increase with the closeness of biological relationships, and because near relatives are closer in biological makeup than more distant relatives, it is almost impossible to separate the "kin altruism" effect from other biological effects.

There is an emerging body of research supporting the conclusion that protective factors such as proper parenting and a supportive home can reduce the probabilities of law violation associated with the various biological risk factors considerably (Larsson, Viding, & Plomin, 2008). Beaver, Wright, and Walsh (2008) have recently developed a gene-based evolutionary explanation for the association between criminal involvement and number of sex partners, but they would likely agree that the probability of *both* of these associated behaviors is likely to be reduced simultaneously by certain forms of socialization. Empathy is a protective factor reducing the risk of antisocial behavior, and there is increasing evidence that it has followed a particular evolutionary pattern that can be encouraged (de Waal, 2008). Current biosocial studies of factors involved in the "resilience" of some inner-city youth in the face of enormous pressures toward law violation remind one of the sociological studies of "self-concept as an insulating factor" reducing the probability of young males becoming involved in delinquency even in extremely high delinquency neighborhoods where such misbehavior was the norm, which were discussed in Chapter 5 (DiRago & Viallant, 2007). Such biosocial variables may even turn out to be contributing factors in the formation of "favorable self-concepts" that are even more important than the influences of the family and other institutions of socialization.

Environmental Toxins

Biosocial criminology is often accused of a conservative bias, but this is not always the case, especially when we consider the implied criticisms of environmental toxins. Here it is joined by the most radical criminological theorists, who argue that environmental damage is perhaps the most serious problem facing the world today (Lynch, Schwendinger, & Schwendinger, 2006).

Fishbein (2003) points out that the frontal lobe deficits that biosocial criminologists have associated with antisocial behavior can often be traced not only to head injury, prenatal drug exposure, childhood deprivation, and chronic drug use, but also to common environmental neurotoxins such as mercury and lead. This area of the brain, which is getting more and more attention from biosocial criminologists, seems to be remarkably sensitive to environmental influences in general, and if the biosocial theorists are correct, environmental toxins pose a serious criminogenic problem.

Environmental toxins are also significant factors in outcomes such as hyperactivity, learning disabilities, and IQ deficits, all risk factors for antisocial behavior identified by biosocial theory (Colburn, Dumanoski, & Myers, 1997; Davis, 2002; Rodericks, 1992). In addition to lead and mercury, which is emitted by coal-burning power plants, the most dangerous include heavy metals and substances such as manganese (Pueschel,

Linakis, & Anderson, 1996). Aside from its damaging effects upon the frontal lobes, exposure to lead in particular is associated with all of these deficits and a host of others including impulsivity, aggression, lack of self-control, and school failure, and a 5 micrograms-per-deciliter increase in blood lead levels across a sample of 376 six-year-olds increased the risk of being arrested for a violent crime in adulthood by 150% (Wright et al., 2008). One common environmental toxin is tobacco smoke, and research has shown an association between prenatal tobacco exposure and emotional instability, physical aggressiveness, social immaturity, and oppositional defiance disorder (ODD), as well as criminal behavior (Regoli et al., 2010). Recent epigenetic research suggests that even when the environmental toxins are eliminated, the offspring of people affected may be damaged for several generations.

Just as some of the most stigmatized illegal drugs such as opium, heroin, and cocaine are so addictive precisely because they mimic certain endogenous processes of endorphins and other substances produced naturally by the body, recent evidence indicates that synthetic chemicals act as "hormone mimics" possessing the ability to alter normal biological processes (Lynch et al., 2006). Thus, early exposure to such chemicals has been linked to elevated estrogen levels in the womb, which has in turn been linked with aggression in male mice (Lynch et al., 2006). Similarly, proximity to hazardous waste production, treatment, storage, and disposal facilities increases the likelihood of brain and general CNS impairment, generating specific disorders such as IQ deficits, learning disabilities, increased aggression, low frustration tolerance, hyperactivity, diminished self-control, impulsivity, and ADHD, plus general disorders such as increased violence, antisocial behavior, and crime (Lynch et al., 2006). Such theoretical integration demonstrates how the biological approach, which tends to focus upon how such factors might produce criminal behavior in individuals, can be broadened to address ecological, race, and class dimensions, thereby showing how society is reacting to its own criminogenic environmental policy (which allows for the proliferation of environmental neurotoxins because it is profitable for some) by punishing the very victims of that policy when factors such as neurological damage make them more susceptible to criminality.

While some biosocial criminologists have moved toward integration of the biological approach with specific sociological and psychological theories such as control theory, conflict theorists and others using critical theory or radical theory have moved toward integration with biological theory by focusing upon environmental toxins. This work has stressed, for example, the fact that minorities, especially African Americans, are much more likely to live in areas proximate to hazardous waste sites than are Whites (Lynch et al., 2006). The same sources also cite research indicating that low-income groups are more likely to live in such areas than are high-income groups, and that both minority groups and low-income communities are likely to be closer to the location of chemical accidents than are White groups or high-income communities. Focusing on schools, for example, Stretsky and Lynch (1999) point out that the higher the percentage of African Americans in a school, the closer that school is to a hazardous waste site.

The Consequences of Biological
Theories: Policy Implications

As we have tried to emphasize through this book, the success of criminological theories may depend more upon the *context* of the times than upon theoretical rigor or research support. This was as true for social Darwinism early in the 20th century as for any ideological preferences reinforced by politically correct sociological jargon late in the same century. We should hope for fewer "true believers" who embrace their theories like religions and more open-minded scientists who regard them as tentative models subject to testing through systematic research, but that may be unlikely when dealing with such a politically charged topic as crime.

Clearly, the biological sciences are riding a wave of popularity. The 1990s saw a boom in biotechnology stocks as many corporations began to seek ways of making profits from the new work in genetic engineering, which promised to provide cures for diseases ranging from baldness to cancer and to allow couples to design their own children. At least one leading figure in human genome research began to seek patents for his work, and concern developed over the possibility that everything from genetically engineered seeds to human body parts might come under corporate control. By the beginning of the new millennium, however, government researchers and corporations seemed to have struck a bargain by which corporations would be allowed considerable profit from genetically engineered "products" provided that they shared credit with government agencies and did not seek exclusive control over all aspects of the new technology.

AN AGENDA FOR RESEARCH AND POLICY

As the biological theorizing became more and more prominent in criminology during the 1980s and 1990s, concern once again turned to the possible policy consequences. Early in the new millennium, Fishbein (2001) published a lengthy summary of "biobehavioral" perspectives in criminology, in which she acknowledged some of the dark side of the history of imputing socially disapproved behavior to biological pathologies as well as some of the pseudoscientific aspects of biological criminology. At the same time, she suggested some means by which such theorizing and research might be used to more progressive ends.

As Fishbein (2001) pointed out, "Before we can begin to design and implement programs and policies based on this science which may have an impact on criminal offending, the relevance and significance of biological perspectives for criminology must be fully evaluated" (p. 98). She suggested that such an evaluation must await four prerequisites: (1) estimation of the incidence of biological disorders among antisocial populations, (2) identification of etiologic or causal mechanisms, (3) assessment of the dynamic interaction among biological and socioenvironmental factors, and (4) determination as to whether improvements in behavior follow large-scale therapeutic manipulations. Each requires a great deal of work.

There is evidence indicating that both biological and social disadvantages may be more prevalent among offender populations than among the general public, but the extent of any differences remains unclear. As in Lombroso's day, there is a need for careful studies comparing offender groups to control groups. If such differences do indeed exist, and if we are able to systematically identify them, then it might be possible to proceed with the second step in policy making based on biosocial research and theorizing. This is, however, much more problematic than it appears. Those familiar with the prison construction boom know that if such institutions are built, they will be filled. In the same way, once a diagnostic category has been developed, it may tend to be extremely over-diagnosed, especially if it seems to offer an easy way out of a problem or if there is money to be made. Thus, one study has shown that more than half of the children diagnosed with ADHD and "treated" with a popular medication were misdiagnosed and inappropriately drugged, a serious danger when these drugs have significant side effects (August, Realmuto, & MacDonald, 1996). With children as young as 5 years old being treated with methyphenidate (Ritalin) or an amphetamine such as Adderall or Dexadrine, drugs that produce side effects such as psychosis, mania, loss of appetite, depression, sleep problems, moodiness and stunting of growth, sales of these drugs now exceed $1 billion annually (Regoli et al., 2010).

Fishbein's second step entails the identification of exact etiologic or causal mechanisms. Just how do these identified genetic or biological factors (in combination with environmental factors) actually influence an individual's likelihood of criminal behavior? Although some interesting correlations have been discovered between antisocial behavior and particular physiological factors, these studies have not been able to identify mechanisms by which such physiological traits may be inherited or to demonstrate which biological systems are being affected by genetic factors. In any attempt to predict risk of reoffending, one should not proceed through "guilt by association," in which, for example, any prisoner up for parole might be denied release because of certain physiological characteristics associated with criminality. As Fishbein (2001) stressed, "Similar to other forms of evidence, the state of biological and genetic research must be thoroughly scientifically tested, reliable, and agreed upon by everyone in the courtroom in order to meet criteria to establish causality" (p. 100). In short, does the physiological characteristic actually cause antisocial behavior, or is it merely associated with it? As we hear so often, correlation is not cause. Perhaps the causal relationship is reversed and the antisocial behavior has produced the physiological characteristic. Low levels of the neurotransmitter serotonin, for example, may predispose one to suicide or to aggressive behavior, but such behavior also may produce lowered levels of serotonin. Here is the "chicken and egg" problem all over again.

As the incidence of biological disorders among antisocial populations is established and the causal mechanisms, if any, are identified, it will be necessary to assess the interaction between biology and the environment. This means taking a biosocial approach in which the biological factors may be examined as "vulnerabilities" that may amplify the antisocial effects of certain environmental influences. Why, for example, do not all children exposed to child abuse become violent as adults? Fishbein (2001) pointed to research suggesting that whether child abuse contributes to violent behavior depends partially on factors such as the presence of brain damage, low serotonin, and an underactive ANS. It will be some time before these interactions, if they do exist, can be established, and even then the results must be expressed in probabilities rather than as simple cause-effect statements.

If the steps just outlined can be accomplished, then it still will be necessary to show that researchers can actually manipulate and control antisocial behavior within the context of biological variables. Treating sex offenders with Depo-Provera (an anti-androgen agent administered to reduce someone's sex drive) offers one example of such attempts to control behavior, although it must be admitted that this is proceeding without complete understanding of the issues outlined in the three steps just described. As Fishbein (2001) noted, manipulating only the biological factors is unlikely to produce the desired changes in antisocial behavior. Even if the causal mechanisms can be identified and the interaction effects can be specified, it still will be necessary for the individual also to receive interventions such as "supportive counseling, behavior modification strategies, and a change in aspects of the environment that are contributing to the behavioral difficulties" (Fishbein, 2001, p. 102).

Fishbein (2003) proceeded to point out that the programs most effective in dealing with behavioral disorders related to the ECF (executive cognitive functioning) deficits discussed above are those focused on improving cognitive ability. She notes that verbal cues, reminders, psychoanalytic practices, and interactional therapies not only tend to be ineffective but may also actually aggravate the problem. According to Fishbein, the programs showing the most favorable outcomes are the cognitive rehabilitation programs (see also Gendreau, Smith, & French, 2006; Lipsey, Chapman, & Landenberger, 2001; MacKenzie, 2006). These emphasize problem solving, social skills, independence, impulse control, self-monitoring, and goal setting, using cues and consequences to slow down negative behavior and draw the subject's attention to the behavior and its consequences, with rewards or penalties used to highlight these consequences.

PREVENTION AND TREATMENT

The biosocial theories may have their greatest policy applications in terms of prevention and treatment programs. Robinson (2009) has pointed to work suggesting that it may soon be possible to alter genetic makeup and/or replace particular genes so as to reduce the likelihood of "antisocial behavior." While they do not seem to have had much influence upon the determination of criminal responsibility in trials, data obtained from brain scans are being used more and more in the sentencing stage, where it is possible to introduce "mitigating circumstances" that the jury may take into account in arriving at an appropriate punishment (Raine, 2002). Here again it is clear that, although there may be an individualistic bias inherent in the biological and biosocial theories, this approach can also serve to advance the cause of those arguing for an emphasis on prevention rather than punishment or for punishment that takes account of circumstances. Although biosocial theory and research tends to ignore white-collar crime—despite its high costs—research focusing upon the contribution of environmental toxins to crime may have some impact even there.

As we have indicated above, the biological and biosocial theories seem to have the most immediate promise in dealing with problems of substance abuse. Citing the alleged effectiveness of such items as methadone and nicotine patches as examples of biologically based treatment strategies capable of dealing with antisocial behavior, Gove and Wilmoth (2003) insist that many other such strategies can be used as successful treatment modalities with other forms of antisocial behavior. The problem, as they see it, is to learn more about the "endogenous reward systems" of the brain that make substance abuse so pleasurable. They maintain that those drawn to substance abuse because they are risk-taking, sensation-seeking types might be provided with alternative pleasures that activate the same reward systems in the brain. This seems like a good idea if only we remember what happened after heroin was so highly touted at the beginning of the 20th century as a solution to the "morphine problem." It will be crucial to establish the exact etiologic or causal processes at work with the "treatment" and to make sure through long-term studies that the "cure" is at least no worse than the problem we are trying to solve. This will present a real challenge in an era when such huge profits are being made from new drugs that often turn out to be not only less effective than the old ones but more dangerous as well.

Some have pointed out that from the perspective of cost-benefit analysis, it may make more sense to intervene at the environmental level to reduce child abuse, environmental toxins, and other "triggers" for the expression of genetic predispositions, especially because many people are exposed to such conditions simultaneously. In the event that neither genetic intervention nor environmental intervention is successful, biosocial theory suggests that behavior be modified by a variety of drugs to regulate neurotransmitter, enzymes, hormones, and the like.

Some parts of the justice system are better positioned than others to develop and employ policies flowing from biological and biosocial theory. In the juvenile justice system, for example, there is the "right to treatment" recognized by the courts, which has the effect of encouraging rehabilitative efforts based on biologically based intervention. The expansion of special "drug courts" in recent years has produced a special court system where treatments based on biological and biosocial approaches seem especially at home.

THE CONSTRUCTION OF CRIME

Biologically oriented criminology often has maintained the aura of science while failing to meet the basic canons of science. One of the problems with biological criminology always has been the criminological naïveté of the biologists or neuroscientists themselves, who often were very good biologists who knew little about the realities of criminal law or criminal behavior. Thus, the search for the biological roots of "criminal behavior" often has gone on without any apparent recognition of the fact that what is criminal behavior is different from one society to another and even within the same society over a span of a few years. Close examination of much of this work shows that even the biological research that traces apparently more specific behavior such as "violence" to biological pathologies actually uses widely differing definitions of violence, for example, treating violence in sports or war as something quite different from gang violence. The "bad violence" apparently is traceable to biological pathologies, but what about the "good violence"? There are, however, indications that this is beginning to change.

Moreover, the tendency of biosocial theories to focus attention entirely upon the offender is very clear. Rowe (2002) begins his book, *Biology and Crime,* for example, with the assertion that, "A book about the biology of crime must be about the biology of criminals—about their traits, physiology, motives, and so on." This is simply not true. "Crime" is a clash between behavior and law, and (in keeping with the thrust of labeling theory) the study of biosocial factors could just as logically focus upon those characterizing the "labelers" or "law constructors." To take one example, some studies outside criminology suggest the possibility that subjects who do not make much use of their right brain hemispheres (for various reasons) are prone to lose both psychological perspective and awareness of social context, resulting in their being more likely to stereotype and stigmatize "evildoers" and leading them into extremely punitive social policies against "outsiders" (Wilson, 1984). On the other hand, other studies (Fox, Bell, & Jones, 1992) indicate the opposite, maintaining that the left hemisphere is the more empathic and socially oriented and suggesting that subjects making less use of their right hemispheres might be *less* likely to advocate harsh crime control policies. The very fact that we have to strain so hard to see the "relevance" of biosocial theory to aspects of criminology other than the biology of criminals shows just how narrow the focus of such theory remains. As was pointed out many years ago by the conflict theorists, theories that insist on locating the problem of "crime" in the offender are really theories of "criminality" rather than of crime per se, unless, of course, we wish to redefine "criminality" as a predisposition toward law violation rather than as the outcome of the criminalization process.

One can readily see the differences in perspective between, for example, the biosocial theorists and the social constructionists. If criminal behavior is a social construction defined by changing laws within a shifting culture, then is it logical to expect it to be predictable through chemical formulas that remain the same across cultures and across time? If criminal behavior were a biological "fact" rather than a changing social construction, then the search for biological or genetic predictors would represent one more struggle in the long history of science, but it is not. Perhaps we can expect biosocial research to prove more successful in establishing such relationships with respect to pathological conditions such as alcoholism (defined as a physically debilitating susceptibility to alcohol addiction) than, for example, with respect to white-collar crimes such as price fixing, environmental pollution, medical fraud, and corporate violations of health and safety laws.

Of course, some biological and biosocial theorists have acknowledged this fact. Thus, for example, Walsh (2003a) admits that the so-called *mala prohibita* crimes (i.e., those that are a matter of changing legal prohibitions) cannot be explained biologically, but he insists that the so-called *mala in se* crimes (i.e., those that are "bad in themselves") can be so explained, and one gets the feeling that he, like other biologically oriented criminologists, considers these the "real" crimes. Rowe (2002) finds it necessary to *redefine* crime from a social/legal category to a biological category so as to produce a proper fit between causal concepts and effect concepts, holding that, "Criminal acts are those acts intended to exploit people belonging to one's own social group in ways that reduce their fitness" (p. 3). This is surely an odd definition of crime. In fact, the definition is *formulated* so as to *imply* evolutionary consequences, meaning that a theoretical framework is built into the definition rather than developed to explain a long-standing problem already defined by common law and legislation. Even the phrase "one's own social group" excludes everything from gang warfare to genocide from criminology. This is very unsatisfying, to say the least. Perhaps we are better off sticking with a vague definition of crime as "antisocial behavior," which is the solution preferred by many biological and biosocial theorists.

In any case, the policy implications of biosocial criminology seem likely to have a significant effect upon the legal system (Heide & Solomon, 2006). The fact is that evolutionary psychology, genetics, and neuroscience make fundamentally different claims about "human nature" than does the criminal law as bequeathed to us by the classical school. The first regards our behavior as essentially determined while the second insists that it is a matter of choice based on free will. The courts have been willing to allow some science into their chambers, but the relationship is a very uneasy one, reflecting two quite different worldviews. Thus, the *Roper v. Simmons* (2005) case that prohibited the execution of juveniles was supported by amicus briefs submitted by the American Medical Association and the American Psychological Association citing brain imaging studies showing the relatively underdeveloped prefrontal lobes of adolescents, but the U.S. Supreme Court may have been looking for a way out of an embarrassing situation that had put the United States almost alone among the nations of the world in executing adolescents.

Just as DNA profiling has entered the criminal justice system through such applications as CODIS, some entrepreneurs are now offering what they insist is a lie detection test much superior to the polygraph—based on fMRI work (Haederle, 2010). In 2007, the MacArthur Foundation committed $10 million to create the Law and Neuroscience Project at the SAGE Center for the Study of Mind, and it is quite possible that such projects will lead to a rethinking of the Model Penal Code, which was developed in 1962 and reflects the input of the psychoanalytic thinking popular in the mid-20th-century United States with virtually no insights from the new biosocial work (Haederle, 2010). It will certainly be difficult to integrate the two worldviews, but we might expect that biosocial data can contribute something to the establishment of certain elements of "diminished capacity" or "mitigating circumstances," both of which are well-established legal concepts.

The best solution here is probably to focus upon certain relatively specific behavioral categories rather than trying to explain "crime" or even "antisocial behavior" in general. As many have pointed out, the problem of "crime" is one of "apples and oranges." Different types of crime are as behaviorally dissimilar as night and day. Some seem to suggest lack of self-control, for example, but others suggest extremely calculated and highly controlled activity. Some may be associated with low IQ in some fashion, but others seem to require a much higher IQ than average. Of course, this is a problem for every theory we have examined to this point, and it was partly this issue that led Sutherland to seek a general explanation based on learning from those who had criminal attitudes and techniques. To some the problem may seem more blatant with the biological and biosocial theories, but the problem surely carries embarrassing implications for many of the psychological and sociological theories as well.

Of course, there are biosocial theorists whose work is much more specific. They are more successful in avoiding the temptation to overgeneralize. Comings (2003), for example, tries to maintain a specific focus in his recent work titled "Conduct Disorder: A Genetic, Orbitofrontal Lobe Disorder That Is the Major Predictor of Adult Antisocial Behavior." Noting that conduct disorder (a *somewhat* specific *behavioral* diagnosis, not a crime) is such a powerful predictor of adult antisocial personality disorder (ASPD) that ASPD requires the presence of preexisting childhood conduct disorder (CD), he cites twin studies indicating the likelihood that CD is predominately a genetic disorder rather than an environmentally caused problem. Despite the title, we can see that he is dealing with the *somewhat* more specific behavioral diagnosis of ASPD rather than the very general notion of "adult antisocial behavior," thus getting rid of some of the "apples" among the "oranges" by at least using a more internally consistent term. Furthermore, Comings is a clinician as well as a molecular geneticist, so he has the advantage of observing and dealing on what he calls a "daily basis" with the actual human behavior outside the laboratory that he is studying under laboratory conditions, thus avoiding some of the naïveté of which we spoke earlier.

On the whole, one cannot help being impressed by the recent energy and diligence of the biosocial theorists. As Rowe (2002) points out, these researchers have obtained approximately the same range of predictive power (i.e., correlation coefficients of .20–.40) with their biological variables as social scientists have with the various social variables long in use. With both sets of variables, success so far consists of explaining as much as perhaps 20% of the variance in the dependent variable. Clearly, much work needs to be done on both fronts.

CHALLENGES AHEAD

This last point calls attention to something that gives pause to many criminologists. For all of the reasons outlined heretofore, it might prove much more difficult to develop effective social policy based on biosocial theory than some imagine. But what if the effort eventually bears fruit? For some, every successful step in biosocial criminology is a dangerous step. Some fear that success in the narrow fields most amenable to genetic or neuroscientific research and theorizing will tend to direct social policy even further away from the crimes of the powerful toward an even more narrow focus on the "antisocial behavior" of less influential segments of the population. Others fear that if the "hard evidence" is ever established with respect to those offenses, it will provide dangerous justifications for those who wish to suppress the particular behaviors that most annoy them and perhaps even to eliminate those now seen as "unfit" to survive.

Some of the biosocial theorists themselves call attention to the problems involved in predicting criminal disposition from genetic evidence, citing the fact that many "false positives" (i.e., individuals who have not only never committed a delinquent or criminal act but who have been *mistakenly* identified as "crime-prone" and "treated" accordingly) may be caught up in the net (Rowe, 2002). These same theorists also point out that outside the criminal justice system, successful identification of genetic markers for criminality might well lead parents trying for the "perfect child" to abort fetuses carrying such markers (Rowe, 2002). This issue has already arisen in the case of markers for serious diseases, and it may be just a matter of time before it surfaces as a means of producing more "domesticated" children. Whether this would be desirable or not is, of course, a value judgment, but it certainly conjures visions of Huxley's *Brave New World*.

If the biological and biosocial approaches are successful, they have the potential to pull us away from the "criminal justice model" toward the "medical model" (Rowe, 2002). Indeed, the entire edifice upon which the idea of "responsibility" rests would have been challenged, along with the concept of "free will." Social policy would be pushed in the direction of a system based on diagnosis and treatment rather than one of adjudication and punishment. This would represent quite a shift from the trend of the past three decades, which has been to restrict the "medical model" by insisting that criminals are moral outcasts rather than impaired citizens, making insanity pleas more difficult in criminal trials, and rejecting the idea of "rehabilitation" along with most of the rest of the medical model as ineffective in corrections. This would be quite a shift indeed.

Still, we must be cautious, because research showing that criminals are biologically different from "the rest of us" has repressive potential as well. Seeing offenders as dangerous due to their very nature may reinforce current policies aimed at incapacitating "career criminals" and "chronic offenders." Although the notion of biological determinism might diminish notions of free will and criminal responsibility, it also could be used to suggest that criminals are incurably wicked and beyond the powers of therapeutic interventions to save. Were this mode of thinking to prevail, it could have disquieting implications, given that the correctional system is disproportionately populated by the nation's poor and minority citizens. Seeing the disadvantaged as biologically deficient risks acquitting "the rest of us" of any responsibility for social inequality and the role it plays in fostering criminogenic conditions.

We began this chapter by remarking on the promising interdisciplinary thrust of recent work such as that of Anthony Walsh (2009), who has made a special effort to comment on most of the theoretical perspectives examined in this book in the light of biosocial criminology. This seems promising in that it acknowledges sociological and psychological theorizing. Unfortunately, however, there is still considerable animus here, perhaps as a result of years of pleading for some "respect" for research in a subfield of criminology that has been at odds with the mainstream for so long. Thus, one of the leading biosocial theorists, Walsh (2009) writes as follows:

> I for one would love to see a form of long-acting implant that calms the violent, shames the psychopath, brings tears to the eyes of the heartless, makes the impulsive contemplate, and causes the man contemplating rape to have immensely graphic visions of pit bulls devouring his nether parts. Yes, this is "treating symptoms, not

causes," but medicine treats symptoms rather than elusive causes all the time. It is the symptoms that cause the patient's pain, and it is the criminal's manifest symptoms that cause his or her victim's suffering, not the supposed "root causes" of those symptoms, whatever they may be. (p. 292)

Perhaps this is an understandable sentiment, but we should remember that even the "violent" are not so different from us, that the "impulsive" are often as good examples of the old truism that, "All work and no play makes Jack a dull boy" as they are of "a dangerous lack of self-control," and that the "supposed 'root causes'" of which Walsh speaks are really the *deeper* causes of the "symptoms." These may be the social conditions studied by sociologists, the labeling behaviors studied by the social psychologists, and the criminalization processes studied by the conflict theorists. Preventive medicine tries to drain the swamp even as it treats the malaria. Otherwise, there is no end to the "long-acting implants," genetic manipulations, and the like.

Conclusion

It is clear that the time has arrived for criminologists to abandon their ideological distaste for biological theorizing; after all, what role biology plays in crime is ultimately an empirical question that requires careful investigation. As noted, scholars' wariness concerning the notion that offenders have different bodies is understandable, given that early theories of the "criminal man" were used to justify eugenics policies that resulted in the institutionalization and sterilization of "defectives," restrictive immigration policies, and calls for racial purity (Bruinius, 2006; Cornwell, 2003). As we have seen, however, more nuanced, complex biosocial perspectives are now being proposed that avoid the more blatant biases found in the work of Lombroso, Hooton, and others of this genre. Contemporary biological theories are not inherently conservative or repressive and, in some cases, can be used to call for progressive intervention (e.g., early childhood programs, protection of inner-city residents from lead exposure).

Still, two challenges will remain as biological theorizing comes to have a more prominent place not only in criminology but also in the explanation of a variety of social problems. First, as Duster (2006) notes, there is a risk that finding biological correlates of wayward conduct will encourage a "reductionist" view of this behavior. A biological paradigm encourages a focus primarily on the individual and on what is inside that person. It takes attention away from larger units of analysis (e.g., why crime is concentrated in certain communities) and away from the historical trends and social contexts in which individuals are enmeshed. It is instructive, observes Duster (2006), that government support for biological research on violence is increasing whereas its sponsorship of social science research is decreasing. This may be because biology promises "harder" findings, but it also may involve the fact that social science research is more critical of existing power structures and sources of inequality.

A biosocial approach need not be reductionist, but there does seem to be such a tendency. Thus, Anthony Walsh (2009), even while ostensibly integrating various sociological theories with the biosocial approach, still insists that, "All social sciences eventually reduce to psychology, psychology to biology, biology to chemistry, chemistry to physics, and physics to mathematics" (p. 295). This is not necessarily the case. As one of the authors of this book wrote more than 30 years ago, approaches such as "general systems theory" take us "beyond reductionism" by positing that each of the levels of study mentioned by Walsh tends to be constructed upon the lower levels but cannot be reduced to those levels because each level tends to be structured in terms of distinct organizational principles of its own, which it is the mission of the respective sciences of sociology, psychology, biology, chemistry, physics, and mathematics to elucidate (Ball, 1978b). Although each level emerges from the level below it, there is no reason to believe that it must be structured in accordance with the principles of the preceding level. This is often called the *principle of emergence*. Whether it holds or not will depend upon research rather than preemptive judgment. Humans have a tendency to be seduced by the latest "gods," which Francis Bacon called the "idols of the mind," regardless of whether these are Marxism, Darwinism, or some other "ism."

The second problem is risk that biological theory—even if done implicitly and with no intention—can be used to socially construct offenders as members of a different class of humans: super-predators whose brains, genes, or constitution are defective. When criminals are depicted as "the other" and the social distance between offenders and the "good people" in society widens, policies to manage the risk posed by this "dangerous class" seem more reasonable. It becomes possible to embrace a "culture of control" (Garland, 2001) that favors measures—such as lengthy prison terms—that protect one's own safety and welfare. In this scenario, theories no longer call attention to the complex ways in which social arrangements are implicated in crime. Instead, the existing order is acquitted of responsibility in inducing criminal conduct because, after all, it is the people with bad biology who are causing all the trouble. On the other hand, it must be acknowledged that biosocial research can also contribute to scenarios in which deviants are treated more leniently by the law, or at least serve as a "cover" for courts willing to do so in the light of changing public opinion, as suggested by the landmark *Roper v Simmons* decision prohibiting execution of those who committed their offenses as juveniles. As is always the case, whether the theories gain acceptance and how they are actually applied will probably depend upon the context of the times.

John H. Laub
1953–
University of Maryland
Author of Life-Course Theory

15

The Development of Criminals

Life-Course Theories

F or years, criminologists concentrated most of their attention on the teenage years and delinquency. Even a brief review of the titles of the field's defining works shows the emphasis on wayward adolescents. There was Shaw and McKay's (1972) *Juvenile Delinquency and Urban Areas;* Reckless, Dinitz, and Murray's (1956) analysis of the "Self-Concept as Insulator Against Delinquency"; Cohen's (1955) *Delinquent Boys;* Cloward and Ohlin's (1960) *Delinquency and Opportunity;* Hirschi's (1969) *Causes of Delinquency;* Matza's (1964) *Delinquency and Drift;* and Schur's (1973) *Radical Nonintervention: Rethinking the Delinquency Problem.* Furthermore, during the past several decades, the vast majority of empirical tests of criminological theories have used samples of juveniles.

But why this early and enduring interest in adolescents? In large part, this interest is tied to the apparent fact that this stage in the life course generates high rates of illegal behavior. When crime and age are plotted on a graph, it is possible to draw an "age-crime curve." Notably, the curve "peaks," or reaches its highest point, at approximately 17 years of age—a bit younger for property crimes, a bit older for violent crimes (Agnew, 2001b; Caspi & Moffitt, 1995). The curve rises steeply between 7 and 17 years of age and then declines thereafter. The peak of the curve is produced by two factors—prevalence and incidence—meaning that criminal acts both are widespread among youths and are committed at a higher rate than at other ages. As

Caspi and Moffitt (1995) noted, "The majority of criminal offenders are teenagers; by the early 20s, the number of active offenders decreases by over 50%; by age 28, almost 85% of former delinquents desist from offending" (p. 493). It is during adolescence, moreover, that youths form gangs that are capable of both facilitating criminal acts and disrupting neighborhood order.

Given these empirical realities, one might conclude that the crime problem is mainly a teenage problem and that the roots of crime lie mainly in what happens to youths during this stage in life. And, in fact, this is precisely what most criminologists surmised. Although there was a vague sense in theoretical writings that what occurred during childhood might be related to what occurs during the teenage years, this insight was not followed up or incorporated into the theories in a systematic way. Instead, the assumption reigned that solving the riddle of crime causation depended on figuring out what unique features of adolescence prompted youths to suddenly break the law. Each theory, of course, could find its plausible candidates. For example, for the Chicago school, it was the gang and the differential association this afforded. For strain theory, it was the status frustration and blocked opportunities that youths confronted. For control theory, it was the weakening attachment to parents and the failure to develop school commitments. For labeling theory, it was being subjected to arrest and the debilitating effects of the juvenile justice system.

This way of thinking was given an added boost by the discovery and sophistication of the "self-report survey," which provided a method for testing which theory could best account for juvenile delinquency. Hirschi's (1969) landmark study, *Causes of Delinquency,* was perhaps most instrumental in demonstrating the power of the self-report study to assess competing delinquency theories. The strategy, of course, was to include on the same survey instrument questions asking youths to report their involvement in delinquent acts and questions measuring different theories. In Hirschi's study, the theories measured and tested against one another were social bond, cultural deviance (or differential association), and strain.

Literally hundreds of studies using this approach have since followed. In part, the extensive use of self-report surveys is due to their convenience. Virtually the entire youth population in a community can be found in its junior and senior high schools. There, youths sit at desks with pencils in hand—potentially ready to complete the questionnaires devised by criminologists. Also, however, the use of self-report surveys was based in part on the assumption, noted previously, that the roots of crime were to be found in adolescence. Equipped with this assumption, the study of delinquency became even more convenient. Criminologists would not have to study the participants over time; instead, they could simply give the participants questionnaires while they were teens. In essence, criminologists could "get away with" *cross-sectional* research.

This way of studying crime, which had crucial ramifications for how theories were developed and tested, was called into question by a simple insight that was present previously but had not been paid much attention: What happens at one stage in life often affects what happens at another stage in life. This reality became apparent when cohort studies, which employed a *longitudinal* research design, showed that involvement in crime during early adolescence was related to criminal conduct during late adolescence and during early adulthood (see, e.g., Wolfgang, Figlio, & Sellin, 1972).

The emergence of the notion of "criminal careers"—that offenders have "careers" much like those who work have careers—also played a role in focusing attention on the time or life-course feature of crime (Blumstein, Cohen, Roth, & Visher, 1986; see also Piquero, Farrington, & Blumstein, 2003, 2007). As Petersilia (1980) observed, "The distinguishing characteristic of criminal career research is its concern with systematic changes in behavior over time" (p. 322). Thus, in this research paradigm, attention was given to when crime begins ("onset"), how long crime lasts ("duration" or "persistence"), how frequently crime is committed ("incidence" or *lambda*), and when crime stops ("desistance"). Scholars added the further insight that the factors affecting each of these dimensions of a criminal career might not be identical—for example, that what might cause someone to begin committing crime (e.g., ineffective parenting) might not be the same as what causes the individual to persist in or desist from criminal involvement (e.g., whether or not employment is obtained).

In this context, the most startling insight to emerge, albeit an insight long known to psychologists interested in human development, was that what happens during *childhood* is related to delinquency during *adolescence*. That is, there appears to be *continuity* or *stability* in antisocial behavior that extends from the early years of life and into adolescence and beyond. If scholars had not attended to this empirical reality, then they were forced to do so by the highly publicized work of Wilson and Herrnstein (1985), *Crime and Human Nature*. "At one time," they noted, "it was widely assumed that the child is father to the man—that there were great continuities in attitude and behavior from infancy on" (p. 241). But this thesis has been ignored in criminology "because childhood was subordinated to gangs and social structure" (p. 241). Most important, research now shows that "misbehavior among young children is powerfully predictive of misbehavior among older ones" (p. 241). As we will see shortly, the thesis of continuity also would be embraced by Gottfredson and Hirschi (1990) and, as represented in their theory, would throw criminology into a crisis. The point at hand was simple but powerful: *If the roots of crime lie in childhood—and not in the juvenile years—then most theories of delinquency and crime must be partially, if not wholly, incorrect.*

By the late 1980s and early 1990s, interest in childhood and in the stability of antisocial behavior across offenders' lives was growing rapidly. Some scholars called for a *developmental criminology* (Le Blanc & Loeber, 1998; Loeber & Le Blanc, 1990; Loeber & Stouthamer-Loeber, 1996), although the term most often used to describe this emerging paradigm is *life-course criminology* (Sampson & Laub, 1992, 1993, 1995). Much of the research in this area was empirical, seeking to map both the predictors of various aspects of offending (e.g., onset, persistence, desistance) and the pathways or sequence of events that directed people into and out of crime. Although invaluable, the empirical thrust of criminology tended to slow the design of life-course *theories* of crime.

Even so, theoretical approaches have increasingly emerged (Farrington, 2005, 2006), and we review the most influential ones here. We start by revisiting the work of Gottfredson and Hirschi (1990) who, as may be recalled from Chapter 6, argued that low self-control consigns individuals to a life of social failure and wayward conduct from childhood onward. We then focus on Patterson's "social-interactional developmental model," which has served as the basis for interventions with troubled families. Moffitt has proposed that criminal development proceeds along two different pathways, one leading to persistent involvement in conduct problems and the second leading youths into and then out of crime. Her theory has been inordinately influential and, as we will see, has generated research and controversy. Perhaps the most significant life-course theory has been proposed by Sampson and Laub. Their work, which has evolved over the past decade or so, explores both continuity and change in offending across time (see Laub & Sampson, 2003; Sampson & Laub, 1993). Notably, their theorizing both borrows from and criticizes Hirschi's control theories. Finally, we consider recent theories of desistance that seek to explain why seemingly career offenders cease their criminal careers and turn in different directions. In this regard, we review the perspectives of Maruna and of Giordano and her colleagues.

Before embarking on this analysis, we first discuss integrated theories of crime. Although not initially formulated as life-course theories, in retrospect they represent early efforts within this paradigm. In particular, using a longitudinal approach, they attribute criminal involvement to sequences of events that transpire across developmental stages (see also Catalano & Hawkins, 1996).

Integrated Theories of Crime

Although different kinds of theoretical integration are possible (Barak, 1998; Messner, Krohn, & Liska, 1989), an *integrated theory* typically is an explanation of crime that attempts to merge the insights from two or more theories into a single framework. There are numerous integrated theories in criminology (Akers & Sellers, 2004; Barak, 1998, p. 193). For many theories, the integration of ideas from different models is made clear. In their classic theory of gang subcultural delinquency, for example, Cloward and Ohlin (1960) told readers that their intent was to merge the insights from Merton's strain theory and from the Chicago school writers who illuminated how criminal behavior is learned through cultural transmission. In other theories, however, the integration is more hidden and is recognized only by those well acquainted enough with criminological theories to see how variables from differing perspectives have been borrowed selectively. Take, for example, Agnew's general strain theory. Many of the factors that he argued "condition" whether strain results in crime are borrowed from control, social learning, and rational choice theories. A number of other theories reviewed in this book could be considered, at least in part, to be integrated theories.

The most noted integrated theories tend to combine elements from theories covered in Chapters 3, 4, 5, and 6—from differential association/ social learning theory, strain theory, and control/social bond theory. This is understandable, because these perspectives have long been at the center of criminology; in fact, Agnew (2001b) contended that these theories are the "dominant theories of delinquency" (p. 117). In what follows, we review two perspectives that represent systematic attempts to combine elements from among these theoretical traditions—Elliott, Ageton, and Canter's (1979) *integrated strain-control paradigm* and Thornberry's (1987) *interactional theory of delinquency.* As noted, these perspectives are important not only because of the attention they have received but also because they were early examples of attempts to study how youths become involved in crime across time. We start by adding a few brief comments on the merits of integrated theorizing.

INTEGRATED THEORIZING

The logic behind integrated theorizing seems, on the surface, to be incontrovertible. Each theoretical tradition captures only a slice of reality. In so doing, each theory alerts us to a potentially important cause of crime. But each theory also has the inherent weakness of not including variables from competing theories that might well be implicated in crime causation. By contrast, integrated theories are not wed to any one perspective. Liberated from this unnecessary allegiance, they are free to incorporate into a single model all factors—regardless of their theoretical origin—that might be causes of criminal conduct.

Although this logic is persuasive, integrated theorizing also has at least two shortcomings (see, e.g., Hirschi, 1989). First, it assumes that criminological knowledge will grow most quickly by trying to bring theories together rather than by having competing perspectives battle it out for explanatory supremacy. Such battles force scholars to sharpen their arguments and to search for innovative ways of showing that their models are more able to account for criminal involvement. Second, integration can lead to sloppy theorizing in which scholars pick a variable they like from one theory, and then a variable they like from another theory, and so on—much like going through a cafeteria line or buffet. The result may be a tray full of ideas that "taste good" but that do not really combine into a "well-balanced meal."

Take, for example, attempts to merge ideas from social bond and social learning theories. These theories *might* be compatible, but they are based on different assumptions about human nature, ask different questions ("Why do they do it?" vs. "Why don't they do it?"), and make different predictions on the effects of a range of variables (e.g., peer groups). These differing assumptions, fundamental questions, and predictions typically are ignored when scholars state their integrated theories. Instead, social bond and social learning variables are merely placed into a causal model, and the potential incompatibility of these two sets of factors is not discussed. In the end, of course, the wisdom of theoretical integration lies in whether these efforts allow us to explain the causes of crime more adequately, although these criteria must be applied to whether integration achieves not only short-term gains in explanatory power but also long-term growth in theoretical knowledge.

ELLIOTT AND COLLEAGUES' INTEGRATED STRAIN-CONTROL PARADIGM

One of the most significant efforts at theoretical integration was set forth by Delbert Elliott, Suzanne Ageton, and Rachelle Canter (Elliott et al., 1979; see also Elliott,

Huizinga, & Ageton, 1985). Although Elliott et al. (1979) labeled their perspective an "integrated strain-control paradigm" (p. 10), it also draws on social learning theory and, in particular, on the importance this perspective places on adolescent peer groups in sustaining criminal involvement.

Beyond suggesting how major theories might be combined, Elliott and colleagues' theory is significant in two other respects. First, the theory suggests that factors from certain theories might be more important at particular stages in life (e.g., childhood vs. adolescence). Second, the theory posits that there may be more than one pathway to delinquency. Notably, issues such as these have reappeared recently in life-course or developmental theories of crime. We revisit these matters later in the chapter.

Elliott et al.'s (1979) model began by focusing on "early socialization outcomes." They noted that the key feature of childhood is whether children establish strong or weak bonds. Thus, the foundation for their model lies within social control theory. In any event, they divided the social bond into two parts: *integration* (the "external" or "social" bond) and *commitment* (the "internal" or "personal" bond).

By *integration*, Elliott et al. (1979) meant the extent to which people are involved in and attached to "conventional groups and institutions, such as the family, school, peer networks, and so on" (p. 12). Participation in these social roles causes the person to be regulated by role expectations and accompanying sanctions. In Elliott et al.'s terms, "Integration is akin to Hirschi's concepts of involvement and commitment" (p. 12). By *commitment*, they meant the individual's "personal attachment to conventional roles, groups, and institutions" (p. 12). This is the extent to which someone "feels morally bound" by social norms. In their view, "commitment is akin to Hirschi's concepts of attachment and belief" (p. 12).

Those who establish strong bonds during childhood and then maintain them in adolescence have a low probability of engaging in delinquency. By contrast, weak bonds in childhood place youngsters on a clear pathway to stable criminal involvement. As these children age into the teen years, they increasingly are exposed to delinquent peer groups. As predicted by social learning theory, these peer groups provide "a positive social setting that is essential for the performance and maintenance of delinquent patterns of behavior over time" (Elliott et al., 1979, p. 14). In short, in the pathway to crime, (1) weak bonds in childhood lead to (2) participation in delinquent peer groups, which in turn (3) results in stable criminal behavior.

Elliott et al. (1979) argued, however, that there is a second pathway to delinquency. Many children who establish strong bonds are set on a pathway that leads them into conventional peer groups and lives of conventional behavior. Even so, some children who initially form strong bonds nonetheless become enmeshed in crime and drug use. This outcome, it should be noted, is inconsistent with traditional social bond theory, which would predict that such bonds should insulate youths from delinquency. But in Elliott and colleagues' model, events can occur in adolescence that create sufficient *strain* on a youth personally or on the social bond to cause the individual's "commitment" to and "integration" into conventional society to "attenuate" (p. 10). One possibility, for example, is that a youth who is highly committed to the goal of success fails to achieve this goal. It also is possible that social disorganization in a juvenile's environment might cause instability and a lack of cohesion in the individual's social group. Consistent with control theory, once strain causes the bonds to weaken, a youth is free to engage in delinquency.

At this point, Elliott et al. (1979) reintroduced social learning theory. A small proportion of youths, they admitted, might experience so much strain from blocked goals that they proceed directly into delinquency, at least on a short-term basis. Others might lose their commitment to success goals—become alienated from conventional ideas of success—and seek some adventure or thrills, for example, through drug use. Mostly, however, youths whose bonds are strained and weakened become involved in delinquent peer groups. Like those who never formed strong bonds, due to these associations, these youngsters are embedded in stable criminality. This pathway to crime, according to Elliott et al., "represents an integration of traditional strain and social learning perspectives" (p. 17). Again, in this second pathway to crime, (1) strong bonds initially insulate the child from conduct problems, but (2) strain attenuates bonds in adolescence, allowing youths (3) to participate in delinquent peer groups and, in turn, (4) to engage in stable criminal behavior.

Elliott et al. (1985) produced data largely supportive of their integrative model. Even so, some important questions remain. Most noteworthy, it is not clear why social learning variables would have effects only during adolescence and only through delinquent peer groups. There is evidence, for example, that peer groups begin to exert socializing influences during childhood (Harris, 1998). Furthermore, Elliott et al. largely saw the family as a socializing agent that inculcates bonds and not, again, as a context in which social learning occurs. Clearly, Akers (2000), the primary contemporary social learning theorist, would dispute this contention. This observation reminds us of the potential weakness of an integrated theory: By selecting variables and placing them in a particular order, integrative theories risk overlooking causal processes that the original theories would see as integral to crime causation.

THORNBERRY'S INTERACTIONAL THEORY

The Theory. According to Terence Thornberry (1987), "Human behavior occurs in social interaction and can therefore be explained by models that focus on interactive processes" (p. 864). In his view, "adolescents interact with other people and institutions," and their "behavioral outcomes are formed by that interactive process" (p. 864). Thus, key causal conditions, such as attachment to parents, are not invariably stable over time but rather may differ depending on whether the nature of the interaction between parents and children changes. Furthermore, delinquents not only are influenced by their social surroundings but also have an impact on others through their behavior—including their delinquent behavior. In short, they are part of an "interactive system" (p. 864). As we will see shortly, this view is significant because it suggests that the relationships between variables in Thornberry's model are not unidirectional but rather interactive or *reciprocal*. It is noteworthy that in a review of existing research, Thornberry (1996) concluded that there is "substantial empirical support for interactional theory's central contention that delinquent behavior is embedded in a set of mutually reinforcing causal networks" (pp. 228–229). That is, the "overwhelming weight of the evidence suggests that many presumed unidirectional causes of delinquency are in fact either products of delinquent behavior or involved in mutually reinforcing causal relationships with delinquent behavior" (p. 229; see also Thornberry, Lizotte, Krohn, Smith, & Porter, 2003).

Thornberry (1987) based his model on the core premise of control theory that the "fundamental cause of delinquency lies in the weakening of social constraints over the conduct of the individual" (p. 865). His basic model is that during childhood, youngsters develop attachments to parents. When this occurs, youths are likely to embrace conventional beliefs, show a commitment to school, avoid association with delinquent peers, reject delinquent values, and resist wayward conduct. When youths fail to develop strong attachments to parents, however, the opposite causal sequence is initiated. They do not develop either strong conventional beliefs or a strong commitment to school. With these controls attenuated, they are free to explore other behavioral options. At this point, they encounter delinquent peers. Consistent with social learning theory, the association with antisocial friends fosters delinquent values and delinquent behavior. In essence, then, Thornberry's model integrates social control and social learning theories.

In an attempt to elaborate this basic causal model, Thornberry (1987) then made two key theoretical insights. First and most important, he argued that the variables in his model have *reciprocal effects*. Take, for example, the variable of peer associations. Weak parental attachment might make delinquent associations more likely, but these associations also can weaken attachment to parents. Similarly, peer associations may cause delinquent behavior, but such behavior also can affect friendship choices. That is, it is possible both for peers to amplify delinquency and for "birds of a feather to flock together." Previous theories generally had ignored the way in which social control and learning variables interacted with one another. In Thornberry's view, however, specifying such interactions is essential because doing so models more closely the actual complexities of social life.

In this regard, Thornberry (1987) noted that this interactional process creates a "behavioral trajectory" that "predicts increasing involvement in delinquency and crime" (p. 883). In this trajectory, "the initially weak bonds lead to high delinquency involvement, the high delinquency involvement further weakens the conventional bonds, and in combination both these effects make it extremely difficult to reestablish bonds to conventional society at later ages" (p. 883). This process is what later life-course theorists would call "cumulative disadvantage"—a set of interlocking conditions that coalesce to progressively deepen a person's criminality. As Thornberry noted, "All the factors tend to reinforce one another over time to produce an extremely high probability of continued deviance" (p. 883).

Second, Thornberry (1987) also realized that the effects of variables differ with a person's stage in the life course. As youths move from early to middle adolescence (15–16 years of age), the effects of parents wane and those of peers and school become more salient. Similarly, during late adolescence, new variables emerge, such as employment, college, military service, and marriage. These now represent a person's "major sources of bonds to conventional society" (p. 881), and they play an important role in determining whether delinquency will continue or desist.

In summary, Thornberry's work alerted us to the fact that criminal behavior emerges in the context of the developmental process in which the person and environment interact with one another. Individuals are simultaneously a product and an architect of their social environment—created by and creating their life situations. Their development is dynamic rather than static, and it can be altered as the conditions of a person's life change. At the same time, there is a tendency for people to become embedded in certain behavioral trajectories. Most disquieting is the interactional process through which weak bonds, delinquent peer associations, and delinquency combine to steadily trap a person in a criminal role. Notably, research exists that is favorable to the core propositions of interactional theory.

Implications for Life-Course Theory. Notably, these themes anticipated the insights of life-course criminology—a contemporary perspective that we examine later in this chapter. Perhaps it is not surprising, then, that Thornberry has transformed interactional theory into a life-course approach and presented longitudinal evidence consistent with this perspective (see Thornberry & Krohn, 2005; Thornberry et al., 2003; see also Kubrin et al., 2009). A special feature of this revised theory is that it seeks to explain why onset into misconduct might occur at three different stages in the life course (Thornberry & Krohn, 2005).

First, there are those who manifest conduct problems in childhood. This early onset is the result of exposure to family disorganization and ineffective parenting, school failure, and association with delinquent peers. These problems are typically exacerbated by living in a community marked by "structural adversity," which places stresses on the family and provides easy access to deviant peer groups. These youngsters tend to persist in their misbehavior because their initial deficits are "extreme" and "interwoven." Their involvement in antisocial behavior further roots them in criminality by causing rejection by parents, prosocial peers, and teachers and increased isolation into delinquent peer groups.

Second, most youths start offending "in mid-adolescence, from about age 12 to age 16" (Thornberry & Krohn, 2005, p. 192). These juveniles are not beset with serious deficits, but rather tend to react to features of adolescence that are "developmentally specific" (p. 193). In particular, these youths are in "the process of establishing age-appropriate autonomy" from "adult authorities," especially parents (p. 193). As a result, parental control weakens and conflict with and alienation from parents may arise. Unsupervised associations with peers arise, who reinforce deviant behaviors. This usually does not involve serious crimes (e.g., robbery, burglary) but rather smoking, drinking, drug experimentation, and vandalism. These adolescents tend to leave crime as they age, in large part because their deficits are not extreme and they have not "knifed off" prosocial adult options.

Third, there are "late bloomers" who wait until adulthood to begin offending. As Thornberry and Krohn (2005) note, it is estimated from previous studies that, somewhat surprisingly, fully "17.2 percent of non-delinquents begin offending in adulthood" (p. 195; see also Eggleston & Laub, 2002). They hypothesize that while growing up, these adults had personal deficits, such as low IQs and poor academic performance, that were "buffered" by supportive parents. As they "begin to leave the protective environment of the family and school," these deficits are no longer mitigated by parents and thus make it difficult to gain employment and to establish enduring intimate relationships such as marriage (p. 196). In turn, these adults become "more vulnerable to the influence of deviant friends" (p. 196) and find their way into crime.

As we will see, Thornberry's interest in explaining differential points of onset into and desistance from delinquency and crime are themes found across life-course theories. In fact, the debate over the nature and causes of continuity and change in offending is perhaps the central theoretical challenge to perspectives in the life-course paradigm.

POLICY IMPLICATIONS

We should note that the policy implications of integrated theories would be largely consistent with social learning, control, and strain theories. Because the building blocks of these theories are families, schools, and peer groups, they would favor interventions that strengthen families and parent–child attachments, increase school commitment, and foster prosocial peer group interactions. However, the tendency of integrated theories to focus on *childhood* and on *development* into crime differentiates them from traditional mainstream theories that were static and limited mainly to adolescence. As a result, integrated theories, like life-course theories, suggest the importance of considering how events during childhood heighten the risk for criminal involvement and how early intervention programs might be used to divert youngsters from lives in crime. We will return to this issue near the close of the chapter.

Life-Course Criminology: Continuity and Change

As noted earlier, the viability of the life-course perspective was enhanced by the recognition that there is *continuity* or *stability* in antisocial conduct from childhood into adolescence and adulthood. However, scholars also observed that the behavior of offenders can *change* or experience *discontinuity*. The centrality of continuity and change to the understanding of crime was captured in Robins's (1978) famous remark that although "adult antisocial behaviour virtually requires childhood antisocial behaviour . . . most antisocial children do not become antisocial adults" (p. 611). Clearly, childhood misconduct predicts later problem behavior, but the relationship is not ironclad. Although assessing the degree of stability is complicated by daunting methodological challenges and by how key concepts are defined and measured (Asendorpf & Valsiner, 1992), it appears that a large minority of antisocial children— perhaps nearly half of those at risk for future criminal involvement—do not develop into serious or chronic criminals during adolescence. Accordingly, a key theoretical issue in life-course criminology is explaining both continuity and change in offending.

To date, life-course theories may be seen as falling into one of four types. First, there are theories arguing that there is only, or mainly, *continuity* in offending (Gottfredson

& Hirschi, 1990). Second, there are theories stating that offending is marked by *either* continuity *or* change (Moffitt, 1993; Patterson, DeBarshy, & Ramsey, 1989). Third, there are theories contending that offending is marked by continuity *and* change (Sampson & Laub, 1993). Finally, there are theories that focus chiefly on change (Giordano, Cernkovich, & Rudolph, 2002; Maruna, 2001). We use this framework of continuity and change as we discuss theories in the sections that follow.

In addition, we will take this opportunity to highlight another controversy that has informed current theorizing: Are continuity and/or change in offending part of a *developmental process?* Some theorists, particularly Moffitt (1993), argue that individuals' movement into and out of crime occurs in predictable stages or patterns. In this view, becoming an offender is a developmental process that unfolds in established ways. These perspectives are thus truly *developmental theories of crime.* In contrast, other theorists, particularly Sampson and Laub (1993; Laub & Sampson, 2003), believe that whether individuals persist or desist in offending is not easily predicted. To be sure, they identify factors that make offending more or less likely, but they also recognize that there is an element of unpredictability—even of chance or good fortune—that can shape where and when an offender might choose to stop breaking the law. Accordingly, crime occurs across and is influenced by key events in the life course, but criminal careers do not unfold in neatly packaged ways. This issue will resurface as we discuss the theories to follow.

Criminology in Crisis: Gottfredson and Hirschi Revisited

Recall that in Chapter 6 we discussed the essentials of Michael Gottfredson and Travis Hirschi's theory of low self-control and crime (also known as the *general theory of crime*). We revisit this perspective here to show its relevance to life-course theory and to show the challenges it posed, on publication, to sociological theories of crime.

In developing their theory, Gottfredson and Hirschi (1990) did what most theorists fail to do: They thought seriously about the nature of crime. In this process, they noted that "competent research regularly shows that the best predictor of crime is prior criminal behavior" (p. 107). Equally important, they argued that traditional theories were deficient because "there has been little systematic integration of the child development and criminological literatures" (p. 102). As a result, "the most influential social scientific theories of crime and delinquency ignore or deny the connection between crime" and such childhood problems as "talking back, yelling, pushing and shoving, insisting on getting one's own way, trouble in school, and poor school performance" (p. 102). But this connection is significant, Gottfredson and Hirschi claimed, because it shows stability or continuity in conduct that shares the same underlying cause—low self-control. That is, the children who manifest conduct problems during childhood are the same ones who manifest delinquency during their teens. This continuity in problem behavior is what led Gottfredson and Hirschi to argue that individual differences in criminal propensities have their origins largely in childhood. It also is why they asserted that "the major 'cause' of low self-control thus appears to be ineffective child-rearing" (p. 97).

Gottfredson and Hirschi's (1990) contention that individual differences in self-control are the chief determinant of stability in waywardness from childhood to adulthood called into question nearly all major theories of crime that, again, emphasized the causal importance of experiences that youngsters had during adolescence. For example, according to Gottfredson and Hirschi, deviant peer group relationships had no causal influence on crime but rather were produced by "birds of a feather flocking together"—youths with low self-control self-selecting to associate with one another. The school failure–delinquency relationship was not causal either; rather, it was due to youths with low self-control failing at school and engaging in delinquency. As may be recalled from Chapter 6, this is the spuriousness thesis: The relationship of the social conditions that criminologists hypothesize cause crime actually is spurious because both the conditions and crime are caused by low self-control (Hirschi & Gottfredson, 1995).

In a very real way, Gottfredson and Hirschi's (1990) *General Theory of Crime* created a crisis in criminology because it made a powerful case that existing theories were simply wrong. These theories ignored the continuity in problem behavior, ignored causal processes during childhood, and rejected the notion that persistent individual differences emerged during childhood and accounted for the stability of such problem behavior. As Gottfredson and Hirschi warned:

> Thus no currently popular criminological theory attends to the stability of individual differences in offending over the life course. We are left with a paradoxical situation: A major finding of criminological research is routinely ignored or denied by criminological theory. After a century of research, crime theories remain inattentive to the fact that people differ in the likelihood that they will commit crimes and that these differences appear early and remain stable over much of the life course. (p. 108)

In short, Gottfredson and Hirschi (1990) proposed a theory of *stability* or *continuity* in offending. They argued that this stability is the result of what has been called "persistent heterogeneity" or stable underlying individual differences that people carry with them across situations at any one time and across their life courses over time. A competing explanation is that the continuity is due to what has been called "state dependence." This explanation sees participation in crime as evoking certain reactions (e.g., stigmatization), changing the offender (e.g., more likely to see crime as beneficial), or changing a life situation (e.g., weakening social bonds) in such a way that continuing in crime becomes more likely. Although the research is not entirely clear on this issue (Nagin & Paternoster, 1991; Sampson & Laub, 1993), there is evidence that persistent individual differences play a role in accounting for continuity in offending (Farrington, 1994b, pp. 531–535; Nagin & Farrington, 1992). At issue, however, is whether low self-control—the central theoretical construct in Gottfredson and Hirschi's theory—is key to the individual difference that accounts for continuity in misconduct.

It should be noted that Gottfredson and Hirschi do not deny that change in offending can occur. For example, they argue that the age–crime curve is "invariant," and thus that criminal involvement will decline with age in virtually all societies.

However, the change they allow for is *intra-individual*—that is, change that occurs within an individual over time. By contrast, they argue for stability in *inter-individual* differences in self-control and participating in offending and other deviant behavior. Thus, once levels of self-control are established in childhood, a person with lower self-control than a second person will always have less self-control and be more involved in wayward activities across the entire life course. In short, between-individual differences in self-control are enduring and are not affected by social or other factors.

Life-course theorists continue to grapple with the issues raised by Gottfredson and Hirschi. Their "general theory" has the advantage of being parsimonious (self-control explains continuity in offending across the life-course). However, although not without value, their theory's simplicity seems to ignore the reality that other factors are involved in causing continuity in offending and that change is as much a part of criminal careers as is stability.

Patterson's Social-Interactional Developmental Model

EARLY-ONSET DELINQUENCY

"Antisocial behavior," Gerald Patterson and colleagues observed, "appears to be a developmental trait that begins early in life and often continues into adolescence and adulthood" (Patterson et al., 1989, p. 329). According to Patterson et al. (1989), "If early forms of antisocial behavior are indeed the forerunners of later antisocial acts, then the task for developmental psychologists is to determine which mechanisms explain stability of antisocial behavior and which control changes over time" (p. 329). A key to understanding this development is that it is "social-interactional." This means that children and their environment are in constant interchange, where the actions of children elicit negative reactions, which in turn prompt children to act in antisocial ways that elicit more counterproductive reactions, and so on. "Each step in this action-reaction sequence," Patterson et al. cautioned, "puts the antisocial child more at risk for long-term social maladjustment and criminal behavior" (p. 329).

Similar to Gottfredson and Hirschi (1990), Patterson and colleagues (1989) linked the start of antisocial conduct to dysfunctional families that are "characterized by harsh and inconsistent discipline, little positive parental involvement with the child, and poor monitoring and supervision" (p. 329). Unlike Gottfredson and Hirschi, they did not propose that such inadequate parenting results in low self-control. In fact, they rejected control theory, including social bond theory, in favor of a social learning explanation. "The social-interactional perspective takes the view that family members directly train the child to perform antisocial behaviors" (p. 330).

In dysfunctional families, coercion is a way of life. The use of coercion is at times positively rewarded. More often, coercion is reinforcing because it enables the child to stop other family members from employing "aversive behaviors," such as hitting, against them. Soon the child "learns to control other family members through coercive means" (Patterson et al., 1989, p. 330). By contrast, prosocial ways of acting and solving problems are not taught or are responded to inappropriately.

As antisocial children move outside the home, they manifest "child conduct problems." They are ill suited for the school environment and are likely to fail academically. Meanwhile, they act aggressively toward other children and suffer rejection from the "normal peer group" (Patterson et al., 1989, p. 330). As they move into late childhood and early adolescence, these youngsters gravitate toward "deviant peer group membership," in part due to their failure in school and rejection by conventional youths and in part due to their antisocial predispositions. These affiliations, in turn, are a means through which deviant conduct is positively reinforced, thereby helping to consolidate these youths' involvement in misconduct.

Notably, Patterson et al. (1989) did not suggest that coercive parents and families are isolated entities. Rather, they attempted to assess why it is that some family systems become dysfunctional. What is it, they asked, that disrupts "family management practices"? They proposed three sets of factors. First, parents who are antisocial—who themselves were raised by antisocial parents—often employ ineffective discipline with their own children. Second, social disadvantage seems to be related to more physical and authoritarian parenting styles as opposed to discipline that is more verbal and cognitive in orientation. Third, when families experience stress—arising, for example, from unemployment, family violence, or family conflict—effective parental management of children may be disrupted.

LATE-ONSET DELINQUENCY

Patterson et al.'s (1989) analysis of "early starters" in antisocial conduct is largely a theory of continuity in offending, in which social-interactional processes increasingly constrain individuals to remain on deviant life courses. However, he subsequently joined with Karen Yoerger to propose "a developmental model for late-onset delinquency" (Patterson & Yoerger, 1997, 2002). This group was hypothesized to have its own developmental pathway to delinquency—an insight that, as we will see, is similar to the model formulated by Moffitt (1993). Patterson and Yoerger use the "marginality hypothesis" to characterize these late-onset offenders (1997, pp. 143–145). Thus, these youngsters grow up in contexts that are marginally disadvantaged, are raised by parents who use marginally effective child-rearing techniques, and have social skills (e.g., for work and relationships) that are marginally developed. As a result, unlike the early-onset youths, their childhood is not marked by extreme early deficits that cause serious antisociality and involvement in delinquency in childhood.

For the late-onset group, the key causal mechanism is not a dysfunctional, coercive family but a deviant peer group. Although more antisocial in childhood than nondelinquents due to more adverse family conditions, they do not enter into delinquency until middle adolescence (about age 14). At this time, parental supervision—not strong to begin with—loses its traction. Freed from direct monitoring by parents, the youths are able to associate with deviant peers. These interactions in turn lead them into criminal acts that are serious enough to trigger formal discipline by school authorities and arrests by the police.

Importantly, Patterson and Yoerger (1997) propose that compared with their early-onset counterparts, late-onset delinquents are less likely to persist and more like to desist from serious offending. Most crucial, as they age, the late-onset group still possesses social skills that will enable them to assume prosocial adult roles (e.g., such as a job or marriage) whose "payoff" exceeds that of deviant activities (p. 141). As Patterson and Yoerger contend, the "early-onset boys are both extremely deviant and extremely unskilled, whereas the late-onset boys are both moderately deviant and moderately skilled. The difference in starting points almost guarantees differences in growth patterns" (p. 141). In short, whereas early-onset youngsters are likely to experience substantial stability in offending across the life course, late-onset youngsters' delinquency is likely to be limited to adolescence (see also Moffitt, 1993). Thus, the developmental pattern of one group is marked by continuity, while that of the second group is marked by change.

INTERVENING WITH FAMILIES

Although they believe that early-onset delinquents face dismal life prospects, Patterson and his colleagues do not contend that these youths are beyond reform, especially if attempts to intervene occur earlier rather than later. For these theorists, the most promising solution to the problem of delinquency lies in early intervention with dysfunctional families. In fact, at the Oregon Social Learning Center, Patterson and his colleagues have implemented "parent management training" with at-risk families. In particular, there is evidence that parent training programs that seek to equip parents with effective ways of disciplining their antisocial children reduce problem behavior (Reid, Patterson, & Snyder, 2002; see also Van Voorhis & Salisbury, 2004).

Two final observations merit attention. First, to the extent that these parental management interventions are effective, they provide support for Patterson et al.'s theory and, more generally, for social learning approaches. That is, when theoretically informed treatments target certain risk factors for change—in this case parenting practices that support coercion and antisocial conduct—and the delinquent behavior is subsequently reduced, this is evidence that the theory has correctly identified a cause of criminal involvement (Cullen, Wright, Gendreau, & Andrews, 2003; Reid et al., 2002). Second, life-course theories that trace the origins of crime to experiences in childhood almost invariably justify early intervention programs. In fact, as we will see, this is the major policy recommendation of the life-course paradigm.

Moffitt's Life-Course-Persistent/ Adolescence-Limited Theory

In one of the more noteworthy developmental theories, Terrie Moffitt argued that offending is marked by either continuity *or* change (Moffitt, 1993; see also Caspi & Moffitt, 1995). The age-group curve is deceptive because the jump in offending during the teenage years *"conceals two qualitatively distinct categories of individuals, each in need of its own distinct theoretical explanation"* (Caspi & Moffitt, 1995, p. 494, emphasis in original). One group is small, consisting of 5% to 10% of the male population; the antisocial proportion of the population for females is even lower. This group manifests antisocial behaviors during childhood and shows *continuity* in misconduct into and beyond adolescence. Moffitt called members of this group *life-course-persistent* offenders (LCPs). A second group is large and includes most youths during their juvenile years. The members of this group evidence little or no antisocial tendencies during childhood but suddenly engage in a range of delinquent acts during adolescence, only to stop offending as they mature into young adulthood. Because their antisocial behavior is restricted to this one stage of development, Moffitt called this group *adolescence-limited* offenders (ALs). Thus, the offending or antisocial conduct of the ALs is marked by *change* or *discontinuity,* not by continuity.

Moffitt thus presents a developmental theory in which, while growing up, virtually all youths take one pathway or the other through adolescence and into adulthood. As a result, her theory suggests that at least when it comes to antisocial behavior, there are *two groups* of people: the LCPs who are the high-rate, chronic offenders and the ALs who move in and out of crime. On reflection, this claim is bold (are there no other groups?) and parsimonious (the development of offenders happens one way or the other)—good reasons, perhaps, for why her theory has garnered considerable attention.

Importantly, Moffitt's ideas about LCPs and ALs emerged in part from perplexing research findings (Piquero, 2011). Trained as a clinical psychologist, she was considering a career in practice or perhaps in a hospital doing clinical neuropsychology. But in 1984, a broken leg from a parachute-jumping accident confined her to a wheelchair in an office as she wrote her dissertation. As chance would have it, she was consigned to spend time with a department visitor. He persuaded her to come to New Zealand and work on a longitudinal study he was directing that was following a birth cohort. She then spent 2 years in New Zealand giving neuropsychological assessments to 1,000 13-year-olds. The initial results were promising in showing, as Moffitt predicted, that neuropsychological deficits were related to criminal involvement. But when the youngsters were followed up a few years later, the results—quite unexpectedly—grew weaker. As Moffitt explains:

> To my horror, as more and more adolescents in the cohort joined the ranks of offending at age 15 and 18, the correlations between risk factors and delinquency became weaker and weaker, until finally I had no findings at all to publish. Sometimes we learn the most when the data do not cooperate with a cherished hypothesis. This was one of those times. The taxonomy of life-course-persistent versus adolescence-limited antisocial behavior grew out of my struggles to understand why the data betrayed me. Why did risk factors characterize kids whose delinquency began before adolescence, but not kids whose delinquency began in mid-adolescence? (quoted in Piquero, 2011, p. 402)

Moffitt thus was led to the realization that the data set contained two types of antisocial youths on different developmental pathways: those affected by neuropsychological deficits who started misbehaving early in life (the LCPs) and those without such deficits who entered delinquency later during adolescence (the ALs). She eventually wrote a classic article, which she submitted to *Criminology*, the main journal in the field. It was rejected but then was later published in the *American Psychologist* (Piquero, 2011).

LIFE-COURSE-PERSISTENT ANTISOCIAL BEHAVIOR

According to Moffitt (1993), "Continuity is the hallmark of the small group of life-course-persistent antisocial persons" (p. 697). They bite and hit others at 4 years of age, they skip school and steal from stores at age 10, they sell drugs and steal vehicles at age 16, they rape and rob at age 20, and they engage in fraud and child abuse at age 30 (p. 695). Within any one developmental stage, they deviate across settings. Thus, they "lie at home, steal from shops, cheat at school, fight at bars, and embezzle at work" (p. 697).

"If some individuals' antisocial behavior is stable from preschool to adulthood," observed Moffitt (1993), "then investigators are compelled to look for its roots early in life, in factors that are present before or soon after birth" (p. 680). A key contribution made by Moffitt is in illuminating the potential life-altering effects of *neuropsychological deficits*. Neural development can be hindered by a range of factors during the prenatal period (e.g., mother's drug use, poor nutrition) and during the postnatal period (e.g., exposure to a toxic agent such as lead, brain injury from child abuse). These deficits can, in turn, affect the psychological traits of a child, especially verbal development and executive functioning (e.g., self-control of impulsivity). "By combining *neuro* with *psychological*," Moffitt noted, "I refer broadly to the extent to which anatomical structures and physiological processes within the nervous system influence psychological characteristics such as temperament, behavioral development, cognitive abilities, or all three" (p. 681, emphasis in original).

Neuropsychological deficits thus burden children with a "high activity level, irritability, poor self-control, and low cognitive ability" (Moffitt, 1993, p. 683), traits that place youngsters at risk for antisocial conduct. "Difficult" children, however, also tend to find themselves in difficult social environments. Their parents are likely to share common traits with their offspring and are likely to reside in adverse neighborhoods. Although any parent would find these children a daunting challenge, mothers and fathers with their own neuropsychological and social deficits are especially ill equipped to respond in productive ways. Instead, they tend to use disciplinary methods that intensify the difficult children's initial problem behaviors and that foster weak parent–child bonds (Caspi & Moffitt, 1995; see also Patterson et al., 1989). "Children's dispositions," observed Moffitt (1993), "may evoke exacerbating responses from the environment and may also render them more vulnerable to criminogenic environments" (p. 682). This is a process of "reciprocal interaction between personal traits and the environmental reaction to them" (p. 684).

Similar to Gottfredson and Hirschi (1990), Moffitt (1993) indicated that antisocial behavior at any one stage in development is, at least in part, a reflection of *contemporary continuity*—that is, of persistent heterogeneity of criminal propensities.

This occurs where an individual becomes involved in trouble because the person "continues to carry into adulthood the same underlying constellation of traits that got him into trouble as a child" (p. 683). Unlike Gottfredson and Hirschi, however, Moffitt argued that stability of antisocial behavior also is fundamentally affected by *cumulative continuity*. Here she illuminated how, over time, actions of youngsters and the reactions to them essentially trap them in a stable antisocial role, thereby "pruning away the options for change" (p. 684).

In this regard, Moffitt (1993) suggested that cumulative continuity in antisocial behavior is fostered by two considerations. First, LCPs have a "restricted behavioral repertoire" because they are afforded few opportunities to learn prosocial alternatives to misbehavior. For example, due to their conduct problems, they often are rejected by normal peers. When this occurs, however, they have fewer chances to "practice conventional social skills," which in turn makes antisocial ways of acting more probable (p. 684). Second, they become "ensnared by consequences of antisocial behavior" (p. 684). For example, they are at risk of becoming pregnant as teenagers, drug addicted, or school dropouts. Events such as these are "*snares* that diminish the probabilities of later success by eliminating opportunities for breaking the chain of cumulative continuity" (p. 684, emphasis in original).

Finally, those who persist in offending into adulthood experience "a bewildering forest of unsavory outcomes" (Caspi & Moffitt, 1995, p. 497). These problems include "drug and alcohol addiction, unsatisfactory employment, unpaid debts, homelessness, drunk driving, violent assault, multiple and unstable relationships with women, spouse battery, abandoned, neglected, or abused children, and psychiatric illnesses" (p. 497).

ADOLESCENCE-LIMITED ANTISOCIAL BEHAVIOR

In contrast to LCP antisocial behavior that occurs across the life course, most youths mainly "get into trouble" during adolescence. In fact, delinquency is so commonplace that it might be considered a normal part of adolescent development (Moffitt, 1993, p. 689). But then it largely halts. As Caspi and Moffitt (1995) noted, the theoretical challenge is to explain this qualitatively distinct type of development—the onset of delinquency "in early adolescence, recovery by young adulthood, widespread prevalence, and lack of continuity" (p. 499). If LCP behavior is to be explained by starting the search for causes during early childhood, then the "clues" to AL behavior are to be found in the unique features of "adolescent development" (p. 499).

For Moffitt, adolescents confront a fundamental developmental problem. By adolescence, they are biologically mature, capable of sexual behavior, and desirous of assuming adult social roles. Modern society is structured in such a way, however, that youths are expected to refrain from sexual relationships and to wait until their late teens to engage in adult behaviors (e.g., smoking, consumption of goods). They suffer, in short, from a "maturity gap" that is a source of dissatisfaction and a potential motivation for misbehavior.

But how can adolescents resolve this unhappy situation and, in a sense, show their maturity? It is instructive that, similar to several other delinquency theories (e.g.,

Messerschmidt, 1993; Tittle, 1995), Moffitt saw delinquency as functional and as "adaptable behavior": It can be a resource used to alleviate problems that youngsters confront. The main function—and reinforcing quality—of antisocial acts for those facing the maturity gap is that they demonstrate autonomy. Thus, "every curfew broken, car stolen, joint smoked, and baby conceived is a statement of independence"—a way of showing maturity (Caspi & Moffitt, 1995, p. 500). By contrast, according to Caspi and Moffitt (1995), "Algebra does not make a statement about independence; it does not assert that a youth is entitled to be taken seriously. Crime does" (p. 500). Furthermore, the very negative consequences that waywardness evokes are additionally reinforcing. If parents are angry or teachers alienated, then so much the better. Defiance is meant to get under their skins.

In short, adolescent antisocial conduct is motivated by the maturity gap and is reinforcing because of its distinctive qualities and the reactions it evokes. But how do those entering this stage in life come to engage in delinquency in the first place? For Moffitt (1993), antisocial conduct is not simply discovered but rather *learned* through a process of *social mimicry*. In part, delinquency can be modeled from older youths. Moffitt proposed, however, that critical sources of delinquent modeling are the LCPs. During childhood, ALs largely rejected LCPs. Now, during adolescence, LCPs emerge as role models because they appear more mature; they flout rules, drink, smoke, sport tattoos, and are more sexually active. They become "magnets" for ALs and "trainers for new recruits" into delinquency (p. 688).

LCPs, of course, continue to manifest antisocial tendencies, including crime, into adulthood. Some ALs also become "ensnared" in their waywardness by the same kind of events—drug abuse, pregnancy, incarceration, and so on—that foster cumulative continuity and embed LCPs in an antisocial life course (Caspi & Moffitt, 1995, p. 500). The vast majority of ALs, however, does not persist in their offending. But why is their misconduct "adolescence limited"? In large part, discontinuity in offending is possible because these youths avoid cumulative continuity and are not burdened by contemporary continuity. In other words, they avoid the "snares" that trap people in crime, and they do not have the kind of persistent individual differences (e.g., low self-control) that cause antisocial acts at every stage of the life course.

Because ALs are psychologically healthy, as they move into adulthood they experience "waning motivation" and are responsive to "shifting contingencies" (Moffitt, 1993, p. 690). Their antisocial motivations diminish because they are able to move into adult roles—marriage, employment, leisure activities, and so on—that render the maturity gap less relevant. Their contingencies shift because delinquency is perceived to elicit more punishment (e.g., an adult arrest record) and to jeopardize the rewards that can now be accrued from conventional options that are within reach. The result is that ALs largely desist from antisocial acts.

ASSESSING MOFFITT'S THEORY

Moffitt's (1993) perspective is theoretically elegant and is shaping criminological research and thinking in important ways. In all likelihood, however, it will continue to receive at least two types of challenges. First and most obvious, the question arises as to whether offenders can be divided neatly into two—and only two—groups (Moffitt, Caspi, Dickson, Silva, & Stanton, 1996). There is some research, for example, suggesting that there might be two different kinds of LCPs: those who commit criminal acts at a high rate and those whose offending is chronic but at a low rate (D'Unger, Land, McCall, & Nagin, 1998; Nagin, Farrington, & Moffitt, 1995). There might even be ALs who participate in crime at high and low levels (D'Unger et al., 1998). Further, in a study of Dutch offenders from age 12 to age 72, Blokland, Nagin, and Nieuwbeerta (2005) detected four groups: those who offended sporadically, those who offended at low rates before desisting, those who offended at moderate rates before desisting, and those who persisted in offending at high rates. Taken together, these results suggest that a two-group theory might be too parsimonious to capture the full complexity of the ways in which development into and out of crime occurs.

Second, scholars such as Gottfredson and Hirschi (1990) would contend that the search for a typology of offenders is a fool's errand. In their view, any differences in levels of offending reflect not a *qualitative* difference between people but merely a *quantitative* difference in the underlying levels of criminal propensities. That is, LCPs and ALs do not really exist but rather are invented when scholars take a distribution of offenders and draw artificial cutoff points in their data. In reality, individual differences in criminal propensities are more spread across the population, leading some youngsters to deviate a little, some to deviate more, some to deviate even more, and some to deviate a great deal. These propensities do not suddenly emerge during adolescence (as the AL concept suggests) but rather persist for *everyone* from childhood, to adolescence, and into adulthood. Future research will be needed to offer a definitive assessment of this controversy.

As might be anticipated, Moffitt has not remained silent on these issues. In several places, she presents evidence that is largely consistent with the causal processes outlined in her theory (see, e.g., Moffitt, 2006a, 2006b; Moffitt, Caspi, Rutter, & Silva, 2001; Piquero & Moffitt, 2005). Moffitt recognizes that virtually any theory will need to be revised as it is tested—in the case of her paradigm, assessed now for well over a decade. For example, Moffitt (2006b) notes that in addition to the LCPs and ALs, there may be low-level chronic offenders and those who abstain from any antisocial conduct. She also calls for more research on the extent to which adolescence-limited offending can have enduring effects in adulthood, making adjustment problematic. And she admits that as LCPs age deep into adulthood, desistance is likely. Even so, Moffitt asserts that her two-group typology explains much of the variation in the development of offending. In particular, it is clear that "after ten years of research, what can be stated with some certainty is that the hypothesized life-course persistent antisocial individual exists, at least during the first three decades of life" (2006b, pp. 301–302).

Sampson and Laub: Social Bond Theory Revisited

In *Crime in the Making*, Robert Sampson and John Laub (1993) proposed to explain crime over the life course through a *theory of age-graded informal social control*

(Sampson & Laub, 1993; see also Laub, Sampson, & Allen, 2001). In so doing, they made a conscious attempt to revisit and revitalize Hirschi's (1969) original social bond theory. Because of his interest in delinquency, Hirschi examined the impact of social bonds on youngsters. By contrast, Sampson and Laub indicated that social bond theory can help to organize our understanding of continuity and change in offending across the entire life course—from childhood, to adolescence, and into adulthood.

Furthermore, Sampson and Laub (1993) made Hirschi's concept of the social bond more sophisticated by introducing the emerging idea of "social capital," which is the "capital" or resources produced by the quality of relationships between people (e.g., social support, contacts for jobs). As social bonds strengthen, social capital rises. Such capital furnishes resources that can be used to solve problems, but a dependence on this capital also means that much can be jeopardized if it is lost. Social control is thus enhanced by this growing sense, rooted in deepening social relationships, that crime is now too costly to commit.

Although not denying the importance of early individual differences in criminal propensities, Sampson and Laub (1993) used social bond theory to show the continued relevance of sociology to developmental criminology. In contrast to Gottfredson and Hirschi (1990), they argued that offending is marked by both continuity *and* change across time, not just by continuity. Social bonds are not spurious; rather, they exert causal influences on behavior (cf. Sampson & Laub, 1995, to Hirschi & Gottfredson, 1995). In contrast to Moffitt (1993), they denied that continuity characterizes a distinct set of offenders and that change characterizes a second distinct set of offenders. Sampson and Laub maintained that the life course is potentially dynamic, meaning that even LCPs can reestablish bonds during adulthood that can divert them from crime.

Ten years later, in their *Shared Beginnings, Divergent Lives,* Laub and Sampson (2003) revised their theory. Although retaining the core components of their social bond perspective, they expanded their analysis of the process of desistance, suggesting that stopping crime was the result of the convergence of several factors and of "human agency." Below, we first discuss Sampson and Laub's original theory and then explore their more recent elaboration of their perspective. In so doing, we will see that their approach represents a critique of both Gottfredson and Hirschi's self-control theory of continuity in offending and of Moffitt's two-group developmental theory of antisocial conduct.

AN AGE-GRADED THEORY OF INFORMAL SOCIAL CONTROL

Embracing a sociological perspective, Sampson and Laub began their causal model by noting that individuals and social control processes—such as those exerted by the family—do not exist in a vacuum but rather exist within a structural context, itself shaped by larger historical and macro-level forces. They contended that what goes on inside the family, for example, is influenced by "structural background factors" such as poverty, residential mobility, family size, employment, and immigrant status (Sampson & Laub, 1994). Unlike most sociologists, however, they also recognized that the individual differences of children—difficult temperaments, early conduct disorders, and so on—also influence those attempting to deliver social control. There are "child effects" on the social environment. These individual differences can elicit social control—sometimes poorly delivered—but also can potentially be affected or constrained by social control.

During the first stages of life, the most salient "social control process" is found in the family, which is an instrument for both direct controls (monitoring) and indirect controls (attachment). In families where discipline is harsh and erratic and where children and parents reject one another, bonds are weak. Delinquency is the likely result. Note as well that family processes "mediate" the effects of structural factors on youth misconduct; that is, it is through families that structural factors produce their effects on delinquency (Laub et al., 2001; Sampson & Laub, 1994). These processes also mediate, in part, individual differences in criminal propensities that emerge during childhood. Consistent with Gottfredson and Hirschi, however, these individual differences appear to have some independent effects on antisocial conduct across the life course. Finally, beyond the family, juvenile delinquency is fostered by weak school attachments and attachments to delinquent peers.

Sampson and Laub (1993) observed that there is "strong continuity in antisocial behavior running from childhood through adulthood across a variety of life domains" (p. 243). They largely attributed this continuity to what Moffitt (1993) called "cumulative continuity," in which crime and reactions to it are in a cycle that embeds offenders in a trajectory of offending (see also Patterson et al., 1989; Thornberry, 1987). For example, Sampson and Laub showed that delinquency weakens adult social bonds by making stable employment and rewarding marriages less likely, which in turn fosters continued criminality. Incarceration, another likely outcome of persistent criminal involvement, helps to stabilize crime by weakening social bonds. And as chances to establish social bonds are "knifed off," the prospects of escaping crime grow even dimmer.

Importantly, however, Sampson and Laub (1990, 1993) argued that if meaningful social bonds are established during adulthood, they can function as a "turning point" that leads offenders into conformity. In short, if weak social bonds underlie continuity, then strong social bonds underlie *change*. It is noteworthy that Sampson and Laub rejected the idea that chronic offenders who establish bonds during adulthood have some previously hidden individual difference that distinguishes them from chronic offenders who fail to establish bonds. Again, this is the idea that all social bonds—no matter at what stage in life—are a mere manifestation of some individual difference that leads to self-selection into productive relationships (Hirschi & Gottfredson, 1995). But for Sampson and Laub (1995), such thinking is reductionistic. Not all associations of bonds to crime are spurious. People's choices certainly affect their social relationships, but at times life is fortuitous. For example, an individual finds a good job, or an individual stumbles into a healthy intimate relationship. When this occurs, social bonds can develop, social capital can be produced, and constraints previously absent from an offender's life can become controlling.

Finally, from this discussion, it should be clear that Sampson and Laub (1993) stressed "the importance of informal social ties and bonds to society *at all ages across the life course*" (p. 17, emphasis added). However, informal social control is "age graded" in that the salience of different types of bonds varies in the life course (e.g., bonds to parents during childhood, bonds to employers during adulthood).

ASSESSING SAMPSON AND LAUB'S LIFE-COURSE THEORY

In *Unraveling Juvenile Delinquency*, Sheldon and Eleanor Glueck presented the results of their longitudinal study comparing matched samples of 500 delinquent and 500 nondelinquent boys (Glueck & Glueck, 1950), whom the Gluecks eventually traced to 32 years of age. In one of the more remarkable stories of modern criminology, in 1987 Sampson and Laub found the Gluecks' "original case files . . . stored away in a dusty, dark sub-basement of the Harvard Law School library" (Laub et al., 2001, p. 98). These files had to be restored and recoded so that they would be amenable to reanalysis through the use of computers. Sampson and Laub went about this painstaking task but reaped the fruits of their labor by producing one of the richest data sets for studying crime across the life course.

Importantly, Sampson and Laub's theory guided their investigation into the Gluecks' data. Sampson and Laub used social bond theory to "make sense" of the diverse types of information collected by the Gluecks' research team. Their theory led them to develop certain constructs and to develop certain empirical measures. At the same time, Sampson and Laub used these data to "put their theory on the line." It is one thing to set forth a theory; it is another thing to risk its falsification by testing it in a rigorous way. The power of Sampson and Laub's perspective is that their analysis of the Gluecks' data supplied persuasive empirical evidence that their theoretical framework had considerable merit (Sampson & Laub, 1990, 1993, 1994). Perhaps most noteworthy, they showed that family social control mediates the effects of both structural and (to a degree) individual traits on delinquency and that *quality* social bonds during adulthood—stable work, a rewarding marriage, and so on—can divert persistent offenders away from crime (Sampson & Laub, 1990, 1993). Furthermore, the existing research is consistent with their contention that social bonds across the life course are related to criminal behavior (Horney, Osgood, & Marshall, 1995; Laub et al., 2001; Laub, Sampson, & Sweeten, 2006).

There are two potential challenges, however, that Sampson and Laub's theory will have to address. First, by embracing social bond theory, Sampson and Laub attributed the crime-reducing effects of quality family life during childhood and of adult conventional relationships to *social control*. It is equally plausible that other social processes are embedded in or triggered by these conditions that also play a role in turning people away from criminal behavior. Take, for example, marriage and employment. These "social bonds" may foster control, but they also may be contexts that lure individuals away from antisocial peers and into contact with prosocial influences (Warr, 1998; Wright & Cullen, 2004). In addition to social control, then, differential association may account for a person's desistance from offending. Future research will need to explore the extent to which social bonds foster conformity through control, social learning, or other social processes such as social support (Cullen, 1994). Notably, in their updated theory, Sampson and Laub addressed this limitation of their perspective and expanded their discussion of the change process.

Second, Sampson and Laub's perspective may well be locked in a continuing competition with Gottfredson and Hirschi's self-control theory and, more generally, with life-course theories attributing continuity in offending to underlying individual differences. (Recall the discussion in Chapter 6 regarding *self-control and social bonds*.) Again, by showing the salience of social bonds across the life course, Sampson and Laub helped to "rescue" sociology from the charge, rendered by Gottfredson and Hirschi, that social processes outside of childhood are irrelevant to the level of criminality that people manifest. In the end, however, the challenge will be to move beyond the rather simplistic debate over whether continuity is caused by self-control or social bonds. Indeed, the more sophisticated message of Sampson and Laub's theory—one consistent with other life-course theories—is that trajectories in crime are caused not by one type of variable or another but rather by the *intersection* or interlocking nature of individual and social conditions. People may, to a degree, be the architects of their lives, but they also become embedded in life courses that dramatically constrain their current choices and likely futures. Disentangling how this complex process unfolds is likely to occupy criminologists' attention during the years ahead. Indeed, these are issues that have occupied Sampson and Laub's attention in their more current work.

REVISING THE AGE-GRADED THEORY OF CRIME

Shared Beginnings, Divergent Lives was prompted by Laub and Sampson's (2003) decision to extend the Gluecks' data set by studying the 500 males defined as delinquents (i.e., remanded to reformatory schools) in the original data set until they were age 70 (they had been followed until age 32 by the Gluecks). In this research, they examined the criminal records of these 500 offenders and also conducted interviews with 52 of the men. These qualitative data were important because they allowed Laub and Sampson to probe more deeply into why these offenders persisted and, in particular, desisted from crime. This supplied them with an enriched understanding of the process of continuity and change. Further, tracing the sample until old age (or death) allowed them to conduct a true *life-course* study. Most previous longitudinal projects had followed respondents only into early adulthood. As a result, these projects were limited in their ability to explore the nature of offending patterns into middle and later adulthood, and they were unable to study how childhood and adolescent experiences predict criminal conduct across the adult years. As we will see, these methodological limitations have important theoretical implications (Sampson & Laub, 2005).

With data from across the entire life course, Laub and Sampson present two key findings. First, it appears that desistance from crime—even among high-rate offenders—is virtually universal. Unless death intervenes first, everyone eventually stops breaking the law. Second, it is difficult to predict when desistance will occur. Events earlier in life—such as childhood risk factors—do not seem to differentiate the point at which crime is surrendered (Laub & Sampson, 2003; Sampson & Laub, 2005).

These findings are a direct challenge to Moffitt's (1993) views on life-course-persistent offenders. It appears that LCPs do not persist in their offending forever (Moffitt [2006b, pp. 300–301] has responded that the males in the Gluecks' sample continued high-rate offending into middle adulthood, "well beyond the age when most young men in their cohort population desisted"). More consequential, desistance does not seem to be part of a neatly unfolding developmental sequence. Once into adulthood, stopping crime is affected by what occurs at this stage in life and not by what took place much earlier in life. In this scenario, change is not predetermined—childhood is not developmental destiny as Moffitt's theory implies—but rather an emergent event that requires its own theoretical explanation.

Laub and Sampson (2003) identify five aspects to the process of desistance during adulthood. First, similar to their earlier work (Sampson & Laub, 1993), they argue that structural turning points—such as marriage and employment—set the stage for change. Unlike Gottfredson and Hirschi, they do not see entry into these social arrangements as simply due to self-selection (e.g., those with more self-control get married and keep jobs). Rather, these events often are fortuitous, a matter of bumping into the right partner or hearing about a good job from a friend.

Second, again consistent with their earlier theorizing, they contend that these structural events create social bonds that increase the informal controls over offenders. They also recognize, however, that these bonds can be a conduit for social support, which in turn can foster attachments and indirect control (see Hirschi, 1969).

Third, departing from a strict control theory, they observe that as offenders move into marriages and jobs, their daily routine activities change from unstructured and focused on deviant locations (e.g., bars) to structured and filled with prosocial responsibilities. As a result, offenders are cut off from deviant associates and other "bad influences."

Fourth, this changed, prosocial lifestyle creates "desistance by default." Offenders wake up one day and find that their lives have been transformed and that criminal activity has become an increasingly distant reality. Much like Hirschi's (1969) bond of commitment, they now have so much invested in their new way of life that it would make little sense to sacrifice everything for a foolish criminal adventure. In this sense, desistance is not typically a born-again, momentary experience but rather a process of deepening prosocial involvement that gradually, but powerfully, weans offenders away from a criminogenic lifestyle.

Fifth and perhaps most important theoretically, Laub and Sampson (2003) assert that the desistance process they describe constrains, but does not fully determine, the choices offenders make. In an effort to bring offenders back into criminology (not unlike the theorists in Chapter 13), they observe that these individuals have a subjective reality and exercise *human agency*. Human agency is a somewhat vague construct, but it implies that offenders have "will" and are active participants in the journey they are taking—either resisting or voluntarily participating in opportunities to desist from crime.

In Laub and Sampson's conceptualization, human agency is not a trait that is carried around (such as low self-control) but rather an emergent property within situations. Facing the temptation of crime, offenders are not mindless participants pushed or pulled to break the law but rather motivate themselves to resist or take advantage of the criminal opportunity that has been presented. The construct of human agency thus adds motivation to control theory. This motivation, however, is not rooted in human nature (as Hirschi's control theories would contend) or inculcated earlier in life (as strain and social learning theories would contend). Rather, the motivation—the will—that offenders exercise is situational and is created by offenders as they exercise "agency" over their lives.

The concept of human agency may accurately describe the nature of how offenders make choices, but is it scientifically useful? The idea of human agency rings true because it captures the realities of crime that offenders convey during interviews. Talking with offenders—as Laub and Sampson (2003) did—inevitably leads to the conclusion that the men in the Gluecks' sample not only were shaped by their personal and social circumstances but also were the architects of their lives. The difficulty, however, is that the very nature of human agency—a type of socially situated free will—makes it virtually impossible to measure. And unless human agency can be operationalized, scholars cannot test empirically whether this is an important causal factor that explains variation in offending or merely reflects self-serving or distorted memories relayed by old criminals about their past lives in crime.

Table 15.1 summarizes the life-course theories discussed to this point. As noted, these perspectives can be understood as providing distinct insights into the issue of continuity and change in offending. Table 15.1 thus is organized to display how prominent life-course theories address these two core dimensions of antisocial and criminal conduct.

Rethinking Crime: Cognitive Theories of Desistance

Whatever the ultimate value of the concept of human agency, Laub and Sampson (2003) are correct in asserting that life-course researchers need to pay more attention to the foreground of crime and, in particular, to what offenders *think about their lives as criminals*. Other perspectives have focused on offenders' thinking and how they decide whether to commit a particular criminal act (e.g., rational choice theory). In the case of life-course theory, however, the issue is not why one offense was chosen but why individuals are, or are not, trapped on a pathway to crime. More similar to labeling theory (see Chapter 7), the theoretical task is to explain how the way offenders conceptualize their identities and life circumstances facilitates either their continued criminal involvement or their desistance from a life in crime. Fortunately, scholars have begun to take up this challenge. Below, we review two prominent attempts to uncover the way "rethinking crime" can allow individuals to desist from persistent offending: Maruna's (2001) *theory of redemption scripts* and Giordano et al.'s (2002) *theory of cognitive transformation*.

Table 15.1   Continuity and Change in Life-Course Theories

Authors and Theory	Continuity	Change
Gottfredson and Hirschi: Low Self-Control Theory—Continuity Only	Yes—people continue to offend because low self-control remains stable.	No—individual differences remain constant across life.
Patterson: Social-Interactional Developmental Model—Continuity or Change	Early-onset group only shows continuity in offending.	Late-onset group only shows change in offending.
Moffitt: Developmental Taxonomy of LCPs and ALs—Continuity or Change	LCPs only show continuity in offending across life.	ALs only show change; offending limited to adolescence.
Sampson and Laub: Age-Graded Social Bond Theory—Continuity and Change	Yes—continuity occurs due to individual propensities and weak social bonds, which combine to create cumulative continuity.	Yes—acquiring quality social bonds can enable an offender to change and thus to desist from crime; human agency also makes change or desistance possible.

MARUNA'S THEORY OF REDEMPTION SCRIPTS

In an interview study of 65 offenders and ex-offenders in Liverpool, England, Shaad Maruna (2001) noted that these individuals all faced dismal futures and were marked by multiple risk factors, including "poverty, child abuse, detachment from the labor force, the stigma of social sanctions, low educational attainment, few legitimate opportunities in the community, serious addictions and dependencies, high-risk personality profiles, and, of course, long-term patterns of criminal behavior" (p. 55). It is clear that this combination of "criminogenic traits," "criminogenic backgrounds," and "criminogenic environments" had led all these people into persistent criminal careers (recall Moffitt's LCPs). Still, some of these offenders had desisted from crime whereas others had not. Maruna discerned that these two groups could not be clearly differentiated by their social, psychological, and criminal histories. Something else seemed to separate the desisters from the persistent offenders.

To find the solution to this puzzle, Maruna argued that it was necessary to move from the "background" to the "phenomenological foreground" of crime (see Katz, 1988). In his detailed analysis of interview data (see also Maruna & Copes, 2005, pp. 277–281), he discovered that desisters and persisters differed markedly in their cognitive understanding of their lives in crime. All of us—not just lawbreakers—seek meaning by having a "life story" or "narrative" that allows us to "make sense" of who we are, what we do, and where we are headed. According to Maruna (2001), the two groups of offenders varied in the stories or "scripts" they used to explain their long-term criminality. Notably, the desisters were able to embrace a life narrative that allowed them to "make good"—"to find reason and purpose in the bleakest of life histories" (pp. 9–10). As we will see shortly, these narratives are called "redemption scripts."

Those seemingly trapped in crime manifested very different views of their lives. According to Maruna (2001), the persistent offenders described themselves as "doomed to deviance" (p. 74). They conceptualized their fate by conveying a "condemnation script" (p. 73) in which they were "condemned" to a life in crime by circumstances beyond their control. Although lacking "any sort of enthusiastic commitment to crime," they "said that they felt powerless to change their behavior because of drug dependency, poverty, a lack of education or skills, or social prejudice. They do not want to offend, they said, but feel that they have no choice" (pp. 74–75). In many ways, their diagnosis of their existing situation was realistic, which might make them "'sadder but wiser' than their contemporaries who struggle to desist" (p. 84).

In contrast, offenders who had desisted—who were "making good"—adopted the "rhetoric of redemption" (p. 85). As long-term offenders, they face the dual challenge of overcoming their dismal circumstances and of accounting for "around a decade" of "selling drugs, stealing cars, and sitting in prison" (p. 85). They accomplish this challenge through a "redemption script" that distorts the "grim realities of their past lives" as surmountable and that redefines who they are and what they have experienced (Maruna & Copes, 2005, p. 280).

Essential to this story is the assertion that their previous criminality was not part of the "real me" or who they really are "deep down" (Maruna, 2001, pp. 88–89). Crime thus was circumstantial and not a reflection of their true nature, which had now surfaced and allowed them to desist from lawbreaking. The redemption script also includes the idea that their past woes have the silver lining of making them a stronger, better person. Indeed, they are now able to serve higher purposes, such as helping juveniles in trouble. They have come to feel that they are in control of their destiny, that they have the self-efficacy to turn in a different direction and stay the course. In the end, this redemption script provides ex-offenders with a "coherent, prosocial identity" that enables them to resist crime and "make good" in society (p. 7).

As might be apparent, Maruna's theory, unlike Sampson and Laub's, does not attribute desistance to structural turning points or to situational human agency (see, however, Maruna & Roy, 2007). For Maruna, offenders who desist experience a fundamental, qualitative cognitive transformation that sustains them in the face of often dire circumstances (cf. Laub & Sampson, 2003, pp. 278–279). As he understands, it still remains to be seen whether redemption scripts precipitate a movement away from crime or largely emerge afterwards and perhaps serve to reinforce a nascent prosocial experiment (Maruna & Copes, 2005, p. 281). Regardless, future life-course researchers will profit from more systematic attention to how offenders and ex-offenders construct their identities and envision their lives.

GIORDANO ET AL.'S THEORY OF COGNITIVE TRANSFORMATION

In their attempt to explain adult desistance from crime, Peggy Giordano, Stephen Cernkovich, and Jennifer Rudolph (2002) anticipated Laub and Sampson's (2003) revised life-course theory. Similar to Laub and Sampson, they criticized traditional social bond theory for being too deterministic. Social bond theory had implicitly assumed that exposure to bonds (e.g., attachment to parents, commitment to school) meant that individuals would automatically be subjected to informal social control and reduce their offending. In Sampson and Laub's (1993) original statement of their perspective, social bonds (e.g., marriage, employment) were seen to function in this way. They were structural "turning points": If an offender was fortunate enough to find a good marital partner, control would be forthcoming and desistance would occur. As we have seen, however, Laub and Sampson (2003) subsequently argued that this scenario leaves out the role of "human agency" in the change process. In a like vein, Giordano et al. (2002) assert that offenders—and especially adult offenders—are "intentional" and "reflective." Accordingly, they "theorize a more reciprocal relationship between actor and environment and reserve a central place for agency in the change process" (p. 999).

Their theory, however, departs from Laub and Sampson's revised perspective in two ways. First, rather than conceptualize marriage and employment as "turning points," Giordano et al. (2002) use the construct of "hooks for change." These events create the possibility, not the inevitability, of change and desistance. In their view, offenders must take advantage of—that is, be "hooked by"—these opportunities to redirect their life-course. This process involves not only the attractiveness of the opportunity but also the offender's agency. As they note, "we want to emphasize the actor's own role in latching onto opportunities presented by the broader environment" (p. 1000). Second, in contrast to Laub and Sampson's amorphous concept of "human agency," Giordano et al. see agency as manifesting itself in a "cognitive transformation" that involves four "cognitive shifts." We should note that by identifying these specific features, Giordano et al. make their theory of desistance amenable to empirical assessment.

For change and thus desistance to transpire, the offender must first develop a "general cognitive openness to change" (Giordano et al., 2002, p. 1000). Once this more global cognitive shift has taken place, offenders must interpret the specific potential hooks for change that they encounter—a budding intimate relationship, a job offer— "as a positive development" (p. 1001). They must "connect" with the opportunity. This second shift also involves defining the "new state of affairs as fundamentally incompatible with continued deviation" (p. 1001).

At this point, Giordano et al. describe a third cognitive shift that overlaps with Maruna's (2001) concept of "redemption scripts" described above. According to Giordano et al., this cognitive transformation involves attempts by offenders "to envision and begin to fashion an appealing and conventional 'replacement self' that can supplant the marginal one that must be left behind" (p. 1001). The key part of this transformation is the development of a new identity that provides direction and acts as a "filter for decision making" (e.g., will this act be consistent with "who I am"?). Notably, hooks for change—again, such as being in a quality marriage—can reinforce this fresh identity and, as Sampson and Laub (1993) observed, provide informal social control. Still, the offender's restructured identity plays an independent and important role in the desistance process. This is particularly the case as offenders encounter "novel situations" on their own or perhaps experience a loss of the original hook for change (e.g., a relationship breaks up, a job is lost; pp. 1001–1002).

Finally, the fourth type of cognitive transformation occurs when offenders reinterpret their past deviant behavior and wayward lifestyle. "The desistance process," observe Giordano et al. (2002), "is relatively complete when the actor no longer sees these same behaviors as positive, viable, or even personally relevant" (p. 1002). This revised worldview now provides an impetus for continued conformity. Further, the transformation is theoretically important because it suggests that the offender's attitudes and thus behavioral motivation have been shifted from antisocial to prosocial. This perspective is more consistent with social learning theory and is inconsistent "with a control position, where motivation is viewed as a relative constant" (p. 1002).

In closing, differences aside, the work of Giordano et al. on cognitive transformation, Maruna on redemption scripts, and Laub and Sampson on human agency converge in showing that offenders are active participants in the paths they take in the life course. This observation is not meant to depreciate the importance of the structural and personal risk factors that initially placed them on—and, for years, have kept them on—a criminal pathway. Still, it appears that exiting a criminal life-course is not simply a natural developmental process or a matter of fortuitously bumping into a good partner or job opportunity. Instead, offenders make choices that allow hooks for change to be exploited or, in other cases, forfeited. The challenge for future theory and research is to demarcate further the meaning of human agency and the precise ways in which offenders rethink their lives and desist from crime.

The Consequences of Theory: Policy Implications

The focus of life-course or developmental theories on childhood as the incubator of serious persistent offending has a direct policy implication: If the roots of crime are in the first years of life, then the most effective way of preventing crime is through *early intervention programs*. These programs tend to concentrate on three areas: parent training, improving the cognitive development of children, and reversing early manifestations of conduct problems (Farrington, 1994a; Tremblay & Craig, 1995). These programs can be delivered during the prenatal period and infancy up through the school years (Farrington & Welsh, 2007; Greenwood, 2006; Loeber & Farrington, 1998). They also can include interventions in criminal justice settings (Cullen & Gendreau, 2000; Lipsey, 2009; Lipsey & Cullen, 2007). The goal of early intervention is to decrease the risk factors for the onset of offending and to increase the protective factors that foster resilience in the face of these criminogenic risks (Tremblay & Craig, 1995). When an antisocial life course is initiated, the task is to intervene so as to divert a youngster from this trajectory.

To provide one example, David Olds and his colleagues developed the "Prenatal and Early Childhood Nurse Home Visitation Program" in Elmira, New York, and then later in Memphis, Tennessee (Olds, 2007; Olds, Hill, & Rumsey, 1998). The program focused on first-time, low-income mothers who were at risk of having antisocial children. During pregnancy and the first 2 years after a child's birth, socially skilled nurses would visit with the mothers every week or two. The nurses focused on mitigating three potential risk factors. First, to reduce neuropsychological deficits associated with antisocial conduct (see Moffitt, 1993), they worked with women during pregnancy to ensure proper prenatal care and to avoid behaviors that would compromise fetal development (e.g., smoking, drug use). Second, to reduce child abuse and neglect, the nurses helped the mothers to learn effective parenting skills. Third, to reduce problems caused by maternal deviance, they assisted mothers in efforts to complete schooling, obtain employment, and avoid unplanned pregnancies. Based on a random experimental study, the data revealed that nurse visitation limited the targeted risk factors and, as hypothesized, lowered subsequent misconduct. As Olds et al. (1998) conclude:

> As described in a 15-year followup of the Elmira nurse visitation program, the long-term effects of the program on children's criminal and antisocial behavior are substantial and have groundbreaking implications for juvenile justice and delinquency prevention. Adolescents whose mothers received nurse home visitation services over a decade earlier were 60 percent less likely than adolescents whose mothers had not received a nurse home visitor to have run away, 55 percent less likely to have been arrested, and 80 percent less likely to have been convicted of a crime, including a violation of probation. (p. 159)

The early intervention strategy for crime prevention is "liberal" politically. Unlike the conservatives' approach, it does not advocate "get tough" policies that leave the underlying causes of persistent antisocial conduct untouched. Indeed, if crime is part of a developmental process, then it is inconsistent to imagine that simple rational choice or deterrence models could explain how a criminal life course is initiated during childhood when cognitive maturity is limited. Life-course theories might supply some credence to incapacitation, given the presence of LCPs and the prediction of continuity in offending over long periods of time. In general, however, theorists in this paradigm eschew prisons, arguing that incarceration may, if anything, only further ensnare offenders in a criminal trajectory (see, e.g., Laub, Sampson, Corbett, & Smith, 1995). But this approach also is not "critical" or leftist. The critique of the existing order—an inequitable political economy—is implicit and not elevated to the point of directing policy. Again, there is a call for programs, not for the redistribution of wealth and power.

The thinking of life-course theorists might prove to be consistent with the times. The political struggle between the Left and the Right will continue, but the nation seems ideologically "in the middle." There seems to be little interest in grand revolutions as there once was during the 1960s (for social equality) or during the 1980s (for conservative values). With William Bennett and Hillary Rodham Clinton both speaking out on behalf of children, saving troubled kids is politically viable (Bennett et al., 1996; Clinton, 1996). In fact, public support for early intervention programs, including juvenile rehabilitation, is extensive (Cullen, Fisher, & Applegate, 2000; Cullen, Vose, Jonson, & Unnever, 2007). There are two other considerations that might make this policy agenda feasible. First, there is growing evidence that early intervention programs "work" to prevent or reduce future offending (Farrington, 1994a; Farrington & Welsh, 2007; Henggeler, 1997; Karoly et al., 1998; Laub et al., 1995; Loeber & Farrington, 1998; Tremblay & Craig, 1995). Second, although more complex to calculate, there also seems to be evidence that early prevention programs tend be cost-effective (Aos, Phipps, Barnoski, & Lieb, 2001; Drake, Aos, & Miller, 2009; Farrington & Welsh, 2007; Karoly et al., 1998; Welsh, 2003; see also Welsh, Farrington, & Sherman, 2001).

Finally, beyond early intervention, life-course research would suggest the need to consider strategies for those who are not "saved" as youngsters but persist in their offending into adulthood. Again, it is tempting to recommend a policy of long-term incarceration for hardened, high-rate criminals. But as Laub and Sampson (2003) note, this approach has unanticipated consequences because imprisonment may increase recidivism by weakening social bonds to core institutions. "Cumulative disadvantage," they observe, "posits that arrest and especially incarceration may spark failure in school, unemployment, and weak community bonds, in turn increasing adult crime" (2003, p. 291). For those whose life course leads them into prison, a special challenge is thus to develop treatment and reentry programs that diminish the attenuation of social bonds (2003, p. 292; see also Cullen & Gendreau, 2000; MacKenzie, 2006; Petersilia, 2003; Travis, 2005).

Conclusion

At this stage in its development, criminology might be said to suffer from an embarrassment of riches. The field is filled with an array of competing theoretical paradigms. Part of this richness in theorizing stems from attempts to revitalize old models in new ways, to integrate traditional approaches into fresh perspectives, and to elaborate ideas that heretofore were underdeveloped within an existing perspective. Part of this richness reflects the efforts of scholars coming to the discipline with different ideologies and with different scholarly training (e.g., economics, psychology). Part of it manifests truly fresh ideas and ways of illuminating the world that redirect theoretical inquiry and empirical investigation (e.g., life-course criminology).

It remains to be seen which theory or theories will dominate criminology during the first decades of the 21st century. During past eras, the hegemony of certain paradigms was clearer because the shifts in the social context and its effect on the criminological imagination were more dramatic. Even so, we would do well to keep two considerations in mind. First, allegiances to contemporary theories will not be a random event; rather, they will continue to be shaped by people's social experiences and corresponding views of the world. Debates over the causes of crime, whether among criminologists or among elected officials, will occur on an intellectual level but will be shaped fundamentally by hidden—or not so hidden—political ideologies. Second, during the time ahead, rapid social changes could coalesce in such a way as to once again nourish certain theories more than others. The prevailing social context, which we take for granted today, inevitably will change and alter how we think—and what we do—about crime.

These observations remind us of the central point of this book: Ideas about crime—or what we call theories—are a product of society that develop in a particular context and then have their consequences for social policy. We hope that this theme has served as a useful framework for understanding criminological theory's stages of development and its impact on the criminal justice system and related prevention strategies. We also hope, however, that our observations on the intimate link among context, theory, and policy will have a personal relevance to the reader. A look to the past illuminates the risk of taking for granted one's existing social reality. Whether criminologists or simply interested citizens, we are immersed in a social context that shapes our beliefs about crime and its control. We cannot fully escape this context, but as reflexive creatures we can explore our biases, think more clearly about crime, and embrace policies less contaminated by our prejudices.

References

Abbott, J. H. (1981). *In the belly of the beast: Letters from prison.* New York: Random House.

Adler, F. (1975). *Sisters in crime: The rise of the new female criminal.* New York: McGraw-Hill.

Adler, F., & Laufer, W. S. (Eds.). (1995). *The legacy of anomie theory* (Advances in Criminological Theory, Vol. 6). New Brunswick, NJ: Transaction.

Adler, J. (2006, March 27). Freud in our midst. *Newsweek,* pp. 43–49.

Agnew, R. (1992). Foundation for a general strain theory of crime and delinquency. *Criminology, 30,* 47–87.

Agnew, R. (2001a). Building on the foundation of general strain theory: Specifying the types of strain most likely to lead to crime and delinquency. *Journal of Research in Crime and Delinquency, 36,* 123–155.

Agnew, R. (2001b). *Juvenile delinquency: Causes and control* (2nd ed.). New York: Oxford University Press.

Agnew, R. (2006a). General strain theory: Current status and directions for further research. In F. T. Cullen, J. P. Wright, & K. R. Blevins (Eds.), *Taking stock: The status of criminological theory* (Advances in Criminological Theory, Vol. 15, pp. 101–123). New Brunswick, NJ: Transaction.

Agnew, R. (2006b). *Pressured into crime: An overview of general strain theory.* Los Angeles: Roxbury.

Agnew, R. (2011). Revitalizing Merton: General strain theory. In F. T. Cullen, C. L. Jonson, A. J. Myer, & F. Adler (Eds.), *The origins of American criminology* (Advances in Criminological Theory, Vol. 16, pp. 137–158). New Brunswick, NJ: Transaction.

Agnew, R., Cullen, F. T., Burton, V. S., Jr., Evans, T. D., & Dunaway, R. G. (1996). A new test of classic strain theory. *Justice Quarterly, 13,* 681–704.

Agnew, R., Piquero, N. L., & Cullen, F. T. (2009). General strain theory and white-collar crime. In S. S. Simpson & D. Weisburd (Eds.), *The criminology of white-collar crime* (pp. 35–60). New York: Springer.

Agnew, R., & White, H. R. (1992). An empirical test of general strain theory. *Criminology, 30,* 475–499.

Ahmed, E., & Braithwaite, V. (2004). "What, me ashamed?" Shame management and school bullying. *Journal of Research in Crime and Delinquency, 41,* 269–294.

Aichhorn, A. (1936). *Wayward youth.* New York: Viking.

Akers, R. L. (1977). *Deviant behavior: A social learning approach* (2nd ed.). Belmont, CA: Wadsworth.

Akers, R. L. (1994). *Criminological theories: Introduction and evaluation.* Los Angeles: Roxbury.

Akers, R. L. (1998). *Social learning and social structure: A general theory of crime and deviance.* Boston: Northeastern University Press.

Akers, R. L. (1999). Social learning and social structure: Reply to Sampson, Morash, and Krohn. *Theoretical Criminology, 3,* 477–493.

Akers, R. L. (2000). *Criminological theories: Introduction, evaluation, and application* (3rd ed.). Los Angeles: Roxbury.

Akers, R. L. (2011). The origins of me and social learning theory: Personal and professional recollections and reflections. In F. T. Cullen, C. L. Jonson, A. J. Myer, & F. Adler (Eds.), *The origins of American criminology* (Advances in Criminological Theory, Vol. 16, pp. 347–366). New Brunswick, NJ: Transaction.

Akers, R. L., & Jensen, G. F. (Eds.). (2003). *Social learning theory and the explanation of crime: A guide for a new century* (Advances in Criminological Theory, Vol. 11). New Brunswick, NJ: Transaction.

Akers, R. L., & Jensen, G. F. (2006). The empirical status of social learning theory of crime and deviance. In F. T. Cullen, J. P. Wright, & K. R. Blevins (Eds.), *Taking stock: The status of criminological theory* (Advances in Criminological Theory, Vol. 15, pp. 37–76). New Brunswick, NJ: Transaction.

Akers, R. L., Krohn, M. D., Lanza-Kaduce, L., & Radosevich, M. (1979). Social learning and deviant behavior: A specific test of general theory. *American Sociological Review, 44,* 636–655.

Akers, R. L., & Sellers, C. S. (2004). *Criminological theories: Introduction, evaluation, and application* (4th ed.). Los Angeles: Roxbury.

Alarid, L. F., Burton, V. S., Jr., & Cullen, F. T. (2000). Gender and crime among felony offenders: Assessing the generality of social control and differential association theories. *Journal of Research in Crime and Delinquency, 37,* 171–199.

Alexander, F., & Healy, W. (1935). *Roots of crime.* New York: Knopf.

Allen, F. A. (1973). Raffaele Garofalo. In H. Mannheim (Ed.), *Pioneers in criminology* (2nd ed., pp. 318–340). Montclair, NJ: Patterson Smith.

Allen, H. E., Latessa, E. J., Ponder, B., & Simonson, C. E. (2007). *Corrections in America: An introduction* (11th ed.). Upper Saddle River, NJ: Prentice Hall.

Allen, M. (1977). James E. Carter and the trilateral commission: A southern strategy. *Black Scholar, 8,* 27.

Alvi, S. (2005). Left realism. In R. A. Wright & J. M. Miller (Eds.), *Encyclopedia of criminology* (Vol. 2, pp. 931–933). New York: Routledge.

American Society of Criminology. (2009). *Criminology and criminal justice policy: Program for the 61st annual meeting of the American Society of Criminology.* Columbus, OH: Author.

Anderson, E. (1999). *Code of the street: Decency, violence, and the moral life of the inner city.* New York: W. W. Norton.

Anderson, K. (1991). Radical criminology and the overcoming of alienation: Perspectives from Marxian and Gandhian humanism. In H. E. Pepinsky & R. Quinney (Eds.), *Criminology as peacemaking* (pp. 14–30). Bloomington: Indiana University Press.

Andrews, D. A. (1980). Some experimental investigations of the principles of differential association through deliberate manipulations of the structure of service systems. *American Sociological Review, 45,* 448–462.

Andrews, D. A., & Bonta, J. (1998). *The psychology of criminal conduct* (2nd ed.). Cincinnati, OH: Anderson.

Andrews, D. A., & Bonta, J. (2003). *The psychology of criminal conduct* (3rd ed.). Cincinnati, OH: Anderson.

Andrews, D. A., & Bonta, J. (2006). *The psychology of criminal conduct* (4th ed.). Cincinnati, OH: Anderson.

Andrews, D. A., & Bonta, J. (2010). Rehabilitative criminal justice policy and practice. *Psychology, Public Policy, and Law, 16,* 39–55.

Anti-Semitic acts were up 12 percent in 1987. (1987, January 27). *New York Times,* p. A13.

Aos, S., Phipps, P., Barnoski, R., & Lieb, R. (2001). The comparative costs and benefits of programs to reduce crime: A review of research findings with implications for Washington state. In B. C. Welsh, D. P. Farrington, & L. W. Sherman (Eds.), *Cost and benefits of preventing crime* (pp. 149–175). Boulder, CO: Westview.

Archibold, R. C. (2010, March 23). California, in financial crisis, opens prison doors. *New York Times.* Retrieved March 24, 2010, from http://www.nytimes.com/2010/03/24/us/24calprisons.html?ref=us

Argentining of America. (1988, February 22). *Newsweek,* pp. 22–45.

Arneklev, B. J., Grasmick, H. G.., Tittle, C. R., & Bursick, R. J., Jr. (1993). Low self-control and imprudent behavior. *Journal of Quantitative Criminology, 9,* 225–247.

Aronowitz, S. (1973). *False promises: The shaping of American working class consciousness.* New York: McGraw-Hill.

Arrigo, B. A. (2003). Postmodern justice and critical criminology: Positional, relational, and provisional science. In M. D. Schwartz & S. E. Hatty (Eds.), *Controversies in critical criminology* (pp. 43–55). Cincinnati, OH: Anderson.

Arrigo, B. A., & Bersot, H. Y. (2010). Postmodern theory. In F. T. Cullen & P. Wilcox (Eds.),

Encyclopedia of criminological theory (pp. 728–732). Thousand Oaks, CA: Sage.

Arrigo, B. A., & Milovanovic, D. (2009). *Revolution in penology: Rethinking the society of captives.* Lanham, MD: Rowman & Littlefield.

Arrigo, B. A., & Williams, C. R. (Eds.). (2006). *Philosophy, crime and criminology.* Urbana and Chicago: University of Illinois Press.

Aseltine, R. H., Jr., Gore, S., & Gordon, J. (2000). Life stress, anger and anxiety, and delinquency: An empirical test of general strain theory. *Journal of Health and Social Behavior, 41,* 256–275.

Asendorpf, J. B., & Valsiner, J. (1992). *Stability and change in development: A study of methodological reasoning.* Newbury Park, CA: Sage.

Ashley-Montagu, M. F. (1984). "Crime and the anthropologist": Forty-three years later. In R. W. Rieber (Ed.), *Advances in forensic psychology* (Vol. 1, pp. 174–175). Norwood, NJ: Ablex.

Assault on the welfare state. (1993, November 21). *Observer,* p. 1 (London).

August, G. J., Realmuto, G. M., & MacDonald, A. W. (1996). Prevalence of ADHD and comorbid disorders among elementary school children screened for disruptive behavior. *Journal of Abnormal Child Psychiatry, 24,* 571–595.

Aulette, J. R., & Michalowski, R. (1993). Fire in Hamlet: A case study of a state-corporate crime. In K. D. Tunnell (Ed.), *Political crime in contemporary America* (pp. 171–206). New York: Garland.

Ball, R. A. (1966). An empirical exploration of the neutralization hypothesis. *Criminologica, 4,* 22–32.

Ball, R. A. (1978a). The dialectical method: Its application to social theory. *Social Forces, 57,* 785–798.

Ball, R. A. (1978b). Sociology and general systems theory. *The American Sociologist, 13,* 65–72.

Ball, R. A. (1979). Toward a dialectical criminology. In M. D. Krohn & R. L. Akers (Eds.), *Crime, law, and sanctions* (pp. 11–26). Beverly Hills, CA: Sage.

Ball, R. A. (in press). Biological and biosocial theory. In C. D. Bryant (Ed.), *The handbook of deviant behavior.* New York: Routledge.

Ball, R. A., Huff, C. R., & Lilly, J. R. (1988). *House arrest and correctional policy: Doing time at home.* Newbury Park, CA: Sage.

Ball, R. A., & Lilly, J. R. (1982). The menace of margarine: The rise and fall of a social problem. *Social Problems, 29,* 488–498.

Ball, R. A., & Lilly, J. R. (1985). Home incarceration: An international alternative to institutional incarceration. *International Journal of Comparative and Applied Criminal Justice, 9,* 85–97.

Barak, G. (Ed.). (1994). *Media, process, and the social construction of crime: Studies in newsmaking criminology.* New York: Garland.

Barak, G. (1998). *Integrating criminologies.* Boston: Allyn & Bacon.

Baran, P. A., & Sweezy, P. M. (1966). *Monopoly capitalism: An essay on the American economic and social order.* New York: Monthly Review Press.

Barkley, R. A., et al. (2002). International consensus statement on ADHD. *Clinical Child and Family Psychology Review, 5,* 89–111.

Barkley, R. A., Fischer, M., Smallish, L., & Fletcher, K. (2004). Young adult follow-up of hyperactive children: Antisocial activities and drug use. *Journal of Child Psychology and Psychiatry, 45,* 195–211.

Barnes, H. E. (1930). Criminology. In E. R. A. Seligmann (Ed.), *Encyclopedia of the social sciences* (Vol. 4). New York: Macmillan.

Baron, S. W. (2003). Self-control, social consequences, and criminal behavior: Street youth and the general theory of crime. *Journal of Research on Crime and Delinquency, 40,* 405–425.

Baron, S. W. (2009). Differential coercion, street youth, and violent crime. *Criminology, 47,* 239–268.

Baron, S. W., & Forde, D. R. (2007). Street youth crime: A test of control balance theory. *Justice Quarterly, 24,* 335–355.

Barr, R., & Pease, K. (1990). Crime placement, displacement, and deflection. In M. Tonry & N. Morris (Eds.), *Crime and justice: A review of research* (Vol. 12, pp. 277–318). Chicago: University of Chicago Press.

Bartol, C. R. (1995). *Criminal behavior: A psychosocial approach* (5th ed.). Englewood Cliffs, NJ: Prentice Hall.

Bartollas, C. (1985). *Juvenile delinquency.* New York: John Wiley.

Basran, G. S., Gill, C., & MacLean, B. D. (1995). *Farmworkers and their children.* Vancouver, BC: Collective Press.

Bauman, A., Gehring, K., & Van Voorhis, P. (2009). Cognitive behavior programming for women and girls. In B. Glick (Ed.), *Cognitive behavioral interventions for at-risk youth* (chpt. 16, pp. 1–30). Kingston, NJ: Civic Research Institute.

Baumer, E. P., & Gustafson, R. (2007). Social organization and instrumental crime: Assessing the empirical validity of classic and contemporary anomie theories. *Criminology, 45,* 617–664.

Bayer, R. (1981). Crime, punishment, and the decline of liberal optimism. *Crime & Delinquency, 27,* 169–190.

Bazemore, G. (1985). Delinquency reform and the labeling perspective. *Criminal Justice and Behavior, 12,* 131–169.

Bazemore, G., & Walgrave, L. (Eds.). (1999). *Restorative juvenile justice: Repairing the harm of youth crime.* Monsey, NY: Willow Tree Press.

Beaver, K. M. (2009). Molecular genetics and crime. In A. Walsh & K. M. Beaver (Eds.), *Biosocial criminology: New directions in theory and research* (pp. 50–72). New York: Routledge.

Beaver, K. M., Wright, J. P., & DeLisi, M. (2008). Delinquent peer group formation: Evidence of a gene × environment correlation. *Journal of Genetic Psychology, 169,* 227–244.

Beaver, K. M., Wright, J. P., DeLisi, M., Daigle, L. E., Swatt, M. L., & Gibson, C. L. (2007). Evidence of a gene × environment interaction in the creation of victimization: Results from a longitudinal study of adolescents. *International Journal of Offender Therapy and Comparative Criminology, 51,* 620–645.

Beaver, K., Wright, J., & Walsh, A. (2008). A gene-based evolutionary explanation for the association between criminal involvement and number of sex partners. *Biodemography and Social Biology, 54,* 47–55.

Beccaria, C. (1963). *On crimes and punishments* (H. Paolucci, Trans.). Indianapolis, IN: Bobbs-Merrill. (Original work published 1764)

Becker, H. S. (1963). *Outsiders: Studies in the sociology of deviance.* New York: Free Press.

Beckwith, J. (1985). Social and political uses of genetics in the United States: Past and present. In F. H. Marsh & J. Katz (Eds.), *Biology, crime, and ethics: A study of biological explanations for criminal behavior* (pp. 316–326). Cincinnati, OH: Anderson.

"Behind bars" amid bad times. (2009). National Public Radio. Retrieved August 14, 2010, from http://www.npr.org/templates/story/story.php?storyId=106550398

Bell, L. (1998). The victimization and revictimization of female offenders. *Corrections Today, 60,* 106–122.

Bellah, R. N., Madsen, R., Sullivan, W. M., Swidler, A., & Tipton, S. M. (1985). *Habits of the heart: Individualism and commitment in American life.* Berkeley: University of California Press.

Bellah, R. N., Madsen, R., Sullivan, W. M., Swidler, A., & Tipton, S. M. (1991). *The good society.* New York: Knopf.

Belsky, J. (1980). Child maltreatment: An ecological integration. *American Psychologist, 35,* 320–335.

Bennett, G. (1987). *Crimewarps: The future of crime in America.* Garden City, NY: Anchor/Doubleday.

Bennett, T. (1998). Crime prevention. In M. Tonry (Ed.), *The handbook of crime and punishment* (pp. 369–402). New York: Oxford University Press.

Bennett, W. J. (1994). *The index of leading cultural indicators: Facts and figures on the state of American society.* New York: Touchstone.

Bennett, W. J., DiIulio, J. J., Jr., & Walters, J. P. (1996). *Body count: Moral poverty and how to win America's war against crime and drugs.* New York: Simon & Schuster.

Benson, M. L. (1985). Denying the guilty mind: Accounting for involvement in a white-collar crime. *Criminology, 23,* 583–608.

Benson, M. L., & Cullen, F. T. (1998). *Combating corporate crime: Local prosecutors at work.* Boston: Northeastern University Press.

Benson, M. L., & Moore, E. (1992). Are white-collar offenders and common criminals the same? An empirical and theoretical critique of a recently proposed general theory of crime. *Journal of Research in Crime and Delinquency, 29,* 251–272.

Benson, M. L., & Simpson, S. S. (2009). *White-collar crime: An opportunity perspective.* New York: Routledge.

Bentham, J. (1948). *An introduction to the principles of morals and legislation* (L. J. Lafleur, Ed.). New York: Hafner.

Berger, P. L., & Luckmann, T. (1966). *The social construction of reality.* Garden City, NY: Anchor.

Bernard, M. A. (1969). Self-image and delinquency: A contribution to the study of female criminality and women's image. *Acta Criminologica: Etudes sur la Conduiter Antisociale, 2,* 71–144.

Bernard, T. J. (1984). Control criticisms of strain theories: An assessment of theoretical and empirical adequacy. *Journal of Research in Crime and Delinquency, 21,* 353–372.

Bernburg, J. G., & Krohn, M. D. (2003). Labeling, life chances, and adult crime: The direct and indirect effects of official intervention in adolescence on crime in early adulthood. *Criminology, 41,* 1287–1318.

Bernburg, J. G., Krohn, M. D., & Rivera, C. J. (2006). Official labeling, criminal embeddedness, and subsequent delinquency: A longitudinal test of labeling theory. *Journal of Research in Crime and Delinquency, 43,* 67–88.

Bersani, B. E., Laub, J. H., & Nieuwbeerta, P. (2009). Marriage and desistance from crime in the Netherlands: Do gender and socio-historical context matter? *Journal of Quantitative Criminology, 25,* 3–24.

The Big Apple. (2010, January 18). Muckraker. Retrieved April 10, 2010, from http://www.barrypopik.com/index.php/new_york_city/entry/muckraker?/

Binder, A., & Geis, G. (1984). Ad populum argumentation in criminology: Juvenile diversion as rhetoric. *Crime & Delinquency, 30,* 309–333.

Bird, A. (2007). Perceptions of epigenetics. *Nature, 447,* 396–398.

Blackwell, B. S. (2000). Perceived sanction threats, gender, and crime: A test and elaboration of power-control. *Criminology, 38,* 439–488.

Blackwell, B. S., & Piquero, A. R. (2005). On the relationship between gender, power control, self-control, and crime. *Journal of Criminal Justice, 33,* 1–17.

Blair, J., Mitchell, D., & Blair, K. (2005). *The psychopath: Emotion and the brain.* Hoboken, NJ: Wiley Blackwell.

Blokland, A. A. J., Nagin, D. S., & Nieuwbeerta, P. (2005). Life span offending trajectories of a Dutch conviction cohort. *Criminology, 43,* 919–954.

Bloom, B., Owen, B., & Covington, S. (2003). *Gender-responsive strategies: Research, practice, and guiding principles for women offenders.* Washington, DC: U.S. Department of Justice, National Institute of Corrections, June.

Blumstein, A. (2000). Disaggregating the violence trends. In A. Blumstein & J. Wallman (Eds.), *The crime drop in America* (pp. 13–44). New York: Cambridge University Press.

Blumstein, A., & Cohen, J. (1973). A theory of the stability of punishment. *Journal of Criminal Law and Criminology, 64,* 198–206.

Blumstein, A., Cohen, J., Roth, J. A., & Visher, C. A. (1986). *Criminal careers and "career criminals"* (Vol. 1). Washington, DC: National Academies Press.

Blumstein, A., & Wallman, J. (Eds.). (2000). *The crime drop in America.* New York: Cambridge University Press.

Bohm, R. M. (1982). Radical criminology: An explication. *Criminology, 19,* 565–589.

Bonger, W. (1969). *Criminality and economic conditions.* Bloomington: Indiana University Press. (Original work published 1916)

Bonta, J., Jesseman, R., Rugge, T., & Cormier, R. (2006). Restorative justice and recidivism: Promises made, promises kept? In D. Sullivan & L. Tiff (Eds.), *Handbook of restorative justice* (pp. 108–120). New York: Routledge.

Bonta, J., Wallace-Capretta, S., & Rooney, J. (1998). *Restorative justice: An evaluation of the Restorative Resolutions Project.* Ottawa, ON: Solicitor General Canada.

Bonta, J., Wallace-Capretta, S., Rooney, J., & McAnoy, K. (2002). An outcome evaluation of a restorative justice alternative to incarceration. *Contemporary Justice Review, 5,* 319–338.

Bordua, D. J. (1961). Delinquent subcultures: Sociological interpretations of gang delinquency. *Annals of the Academy of Political and Social Science, 338,* 119–136.

Born bad? (1997, April 21). *U.S. News & World Report,* cover.

Bottcher, J. (2001). Social practices of gender: How gender relates to delinquency in everyday lives of high-risk youths. *Criminology, 39,* 893–932.

Bottoms, A. E. (1994). Environmental criminology. In M. Maguire, R. Morgan, &

R. Reiner (Eds.), *The Oxford handbook of criminology* (pp. 585–656). New York: Oxford University Press.

Bouffard, L. A., & Piquero, N. L. (2010). Defiance theory and life course explanations of persistent offending. *Crime & Delinquency, 56,* 227–252.

Boulding, E. (1992). *The underside of history: A view of women through time* (Vol. 1). Newbury Park, CA: Sage.

Bowers v. Hardwick, 478 U.S. 186 (1986).

Brady, K. (1989). *Ida Tarbell: Portrait of a muckraker.* Pittsburgh, PA: University of Pittsburgh Press.

Brain defects seen in those who report violent acts. (1985, September 17). *New York Times,* pp. A17, A19.

Braithwaite, J. (1984). *Corporate crime in the pharmaceutical industry.* London: Routledge & Kegan Paul.

Braithwaite, J. (1985). *To punish or persuade: Enforcement of coal mine safety.* Albany: State University of New York Press.

Braithwaite, J. (1989). *Crime, shame and reintegration.* Cambridge, UK: Cambridge University Press.

Braithwaite, J. (1998). Restorative justice. In M. Tonry (Ed.), *The handbook of crime and punishment* (pp. 323–344). New York: Oxford University Press.

Braithwaite, J. (1999). Restorative justice: Assessing optimistic and pessimistic accounts. In M. Tonry (Ed.), *Crime and justice: A review of research* (Vol. 25, pp. 1–27). Chicago: University of Chicago Press.

Braithwaite, J. (2002). *Restorative justice and responsive regulation.* New York: Oxford University Press.

Braithwaite, J., Ahmed, A., & Braithwaite, V. (2006). Shame, restorative justice, and crime. In F. T. Cullen, J. P. Wright, & K. R. Blevins (Eds.), *Taking stock: The status of criminological theory* (Advances in Criminological Theory, Vol. 15, pp. 397–417). New Brunswick, NJ: Transaction.

Brantingham, P. L., & Brantingham, P. J. (1993). Environment, routine, and situation: Toward a pattern theory of crime. In R. V. Clarke & M. Felson (Eds.), *Routine activity and rational choice* (Advances in Criminological Theory, Vol. 5, pp. 259–294). New Brunswick, NJ: Transaction.

Brennan, P. A., Mednick, S. A., & Volavka, J. (1995). Biomedical factors in crime. In J. Q. Wilson & J. Petersilia (Eds.), *Crime.* San Francisco: ICS Press.

Brent, J., & Kraska, P. (2010, November). *Tapping into "tapping out": The barbaric spectacle of underground fighting.* Paper presented at the annual meeting of the American Society of Criminology, Philadelphia .

Brezina, T. (1996). Adapting to strain: An examination of delinquent coping responses. *Criminology, 34,* 213–239.

Brezina, T., Agnew, R., Cullen, F. T., & Wright, J. P. (2004). The code of the street: A quantitative assessment of Elijah Anderson's subculture of violence thesis and its contribution to youth violence research. *Youth Violence and Juvenile Justice, 2,* 303–328.

Briar, S., & Piliavin, I. (1965). Delinquency, situational inducements, and commitment to conformity. *Social Problems, 13,* 35–45.

British scandals jeopardizing party's "back to basics" effort. (1994, January 14). *New York Times,* p. A3.

Britt, C. L., & Gottfredson, M. R. (Eds.). (2003). *Control theories of crime and delinquency* (Advances in Criminological Theory, Vol. 12). New Brunswick, NJ: Transaction.

Brodeur, P. (1985). *Outrageous misconduct: The asbestos industry on trial.* New York: Pantheon.

Brody, G. H., Chen, Y.-F., Murry, V. M., Ge, X., Simons, R. L., Gibbons F. X., Gerrard, M., & Cutrona, C. E. (2006). Perceived discrimination and the adjustment of African American youths: A five-year longitudinal analysis with contextual moderation effects. *Child Development, 77,* 1170–1189.

Brokaw, T. (2007). *Boom? Voices of the sixties.* New York: Random House.

Brotherton, D. C. (2004). What happened to the pathological gang? Notes from a case study of Latin Kings and Queens in New York. In J. Ferrell, K. Hayward, W. Morrison, & M. Presdee (Eds.), *Cultural criminology unleashed* (pp. 263–274). London: Glasshouse.

Brown, M. H. (1979). *Laying waste: The poisoning of America by toxic chemicals.* New York: Pantheon.

Brownfield, D., & Sorenson, A. M. (1993). Self-control and juvenile delinquency:

Theoretical issues and an empirical assessment of selected elements of a general theory of crime. *Deviant Behavior, 14,* 243–264.

Brownmiller, S. (1975). *Against our will: Men, women, and rape.* New York: Simon & Schuster.

Bruinius, H. (2006). *Better for all the world: The secret history of forced sterilization and America's quest for racial purity.* New York: Knopf.

Brushett, R. (2010). Abolitionism. In F. T. Cullen & P. Wilcox (Eds.), *Encyclopedia of criminological theory* (pp. 1–3). Thousand Oaks, CA: Sage.

Bulmer, M. (1984). *The Chicago school of sociology: Institutionalization, diversity, and the rise of sociological research.* Chicago: University of Chicago Press.

Bunker, E. (2000). *Education of a felon.* New York: St. Martin's.

Bureau of Justice Statistics. (2007). *Homicide trends in the U.S.: Intimate homicide.* Washington, DC: U.S. Department of Justice.

Burgess, E. W. (1967). The growth of the city: An introduction to a research project. In R. E. Park, E. W. Burgess, & R. D. McKenzie (Eds.), *The city* (pp. 47–62). Chicago: University of Chicago Press. (Original work published 1925)

Burgess, R. L., & Akers, R. L. (1966). A differential association-reinforcement theory of criminal behavior. *Social Problems, 14,* 128–146.

Burns, J. F. (2010, May 12). Conservatives in Britain retake political power: Coalition government. *New York Times,* pp. A1, A12.

Burton, V. S., Jr., & Cullen, F. T. (1992). The empirical status of strain theory. *Journal of Crime and Justice, 15*(2), 1–30.

Burton, V. S., Jr., Cullen, F. T., Evans, D. T., & Dunaway, R. G. (1994). Reconsidering strain theory: Operationalization, rival theories, and adult criminality. *Journal of Quantitative Criminology, 10,* 213–239.

Bushway, S., Stoll, M. A., & Weiman, D. F. (Eds.). (2007). *Barriers to reentry? The labor market for released prisoners in post-industrial America.* New York: Russell Sage.

Calavita, K., Pontell, H. N., & Tillman, R. H. (1997). *Big money crime: Fraud and politics in the savings and loan crisis.* Berkeley: University of California Press.

California Department of Corrections and Rehabilitation. (2010, January 21). CDCR implements public safety reforms to parole supervision, expanded incentive credits for inmates [Press release.] Retrieved March 24, 2010, from http://www.cdcr.ca.gov/News/2010_Press_Releases/Jan_21.html

Camp, C. G., & Camp, G. M. (1999). *The corrections yearbook 1999: Adult corrections.* Middletown, CT: Criminal Justice Institute.

Camp, G. M., & Camp, C. G. (1987). *The corrections yearbook.* South Salem, NY: Criminal Justice Institute.

Campbell, A. (1981). *Girl delinquents.* Oxford, UK: Basil Blackwell.

Cao, L., Adams, A., & Jensen, V. J. (1997). A test of the Black subculture of violence thesis: A research note. *Criminology, 35,* 367–379.

Cao, L., & Maume, D. J., Jr. (1993). Urbanization, inequality, lifestyle, and robbery: A comprehensive model. *Sociological Focus, 26,* 11–26.

Carter, P. (2004). *Managing offenders, reducing crime: A new approach.* London: Home Office Strategy Unit.

Caspi, A., & Moffitt, T. E. (1995). The continuity of maladaptive behavior: From description to understanding in the study of antisocial behavior. In D. Cicchetti & D. Cohen (Eds.), *Manual of developmental psychology* (pp. 472–511). New York: John Wiley.

Catalano, R. F., Arthur, M. W., Hawkins, J. D., Berglund, L., & Olson, J. J. (1998). Comprehensive community- and school-based interventions to prevent antisocial behavior. In R. Loeber & D. P. Farrington (Eds.), *Serious and violent juvenile offenders: Risk factors and successful interventions* (pp. 248–283). Thousand Oaks, CA: Sage.

Catalano, R. F., & Hawkins, J. D. (1996). The social development model: A theory of antisocial behavior. In J. D. Hawkins (Ed.), *Delinquency and crime: Current theories* (pp. 149–197). New York: Cambridge University Press.

Celis, W. (1991, January 2). Growing talk of date rape separates sex from assault. *New York Times,* p. A1.

Chambliss, W. J. (1964). A sociological analysis of the law of vagrancy. *Social Problems, 12,* 67–77.

Chambliss, W. J. (1969). *Crime and the legal process.* New York: McGraw-Hill.

Chambliss, W. J. (1975). Toward a political economy of crime. *Theory and Society, 2,* 149–170.

Chambliss, W. J. (1984). The Saints and the Roughnecks. In W. J. Chambliss (Ed.), *Criminal law in action* (2nd ed., pp. 126–135). New York: John Wiley.

Chambliss, W. J. (1987). I wish I didn't know now what I didn't know then. *The Criminologist, 12,* 1–9.

Chambliss, W. J., & Seidman, R. T. (1971). *Law, order, and power.* Reading, MA: Addison-Wesley.

Chamlin, M. B., & Cochran, J. K. (1995). Assessing Messner and Rosenfeld's institutional anomie theory: A partial test. *Criminology, 33,* 411–429.

Chapple, C. (2005). Self-control, peer relations, and delinquency. *Justice Quarterly, 22,* 89–106.

Chapple, C., & Hope, T. (2003). An analysis of self-control and criminal versatility of dating violence and gang offenders. *Violence and Victims, 18,* 671–690.

Chen, M. K., & Shapiro, J. M. (2007). Do harsher prison conditions reduce recidivism? A discontinuity-based approach. *American Law and Economic Review, 9,* 1–29.

Cherniak, M. (1986). *The Hawk Nest incident: America's worst industrial disaster.* New Haven, CT: Yale University Press.

Chesney-Lind, M. (1973). Judicial enforcement of the female sex roles: The family court and the female delinquent. *Issues in Criminology, 8*(2), 51–69.

Chesney-Lind, M. (1998). Women in prison: From partial justice to vengeful equity. *Corrections Today, 60,* 130–134.

Chesney-Lind, J., & Faith, K. (2001). What about feminism? Engendering theory-making criminology. In R. Paternoster & R. Bachman (Eds.), *Explaining crime and criminals* (pp. 287–302). Los Angeles: Roxbury.

Chesney-Lind, J., & Pasko, L. (2004). *The female offender* (2nd ed.). Thousand Oaks, CA: Sage.

Chicago Area Project. (2010). Frequently asked questions. Retrieved March 1, 2010, from http://www.chicagoareaproject.org/faq1.htm

Chiricos, T., Barrick, K., Bales, W., & Bontrager, S. (2007). The labeling of convicted felons and its consequences for recidivism. *Criminology, 45,* 547–581.

Christie, N. (1981). *Limits to pain.* Oxford, UK: Martin-Robertson.

Christie, N. (1993). *Crime control as industry: Towards gulags Western style?* London: Routledge.

Christie, N. (1997). *Crime control as industry: Towards gulags, Western style?* (2nd ed.). London: Routledge.

Christie, N. (2001). *Crime control as industry: Towards gulags Western style?* (3rd ed.). London: Routledge.

Clarke, R. V. (1980). "Situational" crime prevention: Theory and practice. *British Journal of Criminology, 20,* 136–147.

Clarke, R. V. (1992). Introduction. In R. V. Clarke (Ed.), *Situational crime prevention: Successful case studies* (pp. 3–36). New York: Harrow & Heston.

Clarke, R. V. (1995). Situational crime prevention. In M. Tonry & D. P. Farrington (Eds.), *Building a safer society: Strategic approaches to crime prevention* (pp. 91–150). New York: Cambridge University Press.

Clarke, R. V., & Cornish, D. B. (2001). Rational choice. In R. Paternoster & R. Bachman (Eds.), *Explaining criminals and crime: Essays in contemporary criminological theory* (pp. 23–42). Los Angeles: Roxbury.

Clarke, R. V., & Felson, M. (1993). Introduction: Criminology, routine activity, and rational choice. In R. V. Clarke & M. M. Felson (Eds.), *Routine activity and rational choice* (Advances in Criminological Theory, Vol. 5, pp. 1–14). New Brunswick, NJ: Transaction.

Clarke, R. V., & Felson, M. (2011). The origins of the routine activity approach and situational crime prevention. In F. T. Cullen, C. L. Jonson, A. J. Myer, & F. Adler (Eds.), *The origins of American criminology* (Advances in Criminological Theory, Vol. 16, pp. 245–260). New Brunswick, NJ: Transaction.

Clear, T. R. (1994). *Harm in American penology: Offenders, victims, and their communities.* Albany: State University of New York Press.

Clear, T. R. (2002). The problem with "addition by subtraction": The prison-crime relationship in low-income communities. In M. Mauer & M. Chesney-Lind (Eds.), *Invisible punishment: The collateral* consequences of mass imprisonment (pp. 181–193). New York: New Press.

Clear, T. R. (2007). *Imprisoning communities: How mass incarceration makes disadvantaged neighborhoods worse.* New York: Oxford University Press.

Clear, T. R., Rose, D. R., Waring, E., & Scully K. (2003). Coercive mobility and crime: A preliminary examination of concentrated incarceration and social disorganization. *Justice Quarterly, 20,* 33–64.

Clinard, M. B. (1952). *The black market: A study of white collar crime.* Montclair, NJ: Patterson Smith. (Reprinted in 1969)

Clinard, M. B. (1957). *Sociology of deviant behavior* (3rd ed.). New York: Holt, Rinehart & Winston.

Clinard, M. B. (1983). *Corporate ethics and crime: The role of middle management.* Beverly Hills, CA: Sage.

Clinard, M. B., & Yeager, P. C. (1980). *Corporate crime.* New York: Free Press.

Clinton, H. R. (1996). *It takes a village: And other lessons children teach us.* New York: Touchstone.

Cloward, R. A. (1959). Illegitimate means, anomie, and deviant behavior. *American Sociological Review, 24,* 164–176.

Cloward, R. A., & Ohlin, L. E. (1960). *Delinquency and opportunity: A theory of delinquent gangs.* New York: Free Press.

Cloward, R. A., & Piven, F. F. (1979). Hidden protest: The channeling of female innovation and resistance. *Signs, 4,* 651–669.

Cochran, J. K., Aleska, V., & Chamlin, M. B. (2006). Self-restraint: A study on the capacity and desire for self-control. *Western Criminological Review, 7,* 27–40.

Cohen, A. K. (1955). *Delinquent boys: The culture of the gang.* New York: Free Press.

Cohen, A. K., & Short, J. F., Jr. (1958). Research in delinquent sub-cultures. *Journal of Social Issues, 14,* 20–37.

Cohen, D. V. (1995). Ethics and crime in business firms: Organizational culture and the impact of anomie. In F. Adler & W. S. Laufer (Eds.), *The legacy of anomie theory* (Advances in Criminological Theory, Vol. 6, pp. 183–206). New Brunswick, NJ: Transaction.

Cohen, L. E., & Felson, M. (1979). Social change and crime rate trends: A routine activities approach. *American Sociological Review, 44,* 588–608.

Cohen, L. E., & Machalek, R. (1988). A general theory of expropriative crime: An evolutionary ecological approach. *American Journal of Sociology, 94,* 465–501.

Cohen, S. (1971). *Images of deviance.* Harmondsworth, UK: Penguin.

Cohen, S. (1980). *Folk devils and moral panics.* London: Macgibbon & Kee. (Original work published 1972)

Cohen, S. (1981). Footprints in the sand. In M. Fitzgerald, G. McLenna, & K. Pease (Eds.), *Crime and society: Readings in history and theory* (pp. 220–267). London: Routledge & Kegan Paul.

Cohen, S. (1986). Community control: To demystify or to reaffirm? In H. Bianchi & van Swaaningen, R. (Eds.), *Abolitionism: Towards a non repressive approach to crime* (pp. 127–132). Amsterdam: Free University Press.

Cohen, S. (1988). *Against criminology.* New Brunswick, NJ: Transaction.

Colburn, T., Dumanoski, D., & Myers, J. P. (1997). *Our stolen future.* New York: Plume.

Cole, S. (1975). The growth of scientific knowledge: Theories of deviance as a case study. In L. A. Coser (Ed.), *The idea of social structure: Papers in honor of R. K. Merton* (pp. 175–200). New York: Harcourt Brace Jovanovich.

Coleman, J. W. (1985). *The criminal elite: The sociology of white collar crime.* New York: St. Martin's.

Coleman, J. W. (1992). The theory of white-collar crime: From Sutherland to the 1990s. In K. Schlegel & D. Weisburd (Eds.), *White-collar crime reconsidered* (pp. 53–77). Boston: Northeastern University Press.

Coles, R. (1993). *The call of service: A witness to idealism.* Boston: Houghton Mifflin.

Collins, G. (2009). *When everything changes: The amazing journey of American women from 1960 to the present.* New York: Little, Brown.

Colvin, M. (2000). *Crime and coercion: An integrated theory of chronic criminality.* New York: St. Martin's.

Colvin, M., Cullen, F. T., & Vander Ven, T. (2002). Coercion, social support, and crime: An emerging theoretical consensus. *Criminology, 40,* 19–42.

Colvin, M., & Pauly, J. (1983). A critique of criminology: Toward an integrated structural-Marxist theory of delinquency production. *American Journal of Sociology, 89,* 513–551.

Comings, D. E. (2003). Conduct disorder: A genetic, orbitofrontal lobe disorder and the major predictor of adult antisocial behavior. In A. Walsh & L. Ellis (Eds.), *Biosocial criminology: Challenging environmentalism's supremacy* (pp. 145–164). Hauppauge, NY: Nova Science.

Conklin, J. E. (1977). *"Illegal but not criminal": Business crime in America.* Englewood Cliffs, NJ: Prentice Hall.

Considine, J. (1995). *Restorative justice: Healing the effects of crime.* Lyttleton, New Zealand: Ploughshares.

Cooley, C. H. (1902). *Human nature and social order.* New York: Scribner.

Cooley, C. H. (1909). *Social organization.* New York: Scribner.

Cooley, C. H. (1922). *Human nature and social order* (Rev. ed.). New York: Scribner.

Cornish, D., & Clarke, R. V. (1986). *The reasoning criminal: Rational choice perspectives on offending.* New York: Springer.

Cornwell, J. (2003). *Hitler's scientists: Science, war, and the devil's pact.* New York: Penguin.

Crack down: Reagan declares war on drugs and proposes tests for key officials. (1986, August 18). *Time,* pp. 12–13.

Cressey, D. R. (1950). The criminal violation of financial trust. *American Sociological Review, 15,* 738–743.

Cressey, D. R. (1953). *Other people's money: A study in the social psychology of embezzlement.* Glencoe, IL: Free Press.

Croall, H. (1992). *White collar crime.* Buckingham, UK: Open University Press.

Cullen, F. T. (1984). *Rethinking crime and deviance theory: The emergence of a structuring tradition.* Totowa, NJ: Rowman & Allanheld.

Cullen, F. T. (1988). Were Cloward and Ohlin strain theorists? Delinquency and opportunity revisited. *Journal of Research in Crime and Delinquency, 25,* 214–241.

Cullen, F. T. (1994). Social support as an organizing concept for criminology: Presidential address to the Academy of Criminal Justice Sciences. *Justice Quarterly, 11,* 527–559.

Cullen, F. T. (2009). Preface. In A. Walsh & K. M. Beaver (Eds.), *Biosocial criminology: New directions in theory and research* (pp. xv–xvii). New York: Routledge.

Cullen, F. T. (2010). Elliott Currie: In tribute to a life devoted to confronting crime. *Criminology and Public Policy, 9,* 19–27.

Cullen, F. T., Cavender, G., Maakestad, W. J., & Benson, M. L. (2006). *Corporate crime under attack: The fight to criminalize business violence* (2nd ed.). Cincinnati, OH: LexisNexis/Anderson.

Cullen, F. T., Clark, G. A., & Wozniak, J. F. (1985). Explaining the get tough movement: Can the public be blamed? *Federal Probation, 49,* 16–24.

Cullen, F. T., & Cullen, J. B. (1978). *Toward a paradigm of labeling theory.* Lincoln: University of Nebraska Press.

Cullen, F. T., Eck, J. E., & Lowenkamp, C. T. (2002). Environmental corrections: A new paradigm for effective probation and parole supervision. *Federal Probation, 66(2),* 28–37.

Cullen, F. T., Fisher, B. S., & Applegate, B. K. (2000). Public opinion about crime and punishment. In M. Tonry (Ed.), *Crime and justice: A review of research* (Vol. 27, pp. 1–79). Chicago: University of Chicago Press.

Cullen, F. T., & Gendreau, P. (2000). Assessing correctional rehabilitation: Policy, practice, and prospects. In J. Horney (Ed.), *Criminal justice 2000: Vol. 3. Policies, processes, and decisions in the criminal justice system* (pp. 109–175). Washington, DC: National Institute of Justice.

Cullen, F. T., & Gendreau, P. (2001). From nothing works to what works: Changing professional ideology in the 21st century. *Prison Journal, 81,* 313–338.

Cullen, F. T., Gendreau, P., Jarjoura, G. R., & Wright, J. P. (1997). Crime and the bell curve: Lessons from intelligent criminology. *Crime & Delinquency, 43,* 387–411.

Cullen, F. T., & Gilbert, K. E. (1982). *Reaffirming rehabilitation.* Cincinnati, OH: Anderson.

Cullen, F. T., Hartman, J. L., & Jonson, C. L. (2009). Bad guys: Why the public supports punishing white-collar offenders. *Crime, Law and Social Change, 51,* 31–44.

Cullen, F. T., & Jonson, C. L. (2011). *Correctional theory: Context and consequences.* Thousand Oaks, CA: Sage.

Cullen, F. T., Link, B. G., & Polanzi, C. W. (1982). The seriousness of crime revisited: Have attitudes toward white-collar crime changes? *Criminology, 20,* 82–102.

Cullen, F. T., Maakestad, W. J., & Cavender, G. (1987). *Corporate crime under attack: The Ford Pinto case and beyond.* Cincinnati, OH: Anderson.

Cullen, F. T., & Messner, S. M. (2007). The making of criminology revisited: An oral history of Merton's anomie paradigm. *Theoretical Criminology, 11,* 5–37.

Cullen, F. T., Pealer, J. A., Santana, S. A., Fisher, B. S., Applegate, B. K., & Blevins, K. R. (2007). Public support for faith-based correctional programs: Should sacred places serve civic purposes? *Journal of Offender Rehabilitation, 43,* 29–46.

Cullen, F. T., Pratt, T. C., Micelli, S., & Moon, M. M. (2002). Dangerous liaison? Rational choice theory as the basis for correctional intervention. In A. R. Piquero & S. G. Tibbetts (Eds.), *Rational choice and criminal behavior: Recent research and future challenges* (pp. 279–296). Philadelphia: Taylor & Francis.

Cullen, F. T., Sundt, J. L., & Wozniak, J. F. (2000). The virtuous prison: Toward a restorative rehabilitation. In H. N. Pontell & D. Shichor (Eds.), *Contemporary issues in crime and criminal justice: Essays in honor of Gilbert Geis* (pp. 265–286). Thousand Oaks, CA: Sage.

Cullen, F. T., Vose, B. A., Jonson, C. L., & Unnever, J. D. (2007). Public support for early intervention: Is child saving a "habit of the heart"? *Victims and Offenders, 2,* 109–124.

Cullen, F. T., Unnever, J. D., Wright, J. P., & Beaver, K. M. (2008). Parenting and self-control. In E. Goode (Ed.), *Crime and Criminality: Evaluating the general theory of crime* (pp. 61–74). Stanford, CA: Stanford University Press.

Cullen, F. T., Wright, J. P., & Applegate, B. K. (1996). Control in the community: The limits of reform? In A. T. Harland (Ed.), *Choosing correctional interventions that work: Defining the demand and evaluating the supply* (pp. 69–116). Thousand Oaks, CA: Sage.

Cullen, F. T., Wright, J. P., & Chamlin, M. B. (1999). Social support and social reform: A progressive crime control agenda. *Crime & Delinquency, 45,* 188–207.

Cullen, F. T., Wright, J. P., Gendreau, P., & Andrews, D. A. (2003). What correctional treatment can tell us about criminological theory: Implications for social learning theory. In R. L. Akers & G. F. Jensen (Eds.), *Social learning theory and the explanation of crime: A guide for the new century* (Advances in Criminological Theory, Vol. 11, pp. 339–362). New Brunswick, NJ: Transaction.

Cullen, J. B., Parboteeah, K. P., & Hoegl, M. (2004). Cross-national differences in managers' willingness to justify ethically suspect behaviors: A test of institutional anomie theory. *Academy of Management Journal, 47,* 411–421.

Curran, D. J., & Renzetti, C. M. (1994). *Theories of crime.* Boston: Allyn & Bacon.

Currie, E. (1974). [Book review of *The New Criminology*]. *Issues in Criminology, 9,* 123–142.

Currie, E. (1985). *Confronting crime: An American challenge.* New York: Pantheon.

Currie, E. (1989). Confronting crime: Looking toward the twenty-first century. *Justice Quarterly, 6,* 5–25.

Currie, E. (1993). *Reckoning: Drugs, cities, and the American future.* New York: Hill & Wang.

Currie, E. (1997). Market, crime, and community: Toward a mid-range theory of post-industrial violence. *Theoretical Criminology, 1,* 147–172.

Currie, E. (1998a). Crime and market society: Lessons from the United States. In P. Walton & J. Young (Eds.), *The new criminology revisited* (pp. 130–142). London: St. Martin's.

Currie, E. (1998b). *Crime and punishment in America.* New York: Metropolitan Books.

Currie, E. (2004). *The road to whatever: Middle-class culture and the crisis of adolescence.* New York: Henry Holt.

Currie, E. (2007). Against marginality: Arguments for a public criminology. *Theoretical Criminology, 11,* 175–190.

Currie, E. (2009). *The roots of danger: Violent crime in global perspective.* Upper Saddle River, NJ: Prentice Hall.

Dahrendorf, R. (1958). *Class and class conflict in industrial society.* Stanford, CA: Stanford University Press.

Dahrendorf, R. (1968). Toward a theory of social conflict. *Journal of Conflict Resolution, 2,* 170–183.

Daigle, L. E., Cullen, F. T., & Wright, J. P. (2007). Gender differences in the predictors of juvenile delinquency: Assessing the generality-specificity debate. *Youth Violence and Juvenile Justice, 5,* 254–286.

Dallier, D. J. (2011). Michalowski, Raymond J. and Ronald C. Kramer: State-corporate crime. In F. T. Cullen & P. Wilcox (Eds.), *Encyclopedia of criminological theory* (pp. 628–631). Thousand Oaks, CA: Sage.

Daly, K. (1998). Gender, crime and criminology. In M. Tonry (Ed.), *The handbook of crime and justice* (pp 85–108). Oxford, UK: Oxford University Press.

Daly, K. (2010). Feminist perspectives in criminology: A review with Gen Y in mind. In E. McLaughlin & T. Newburn (Eds.), *The Sage handbook of criminological theory* (pp. 225–246). London: Sage.

Daly, K., & Chesney-Lind, M. (1988). Feminism and criminology. *Justice Quarterly, 5,* 497–535.

Damasio, A. R. (1994). Descartes' error and the future of human life. *Scientific American, 271,* 144.

Darwin, C. (1871). *The descent of man.* London: John Murray.

Darwin, C. (1872). *The expression of emotions in man and animals.* London: John Murray.

Darwin, C. (1981). *Origin of species* (Rev. ed.). Danbury, CT: Grolier. (Original work published 1859)

Davis, A. (2003). *Are prisons obsolete?* New York: Seven Stories Press.

Davis, A. (2005). *Abolition democracy: Beyond prisons, torture and empire.* New York: Seven Stories Press.

Davis, D. (2002). *When smoke ran like water: Tales of environmental deception and the battle against pollution.* New York: Basic Books.

Davis, K. (1948). *Human society.* New York: Macmillan.

Dawkins, R. (1976). *The selfish gene.* New York: Oxford University Press.

Day, J. J., & Carelli, R. M. (2007). The nucleus accumbens and Pavlovian reward learning. *The Neuroscientist, 13,* 148–159.

de Waal, F. B. M. (2008). Putting the altruism back into altruism: The evolution of empathy. *Annual Review of Psychology, 59,* 279–300.

Dean, M. (2005, December 14). Shame of Blair's market madness. *The Guardian,* Opinion Section.

Death row appeals assailed. (1988, January 28). *New York Times,* p. A7.

Defective gene tied to form of manic-depressive illness. (1987, February 26). *New York Times,* p. A1.

DeKeseredy, W. S. (2010). Left realism. In F. T. Cullen & P. Wilcox (Eds.), *Encyclopedia of criminological theory* (pp. 546–550). Thousand Oaks, CA: Sage.

DeKeseredy, W. S. (in press). *Contemporary critical criminology.* New York: Routledge.

DeKeseredy, W. S., Alvi, S., & Schwartz, M. D. (2006). Left realism revisited. In W. S. DeKeseredy & B. Perry (Eds.), *Advanced critical criminology: Theory and application* (pp. 19–42). Lanham, MD: Lexington Books.

DeKeseredy, W. S., Alvi, S., Schwartz, M. D., & Tomaszewski, E. A. (2003). *Under siege: Poverty and crime in a public housing community.* Lanham, MD: Lexington Books.

DeKeseredy, W. S., & Schwartz, M. D. (2010). Friedman economic policies, social exclusion, and crime: Toward a gendered left realist subcultural theory. *Crime, Law and Social Change, 54,* 159–170.

DeKeseredy, W. S., Schwartz, M. D., Fagen, D., & Hall, M. (2006). Separation/divorce sexual assault: The contribution of male peer support. *Feminist Criminology, 1,* 228–250.

Deinstitutionalization: Special report. (1975, November–December). *Corrections Magazine.*

Dewan, S. (2009, August 9). The real murder mystery? It's the low crime rate. *New York Times,* p. WK4.

DiCristina, B. (1995). *Method in criminology: A philosophical primer.* New York: Harrow & Heston.

DiIulio, J. J., Jr. (1994, Fall). The question of Black crime. *Public Interest, 117,* 3–32.

DiIulio, J. J., Jr. (1995, November 27). The coming of super-predators. *Weekly Standard,* pp. 23–28.

Dinitz, S., Reckless, W. C., & Kay, B. (1958). A self gradient among potential delinquents. *Journal of Criminal Law, Criminology, and Police Science, 49,* 230–233.

Dinitz, S., Scarpitti, F. R., & Reckless, W. C. (1962). Delinquency vulnerability: A cross group and longitudinal analysis. *American Sociological Review, 27,* 515–517.

Dionne, E. J., Jr. (1996). *They only look dead: Why progressives will dominate the next political era.* New York: Simon & Schuster.

Dionne, E. J., Jr., & Chen, M. H. (Eds.). (2001). *Sacred places, civic purposes: Should government help faith-based charity?* Washington, DC: Brookings Institution.

DiRago, A. C., & Viallant, G. E. (2007). Resilience in inner city youth: Childhood predictors of occupational status across the lifespan. *Journal of Youth and Adolescence, 36,* 61–70.

Donovan, K. M., & Klahm, C. F. (2009). Prosecuting science: The rational defence of mandatory DNA databases. *The Howard Journal of Criminal Justice, 48,* 411–413.

Dowie, M. (1977, September–October). Pinto madness. *Mother Jones, 2,* pp. 18–32.

Downes, D., & Rock, P. (1988). *Understanding deviance* (2nd ed.). Oxford, UK: Clarendon.

Drake, E. K., Aos, S., & Miller, M. G. (2009). Evidence-based public policy options to reduce crime and criminal justice costs: Implications in Washington state. *Victims and Offenders, 4,* 170–196.

Drug test called costly, often useless. (1986, June 21). *Washington Post,* p. A15.

Dugdale, R. L. (1877). *The jukes.* New York: Putnam.

Dunbar, R. I. M. (2007). Male and female brain evolution is subject to contrasting selection pressures in primates. *BioMedCentral Biology, 5,* 1–3.

D'Unger, A. V. (2005). Feminist theories of criminal behavior. In R. A. Wright & J. M. Miller (Eds.), *Encyclopedia of criminology* (Vol. 1, pp. 559–565). New York: Routledge.

D'Unger, A. V., Land, K. C., McCall, P. L., & Nagin, D. S. (1998). How many latent classes of delinquent/criminal careers? Results from mixed Poisson regression analyses. *American Journal of Sociology, 103,* 1593–1630.

Durkheim, E. (1933). *The division of labor in society.* Glencoe, IL: Free Press.

Durkheim, E. (1951). *Suicide: A study in sociology* (J. A. Spaulding & G. Simpson, Trans.). New York: Free Press. (Original work published 1897)

Durkheim, E. (1964). *The division of labor.* London: Free Press.

Duster, T. (2006). Comparative perspectives and competing explanations: Taking on the new configured reductionist challenge to sociology. *American Sociological Review, 71,* 1–15.

Eck, J. E. (1995). Examining routine activity theory: A review of two books. *Justice Quarterly, 12,* 783–797.

Eck, J. E. (1997). Preventing crime at places. In L. W. Sherman, D. Gottfredson, D. MacKenzie, J. Eck, P. Reuter, & S. Bushway (Eds.), *Preventing crime: What works, what doesn't, what's promising—A report to the United States Congress.* Washington, DC: Office of Justice Programs.

Eck, J. E. (1998). Preventing crime by controlling drug dealing on private rental property. *Security Journal, 11,* 37–43.

Eck, J. E. (2003). Police problems: The complexity of problem theory, research and evaluation. In J. Knutsson (Ed.), *Problem-oriented policing: From innovation to mainstream* (Crime Prevention Studies, Vol. 15, pp. 79–113). Monsey, NJ: Criminal Justice Press.

Eck, J. E., & Maguire, E. R. (2000). Have changes in policing reduced crime? An assessment of the evidence. In A. Blumstein & J. Walman (Eds.), *The crime drop in America* (pp. 207–265). New York: Cambridge University Press.

Eck, J. E., & Spelman, W. (1987). Who ya gonna call: The police as problem-busters. *Crime & Delinquency, 33,* 31–52.

Eck, J. E., & Weisburd, D. (1995). Crime places in crime theory. In J. E. Eck & D. Weisburd (Eds.), *Crime places: Crime prevention studies* (Vol. 4, pp. 1–33). Monsey, NY: Willow Tree Press.

[Editorial]. (1985, October 28). *Cincinnati Enquirer,* p. C12.

Edwards, R. C., Reich, M., & Weisskopf, T. E. (1972). *The capitalist system: A radical analysis of American society.* Englewood Cliffs, NJ: Prentice Hall.

Edwin Meese lifts his lance. (1986, November 3). *Newsweek,* p. 9.

Eggleston, E. P., & Laub, J. H., (2002). The onset of adult offending: A neglected dimension of the criminal career. *Journal of Criminal Justice, 30,* 603–622.

The eighties are over. (1988, January 4). *Newsweek,* pp. 40–44.

Elliott, D. S., Ageton, S. S., & Canter, R. J. (1979). An integrated theoretical perspective on delinquent behavior. *Journal of Research on Crime and Delinquency, 16,* 3–27.

Elliott, D. S., Huizinga, D., & Ageton, S. S. (1985). *Explaining delinquency and drug use.* Beverly Hills, CA: Sage.

Ellis, E. (1977). The decline and fall of sociology: 1977–2000. *American Sociologist, 12,* 30–41.

Ellis, H. (1913). *The criminal* (4th ed.). New York: Scribner.

Ellis, L. (1987). Criminal behavior and r/K selection: An extension of gene-based evolutionary theory. *Deviant Behavior, 8*(1), 149–176.

Ellis, L. (1989). Evolutionary and neurochemical causes of sex differences in victimizing behavior: Toward a unified theory of criminal behavior. *Social Science Information, 28,* 605–636.

Ellis, L. (2003a). Genes, crime, and the evolutionary neuroandrogenic theory. In A. Walsh & L. Ellis (Eds.), *Biosocial criminology: Challenging environmentalism's supremacy* (pp. 13–14). Hauppauge, NY: Nova Science Publishers.

Ellis, L. (2003b). So you want to be a biosocial criminologist? Advice from the underground. In A. Walsh L. Ellis (Eds.), *Biosocial criminology: Challenging environmentalism's supremacy* (pp. 249–256). Hauppauge, NY: Nova Science Publishers.

Ellis, L. (2005). Biological perspectives on crime. In S. Guarino-Ghezzi & A. J. Trevino (Eds.), *Understanding crime* (pp. 143–174). Cincinnati, OH: Anderson.

Ellis, L., Beaver, K. M., & Wright, J. P. (2009). *Handbook of crime correlates.* San Diego, CA: Elsevier.

Ellis, L., & Hoffman, H. (Eds.). (1990). *Crime in biological, social and moral contexts.* New York: Praeger.

Ellis, L., & Walsh, A. (1999). Criminologists' opinions about causes and theories of crime. *The Criminologist, 24,* 3–6.

Ellis, L., & Walsh, A. (2000). *Criminology: A global perspective.* Boston: Allyn & Bacon.

Empey, L. T. (1979). Foreword: From optimism to despair—New doctrines in juvenile justice. In C. A. Murray & L. A. Co, Jr. (Eds.), *Beyond probation: Juvenile corrections and the chronic delinquent* (pp. 9–26). Beverly Hills, CA: Sage.

Empey, L. T. (1982). *American delinquency: Its meaning and construction* (Rev. ed.). Homewood, IL: Dorsey.

Empey, L. T., & Erickson, M. L. (1972). *The Provo experiment: Evaluating community control of delinquency.* Lexington, MA: Lexington Books.

Empey, L. T., & Lubeck, S. (1971). *The Silverlake experiment: Testing delinquency theory and community intervention.* Chicago: Aldine.

Engel, R. S., & Calnon, J. M. (2004). Examining the influence of drivers' characteristics during traffic stops with police: Results form a national survey. *Justice Quarterly, 21,* 49–90.

Engel, R. S., Calnon, J. M., & Bernard, T. J. (2002). Theory and racial profiling: Shortcomings and future directions in research. *Justice Quarterly, 19,* 249–273.

Erikson, K. T. (1966). *Wayward Puritans: A study in sociology of deviance.* New York: John Wiley.

Ermann, M. D., & Lundman, R. J. (Eds.). (1978). *Corporate and governmental deviance: Problems of organizational behavior in contemporary society.* New York: Oxford University Press.

Estrabrook, A. H. (1916). *The Jukes in 1915.* Washington, DC: Carnegie Institute.

Etzioni, A. (1993). *The spirit of community: Rights, responsibilities, and the communitarian agenda.* New York: Crown.

Evans, T. D., Cullen, F. T., Burton, V. S., Jr., Dunaway, R. G., & Benson, M. L. (1997). The social consequences of self-control: Testing the general theory of crime. *Criminology, 35,* 475–500.

Evans, T. D., Cullen, F. T., Dunaway, R. G., & Burton, V. S., Jr. (1995). Religion and crime reexamined: The impact of religion, secular controls, and social ecology on adult criminality. *Criminology, 33,* 195–224.

Eysenck, H. J. (1964). *Crime and personality.* Boston: Houghton Mifflin.

Farrington, D. P. (1994a). Early developmental prevention of juvenile delinquency. *Criminal Behaviour and Mental Health, 4,* 209–227.

Farrington, D. P. (1994b). Human development and criminal careers. In M. Maguire, R. Morgan, & R. Reiner (Eds.), *The Oxford handbook of criminology* (pp. 511–584). New York: Oxford University Press.

Farrington, D. P. (1997). The relationship between low resting heart rate and violence. In A. Raine, P. Brennan, D. P. Farrington, & S. Mednick (Eds.), *Biosocial bases of violence* (pp. 89–106). New York: Plenum.

Farrington, D. P. (Ed.). (2005). *Integrated developmental and life-course theories of offending* (Advances in Criminological Theory, Vol. 14). New Brunswick, NJ: Transaction.

Farrington, D. P. (2006). Building developments and life-course theories of offending. In F. T. Cullen, J. P. Wright, & K. R. Blevins (Eds.), *Taking stock: The status of criminological theory* (Advances in Criminological Theory, Vol. 15, pp. 335–364). New Brunswick, NJ: Transaction.

Farrington, D. P., & Welsh, B. C. (2007). *Saving children from a life in crime: Early risk factors and effective interventions.* New York: Oxford University Press.

Featherstone, M. (1988). In pursuit of the postmodern: An introduction. *Theory, Culture & Society, 5*(2–3), 195–215.

Federal Bureau of Investigation. (2010). *Crime in the United States, 2008.* Retrieved March 2, 2010, from http://www.fbi.gov/ucr/cius2008/index.html

Feeley, M. M. (2010). Elliott Currie's contribution to public criminology: An appreciation and a lament. *Criminology and Public Policy, 9,* 11–17.

Felson, M. (1995). Those who discourage crime. In J. E. Eck & D. Weisburd (Eds.), *Crime and place: Crime prevention studies* (Vol. 4, pp. 53–66). Monsey, NY: Criminal Justice Press.

Felson, M. (1998). *Crime and everyday life* (2nd ed.). Thousand Oaks, CA: Pine Forge.

Felson, M. (2002). *Crime and everyday life* (3rd ed.). Thousand Oaks, CA: Sage.

Felson, M., & Boba, R. (2010). *Crime and everyday life* (4th ed.). Thousand Oaks, CA: Sage.

Felson, R. B., & Haynie, D. L. (2002). Pubertal development, social factors, and delinquency among adolescent boys. *Criminology, 40,* 967–988.

Ferrell, J. (1996). *Crimes of style: Urban graffiti and the politics of criminality.* Boston: Northeastern University Press.

Ferrell, J. (1998). Stumbling toward a critical criminology (and into the anarchy and imagery of postmodernism). In J. I. Ross (Ed.), *Cutting the edge* (pp. 63–76). Westport, CT: Praeger.

Ferrell, J. (2005). Cultural criminology. In R. A. Wright & J. M. Miller (Eds.), *Encyclopedia of criminology* (Vol. 1, pp. 347–351). New York: Routledge.

Ferrell, J. (2010). Cultural criminology. In F. T. Cullen & P. Wilcox (Eds.), *Encyclopedia of criminological theory* (pp. 249–253). Thousand Oaks, CA: Sage.

Ferrell, J., Hayward, K., Morrison, W., & Presdee, M. (Eds.). (2004a). *Cultural criminology unleashed.* London: Glasshouse.

Ferrell, J., Hayward, K., Morrison, W., & Presdee, M. (2004b). Fragments of a manifesto: Introducing *Cultural criminology unleashed.* In J. Ferrell, K. Hayward, W. Morrison, & M. Presdee (Eds.), *Cultural criminology unleashed* (pp. 1–9). London: Glasshouse.

Ferrell, J., Hayward, K., & Young, J. (2008). *Cultural criminology: An invitation.* London: Sage.

Ferri, E. (1929–1930). *Sociologia criminale* (5th ed., 2 vols.). Turin, Italy: UTET.

Fighting narcotics is everyone's issue now. (1986, August 10). *New York Times,* p. A25.

Finckenauer, J. O. (1982). *Scared straight! And the panacea phenomenon.* Englewood Cliffs, NJ: Prentice Hall.

Fishbein, D. H. (1990). Biological perspectives in criminology. *Criminology, 28,* 27–72.

Fishbein, D. H. (1997). Biological perspectives in criminology. In S. Henry & W. Einstadter (Eds.), *The criminology theory reader.* New York: New York University Press.

Fishbein, D. H. (2001). *Biobehavioral perspectives in criminology.* Belmont, CA: Wadsworth.

Fishbein, D. H. (2003). Neurophysiological and emotional regulatory processes in antisocial behavior. In A. Walsh & L. Ellis (Eds.), *Biosocial criminology: Challenging environmentalism's supremacy* (pp. 185–208). Hauppauge, NY: Nova Science Publishers.

Fisher, B. S., Cullen, F. T., & Turner, M. G. (2001). *The sexual victimization of college women: Findings from two national-level studies.* Washington, DC: National Institute of Justice and Bureau of Justice Statistics.

Fisher, B. S., Daigle, L. E., & Cullen, F. T. (2010). *Unsafe in the ivory tower: The sexual victimization of college women.* Thousand Oaks, CA: Sage.

Fisher, B. S., Sloan, J. J., Cullen, F. T., & Lu, C. (1998). Crime in the ivory tower: The level and sources of student victimization. *Criminology, 36,* 671–710.

Flanagan, T. J. (1987). Change and influence in popular criminology: Public attributions of crime causation. *Journal of Criminal Justice, 15,* 231–243.

Fox, N., Bell, M. A., & Jones, N. A. (1992). Individual differences in response to stress and cerebral asymmetry. *Developmental Neuropsychology, 8,* 161–184.

Frank, N. (1985). *Crimes against health and safety.* New York: Harrow and Heston.

Frank, T. (1997). *The conquest of cool: Business culture, counterculture and the rise of hip consumerism.* Chicago: University of Chicago Press.

Franklin, H. B. (1998). *Prison writing in 20th-century America.* New York: Penguin.

Frazier, C. E., & Cochran, J. K. (1986). Official intervention, diversion from the juvenile justice system, and dynamics of human services work: Effects of a reform goal based on labeling theory. *Crime & Delinquency, 32,* 157–176.

Freud, S. (1920). *A general introduction to psychoanalysis.* New York: Boni & Liveright.

Freud, S. (1927). *The ego and the id.* London: Hogarth.

Freud, S. (1930). *Civilization and its discontents.* New York: Cape & Smith.

Friedan, B. (1963). *The feminine mystique.* New York: W. W. Norton.

Friedlander, K. (1949). Latent delinquency and ego development. In K. R. Eissler (Ed.), *Searchlights on delinquency* (pp. 205–215). New York: International University Press.

Friedrichs, D. O. (1979). The law and legitimacy crisis: A critical issue for criminal justice. In R. G. Iacovetta & D. H. Chang (Eds.), *Critical issues in criminal justice* (pp. 290–311). Durham, NC: Carolina Academic Press.

Friedrichs, D. O. (1996). *Trusted criminals: White collar crime in contemporary society.* Belmont, CA: Wadsworth.

Friedrichs, D. O. (2009). Critical criminology. In J. M. Miller (Ed.), *21st century criminology: A reference handbook* (Vol. 1, pp. 210–218). Thousand Oaks, CA: Sage.

Friedrichs, D. O. (2010). Integrated theories of white-collar crime. In F. T. Cullen & P. Wilcox (Eds.), *Encyclopedia of criminological theory* (pp. 479–486). Thousand Oaks, CA: Sage.

Friedrichs, D. O., & Schwartz, M. D. (2008). Low self-control and high organizational control: The paradoxes of white-collar crime. In E. Goode (Ed.), *Out of control? Assessing the general theory of crime* (pp. 145–159). Stanford, CA: Stanford University Press.

Fuller, J. R. (1998). *Criminal justice: A peacemaking perspective.* Boston: Allyn & Bacon.

Fuller, J. R., & Wozniak, J. F. (2006). Peacemaking criminology: Past, present, and future. In F. T. Cullen, J. P. Wright, & K. B. Blevins (Eds.), *Taking stock: The status of criminological theory* (Advances in Criminological Theory, Vol. 15, pp. 251–273). New Brunswick, NJ: Transaction.

Fumento, M. (2003, February 3). Trick question: A liberal "hoax" turns out to be true. *The New Republic,* pp. 18–21.

Gabbidon, S. (2007). *Criminological perspectives on race and crime.* New York: Routledge.

Galliher, J. F., & Walker, A. (1977). The puzzle of the social origins of the Marijuana Tax Act of 1937. *Social Problems, 24,* 367–376.

Galvan, A., Hare, T. A., Parra, C. E., Penn, J., Voss, H., Glover, G., & Casey, B. J. (2006). Earlier development of the accumbens relative to orbitofrontal cortex might underlie risk-taking behavior in adolescents. *Journal of Neuroscience, 26,* 6885–6892.

Gamble, A. (1989). Privatization, Thatcherism, and the British state. *Journal of Law and Society, 16,* 1–20.

Garfinkel, H. (1956). Conditions of successful degradation ceremonies. *American Journal of Sociology, 61,* 420–424.

Garland, D. (2001). *The culture of control: Crime and social order in contemporary society.* Oxford, UK: Oxford University Press.

Garofalo, J. (1987). Reassessing the lifestyle model of criminal victimization. In M. R. Gottfredson & T. Hirschi (Eds.), *Positive criminology* (pp. 23–42). Newbury Park, CA: Sage.

Garofalo, R. (1885). *Criminology.* Naples, Italy: N.P.

Gatti, U., Tremblay, R. E., & Vitaro, F. (2009). Iatrogenic effect of juvenile justice. *Child Psychology and Psychiatry, 50,* 991–998.

Gaukroger, S. (2006). *The emergence of a scientific culture: Science and the shape of modernity 1210–1685.* New York: Oxford University Press.

Gaylord, M. S., & Galliher, J. F. (1988). *The criminology of Edwin Sutherland*. New Brunswick, NJ: Transaction.

Gehring, K. S., Van Voorhis, P., & Bell, V. (2010). "What works" for female probationers? An evaluation of the Moving On Program. *Women, Girls, and Criminal Justice, 11*, 6–10.

Geis, G. (2000). On the absence of self-control as the basis for a general theory of crime: A critique. *Theoretical Criminology, 4*, 35–53.

Geis, G. (2007). *White-collar and corporate crime*. Upper Saddle River, NJ: Pearson Prentice Hall.

Geis, G. (2010). Sutherland, Edwin, H.: White-collar crime. In F. T. Cullen & P. Wilcox (Eds.), *Encyclopedia of criminological theory* (pp. 910–915). Thousand Oaks, CA: Sage.

Geis, G., & Goff, C. (1983). Introduction. In E. H. Sutherland (Ed.), *White collar crime: The uncut version* (pp. ix–xiii). New Haven, CT: Yale University Press.

Geis, G., & Goff, C. (1986). Edwin H. Sutherland's white-collar crime in America: An essay in historical criminology. *Criminal Justice History, 7*, 1–31.

Gelsthorpe, L. (1988). Feminism and criminology in Britain. *British Journal of Criminology, 28*, 93–110.

Gendreau, P., & Goggin, C. (2000). *Comments on restorative justice programmes in New Zealand*. Unpublished manuscript, Centre for Criminal Justice Studies, University of New Brunswick at Saint John.

Gendreau, P., Goggin, C., Cullen, F. T., & Andrews, D. A. (2000, May). The effects of community sanctions and incarceration on recidivism. *Forum on Corrections Research*, pp. 10–13.

Gendreau, P., Goggin, C., & Fulton, B. (2000). Intensive supervision in probation and parole. In C. R. Hollin (Ed.), *Handbook of offender assessment and treatment* (pp. 195–204). Chichester, UK: Wiley.

Gendreau, P., Smith, P., & French, S. (2006). The theory of effective correctional intervention: Empirical status and future directions. In F. T. Cullen, J. P. Wright, & K. R. Blevins (Eds.), *Taking stock: The status of criminological theory* (Advances in Criminological Theory, Vol. 15, pp. 419–446). New Brunswick, NJ: Transaction.

Gerra, G., Garofino, L., Castaldini, L., Rovetto, F., Zamovic, A., Moi, G., Bussandri, M., Branchi, B., Brambilla, F., Friso, G., & Donnini, C. (2005). Serotonin transporter promoter polymorphism genotype is associated with temperament, personality, and illegal drug use among adolescents. *Journal of Neural Transmission, 112*, 1435–1463.

Gettleman, J. (2010, January 4). Americans' role seen in Uganda anti-gay push. *New York Times*, p. A1.

Gibbons, D. C. (1979). *The criminological enterprise: Theories and perspectives*. Englewood Cliffs, NJ: Prentice Hall.

Gibbons, D. C. (1994). *Talking about crime and criminals: Problems and issues in theory development in criminology*. Englewood Cliffs, NJ: Prentice Hall.

Gibbons, D. C., & Garabedian, P. (1974). Conservative, liberal, and radical criminology: Some trends and observations. In C. E. Reasons (Ed.), *The criminologist: Crime and the criminal* (pp. 51–65). Pacific Palisades, CA: Goodyear.

Gibbons, F. X., Gerrard, M., Cleveland, M. J., Wills, T. A., & Brody, G. (2004). Perceived discrimination and substance use in African American parents and their children: A panel study. *Journal of Personality and Social Psychology, 86*, 517–529.

Gibbs, J. C. (2009, November). *Looking at terrorism through Left Realists lenses*. Paper presented at the annual meeting of the American Society of Criminology.

Gibbs, J. P. (1975). *Crime, punishment, and deterrence*. New York: Elsevier.

Gibbs, J. P. (1985). Review essay. *Criminology, 23*, 381–388.

The Gideon case 25 years later. (1988, March 16). *New York Times*, p. A27.

Gilfus, M. E. (1992). From victims to survivors to offenders: Women's routes of entry and immersion into street crime. *Women and Criminal Justice, 4*, 63–89.

Giordano, P. C., Cernkovich, S. A., & Rudolph, J. L. (2002). Gender, crime, and desistance: Toward a theory of cognitive transformation. *American Journal of Sociology, 107*, 990–1064.

Giordano, P. C., Kerbel, S., & Dudley, S. (1981). The economics of female criminality: An analysis of police blotters, 1890–1975. In L. H. Bower (Ed.), *Women and crime in America* (pp. 65–82). New York: Macmillan.

Gitlin, T. (1989). *The sixties: Years of hope, days of rage.* New York: Bantam.

Glueck, S., & Glueck, E. (1950). *Unraveling juvenile delinquency.* New York: Commonwealth Fund.

God and money: Sex scandal, greed, and lust for power split the TV preaching world. (1987, April 6). *Newsweek,* pp. 16–22.

Goddard, H. H. (1912). *The Kallikak family.* New York: Macmillan.

Goddard, H. H. (1914). *Feeblemindedness: Its causes and consequences.* New York: Macmillan.

Goddard, H. H. (1921). Feeblemindedness and delinquency. *Journal of Psycho-Asthenics, 25,* 168–176.

Goddard, H. H. (1927). Who is a moron? *Scientific Monthly, 24,* 41–46.

Goetz case: Commentary on nature of urban life. (1987, June 19). *New York Times,* p. A11.

Goldstein, H. (1979). Improving policing: A problem-oriented approach. *Crime & Delinquency, 25,* 234–258.

Goode, E. (Ed.). (2008). *Out of control: Assessing the general theory of crime.* Stanford, CA: Stanford University Press.

Goodstein, L. (1992). Feminist perspectives and the criminal justice curriculum. *Journal of Criminal Justice Education, 3,* 154–181.

Gordon, D. M. (1971). Class and the economics of crime. *Review of Radical Political Economy, 3,* 51–75.

Gordon, R. A. (1987). SES versus IQ in the race-IQ delinquency model. *International Journal of Sociology and Social Policy, 7,* 29–96.

Goring, C. (1913). *The English convict: A statistical study.* London: Her Majesty's Stationery Office.

Gottfredson, D. G., Wilson, D. B., & Najaka, S. S. (2002). School-based crime prevention. In L. W. Sherman, D. P. Farrington, B. C. Welsh, & D. L. MacKenzie (Eds.), *Evidence-based crime prevention* (pp. 56–164). London: Routledge.

Gottfredson, M. R. (2006). The empirical status of control theory in criminology. In F. T. Cullen, J. P. Wright, & K. R. Blevins (Eds.), *Taking stock: The status of criminological theory* (Advances in Criminological Theory,

Vol. 15, pp. 77–100). New Brunswick, NJ: Transaction.

Gottfredson, M. R. (2011). In pursuit of a general theory of crime. In F. T. Cullen, C. L. Jonson, A. J. Myer, & F. Adler (Eds.), *The origins of American criminology* (Advances in Criminological Theory, Vol. 16, pp. 333–346). New Brunswick, NJ: Transaction.

Gottfredson, M. R., & Hirschi, T. (1990). *A general theory of crime.* Stanford, CA: Stanford University Press.

Gould, S. J. (1981). *The mismeasure of man.* New York: W. W. Norton.

Gouldner, A. W. (1973). Foreword. In I. Taylor, P. Walton, & J. Young, *The new criminology: For a social theory of deviance* (pp. ix–xiv). London: Routledge & Kegan Paul.

Gove, W., & Wilmoth, C. K. (2003). The neurophysiology of motivation and habitual criminal behavior. In A. Walsh & L. Ellis (Eds.), *Biosocial criminology: Challenging environmentalism's supremacy* (pp. 227–245). Hauppauge, NY: Nova Science Publishers.

Gove, W. R. (Ed.). (1975). *The labeling of deviance: Evaluating a perspective.* Beverly Hills, CA: Sage.

Gove, W. R. (Ed.). (1980). *The labeling of deviance: Evaluating a perspective* (2nd ed.). Beverly Hills, CA: Sage.

Gow, D. (2006, January 19). Sans courage [Editorial]. *The Guardian.*

Graham, K., & Wells, S. (2003). "Somebody's gonna get their head kicked in tonight!" Aggression among young males in bars—A question of values? *British Journal of Criminology, 43,* 546–566.

Grasmick, H. G., & Bursik, R. J., Jr. (1990). Conscience, significant others, and rational choice: Extending the deterrence model. *Law and Society Review, 24,* 837–861.

Grasmick, H. G., Hagan, J., Blackwell, B. S., & Arneklev, B. J. (1996). Risk preferences and patriarch: Extending power-control theory. *Social Forces, 75,* 177–199.

Grasmick, H. G., Tittle, C. R., Bursik, R. J., Jr., & Arneklev, B. J. (1993). Testing the core empirical implications of Gottfredson and Hirschi's general theory of crime. *Journal of Research in Crime and Delinquency, 30,* 5–29.

Greenberg, D. F. (1977). Delinquency and the age structure of society. *Contemporary Crises, 1,* 189–233.

Greenberg, D. F. (1981). Introduction. In D. F. Greenberg (Ed.), *Crime and capitalism* (pp. 1–35). Palo Alto, CA: Mayfield.

Greenwood, P. W. (2006). *Changing lives: Delinquency prevention as crime-control policy.* Chicago: University of Chicago Press.

Griffin, S. (1971, September). Rape: The all-American crime. *Ramparts,* pp. 26–35.

Grossman, C. L. (2001, February 26). DiIluio keeps the faith. *USA Today,* p. D1.

Grossman, C. L. (2009, March 9). Most religious groups in USA have lost ground, survey finds. *USA Today,* p. D3.

Grosz, E. (1994). *Volatile bodies: Toward a corporeal feminism.* St. Leonards, NSW, Australia: Allen & Unwin.

Guerette, R. T. (2009). The pull, push, and expansion of situational crime prevention evaluation: An appraisal of thirty-seven years of research. In J. Knutsson & N. Tilley (Eds.), *Evaluating crime reduction initiatives* (Crime Prevention Studies, Vol. 24, pp. 29–58). Monsey, NY: Criminal Justice Press.

Gullo, K. (2001, February 19). Number in prison surged under Clinton. *Cincinnati Enquirer,* p. A1.

Guo, G., Roettger, M. E., & Shih, J. C. (2007). Contributions of the DAT1 and DRD2 genes to serious and violent delinquency among adolescents and young adults. *Human Genetics, 121,* 125–136.

Habermas, J. (1970). *Knowledge and human interests.* Boston: Beacon.

Habermas, J. (1971). *Toward a rational society.* Boston: Beacon.

Haederle, M. (2010, March–April). Trouble in mind: Will the new neuroscience undermine our legal system? *Miller McCune,* pp. 70–79.

Hagan, J. (1973). Labeling and deviance: A case study in the "sociology of the interesting." *Social Problems, 20,* 447–458.

Hagan, J. (1989). *Structural criminology.* New Brunswick, NJ: Rutgers University Press.

Hagan, J., Gillis, A. R., & Simpson, J. (1990). Clarifying and extending power-control theory. *American Journal of Sociology, 95,* 1024–1037.

Hagan, J., & Kay, F. (1990). Gender and delinquency in white-collar families: A power-control perspective. *Crime & Delinquency, 36,* 391–407.

Hagan, J., & McCarthy, B. (1997). *Mean streets: Youth crime and homelessness.* Cambridge, UK: Cambridge University Press.

Hahn, P. (1998). *Emerging criminal justice: Three pillars for a proactive justice system.* Thousand Oaks, CA: Sage.

Hail liberty: A birthday party album [of pictures]. (1986, July 14). *Time.*

Hall, R. M., & Sandler, B. R. (1985). A chilly climate in the classroom. In A. G. Sargent (Ed.), *Beyond sex roles* (pp. 503–511). St. Paul, MN: West.

Hall, S., & Winlow, S. (2004). Barbarians at the gate: Crime and violence in the breakdown of the pseudo-pacification process. In J. Ferrell, K. Hayward, W. Morrison, & M. Presdee (Eds.), *Cultural criminology unleashed* (pp. 275–286). London: Glasshouse.

Hall, S., Winlow, S., & Ancrum, C. (2008). *Criminal identities and consumer culture: Crime, exclusion and the new culture of narcissism.* Devon, UK: Willan.

Hallsworth, S. (2006). Cultural criminology unleashed [Book review]. *Criminology & Criminal Justice, 6*(1), 147–149.

Hanley, S., & Nellis, M. (2001). Crime, punishment, and community in England and Wales. In C. Jones-Finer (Ed.), *Comparing the social policy experience of Britain and Taiwan.* Aldershot, UK: Ashgate.

Harcourt, B. E. (2001). *Illusions of order: The false promise of broken windows policing.* Cambridge, MA: Harvard University Press.

Harris, J. R. (1995). Where is the child's environment? A group socialization theory of development. *Psychological Review, 102,* 458–489.

Harris, J. R. (1998). *The nurture assumption: Why children turn out the way they do.* New York: Free Press.

Harrison, P. M., & Beck, A. J. (2005). *Prisoners in 2004.* Washington, DC: U.S. Department of Justice, Bureau of Justice Statistics.

Hawkins, F. H. (1931). Charles Robert Darwin. In *Encyclopedia of the social sciences* (Vol. 5, pp. 4–5). New York: Macmillan.

Hawkins, G. (1976). *The prison: Policy and practice.* Chicago: University of Chicago Press.

Hawkins, J. D., & Herrenkohl, T. I. (2003). Prevention in the school years. In

D. P. Farrington & J. W. Coid (Eds.), *Early prevention of adult antisocial behavior* (pp. 265–291). Cambridge, UK: Cambridge University Press.

Hawley, A. H. (1950). *Human ecology: A theory of community structure*. New York: Ronald Press.

Hawley, F. F., & Messner, S. F. (1989). The southern violence construct: A review of arguments, evidence, and normative context. *Justice Quarterly, 6,* 481–511.

Hay, C. (2001). An exploratory test of Braithwaite's reintegrative shaming theory. *Journal of Research in Crime and Delinquency, 38,* 132–153.

Hayward, K. J., & Young, J. (2004). Cultural criminology: Some notes on the script. *Theoretical Criminology, 8,* 250–273.

Hayward, K. J., & Young, J. (2005, November). *Cultural criminology: Sharpening the focus.* Paper presented at the annual meetings of the American Society of Criminology, Toronto.

Hebdige, D. (1979). *Subculture: The meaning of style.* London: Methuen.

Heide, K. M., & Solomon, E. P. (2006). Biology, childhood trauma, and murder: Rethinking justice. *International Journal of Law and Psychiatry, 29,* 220–233.

Heidensohn, F. M. (1968). The deviance of women: A critique and an enquiry. *British Journal of Sociology, 19,* 160–176.

Heidensohn, F. M. (1985). *Women and crime: The life of the female offender.* New York: New York University Press.

Henggeler, S. W. (1997). *Treating serious antisocial behavior in youth: The MST approach.* Washington, DC: U.S. Department of Justice.

Henry, F., & Tator, C. (2002). *Discourse of domination: Racial bias in the Canadian English-Language press.* Toronto: University of Toronto.

Henry, S., & Milovanovic, D. (2005). Postmodernism and constitutive theories of criminal behavior. In R. A. Wright & J. M. Miller (Eds.), *Encyclopedia of criminology* (Vol. 2, pp. 1245–1249). New York: Routledge.

Herrnstein, R. J., & Murray, C. (1994). *The bell curve: Intelligence and class structure in American life.* New York: Free Press.

Heusenstamm, F. K. (1975). Bumper stickers and the cops. In D. J. Steffensmeier & R. M. Terry (Eds.), *Examining deviance experimentally: Selected readings* (pp. 251–255). Port Washington, NY: Alfred.

Hill, G., & Atkinson, M. P. (1988). Gender, familial control, and delinquency. *Criminology, 26,* 127–151.

Hills, S. L. (Ed.). (1987). *Corporate violence: Injury and death for profit.* Totowa, NJ: Rowman & Littlefield.

Hines, D. A. (2009). Domestic violence. In M. Tonry (Ed.), *The Oxford handbook of crime and public policy* (pp. 115–139). New York: Oxford University Press.

Hinkle, R. C., & Hinkle, G. J. (1954). *The development of modern sociology.* New York: Random House.

Hirschi, T. (1969). *Causes of delinquency.* Berkeley: University of California Press.

Hirschi, T. (1975). Labeling theory and juvenile delinquency: An assessment of the evidence. In W. R. Gove (Ed.), *The labeling of deviance: Evaluating a perspective* (pp. 181–201). Beverly Hills, CA: Sage.

Hirschi, T. (1983). Crime and the family. In J. Q. Wilson (Ed.), *Crime and public policy* (pp. 53–68). San Francisco: Institute for Contemporary Studies.

Hirschi, T. (1989). Exploring alternatives to integrated theory. In S. F. Messner, M. D. Krohn, & A. E. Liska (Eds.), *Theoretical integration in the study of deviance and crime: Problems and prospects* (pp. 37–49). Albany: State University of New York Press.

Hirschi, T. (2004). Self-control and crime. In R. F. Baumeister & K. D. Vohs (Eds.), *Handbook of self-regulation: Research, theory, and applications* (pp. 537–552). New York: Guilford Press.

Hirschi, T., & Gottfredson, M. R. (1995). Control theory and the life-course perspective. *Studies on Crime and Crime Prevention, 4,* 131–142.

Hirschi, T., & Hindelang, M. J. (1977). Intelligence and delinquency: A revisionist review. *American Sociological Review, 42,* 571–586.

Hochstedler, E. (Ed.). (1984). *Corporations as criminals.* Beverly Hills, CA: Sage.

Hoffman, J. P., & Miller, A. S. (1998). A latent variable analysis of general strain theory. *Journal of Quantitative Criminology, 14,* 83–110.

Hoffman, J. P., & Su, S. S. (1997). The conditional effects of stress on delinquency and drug use: A strain theory assessment of sex differences. *Journal of Research in Crime and Delinquency, 34,* 46–78.

Hofstadter, R. (1955a). *Age of reform.* New York: Knopf.

Hofstadter, R. (1955b). *Social Darwinism in American thought.* Boston: Beacon.

Hofstadter, R. (1963). *The Progressive movement: 1900 to 1915.* New York: Touchstone.

Holzer, H. J., Raphael, S., & Stoll, M. A. (2004). Will employers hire former offenders? Employer preferences, background checks, and their determinants. In M. Pattillo, D. Weiman, & B. Western (Eds.), *Imprisoning America: The social effects of mass incarceration* (pp. 205–243). New York: Russell Sage.

Home Office. (2004). *Offender management caseload statistics, 2004.* London: The Stationery Office.

Hooton, E. A. (1939). *Crime and the man.* Cambridge, MA: Harvard University Press.

Horney, J., Osgood, D. W., & Marshall, I. H. (1995). Criminal careers in the short term: Intra-individual variability in crime and its relation to local life circumstances. *American Sociological Review, 60,* 655–673.

Howard, J. (1973). *The state of the prisons* (4th ed.). Montclair, NJ: Patterson Smith. (Original work published 1792)

Howard League for Penal Reform. (2005, January 18). *England and Wales lead the pack on European imprisonment rates* [Press release].

Hudson, B. (1998). Restorative justice: The challenge of sexual and racial violence. *Journal of Law and Society, 25,* 237–256.

Huff, C. R., & Scarpitti, F. R. (2011). The origins and development of containment theory: Walter C. Reckless and Simon Dinitz. In F. T. Cullen, C. L. Jonson, A. J. Myer, & F. Adler (Eds.), *The origins of American criminology* (Advances in Criminological Theory, Vol. 16, pp. 277–294). New Brunswick, NJ: Transaction.

Hughes, E. C. (1945). Dilemmas and contradictions of statuses. *American Journal of Sociology, 50,* 353–359.

Hulse, C. (2010, February 27). Bill to extend jobless benefits hits a partisan roadblock. *New York Times,* p. A12.

Human Rights Watch. (1998). Women raped in prison face retaliation; Michigan failing to protect inmates, says rights group.

Retrieved June 20, 2006, from http://www.commondreams.org/pressreleases/Sept98/092198c.htm

Hunt, M. M. (1961, January 28). How does it come to be so? Profile of Robert K. Merton. *The New Yorker,* pp. 39–64.

Immarigeon, R., & Daly, K. (1997, December). Restorative justice: Origins, practices, contexts, and challenges. *ICCA Journal of Community Corrections, 8,* 13–18.

Inciardi, J. A. (Ed.). (1980). *Radical criminology: The coming crises.* Beverly Hills, CA: Sage.

Inciardi, J. A. (1986). *The war on drugs: Heroin, cocaine, crime, and public policy.* Palo Alto, CA: Mayfield.

Irwin, J. (1970). *The felon.* Englewood Cliffs, NJ: Prentice Hall.

Irwin, J. (1980). *Prisons in turmoil.* Boston: Little, Brown.

Irwin, J. (1985). *The jail.* Berkeley: University of California Press.

Irwin, J. (2005). *The warehouse prison: Disposal of the new dangerous class.* Los Angeles: Roxbury.

Irwin, J., & Austin, J. (1994). *It's about time: America's imprisonment binge.* Belmont, CA: Wadsworth.

Irwin, J., & Cressey, D. (1962). Thieves, convicts, and inmate culture. *Social Problems, 2,* 142–155.

Ivar Kreuger. (2010). Retrieved April 11, 2010, from http://en.wikipedia.org/wiki/Ivar_Kreuger

Jacobs, J. (1961). *The death and life of great American cities.* New York: Random House.

Jaggar, A. M. (1983). *Feminist politics and human nature.* Totowa, NJ: Rowman & Allanheld.

Jaggar, A. M., & Rothenberg, P. (Eds.). (1984). *Feminist frameworks.* New York: McGraw-Hill.

Jeffery, C. R. (1977). *Crime prevention through environmental design* (2nd ed.). Beverly Hills, CA: Sage.

Jenkins, P. (1987). *Mrs. Thatcher's revolution: The ending of the socialist era.* London: Pan Books.

Jensen, G. F., & Thompson, K. (1990). What's class got to do with it? A further examination of power-control theory. *American Journal of Sociology, 95,* 1009–1023.

Jesilow, P., Pontell, H. N., & Geis, G. (1993). *Prescription for profit: How doctors defraud Medicaid.* Berkeley: University of California Press.

Johnson, B. R. (2004). Religious programs and recidivism among former inmates in Prison Fellowship programs: A long-term follow-up study. *Justice Quarterly, 21,* 329–354.

Jonson, C. L. (2010). *The impact of imprisonment on reoffending: A meta-analysis.* Unpublished doctoral dissertation, University of Cincinnati, Cincinnati, OH.

Jonson, C. L., & Geis, G. (2010). Cressey, Donald R.: Embezzlement and white-collar crime. In F. T. Cullen & P. Wilcox (Eds.), *Encyclopedia of criminological theory* (pp. 223–230). Thousand Oaks, CA: Sage.

Josephson, E., & Josephson, M. (1962). *Man alone: Alienation in modern society.* New York: Dell.

Justice Policy Institute. (2010, February). *The Obama administration's budget: More policing, prisons, and punitive policies.* Washington, DC: Author.

Kalb, C. (2006, March 27). The therapist as scientist. *Newsweek,* pp. 50–51.

Kamin, L. L. (1985, April). Is crime in the genes? The answer may depend on who chooses the evidence. *Scientific American,* pp. 22–25.

Karjane, H. M., Fisher, B. S., & Cullen, F. T. (2005). *Sexual assault on campus: What colleges and universities are doing about it.* Washington, DC: National Institute of Justice and Bureau of Justice Statistics.

Karoly, L. A., Greenwood, P. W., Everingham, S. S., Hoube, J., Kilburn, M. R., Rydell, C. P., Sanders, M., & Chiesa, J. (1998). *Investing in our children: What we know and don't know about the costs and benefits of early childhood interventions.* Santa Monica, CA: RAND.

Katayama, L. (2005). Reforming California's prisons: An interview with Jackie Speier. Retrieved June 20, 2006, from http://www.motherjones.com/news/qa/2005/07/jackie_speier.html

Katz, J. (1980). The social movement against white-collar crime. In E. Bitner & S. Messinger (Eds.), *Criminology review yearbook* (Vol. 2, pp. 161–184). Beverly Hills, CA: Sage.

Katz, J. (1988). *Seductions of crime: Moral and sensual attractions of doing evil.* New York: Basic Books.

Katz, J., & Abel, C. F. (1984). The medicalization of repression: Eugenics and crime. *Contemporary Crises, 8,* 227–241.

Katz, R. (2000). Explaining girls' and women's crime and desistance in the context of their victimization experiences. *Violence Against Women, 6,* 633–660.

Keane, C., Maxim, P. S., & Teevan, J. J. (1993). Drinking and driving, self-control, and gender. *Journal of Research in Crime and Delinquency, 30,* 30–46.

Kelling, G. L., & Coles, C. M. (1996). *Fixing broken windows: Restoring order and reducing crime in our communities.* New York: Simon & Schuster.

Kempf, K. L. (1993). The empirical status of Hirschi's control theory. In F. Adler & W. S. Laufer (Eds.), *New directions in criminological theory* (Advances in Criminological Theory, Vol. 4, pp. 143–185). New Brunswick, NJ: Transaction.

Kennedy, P. (1987, August). The (relative) decline of America. *The Atlantic,* pp. 29–38.

Kitsuse, J. I. (1964). Societal reaction to deviant behavior: Problems of theory and method. In H. Becker (Ed.), *The other side* (pp. 87–102). New York: Free Press.

Klein, D. (1973). The etiology of female crime: A review of the literature. *Issues in Criminology, 8,* 3–30.

Klein, M. W. (1979). Deinstitutionalization and diversion of juvenile offenders: A litany of impediments. In N. Morris & M. Tonry (Eds.), *Crime and justice: An annual review of research* (Vol. 1, pp. 145–201). Chicago: University of Chicago Press.

Klein, M. W. (1986). Labeling theory and delinquency policy: An experimental test. *Criminal Justice and Behavior, 13,* 47–79.

Klein, M. W. (2007). *The shock doctrine: The rise of disaster capitalism.* Toronto: Knopf.

Knopp, F. H. (1991). Community solutions to sexual violence: Feminist/abolitionist perspectives. In H. Pepinsky & R. Quinney (Eds.), *Criminology as peacemaking* (pp. 181–193). Bloomington: Indiana University Press.

Kobrin, S. (1959). The Chicago Area Project: A 25-year assessment. *Annals of the American Academy of Political and Social Sciences, 332,* 19–29.

Kornhauser, R. R. (1978). *Social sources of delinquency: An appraisal of analytical models.* Chicago: University of Chicago Press.

Kramer, R. C. (1992). The Space Shuttle *Challenger* explosion: A case study in state-corporate crime. In K. Schlegel & D. Weisburd (Eds.), *White-collar crime reconsidered* (pp. 214–243). Boston: Northeastern University Press.

Kramer, R. C., Michalowski, R. J., & Kauzlarich, D. (2003). The origin and development of the concept and theory of state-corporate crime. *Crime & Delinquency, 48,* 263–282.

Kretschmer, E. (1925). *Physique and character* (W. J. Sprott, Trans.). New York: Harcourt Brace.

Krisberg, B., & Austin, J. (1978). *The children of Ishmael: Critical perspectives on juvenile justice.* Palo Alto, CA: Mayfield.

Kruttschnitt, C., & Green, D. (1984). The sex-sanctioning issue: Is it history? *American Sociological Review, 49,* 541–551.

Kubrin, C. E., Stucky, T. D., & Krohn, M. D. (2009). *Researching theories of crime and deviance.* New York: Oxford University Press.

Kurki, L. (2000). Restorative and community justice in the United States. In M. Tonry (Ed.), *Crime and justice: A review of research* (Vol. 27, pp. 235–303). Chicago: University of Chicago Press.

Lacan, J. (2006). *Ecrits.* London: W. W. Norton.

Lacey, N. (1997). On the subject of sexing the subject. In N. Naffine & R. J. Owens (Eds.), *Sexing the subject of law* (pp. 65–76). Sydney, Australia: LBC Information Services.

Lack of figures on racial strife fueling dispute. (1987, April 5). *New York Times,* p. A13.

Larsson, H., Viding, E., & Plomin, R. (2008). Callous-unemotional traits and antisocial behavior: Genetic, environmental, and early parenting characteristics. *Criminal Justice and Behavior, 35,* 197–211.

Lasch, C. (1978). *The culture of narcissism.* New York: W. W. Norton.

Latessa, E. J. (1987). The effectiveness of intensive supervision with high risk probationers. In B. R. McCarthy (Ed.), *Intermediate punishments: Intensive supervision, home confinement, and electronic surveillance* (pp. 99–112). Monsey, NY: Criminal Justice Press.

Latimer, J., Dowden, C., & Muise, D. (2005). The effectiveness of restorative justice practices: A meta-analysis. *Prison Journal, 85,* 127–144.

Laub, J. H. (1983). *Criminology in the making: An oral history.* Boston: Northeastern University Press.

Laub, J. H. (2002). Introduction: The life and work of Travis Hirschi. In T. Hirschi, *The craft of criminology: Selected papers* (pp. xi–xlix). New Brunswick, NJ: Transaction.

Laub, J. H., & Sampson, R. J. (1991). The Sutherland-Glueck debate: On the sociology of criminological knowledge. *American Journal of Sociology, 6,* 1402–1440.

Laub, J. H., & Sampson, R. J. (2003). *Shared beginnings, divergent lives: Delinquent boys to age 70.* Cambridge, MA: Harvard University Press.

Laub, J. H., Sampson, R. J., & Allen, L. C. (2001). Explaining crime over the life course: Toward a theory of age-graded informal social control. In R. Paternoster & R. Bachman (Eds.), *Explaining criminals and crime: Essays in contemporary criminological theory* (pp. 97–112). Los Angeles: Roxbury.

Laub, J. H., Sampson, R. J., Corbett, R. P., Jr., & Smith, J. S. (1995). The public policy implications of a life-course perspective on crime. In H. Barlow (Ed.), *Crime and public policy: Putting theory to work* (pp. 91–106). Boulder, CO: Westview.

Laub, J. H., Sampson, R. J., & Sweeten, G. (2006). Assessing Sampson and Laub's life-course theory of crime. In F. T. Cullen, J. P. Wright, & K. R. Blevins (Eds.), *Taking stock: The status of criminological theory* (Advances in Criminological Theory, Vol. 15, pp. 313–333). New Brunswick, NJ: Transaction.

Lea, J. (1998). Criminology and postmodernity. In P. Walton & J. Young (Eds.), *The new criminology revisited* (pp. 163–198). New York: St. Martin's.

Lea, J. (2005). *Terrorism, crime and the collapse of civil liberties.* Lecture presented to the *Criminology Society,* Middlesex University, in April. Retrieved June 20, 2006, from http://www.bunker8.pwp.blueyonder.co.uk/misc/terror.htm

Le Blanc, M., & Loeber, R. (1998). Developmental criminology updated. In M. Tonry (Ed.), *Crime and justice: A review of research* (Vol. 23, pp. 115–198). Chicago: University of Chicago Press.

Lederer, G. (1961). *A nation of sheep.* New York: Fawcett.

Lemert, E. M. (1951). *Social pathology.* New York: McGraw-Hill.

Lemert, E. M. (1972). *Human deviance, social problems, and social control* (2nd ed.). Englewood Cliffs, NJ: Prentice Hall.

Levitt, S. D. (2002). Deterrence. In J. Q. Wilson & J. Petersilia (Eds.), *Crime: Public policies for crime control* (pp. 435–450). Oakland, CA: ICS Press.

Levrant, S., Cullen, F. T., Fulton, B., & Wozniak, J. F. (1999). Reconsidering restorative justice: The corruption of benevolence revisited? *Crime & Delinquency, 45,* 3–27.

Lewis, M. (2010). *The big short: Inside the doomsday machine.* New York: W. W. Norton.

Liao, D., Hong, C., Shih, H., & Tsai, S. (2004). Possible association between serotonin transporter promoter region polymorphism and extremely violent behavior in Chinese males. *Neuropsychobiology, 50,* 284–287.

Liederbach, J., Cullen, F. T., Sundt, J. L., & Geis, G. (2001). The criminalization of physician violence: Social control in transformation? *Justice Quarterly, 18,* 141–170.

Lilly, J. R. (1992). Selling justice: Electronic monitoring and the security industry. *Justice Quarterly, 9,* 493–504.

Lilly, J. R. (2003). *La Face Cachee Des GI's: Les viols commis par des soldats americains en France, en Angleterre et en Allemagne pendant la Seconde Guerre mondiale.* Paris: Payot.

Lilly, J. R. (2004). *Stupri Di Guerra: Le Violenze Commesse Dai Soldati Americani in Gran Bretagna, Francia e Germania 1942–1945.* Milano: Mursia.

Lilly, J. R. (2006a). Issues beyond empirical EM reports. *Criminology & Public Policy, 5,* 501–510.

Lilly, J. R. (2006b). Surveillance electronique et politiques penales aux Etats-Unis: l'etat des lieux en 2005. In R. Levy & X. Lameyre (Eds.), *Poursuivre et puir sans emprisonner* (pp. 25–50). Paris: Livre Broche.

Lilly, J. R. (2007). *Taken by force: Rape and American soldiers in the European theatre of operations during World War II.* London: Palgrave.

Lilly, J. R. (2010). Taylor, Ian Paul Walton, and Jock Young. The new criminology. In F. T. Cullen & P. Wilcox (Eds.), *Encyclopedia of criminological theory* (pp. 936–940). Thousand Oaks, CA: Sage.

Lilly, J. R., & Ball, R. A. (1993). Selling justice: Will electronic monitoring last? *Northern Kentucky Law Review, 20,* 505–530.

Lilly, J. R., & Deflem, M. (1993, June). Penologie en profit. *Delinkt en Delinkent,* pp. 551–557.

Lilly, J. R., & Jeffrey, W., Jr. (1979). On the state of criminology: A review of a classic. *Crime & Delinquency, 25,* 95–103.

Lilly, J. R., & Knepper, P. (1992). An international perspective on the privatization of corrections. *Howard Journal of Criminal Justice, 31,* 174–191.

Lilly, J. R., & Knepper, P. (1993). The corrections-commercial complex. *Crime & Delinquency, 39,* 150–166.

Lilly, J. R., & Marshall, P. (2000). Rape—Wartime. In C. B. Bryant (Ed.), *Encyclopedia of criminology and deviant behavior* (Vol. 3, pp. 318–322). Philadelphia: Taylor & Francis.

Lilly, J. R., & Nellis, M. (2001). Home detention curfew and the future of tagging. *Prison Service Journal, 135,* 59–64.

Link, B. G., Cullen, F. T., Frank, J., & Wozniak, J. F. (1987). The social rejection of former mental patients: Understanding why labels matter. *American Journal of Sociology, 92,* 1461–1500.

Lipset, S. M., & Schneider, W. (1983). *The confidence gap: Business, labor, and government in the public mind.* New York: Free Press.

Lipsey, M. W. (2009). The primary factors that characterize effective interventions with juvenile offenders: A meta-analytic overview. *Victims and Offenders, 4,* 124–147.

Lipsey, M. W., Chapman, G. L., & Landenberger, N. A. (2001). Cognitive-behavior programs for offenders. *Annals of the American Academy of Political and Social Science, 578,* 144–157.

Lipsey, M. W., & Cullen, F. T. (2007). The effectiveness of correctional rehabilitation: A review of systematic reviews. *Annual Review of Law and Social Sciences, 3,* 297–320.

Liptak, A. (2006, March 2). Prisons often shackle pregnant inmates in labor. *New York Times*, pp. A1, A16.

Liska, A. E. (1981). *Perspectives on deviance.* Englewood Cliffs, NJ: Prentice Hall.

Listwan, S. J., Cullen, F. T., & Latessa, E. J. (2006). *How to prevent prisoner reentry from failing: Insights from evidence-based corrections. Federal Probation, 70*(3), 19–25.

Living in terror. (1994, July 4). *Newsweek*, pp. 26–33.

Loeber, R., & Farrington, D. P. (Eds.). (1998). *Serious and violent juvenile offenders: Risk factors and successful interventions.* Thousand Oaks, CA: Sage.

Loeber, R., & Le Blanc, M. (1990). Toward a developmental criminology. In M. Tonry & N. Morris (Eds.), *Crime and justice: A review of research* (Vol. 12, pp. 375–473). Chicago: University of Chicago Press.

Loeber, R., & Stouthamer-Loeber, M. (1996). The development of offending. *Criminal Justice and Behavior, 23,* 12–24.

Lofland, L. H. (1973). *A world of strangers: Order and action in urban public space.* New York: Basic Books.

Lombroso, C. (1876). *On criminal man.* Milan, Italy: Hoepli.

Lombroso, C. (1920). *The female offender.* New York: Appleton. (Original work published 1903)

Lombroso-Ferrero, G. (1972). *Criminal man according to the classification of Cesare Lombroso.* Montclair, NJ: Patterson Smith.

Longshore, D., Chang, E., & Messina, N. (2005). Self-control and social bonds: A combined control perspective. *Journal of Quantitative Criminology, 21,* 419–437.

Lopez-Rangel, E., & Lewis, M. E. S. (2006). Loud and clear evidence for gene silencing by epigenetic mechanisms in autism spectrum and related neurodevelopmental disorders. *Clinical Genetics, 69,* 21–22.

Losel, F., & Bender, D. (2003). Protective factors and resilience. In D. P. Farrington & J. W. Coid (Eds.), *Early prevention of adult antisocial behavior* (pp. 130–204). Cambridge, UK: Cambridge University Press.

Lowenkamp, C. T., Cullen, F. T., & Pratt, T. C. (2003). Replicating Sampson and Grove's test of social disorganization theory: Revisiting a criminological classic. *Journal*

of Research in Crime and Delinquency, 40, 351–373.

Lyall, S. (2008, December 5). European court rules against Britain's policy of keeping DNA database of suspects. *New York Times,* p. A19.

Lynam, D. R., Milich, R., Zimmerman, R., Novak, S. P., Logan, T. K., Martin, C., Leukefield, C., & Clayton, R. (1999). Project DARE: No effects at 10-year follow-up. *Journal of Consulting and Clinical Psychology, 67,* 590–593.

Lynch, J. P. (2002). Crime in international perspective. In J. Q. Wilson & J. Petersilia (Eds.), *Crime: Public policies for crime control* (pp. 5–41). Oakland, CA: ICS Press.

Lynch, J. P., & Sabol, W. J. (2004). Effects of incarceration on informal social control in communities. In M. Pattillo, D. Weiman, & B. Western (Eds.), *Imprisoning America: The social effects of mass incarceration* (pp. 135–164). New York: Russell Sage.

Lynch, M. J. (1999). Beating a dead horse: Is there any basic empirical evidence for the deterrent effect of imprisonment? *Crime, Law, and Social Change, 31,* 347–362.

Lynch, M. J. (2000). The power of oppression: Understanding the history of criminology as a science of oppression. *Critical Criminology, 9,* 144–152.

Lynch, M. J., Cole, S. A., McNally, R., & Jordan, K. (2008). *Truth machine: The contentious history of DNA fingerprinting.* Chicago: University of Chicago Press.

Lynch, M. J., Schwendinger, H., & Schwendinger, J. (2006). The state of empirical research in radical criminology. In F. T. Cullen, J. P. Wright, & K. R. Blevins (Eds.), *Taking stock: The status of criminological theory* (Advances in Criminological Theory, Vol. 15, pp. 191–215). New Brunswick, NJ: Transaction.

Lynch, M. J., & Stretesky, P. B. (2006). The new radical criminology and the same old criticisms. In S. Henry & M. M. Lanier (Eds.), *The essential criminology reader.* Boulder, CO: Westview.

MacKenzie, D. L. (2000). Evidence-based corrections: Identifying what works. *Crime & Delinquency, 46,* 457–471.

MacKenzie, D. L. (2006). *What works in corrections: Reducing the criminal activities of offenders and delinquents*. New York: Cambridge University Press.

MacKenzie, D. L., Ellis, L., Simpson, S., et al. (1994). *Female offenders in boot camp*. College Park: University of Maryland Press.

Maddan, S., Walker, J. T., & Miller, J. M. (2008). Does size really matter? A reexamination of Sheldon's somatypes and criminal behavior. *Social Science Journal, 45,* 330–344.

Madge, J. (1962). *The origins of scientific sociology*. New York: Free Press.

Maher, L. (1997). *Sexed work: Gender, race and resistance in a Brooklyn drug market*. Oxford, UK: Clarendon.

Maher, L., & Curtis, R. (1992). Women on the edge: Crack cocaine and the changing contexts of street-level sex work in New York City. *Crime, Law and Social Change, 18,* 221–258.

Makkai, T., & Braithwaite, J. (1991). Criminological theories and regulatory compliance. *Criminology, 29,* 191–220.

Mankoff, M. (1971). Societal reaction and career deviance: A critical analysis. *Sociological Quarterly, 12,* 204–218.

Manning, M. A., & Hoyme, H. E. (2007). Fetal alcohol syndrome disorders: A practical clinical approach to diagnosis. *Neuroscience and Biobehavioral Review, 31,* 230–238.

Manning, P. (1998). Media loops. In F. Bailey & D. Hale (Eds.), *Popular culture, crime, and justice* (pp. 25–39). Belmont, CA: Wadsworth.

Many rape victims finding justice through civil courts. (1991, September 20). *New York Times,* p. A1.

Manza, J., & Uggen, C. (2006). *Locked out: Felon disenfranchisement and American democracy*. New York: Oxford University Press.

Marcus, B. (2004). Self-control in the general theory of crime: Theoretical implications of a measurement problem. *Theoretical Criminology, 8,* 33–55.

Marcuse, H. (1960). *Reason and revolution*. Boston: Beacon.

Marcuse, H. (1964). *One dimensional man*. Boston: Beacon.

Marcuse, H. (1972). *Counter-revolution and revolt*. Boston: Beacon.

Martindale, D. (1960). *The nature and types of sociological theory*. Cambridge, MA: Houghton Mifflin.

Maruna, S. (2001). *Making good: How ex-convicts reform and rebuild their lives*. Washington, DC: American Psychological Association.

Maruna, S., & Copes, H. (2005). What have we learned from five decades of neutralization research? In M. Tonry (Ed.), *Crime and justice: A review of research* (Vol. 32, pp. 221–320). Chicago: University of Chicago Press.

Maruna, S., & Roy, K. (2007). Amputation or reconstruction? Notes on the concept of "knifing off" and desistance from crime. *Journal of Contemporary Criminal Justice, 23,* 104–124.

Marx, K., & Engels, F. (1992). *Communist manifesto*. New York: Bantam Books. (Original work published 1848)

Massey, D. R. (2007). *Categorically unequal: The American stratification system*. New York: Russell Sage.

Mathiesen, T. (1974). *The politics of abolition*. New York: Halstead.

Matsueda, R. L., Kreager, D. A., & Huizinga, D. (2006). Deterring delinquents: A rational model of theft and violence. *American Sociological Review, 71,* 95–122.

Matthews, R. (2005). Cultural criminology unleashed [Book review]. *British Journal of Criminology, 45,* 419–420.

Matthews, R. (2009). Beyond "so what?" criminology. *Theoretical Criminology, 13,* 341–362.

Matthews, R., & Young, J. (1986). Editors' introduction. In R. Matthews & J. Young (Eds.), *Confronting crime* (p. 1). London: Sage.

Matthews, R., & Young, J. (Eds.). (1992). *Issues in realist criminology*. London: Sage.

Matza, D. (1964). *Delinquency and drift*. New York: John Wiley.

Matza, D. (1969). *Becoming deviant*. Englewood Cliffs, NJ: Prentice Hall.

Mauer, M. (1999). *Race to incarcerate*. New York: New Press.

Mauer, M., & Chesney-Lind, M. (Eds.). (2002). *Invisible punishment: The collateral consequences of mass imprisonment*. New York: New Press.

Maume, D. J., Jr. (1989). Inequality and metropolitan rape rates: A routine activity approach. *Justice Quarterly, 6,* 513–527.

Maume, M. O., & Lee, M. R. (2003). Social institutions and violence: A sub-national test of institutional-anomie theory. *Criminology, 41,* 1137–1172.

Mazerolle, P. (1998). Gender, general strain, and delinquency: An empirical examination. *Justice Quarterly, 15,* 65–91.

Mazerolle, P., Burton, V. S., Jr., Cullen, F. T., Evans, T. D., & Payne, G. L. (2000). Strain, anger, and delinquent adaptations: Specifying general strain theory. *Journal of Criminal Justice, 28,* 89–101.

Mazerolle, P., & Maahs, J. (2000). General strain and delinquency: An alternative examination of conditioning influences. *Justice Quarterly, 17,* 753–778.

Mazerolle, P., & Piquero, A. (1997). Violent responses to strain: An examination of conditioning influences. *Violence and Victims, 12,* 323–343.

McCarthy, B., & Hagan, J. (1992). Mean streets: The theoretical significance of situational delinquency among homeless youths. *American Journal of Sociology, 98,* 597–627.

McCarthy, B., Hagan, J., & Woodward, T. S. (1999). In the company of women: Structure and agency in a revised power-control theory of gender and delinquency. *Criminology, 37,* 761–788.

McDermott, M. J. (1992). The personal is empirical: Feminism, research methods, and criminal justice education. *Journal of Criminal Justice Education, 3,* 237–249.

McGarrell, E. F., & Hipple, N. K. (2007). Family group conferencing and re-offending among first-time juvenile offenders: The Indianapolis experiment. *Justice Quarterly, 24,* 221–246.

Mead, G. H. (1934). *Mind, self, and society* (C. W. Morris, Ed.). Chicago: University of Chicago Press.

Mead, M. (1928). *Growing up in Samoa: A psychological study of primitive youth for Western civilization.* Oxford, UK: Morrow.

Mealey, L. (1995). The sociobiology of sociopathy: An integrated evolutionary model. *Behavioral and Brain Sciences, 18,* 523–599.

Mednick, S. A. (1977). A biological theory of the learning of law-abiding behavior. In S. A. Mednick & K. O. Christiansen (Eds.), *Biosocial bases of criminal behavior* (pp. 1–8). New York: Gardner.

Mednick, S. A., & Christiansen, K. O. (Eds.). (1977). *Biosocial bases of criminal behavior.* New York: Gardner.

Mednick, S. A., Gabrielli, W., & Hutchings, B. (1984). Genetic influences in criminal convictions: Evidence from an adoption cohort. *Science, 224,* 891–894.

Mednick, S. A., Gabrielli, W., & Hutchings, B. (1987). Genetic factors in the etiology of criminal behavior. In S. A. Mednick, T. E. Moffitt, & S. A. Stack (Eds.), *The causes of crime: New biological approaches* (pp. 74–91). New York: Cambridge University Press.

Mednick, S. A., Moffitt, T. E., & Stack, S. A. (Eds.). (1987). *The causes of crime: New biological approaches.* New York: Cambridge University Press.

Mednick, S. A., Pollack, V., Volavka, J., & Gabrielli, W. F. (1982). Biology and violence. In M. E. Wolfgang & N. A. Weiner (Eds.), *Criminal violence* (pp. 21–79). Beverly Hills, CA: Sage.

Mednick, S. A., & Shoham, G. (1979). *New paths in criminology.* Lexington, MA: Lexington Books.

Mednick, S. A., Volavka, J., Gabrielli, W. F., & Itil, T. M. (1981). EEG as a predictor of antisocial behavior. *Criminology, 19,* 219–229.

Meese: Execute teen-age killers. (1985, September 5). *Cincinnati Enquirer,* pp. A1, A6.

Meese says court doesn't make law. (1986, October 23). *New York Times,* pp. A1, A20.

Meese seen as ready to challenge rule on telling suspects of rights. (1987, January 22). *New York Times,* p. A13.

Meier, R. F., & Miethe, T. D. (1993). Understanding theories of criminal victimization. In M. Tonry (Ed.), *Crime and justice: A review of research* (Vol. 17, pp. 459–499). Chicago: University of Chicago Press.

Merton, R. K. (1938). Social structure and anomie. *American Sociological Review, 3,* 672–682.

Merton, R. K. (1957). Priorities in scientific discovery: A chapter in the sociology of science. *American Sociological Review, 22,* 635–659.

Merton, R. K. (1959). Social conformity, deviation, and opportunity structures: A comment on the contributions of Dubin and Cloward. *American Sociological Review, 24,* 177–189.

Merton, R. K. (1964). Anomie, anomia, and social interaction: Contexts of deviant behavior. In M. B. Clinard (Ed.), *Anomie and deviant behavior* (pp. 213–242). New York: Free Press.

Merton, R. K. (1968). *Social theory and social structure.* New York: Free Press.

Merton, R. K. (1984). "Crime and the anthropologist": An historical postscript. In R. W. Rieber (Ed.), *Advances in forensic psychology* (Vol. 1, pp. 171–173). Norwood, NJ: Ablex.

Merton, R. K. (1995). Opportunity structure: The emergence, diffusion, and differentiation of a sociological concept, 1930s–1950s. In F. Adler & W. S. Laufer (Eds.), *The legacy of anomie theory* (Advances in Criminological Theory, Vol. 6, pp. 3–78). New Brunswick, NJ: Transaction.

Merton, R. K., & Ashley-Montagu, M. F. (1940). Crime and the anthropologist. *American Anthropologist, 42,* 384–408.

Messerschmidt, J. W. (1986). *Capitalism, patriarchy, and crime: Toward a socialist feminist criminology.* Totowa, NJ: Rowman & Littlefield.

Messerschmidt, J. W. (1993). *Masculinities and crime: Critique and reconceptualization of theory.* Totowa, NJ: Rowman & Littlefield.

Messner, S. F. (1988). Merton's "social structure and anomie": The road not taken. *Deviant Behavior, 9*(1), 33–53.

Messner, S. F., Krohn, M. D., & Liska, A. E. (Eds.). (1989). *Theoretical integration in the study of deviance and crime: Problems and prospects.* Albany: State University of New York Press.

Messner, S. F., & Rosenfeld, R. (1994). *Crime and the American dream.* Belmont, CA: Wadsworth.

Messner, S. F., & Rosenfeld, R. (1997). Political restraint of the market and levels of criminal homicide: A cross-national application of institutional anomie theory. *Social Forces, 75,* 1393–1416.

Messner, S. F., & Rosenfeld, R. (2001). *Crime and the American dream* (3rd ed.). Belmont, CA: Wadsworth.

Messner, S. F., & Rosenfeld, R. (2006). The present and future of institutional-anomie theory. In F. T. Cullen, J. P. Wright, & K. R. Blevins (Eds.), *Taking stock: The status of criminological theory* (Advances in Criminological Theory, Vol. 15, pp. 127–148). New Brunswick, NJ: Transaction.

Michalowski, R. J., & Bolander, E. W. (1976). Repression and criminal justice in capitalist America. *Sociological Inquiry, 46,* 95–106.

Miles, R. (1989). *The women's history of the world.* London: Paladin.

Milibrand, R. (1969). *The state in capitalist society.* New York: Basic Books.

Miller, A. D., & Ohlin, L. E. (1985). *Delinquency and community: Creating opportunities and controls.* Beverly Hills, CA: Sage.

Miller, J. (2008). *Getting played: African American girl, urban inequality, and gendered violence.* New York: New York University Press.

Miller, J., & Mullins, C. W. (2006). The status of feminist theories in criminology. In F. T. Cullen, J. P. Wright, & K. R. Blevins (Eds.), *Taking stock: The status of criminological theory* (Advances in Criminological Theory, Vol. 15, pp. 217–249). New Brunswick, NJ: Transaction.

Miller, J. G. (1991). *Last one over the wall: The Massachusetts experiment in closing reform schools.* Columbus: Ohio State University Press.

Miller, L. (2010, January 25). Why God hates Haiti: The frustrating theology of suffering. *Newsweek,* p. 14.

Miller, S. L., & Burack, C. (1993). A critique of Gottfredson and Hirschi's general theory of crime: Selective (in)attention to gender and power positions. *Women and Criminal Justice, 4,* 115–134.

Miller, W. B. (1958). Lower class culture as a generating milieu of gang delinquency. *Journal of Social Issues, 14,* 5–19.

Milovanovic, D. (1995). Dueling paradigms: Modernist versus postmodernist. *Humanity and Society, 19,* 1–22.

Milovanovic, D. (2002). *Critical criminology at the edge: Postmodern perspectives, applications, and integrations.* Westport, CT: Praeger.

Mintz, M. (1985). *At any cost: Corporate greed, women, and the Dalkon Shield.* New York: Pantheon.

Miranda v. Arizona, 384 U.S. 436 (1966).

Mobley, A. (2003). Convict criminology: The two-legged data dilemma. In J. I. Ross & S. C. Richards (Eds.), *Convict criminology* (pp. 209–223). Belmont, CA: Thomson.

Moffitt, T. E. (1983). The learning theory model of punishment: Implications for delinquency deterrence. *Criminal Justice and Behavior, 10,* 131–158.

Moffitt, T. E. (1993). Adolescence-limited and life-course–persistent antisocial behavior: A developmental taxonomy. *Psychological Review, 100,* 674–701.

Moffitt, T. E. (2005). The new look at behavioral genetics in developmental psychopathology: Gene-environment interplay in antisocial behaviors. *Psychological Bulletin, 131,* 533–554.

Moffitt, T. E. (2006a). Life-course-persistent versus adolescence-limited antisocial behavior. In D. Cicchetii & D. J. Cohen (Eds.), *Developmental psychopathology: Vol. 3. Risk, disorder, and adaptation* (2nd ed., pp. 570–598). Hoboken, NJ: John Wiley.

Moffitt, T. E. (2006b). A review of research on the taxonomy of life-course persistent versus adolescence-limited antisocial behavior. In F. T. Cullen, J. P. Wright, & K. R. Blevins (Eds.), *Taking stock: The status of criminological theory* (Advances in Criminological Theory, Vol. 15, pp. 277–311). New Brunswick, NJ: Transaction.

Moffitt, T. E., Caspi, A., Dickson, N., Silva, P., & Stanton, W. (1996). Childhood-onset antisocial conduct problems in males: Natural history from ages 3 to 18. *Development and Psychopathology, 8,* 399–424.

Moffitt, T. E., Capsi, A., Rutter, M., & Silva, P. A. (2001). *Sex differences in antisocial behavior: Conduct disorder, delinquency, and violence in the Dunedin Longitudinal Study.* Cambridge, UK: Cambridge University Press.

Mohr, G. J., & Gundlach, R. H. (1929–1930). A further study of the relation between physique and performance in criminals. *Journal of Abnormal and Social Psychology, 24,* 36–50.

Mohr, J. M. (2008–2009). Oppression by scientific method: The use of science to "other" sexual minorities. *Journal of Hate Studies, 7,* 21–45.

Mokhiber, R. (1988). *Corporate crime and violence: Big business, power and the abuse of the public trust.* San Francisco: Sierra Club Books.

Monachesi, E. (1973). Cesare Beccaria. In H. Mannheim (Ed.), *Pioneers in criminology* (2nd ed., pp. 36–50). Montclair, NJ: Patterson Smith.

Morash, M. (1982). Juvenile reaction to labels: An experiment and an exploratory study. *Sociology and Social Research, 67,* 76–88.

More jails are charging inmates for their stay. (1987, November 19). *USA Today,* p. A8.

Morse, J. (2010, January 2). More women rob banks. *Cincinnati Enquirer,* p. 1.

Morris, A. (1987). *Women, crime, and criminal justice.* New York: Blackwell.

Morris, N., & Hawkins, G. (1970). *The honest politician's guide to crime control.* Chicago: University of Chicago Press.

Morrison, W. (2004). Lombroso and the birth of criminological positivism: Scientific mastery or cultural artifice? In J. Ferrel, K. Hayward, W. Morrison, & M. Presdee (Eds.), *Cultural criminology unleashed* (pp. 67–80). London: Glasshouse.

Morrison, W. (2006). *Criminology, civilisation and the new world order.* Oxford, UK: Routledge.

Moynihan, D. P. (1969). *Maximum feasible misunderstanding: Community action in the war on poverty.* New York: Free Press.

Mullins, C. W., & Miller, J. (2008). Temporal, situational and interactional features of women's violence conflicts. *The Australian and New Zealand Journal of Criminology, 41,* 36–62.

Mullins, C. W., & Wright, R. (2003). Gender, social networks, and residential burglary. *Criminology, 42,* 813–840.

Mullins, C. W., Wright, R., & Jacobs, B. A. (2004). Gender, streetlife and criminal retaliation. *Criminology, 42,* 911–940.

Muncie, J. (1998). Reassessing competing paradigms in criminological theory. In P. Walton & J. Young (Eds.), *The new criminology revisited* (pp. 221–233). London: St. Martin's.

Murray, C. (1984). *Losing ground: American social policy, 1950–1980.* New York: Basic Books.

Mutchnick, R. J., Martin, R., & Austin, W. T. (2009). *Criminological thought: Pioneers past and present.* Upper Saddle River, NJ: Prentice Hall.

Nader, R. (1965). *Unsafe at any speed: The designed-in dangers of the American automobile.* New York: Grossman.

Nagin, D. S. (1998). Criminal deterrence research at the outset of the twenty-first century. In M. Tonry (Ed.), *Crime and justice: A review of research* (Vol. 23, pp. 1–42). Chicago: University of Chicago Press.

Nagin, D. S., Cullen, F. T., & Jonson, C. L. (2009). Imprisonment and reoffending. In M. Tonry (Ed.), *Crime and justice: A review of research* (Vol. 38, pp. 115–200). Chicago: University of Chicago Press.

Nagin, D. S., & Farrington, D. P. (1992). The stability of criminal potential from childhood to adulthood. *Criminology, 30,* 235–260.

Nagin, D. S., Farrington, D. P., & Moffitt, T. E. (1995). Life-course trajectories of different types of offenders. *Criminology, 33,* 139.

Nagin, D. S., & Paternoster, R. (1991). On the relationship of past to future participation in delinquency. *Criminology, 29,* 163–189.

Nagin, D. S., & Paternoster, R. (1993). Enduring individual differences and rational choice theories of crime. *Law and Society Review, 27,* 467–496.

Nagin, D. S., & Paternoster, R. (2010). Pogarsky, Greg: Behavioral economics and crime. In F. T. Cullen & P. Wilcox (Eds.), *Encyclopedia of criminological theory* (pp. 716–720). Thousand Oaks, CA: Sage.

Nagin, D. S., & Pogarsky, G. (2001). Integrating celerity, impulsivity, and extralegal sanction threats into a model of general deterrence: Theory and evidence. *Criminology, 39,* 865–889.

Nagin, D. S., & Pogarsky, G. (2003). An experimental investigation of deterrence: Cheating, self-serving bias, and impulsivity. *Criminology, 41,* 167–191.

Naked truth returns from Boren's past. (1995, January 26). *Daily Oklahoman,* p. A10.

Naked truth revealed about Ivy League schools. (1995, January 16). *The Times* (London).

National Advisory Commission on Criminal Justice Standards and Goals. (1975). *A national strategy to reduce crime.* New York: Avon.

Nebraskastudies.org. (2010). The roots of progressivism: The populists. Retrieved April 11, 2010, from http://www.nebraskastudies.org/0600/navigation/b0601_00.html

Nellis, M. (2003a). Crime, punishment and community cohesion. In E. McLauren (Ed.), *Developing community cohesion: Understanding the issues, delivering solutions* (pp. 40–42). London: Runnymede Trust.

Nellis, M. (2003b). Electronic monitoring and the future of the Probation Service. In W. H. Chui & M. Nellis (Eds.), *Moving probation forward: Evidence, arguments and practice* (pp. 245–260). Harlow, UK: Longman.

Nellis, M. (2003c). News media, popular cultural and the electronic monitoring of offenders. *Howard Journal of Criminal Justice, 42*(1), 1–31.

Nellis, M. (2004). They don't even know we're there: The electronic monitoring of offenders in England and Wales. In K. Ball & F. Webster (Eds.), *The intensification of surveillance: Crime, terrorism and warfare in the information age* (pp. 62–89). London: Pluto.

Nellis, M. (2006). NOMS, contestability and the process of technocorrectional innovation. In M. Hough, R. Allen, & U. Padel (Eds.), *Reshaping probation and prisons: The new offender management framework* (pp. 49–67). Bristol, UK: Policy Press.

Nellis, M., & Lilly, J. R. (2000). Accepting the tag: Probation officers and home detention curfew. *Vista, 6,* 68–80.

Nellis, M., & Lilly, J. R. (2004). GPS tracking: What America and England might learn from each other. *Journal of Offender Monitoring, 17*(2), 5–7, 23–27.

Nellis, M., & Lilly, J. R. (2010). Electronic monitoring. In B. S. Fisher & S. P. Lab (Eds.), *Encyclopedia of victimology and crime prevention* (Vol. 1, pp. 360–363). Thousand Oaks, CA: Sage.

Newburn, T. (2007). "Tough on crime": Penal policy in England and Wales. In M. Tonry (Ed.), *Crime, punishment, and politics in comparative* perspective (Crime and Justice: A Review of Research, Vol. 36, pp. 425–470). Chicago: University of Chicago Press.

Newman, O. (1972). *Defensible space: Crime prevention through urban design.* New York: Collier.

Nieuwbeerta, P., Nagin, D. S., & Blokland, A. A. (2009). The relationship between first imprisonment and criminal career development: A matched samples comparison. *Journal of Quantitative Criminology, 25,* 227–257.

Novak, K. J. (2004). Disparity and racial profiling in traffic enforcement. *Police Quarterly, 7,* 65–96.

Nye, F. I. (1958). *Family relationships and delinquency behavior.* New York: John Wiley.

O'Brien, M. (2005). What is *cultural* about cultural criminology? *British Journal of Criminology, 45,* 599–612.

Officials to probe brain use. (1985, October 6). *Cincinnati Enquirer,* p. C12.

Ogle, R. S., Maier-Katkin, J., & Bernard, T. J. (1995). Homicidal behavior among women. *Criminology, 33,* 173–193.

Olasky, M. (1992). *The tragedy of American compassion.* Washington, DC: Regnery.

Olds, D. L. (2007). Preventing crime with prenatal and infancy support of parents: The Nurse-Family Partnership. *Victims and Offenders, 2,* 205–225.

Olds, D. L., Hill, P., & Rumsey, E. (1998). *Prenatal and early childhood nurse home visitation.* Washington, DC: U.S. Department of Justice, Office of Juvenile Justice and Delinquency Prevention.

O'Neill, M. (2004). Crime, culture and visual methodologies: Ethno-mimesis as performative praxis. In J. Ferrell, K. Hayward, W. Morrison, & M. Presdee (Eds.), *Cultural criminology unleashed* (pp. 219–229). London: Glasshouse.

Orland, L. (1980). Reflections on corporate crime: Law in search of theory and scholarship. *American Criminal Law Review, 17,* 501–520.

Osgood, D. W. (2011). Osgood, D. Wayne, Janet K. Wilson, Jerald G. Bachman, Patrick M. O'Malley, and Lloyd D. Johnston: Routine activities and individual deviant behavior. In F. T. Cullen & P. Wilcox (Eds.), *Encyclopedia of criminological theory* (pp. 675–679). Thousand Oaks, CA: Sage.

Osgood, D. W., Wilson, J. K., Bachman, J. G., O'Malley, P. M., & Johnston, L. D. (1996). Routine activities and individual deviant behavior. *American Sociological Review, 61,* 635–655.

Oshinsky, D. M. (1996). *"Worse than slavery": Parchman Farm and the ordeal of Jim Crow justice.* New York: Free Press.

Pager, D. (2003). The mark of a criminal record. *American Journal of Sociology, 108,* 937–975.

Pager, D. (2007). *Marked: Race, crime, and finding work in an era of mass incarceration.* Chicago: University of Chicago Press.

Pager, D., & Quillian, L. (2005). Walking the talk? What employers say versus what they do. *American Sociological Review, 70,* 355–380.

Palamara, F., Cullen, F. T., & Gersten, J. C. (1986). The effect of police and mental health intervention on juvenile deviance: Specifying contingencies in the impact of formal reaction. *Journal of Health and Social Behavior, 27,* 90–105.

Palen, J. J. (1981). *The urban world* (3rd ed.). New York: McGraw-Hill.

Panel on High-Risk Youth. (1993). *Losing generations: Adolescents in high-risk settings.* Washington, DC: National Academies Press.

Paolucci, H. (1963). Preface. In C. Beccaria, *On crimes and punishments* (H. Paolucci, Trans.). Indianapolis, IN: Bobbs-Merrill. (Original work published 1764)

Parsons, T. (1951). *The social system.* New York: Free Press.

Passas, N. (1990). Anomie and corporate deviance. *Contemporary Crises, 14,* 157–178.

Passas, N. (2010). Anomie and white-collar crime. In F. T. Cullen & P. Wilcox (Eds.), *Encyclopedia of criminological theory* (pp. 56–58). Thousand Oaks, CA: Sage.

Passas, N., & Agnew, R. (Eds.). (1997). *The future of anomie theory.* Boston: Northeastern University Press.

Paternoster, R. (1987). The deterrent effect of the perceived certainty and severity of punishment: A review of the evidence and issues. *Justice Quarterly, 4,* 197–217.

Paternoster, R., Brame, R., Bachman, R., & Sherman, L. W. (1997). Do fair procedures matter? The effect of procedural justice on spouse assault. *Law and Society Review, 31,* 163–204.

Paternoster, R., & Iovanni, L. (1989). The labeling perspective and delinquency: An elaboration of the theory and assessment of the evidence. *Justice Quarterly, 6,* 359–394.

Paternoster, R., & Mazerolle, P. (1994). General strain theory and delinquency: A replication and extension. *Journal of Research in Crime and Delinquency, 31,* 235–263.

Paternoster, R., & Simpson, S. (1993). A rational choice theory of corporate crime. In R. V. Clarke & M. Felson (Eds.), *Routine activity and rational choice* (Advances in Criminological Theory, Vol. 5, pp. 37–58). New Brunswick, NJ: Transaction.

Paternoster, R., & Simpson, S. (1996). Sanction threats and appeals to morality: Testing a rational choice model of corporate crime. *Law and Society Review, 30,* 549–583.

Patterson, G. R., DeBarshy, B. D., & Ramsey, E. (1989). A developmental perspective on antisocial behavior. *American Psychologist, 44,* 329–335.

Patterson, G. R., & Yoerger, K. (1997). A developmental model for late-onset delinquency. In D. W. Osgood (Ed.), *Motivation and delinquency: Nebraska Symposium on Motivation* (Vol. 44, pp. 119–177). Lincoln: University of Nebraska Press.

Patterson, G. R., & Yoerger, K. (2002). A developmental model for early- and late-onset delinquency. In J. B. Reid, G. R. Patterson, & J. Synder (Eds.), *Antisocial behavior in children and adolescents: A developmental analysis and model for intervention* (pp. 147–172). Washington, DC: American Psychological Association.

Patterson, J. T. (1996). *Grand expectations: The United States, 1945–1974.* New York: Oxford University Press.

Pattillo, M., Weiman, D., & Western, B. (Eds.). (2004). *Imprisoning America: The social effects of mass incarceration.* New York: Russell Sage.

Peacock, A. (2003). *Libby, Montana: Asbestos and the deadly silence of an American corporation.* Boulder, CO: Johnson.

Pearce, F. (1976). *Crimes of the powerful: Marxism, crime and deviance.* London: Pluto.

Pepinsky, H. (1976). *Crime and conflict: A study in law and society.* New York: Academic Press.

Pepinsky, H. (1999). Peacemaking primer. In B. A. Arrigo (Ed.), *Social justice: Criminal justice* (pp. 52–70). Belmont, CA: Wadsworth.

Pepinsky, H., & Quinney, R. (Eds.). (1991). *Criminology as peacemaking.* Bloomington: Indiana University Press.

Persell, C. H. (1984). An interview with Robert K. Merton. *Teaching Sociology, 11,* 355–386.

Peters, M. (2006, January 19). Second careers and the third age: You're only as old as your new job. *The Guardian* (Business). Retrieved June 20, 2006, from http://business .guardian.co.uk/story/0,1689574,00.html

Petersilia, J. (1980). Criminal career research: A review of recent evidence. In N. Morris & M. Tonry (Eds.), *Crime and justice: An annual review of research* (Vol. 2, pp. 321–379). Chicago: University of Chicago Press.

Petersilia, J. (2003). *When prisoners come home: Parole and prison reentry.* New York: Oxford University Press.

Petersilia, J. (2008). California's correctional paradox of excess and deprivation. In M. Tonry (Ed.), *Crime and justice: A review of research* (Vol. 37, pp. 207–278). Chicago: University of Chicago Press.

Petersilia, J., & Turner, S. (with Peterson, J.). (1986). *Prison versus probation in California: Implications for crime and offender recidivism.* Santa Monica, CA: RAND.

Peterson, R. (2008). Foreword. In J. Miller, *Getting played* (pp. ix–xii). New York: New York University Press.

Petrosino, A., Turpin-Petrosino, C., & Guckenbuerg, S. (2010). *Formal system processing of juveniles: Effects on delinquency.* Oslo, Norway: The Campbell Collaboration.

Pew Center on the States. (2010). *Prison count 2010: State prison population declines for the first time in 38 years.* Washington, DC: Author.

Pew Charitable Trusts. (2008). *One in 100: Behind bars in America 2008.* Washington, DC: Author.

Pfohl, S. J. (1977). The "discovery of child abuse." *Social Problems, 24,* 310–323.

Pfohl, S. J. (1985). *Images of deviance and social control. A sociological history.* New York: McGraw-Hill.

Piliavin, I., & Briar, S. (1964). Police encounters with juveniles. *American Journal of Sociology, 70,* 206–214.

Piquero, A. R. (2011). Understanding the development of antisocial behavior: Terrie Moffitt. In F. T. Cullen, C. L. Jonson, A. J. Myer, & F. Adler (Eds.), *The origins of American criminology* (Advances in Criminological Theory, Vol. 16, pp. 397–408). New Brunswick, NJ: Transaction.

Piquero, A. R., & Bouffard, J. A. (2007). Something old, something new: A preliminary investigation of Hirschi's redefined self-control. *Justice Quarterly, 24,* 1–27.

Piquero, A. R., Farrington, D. P., & Blumstein, A. (2003). The criminal career paradigm. In M. Tonry (Ed.), *Crime and justice: A review of research* (Vol. 30, pp. 359–506). Chicago: University of Chicago Press.

Piquero, A. R., Farrington, D. P., & Blumstein, A. (2007). *Key issues in criminal career research: New analyses of the Cambridge Study in Delinquent Development.* New York: Cambridge University Press.

Piquero, A. R., & Hickman, M. (1999). An empirical test of Tittle's control balance theory. *Criminology, 37,* 319–341.

Piquero, A. R., & Moffitt, T. E. (2005). Explaining the facts of crime: How the developmental taxonomy replies to Farrington's invitation. In D. P. Farrington (Ed.), *Integrated developmental and life-course theories of offending* (Advances in Criminological Theory, Vol. 14, pp. 51–72). New Brunswick, NJ: Transaction.

Piquero, A. R., & Piquero, N. L. (1998). On testing institutional-anomie theory with varying specifications. *Studies on Crime and Crime Prevention, 7,* 61–84.

Piquero, N. L., Exum, M. L., & Simpson, S. S. (2005). Integrating the desire-for-control and rational choice in a corporate crime context. *Justice Quarterly, 22,* 252–280.

Piquero, N. L., & Piquero, A. R. (2006). Control balance and exploitative corporate crime. *Criminology, 44,* 397–430.

Pivar, D. J. (1973). *Purity crusade: Sexual morality and social control, 1868–1900.* Westport, CT: Greenwood.

Platt, A. M. (1969). *The child savers: The invention of delinquency.* Chicago: University of Chicago Press.

Pogarsky, G., Kim, K., & Paternoster, R. (2005). Perceptual change in the National Youth Survey: Lessons for deterrence theory and offender decision-making. *Justice Quarterly, 22,* 1–29.

Pogarsky, G., & Piquero, A. R. (2003). Can punishment encourage offending? Investigating the "resetting effect." *Journal of Research in Crime and Delinquency, 40,* 95–120.

Pollak, O. (1950). *The criminality of women.* Philadelphia: University of Pennsylvania Press.

Pound, R. (1942). *Social control through law.* New Haven, CT: Yale University Press.

Pratt, T. C. (2001). *Assessing the relative effects of macro-level predictors of crime: A meta-analysis.* Unpublished doctoral dissertation, University of Cincinnati, Cincinnati, OH.

Pratt, T. C. (2009). *Addicted to incarceration: Corrections policy and the politics of misinformation in the United States.* Thousand Oaks, CA: Sage.

Pratt, T. C., & Cullen, F. T. (2000). The empirical status of Gottfredson and Hirschi's general theory of crime. *Criminology, 38,* 931–964.

Pratt, T. C., & Cullen, F. T. (2005). Assessing macro-level predictors and theories of crime: A meta-analysis. In M. Tonry (Ed.), *Crime and justice: A review of research* (Vol. 32, pp. 373–450). Chicago: University of Chicago Press.

Pratt, T. C., Cullen, F. T., Blevins, K. R., Daigle, L. E., & Madensen, T. D. (2006). The empirical status of deterrence theory: A meta-analysis. In F. T. Cullen, J. P. Wright, & K. R. Blevins (Eds.), *Taking stock: The status of criminological theory* (Advances in Criminological Theory, Vol. 15, pp. 367–395). New Brunswick, NJ: Transaction.

Pratt, T. C., Cullen, F. T., Blevins, K. R., Daigle, L. E., & Unnever, J. D. (2002). The relationship of attention deficit hyperactivity disorder to crime and delinquency: A meta-analysis. *International Journal of Police Science and Management, 4,* 344–360.

Pratt, T. C., Cullen, F. T., Sellers, C. S., Winfree, L. T., Jr., Madensen, T. D., Daigle, L. E., Fearn, N. E., & Gau, J. M. (2010). The empirical status of social learning theory: A meta-analysis. *Justice Quarterly, 27,* 765–802.

Pratt, T. C., Turner, M. G., & Piquero, A. R. (2004). Parental socialization and community context: A longitudinal analysis of the structural sources of low self-control. *Journal of Research in Crime and Delinquency, 41,* 219–243.

The president's angry apostle. (1986, October 6). *Newsweek,* p. 27.

President's Commission on Law Enforcement and Administration of Justice. (1968). *Challenge of crime in a free society.* New York: Avon.

Provine, W. B. (1973). Geneticists and the biology of race crossing. *Science, 182,* 790.

Public humiliation. (1986, May 20). ABC News *Nightline.*

Pueschel, S. M., Linakis, J. G., & Anderson, A. C. (Eds.). (1996). *Lead poisoning in childhood.* Baltimore: Paul H. Brookes.

Putnam, R. D. (2000). *Bowling alone: The collapse and revival of American community.* New York: Simon & Schuster.

Quinney, R. (1969). *Crime and justice in society.* Boston: Little, Brown.

Quinney, R. (1970a). *The problem of crime.* New York: Dodd, Mead.

Quinney, R. (1970b). *The social reality of crime.* Boston: Little, Brown.

Quinney, R. (1974a). *Criminal justice in America: A critical understanding.* Boston: Little, Brown.

Quinney, R. (1974b). *Critique of the legal order: Crime control in capitalist society.* Boston: Little, Brown.

Quinney, R. (1977). *Class, state, and crime: On the theory and practice of criminal justice.* New York: David McKay.

Quinney, R. (1980). *Class, state, and crime: On the theory and practice of criminal justice* (2nd ed.). New York: David McKay.

Radzinowicz, L. (1966). *Ideology and crime: A study of crime in its social and historical context.* New York: Columbia University Press.

Rafter, N. (2008). *The criminal brain: Understanding biological theories of crime.* New York: New York University Press.

Raine, A. (1993). *The psychopathology of crime.* San Diego, CA: Academic Press.

Raine, A. (2002). The biological basis of crime. In J. Q. Wilson & J. Petersilia (Eds.), *Crime: Public policies for crime control* (pp. 43–74). Oakland, CA: ICS Press.

Rand, M. R. (2009). *Criminal victimization, 2008.* Washington, DC: U.S. Department of Justice, Bureau of Justice Statistics.

Reckless, W. C. (1943). *The etiology of criminal and delinquent behavior.* New York: Social Science Research Council.

Reckless, W. C. (1961). *The crime problem* (3rd ed.). New York: Appleton-Century-Crofts.

Reckless, W. C. (1967). *The crime problem* (4th ed.). New York: Meredith.

Reckless, W. C., & Dinitz, S. (1967). Pioneering with self-concept as a vulnerability factor in delinquency. *Journal of Criminal Law, Criminology, and Police Science, 58,* 515–523.

Reckless, W. C., Dinitz, S., & Kay, B. (1957). The self component in potential delinquency and nondelinquency. *American Sociological Review, 22,* 566–570.

Reckless, W. C., Dinitz, S., & Murray, E. (1956). Self concept as insulator against delinquency. *American Sociological Review, 21,* 744–764.

Redl, F., & Wineman, D. (1951). *Children who hate.* New York: Free Press.

Regoli, R. M., & Hewitt, J. D. (1997). *Delinquency and society* (3rd ed.). New York: McGraw-Hill.

Regoli, R. M., Hewitt, J. D., & DeLisi, M. (2010). *Delinquency and society* (8th ed.). Sudbury, MA: Jones and Bartlett.

Reich, R. B. (Ed.). (1988). *The power of public ideas.* Cambridge, MA: Harvard University Press.

Reich, R. B. (2001, February 12). The new economy as a decent society. *American Prospect,* pp. 20–23.

Reid, J. B., Patterson, G. R., & Snyder, J. (Eds.). (2002). *Antisocial behavior in children and adolescents: A developmental analysis and model for intervention.* Washington, DC: American Psychological Association.

Reiman, J. H. (1979). *The rich get richer and the poor get prison.* New York: John Wiley.

Reiss, A. J. (1949). *The accuracy, efficiency, and validity of a prediction instrument.* Unpublished doctoral dissertation, University of Chicago.

Reiss, A. J. (1951). Delinquency as the failure of personal and social controls. *American Sociological Review, 16,* 196–207.

Rennie, Y. (1978). *The search for criminal man: A conceptual history of the dangerous offender.* Lexington, MA: Lexington Books.

Rennison, C. M. (2003). *Intimate partner violence 1993–2001.* Washington, DC: U.S. Department of Justice, Bureau of Justice Statistics.

Renzetti, C. M. (1993). On the margins of the malestream (or they still don't get it, do they?): Feminist analyses in criminal justice education. *Journal of Criminal Justice Education, 4,* 219–234.

Report traces 45 cases of attacks on minorities. (1987, February 15). *New York Times,* p. A22.

Reports of bias attacks on the rise in New York. (1987, September 23). *New York Times,* p. A19.

Reynolds, M. (1996). *Crime and punishment in Texas: An update.* Dallas, TX: National Center for Policy Analysis.

Reynolds, M. (2000, November 13). Crime and punishment. *Newsweek,* p. 46.

Rich, F. (2009, March 9). The culture warriors get laid off. *New York Times,* p. WK12.

Richards, S. C. (1998). Critical and radical perspectives on community punishment: Lessons from darkness. In J. I. Ross (Ed.), *Cutting edge* (pp. 122–144). Westport CT: Praeger.

Richards, S. C., & Ross, J. I. (2001). Introducing the new school of convict criminology. *Social Justice, 28*(1), 177–190.

Richards, S. C., & Ross, J. I. (2005). Convict criminology. In R. A. Wright & J. M. Miller (Eds.), *Encyclopedia of criminology* (Vol. 1, pp. 232 –236). New York: Routledge.

Richie, B. E. (1996). *Compelled to crime: The gender of entrapment of battered Black women.* New York: Routledge.

Ridenour, A. (1999, July 20). Context of silliest lawsuits ever. *Cincinnati Enquirer,* p. A6.

Roberts, J. V., & Stalans, L. J. (2000). *Public opinion, crime, and criminal justice.* Boulder, CO: Westview.

Robertson, P. (2010). Pat Robertson: Haiti "cursed" by "pact to the devil." Retrieved January 26, 2010, from http://www.huffingtonpost.com/2010/01/13/patrobertson-haiti-curse_n_422099.html

Robins, L. N. (1978). Sturdy childhood predictors of adult antisocial behavior: Replications from longitudinal studies. *Psychological Medicine, 8,* 611–622.

Robinson, M. (2009). No longer taboo: Crime prevention implications of biosocial criminology. In A. Walsh & K. M. Beaver (Eds.), *Biosocial criminology: New directions in theory and research* (pp. 243–263). New York: Routledge.

Rock, P. (1977). Review symposium on women, crime, and criminology. *British Journal of Criminology, 17,* 393–395.

Rodericks, J. V. (1992). *Calculated risks: The toxicity and public health risks of chemicals in our environment.* Cambridge, UK: Cambridge University Press.

Roe v. Wade, 410 U.S. 113 (1973).

Roper v. Simmons 543 U.S. 551 (2005).

Rose, D. R., & Clear, T. R. (1998). Incarceration, social capital, and crime: Implications for social disorganization theory. *Criminology, 36,* 441–479.

Rose, N. (1969). *Inventing our selves: Psychology, power, and personhood.* Cambridge, UK: Cambridge University Press.

Rosenbaum, D. P. (2007). Just say no to D.A.R.E. *Criminology and Public Policy, 6,* 815–824.

Rosenfeld, R. (1989). Robert Merton's contribution to the sociology of deviance. *Sociological Inquiry, 59,* 453–466.

Rosenfeld, R. (2009). Homicide and serious assaults. In M. Tonry (Ed.), *The Oxford handbook of crime and public policy* (pp. 25–50). New York: Oxford University Press.

Rosenfeld, R., & Messner, S. F. (2011). The intellectual origins of institutional-anomie theory. In F. T. Cullen, C. L. Jonson, A. J. Myer, & F. Adler (Eds.), *The origins of American criminology* (Advances in Criminological Theory, Vol. 16, pp. 121–135). New Brunswick, NJ: Transaction.

Rosoff, S., Pontell, H., & Tillman, R. (2007). *Profit without honor: White-collar crime and the looting of America* (4th ed.). Upper Saddle River, NJ: Pearson Prentice Hall.

Ross, E. A. (1907). *Sin and society: An analysis of latter-day iniquity.* New York: Harper and Row. (Reprinted in 1973)

Ross, J. I., & Richards, S. C. (Eds.). (2003a). *Convict criminology* (pp. 2–14). Belmont, CA: Thomson.

Ross, J. I., & Richards, S. C. (2003b). What is the new school of convict criminology? In J. I. Ross & S. C. Richards (Eds.), *Convict criminology* (pp. 2–14). Belmont, CA: Thomson.

Ross, J. I., & Richards, S. C. (2009). *Beyond bars: Rejoining society after prison.* Exton, PA: Alpha Publishing House.

Rossmo, D. K. (1995). Place, space, and police investigations: Hunting serial violent criminals. In J. E. Eck & D. Weisburd (Eds.), *Crime places: Crime prevention studies* (Vol. 4, pp. 217–235). Monsey, NY: Willow Tree Press.

Rothman, D. J. (1971). *The discovery of the asylum: Social order and disorder in the new republic.* Boston: Little, Brown.

Rothman, D. J. (1978). The state as parent: Social policy in the Progressive era. In W. Gaylin, I. Glasser, S. Marcus, & D. Rothman (Eds.), *Doing good: The limits of benevolence* (pp. 69–96). New York: Pantheon.

Rothman, D. J. (1980). *Conscience and convenience: The asylum and its alternatives in progressive America.* Boston: Little, Brown.

Rowe, D. C. (1983). Biometrical genetic models of self-reported delinquent behavior: A twin study. *Behavioral Genetics, 13,* 473–489.

Rowe, D. C. (1986). Genetic and environmental components of antisocial behavior: A study of 265 twins. *Criminology, 24,* 513–534.

Rowe, D. C. (2002). *Biology and crime.* Los Angeles, CA: Roxbury.

Rubia, K., Halari, R., Smith, A. B., Mohammed, M., Scott, S., Giampietro, V., Taylor, E., & Brammer, M. J. (2008). Dissociated functional brain abnormalities of inhibition in boys with pure conduct disorder and in boys with pure ADHD. *American Journal of Psychiatry, 165,* 889–897.

Rubington, E., & Weinberg, M. S. (1971). *The study of social problems.* New York: Oxford University Press.

Ruggiero, V., South, N., & Taylor, I. (1998). Introduction: Towards a European criminological community. In V. Ruggiero, N. South, & I. Taylor (Eds.), *The new European criminology: Crime and social order in Europe* (pp. 1–15). London: Routledge.

Rushton, J. P. (1990). Race and crime: A reply to Roberts and Gabor. *Canadian Journal of Criminology, 32,* 315–334.

Rushton, J. P. (1995). Race and crime: International data for 1989–1990. *Psychological Reports, 76,* 307–312.

Rutter, M. (2006). *Genes and behavior: Nature-nurture interplay.* Malden, MA: Blackwell.

Rutter, M. (2007). Gene-environment interdependence. *Developmental Science 10,* 12–18.

Samenow, S. E. (1984). *Inside the criminal mind.* New York: Times Books.

Samenow, S. E. (1989). *Before it's too late: Why some kids get into trouble—And what parents can do about it.* New York: Times Books.

Sampson, R. J. (1986a). Crime in the cities: The effects of formal and informal social control. In A. J. Reiss & M. Tonry (Eds.), *Community and crime* (Crime and Justice: A Review of Research, Vol. 8, 271–311). Chicago: University of Chicago Press.

Sampson, R. J. (1986b). Effects of socioeconomic context on official reaction to juvenile delinquency. *American Sociological Review, 51,* 876–885.

Sampson, R. J. (1999). Techniques of research neutralization. *Theoretical Criminology, 3,* 438–451.

Sampson, R. J. (2006). Collective efficacy: Lessons learned and directions for future inquiry. In F. T. Cullen, J. P. Wright, & K. R. Blevins (Eds.), *Taking stock: The status of criminological theory* (Advances in Criminological Theory, Vol. 15, pp. 149–167). New Brunswick, NJ: Transaction.

Sampson, R. J. (2011). Communities and crime revisited: Intellectual trajectory of a Chicago school education. In F. T. Cullen, C. L. Jonson, A. J. Myer, & F. Adler (Eds.), *The origins of American criminology* (Advances in Criminological Theory, Vol. 16, pp. 63–85). New Brunswick, NJ: Transaction.

Sampson, R. J., & Groves, W. B. (1989). Community structure and crime: Testing social-disorganization theory. *American Journal of Sociology, 94,* 774–802.

Sampson, R. J., & Laub, J. H. (1990). Crime and deviance over the life course: The salience of adult social bonds. *American Sociological Review, 44,* 609–627.

Sampson, R. J., & Laub, J. H. (1992). Crime and deviance in the life course. *Annual Review of Sociology, 18,* 63–84.

Sampson, R. J., & Laub, J. H. (1993). *Crime in the making: Pathways and turning points through life.* Cambridge, MA: Harvard University Press.

Sampson, R. J., & Laub, J. H. (1994). Urban poverty and the family context of delinquency: A new look at structure and process in a classic study. *Child Development, 65,* 523–540.

Sampson, R. J., & Laub, J. H. (1995). Understanding variability in lives through time: Contributions of life-course criminology. *Studies on Crime and Crime Prevention, 4,* 143–158.

Sampson, R. J., & Laub, J. H. (2005). A life-course view of the development of crime. *Annals of the American Academy of Political and Social Science, 602,* 12–45.

Sampson, R. J., Morenoff, J. D., & Earls, F. (1999). Beyond spatial capital: Spatial dynamics of collective efficacy for children. *American Sociological Review, 64,* 633–660.

Sampson, R. J., & Raudenbush, R. W. (1999). Systematic social observation of public spaces: A new look at disorder in urban neighborhoods. *American Journal of Sociology, 3,* 603–651.

Sampson, R. J., Raudenbush, S. W., & Earls, F. (1997, August 15). Neighborhoods and violent crime: A multilevel study of collective efficacy. *Science, 277,* 918–924.

Sampson, R. J., & Wilson, W. J. (1995). Toward a theory of race, crime, and urban inequality. In J. Hagan & R. D. Peterson (Eds.), *Crime and inequality* (pp. 37–54). Stanford, CA: Stanford University Press.

Scarpitti, F. R., Murray, E., Dinitz, S., & Reckless, W. C. (1960). The good boys in a high delinquency area: Four years later. *American Sociological Review, 25,* 922–926.

Scarr, S., & McCartney, K. (1983). How people make their own environments: A theory of genotype-environment effects. *Child Development, 54,* 424–435.

Scheff, T. J. (1966). *Being mentally ill.* Chicago: Aldine.

Scheingold, S. A. (1984). *The politics of law and order: Street crime and public policy.* New York: Longman.

Schiff, M. F. (1999). The impact of restoration intervention on juvenile offenders. In G. Bazemore & L. Walgrave (Eds.), *Restorative juvenile justice: Repairing the harm of youth crime* (pp. 327–356). Monsey, NY: Willow Tree Press.

Schlossman, S., Zellman, G., & Shavelson, R. (with Sedlak, M., & Cobb, J.). (1984). *Delinquency prevention in South Chicago: A fifty-year assessment of the Chicago Area Project.* Santa Monica, CA: RAND.

Schuessler, K. (Ed.). (1973). Introduction. In E. H. Sutherland (Ed.), *On analyzing crime* (pp. ix–xvi). Chicago: University of Chicago Press.

Schur, E. M. (1965). *Crimes without victims: Deviant behavior and public policy.* Englewood Cliffs, NJ: Prentice Hall.

Schur, E. M. (1973). *Radical nonintervention: Rethinking the delinquency problem.* Englewood Cliffs, NJ: Prentice Hall.

Schur, E. M., & Bedeau, H. A. (1974). *Victimless crimes: Two sides of a controversy.* Englewood Cliffs, NJ: Prentice Hall.

Schutz, A. (1962). *The problem of social reality.* The Hague, The Netherlands: Martinus Nijhoff.

Schwartz, M. D. (1991). The future of critical criminology. In B. D. MacLean (Ed.), *New Directions in critical criminology* (pp. 119–124). Vancouver, BC: Collective Press.

Schwartz, M. D., & Friedrichs, D. O. (1994). Postmodern thought and criminological discontent. *Criminology, 32,* 221–246.

Schwarzenegger, A. (2010, January 6). Governor Schwarzenegger delivers 2010 State of the State Address. Sacramento, CA: Office of the Governor. Retrieved January 7, 2010, from http://gov.ca.gov/speech/14118/

Scientists urge high project to chart all human genes. (1988, February 12). *New York Times,* pp. A1, A13.

Sellers, C. S. (1999). Self-control and intimate violence: An examination of the scope and specification of the general theory of crime. *Criminology, 37,* 375–404.

Sellin, T. (1938). *Culture, conflict, and crime.* New York: Social Science Research Council.

Sellin, T. (1973). Enrico Ferri. In H. Mannheim (Ed.), *Pioneers in criminology* (2nd ed., pp. 361–384). Montclair, NJ: Patterson Smith.

Seringhaus, M. (2010, March 14). To stop crime, share your genes. *New York Times,* p. A21.

Sex in the 80s: The revolution is over. (1984, April 9). *Time,* pp. 74–78.

Shannon, L. W. (1982). *Assessing the relationship of adult criminal careers to juvenile careers: A summary.* Washington, DC: U.S. Department of Justice.

Shapiro, S. P. (1984). *Wayward capitalists: Target of the Security and Exchange Commission.* New Haven, CT: Yale University Press.

Shapland, J., Atkinson, A., Atkinson, H., Dignan, J., Edwards, L., Hibbert, J., Howes, J. J., Robinson, G., & Sorsby, A. (2008). *Does restorative justice affect reconviction? The fourth report from the evaluation of three schemes.* London: Ministry of Justice.

Shaw, C. R. (1930). *The jack-roller: A delinquent boy's own story.* Chicago: University of Chicago Press.

Shaw, C. R. (with Moore, M. E.). (1931). *The natural history of a delinquent career.* Chicago: University of Chicago Press.

Shaw, C. R. (with McKay, H. D., & MacDonald, J. F.). (1938). *Brothers in crime.* Chicago: University of Chicago Press.

Shaw, C. R., & McKay, H. D. (1972). *Juvenile delinquency and urban areas.* Chicago: University of Chicago Press.

Shaw, P., Gormack, M., Lerch, J., Addington, A., Seal, J., Greenstein, D., Sharp, W., Evans, A., Giedd, J. N., Castellanos, F. X., & Rapoport, J. L. (2007). Polymorphisms of DRD4, clinical outcome, and cortical structure in ADHD. *Archives of General Psychiatry, 64,* 921–931.

Sheldon, W. H. (1949). *Varieties of delinquent youth: An introduction to constitutional psychiatry.* New York: Harper.

Sheptycki, J. (2010). Edward Sutherland (1883–1950). In K. Hayward, S. Maruna, & J. Mooney (Eds.), *Fifty key thinkers in criminology* (pp. 63–71). London: Routledge.

Sherman, L. W. (with Schmidt, J. D., & Rogan, D. P.). (1992). *Policing domestic violence: Experiments and dilemmas.* New York: Free Press.

Sherman, L. W. (1993). Defiance, deterrence, and irrelevance: A theory of the criminal sanction. *Journal of Research in Crime and Delinquency, 30,* 445–473.

Sherman, L. W. (1997). Policing for crime prevention. In L. W. Sherman, D. Gottfredson, D. MacKenzie, J. Eck, P. Reuter, & S. Bushway (Eds.), *Preventing crime: What works, what doesn't, what's promising—A report to the United States Congress.* Washington, DC: Office of Justice Programs.

Sherman, L. W. (2000, February 24). *The defiant imagination: Consilience and the science of sanctions.* Lecture delivered at University of Pennsylvania, Philadelphia.

Sherman, L. W., Farrington, D. P., Welsh, B. C., & MacKenzie, D. L. (Eds.). (2002). *Evidence-based crime prevention.* London: Routledge.

Sherman, L. W., Gartin, P. R., & Buerger, M. E. (1989). Hot spots of predatory crime: Routine activity and the criminology of place. *Criminology, 27,* 27–55.

Sherman, L. W., Schmidt, J. D., Rogan, D. P., Smith, D. A., Gartin, P. R., Cohn, E. G., Collins, J., & Bacich, A. R. (1992). The variable effects of arrests on criminal careers: The Milwaukee domestic violence experiment. *Journal of Criminal Law and Criminology, 83,* 137–169.

Sherman, L. W., & Strang, H. (2007). *Restorative justice: The evidence.* London: The Smith Institute.

Sherman, M., & Hawkins, G. (1981). *Imprisonment in America: Choosing the future.* Chicago: University of Chicago Press.

Short, J. F., Jr., & Strodtbeck, F. L. (1965). *Group process and gang delinquency.* Chicago: University of Chicago Press.

Shover, N. (1996). *Great pretenders: Pursuits and careers of persistent thieves.* Boulder, CO: Westview.

Shover, N., & Hochstetler, A. (2006). *Choosing white-collar crime.* New York: Cambridge University Press.

Shover, N., & Hunter, B. W. (in press). Blue-collar, white-collar: Crimes and mistakes. In W. Bernasco (Ed.), *Offender on offending: Learning about crime from criminals.* Collompton, UK: Willan.

Shover, N., & Scroggins, J. (2009). Organizational crime. In M. Tonry (Ed.), *The Oxford handbook of crime and public policy* (pp. 273–303). New York: Oxford University Press.

Simon, D. R., & Eitzen, S. D. (1986). *Elite deviance* (2nd ed.). Boston: Allyn & Bacon.

Simon, J. (1993). *Poor discipline: Parole and the social control of the underclass.* Chicago: University of Chicago Press.

Simon, R. J. (1975). *Women and crime.* Lexington, MA: Lexington Books.

Simons, R. L., Chen, Y.-F., Stewart, E. A., & Brody, G. H. (2003). Incidents of discrimination and risk for delinquency: A longitudinal test of strain theory with an African American sample. *Justice Quarterly, 20,* 27–854.

Simons, R. L., Simons, L. G., Burt, C. H., Drummund, H., Stewart, E., Brody, G. H., Gibbons, F. X., & Cutrona, C. (2006). Supportive parenting moderates the effect of discrimination upon anger, hostile view of relationships, and violence among African American boys. *Journal of Health and Social Behavior, 47,* 373–389.

Simpson, S. S. (1991). Caste, class, and violent crime: Explaining difference in female offending. *Criminology, 29,* 115–135.

Simpson, S. S. (2002). *Corporate crime, law, and social control.* New York: Cambridge University Press.

Simpson, S. S., & Piquero, N. L. (2002). Low self-control, organizational theory, and corporate crime. *Law and Society Review, 36,* 509–548.

Simpson, S. S., Piquero, N. L., & Paternoster, R. (2002). Rationality and corporate offending decisions. In A. R. Piquero & S. G. Tibbetts (Eds.), *Rational choice and criminal behavior: Recent research and future challenges* (pp. 25–30). New York: Routledge.

Sinclair, U. (1906). *The jungle.* New York: Signet.

Singer, S. I., & Levine, M. (1988). Power-control theory, gender, and delinquency. *Criminology, 26,* 627–648.

Skolnick, J. H. (1997, January–February). Tough guys. *American Prospect,* pp. 86–91.

Smart, C. (1976). *Women, crime, and criminology: A feminist critique.* London: Routledge & Kegan Paul.

Smith, J., & Fried, W. (1974). *The uses of the American prison: Political theory and penal practice.* Lexington, MA: Lexington Books.

Smith, N. C., Simpson, S. S., & Huang, C.-Y. (2007). Why managers fail to do the right thing: An empirical study of unethical and illegal conduct. *Business Ethics Quarterly, 17,* 633–667.

Smith, P. (2006). *The effects of incarceration on recidivism: A longitudinal examination of program participation and institutional adjustment in federally sentenced adult male offenders.* Unpublished doctoral dissertation, University of New Brunswick, Canada.

Smithsonian destroys nude photo collection. (1995, January 28). *Plain Dealer,* p. A8. (Cleveland, OH).

Snodgrass, J. (1976). Clifford R. Shaw and Henry D. McKay: Chicago criminologists. *British Journal of Criminology, 16,* 1–19.

Snodgrass, J. D. (1972). *The American criminological tradition: Portraits of the men and ideology in a discipline.* Unpublished doctoral dissertation, University of Pennsylvania.

Snyder, R. C. (2008). What is third-wave feminism: A new direction. *Signs: Journal of Women in Culture and Society, 34,* 175–196.

Spelman, W. (2000). What recent studies do (and don't) tell us about imprisonment and crime. In M. Tonry (Ed.), *Crime and justice: A review of research* (Vol. 27, pp. 419–494). Chicago: University of Chicago Press.

Spiegler, M. D., & Guevremont, D. C. (1998). *Contemporary behavior therapy* (3rd ed.). Pacific Grove, CA: Brooks/Cole.

Spitzer, S. (1976). Toward a Marxian theory of deviance. *Social Problems, 22,* 638–651.

Spohn, C., & Holleran, D. (2002). The effect of imprisonment on recidivism rates of felony offenders: A focus on drug offenders. *Criminology, 40,* 329–357.

Stadler, W. A. (2010). *White-collar offenders and the prison experience: An empirical examination of the "special sensitivity" to imprisonment hypothesis.* Unpublished doctoral dissertation, University of Cincinnati.

Stanko, E. A. (1992). Intimidating education: Sexual harassment in criminology. *Journal of Criminal Justice Education, 3,* 331–340.

Starr, J. M. (1985). Cultural politics in the 1960s. In J. M. Starr (Ed.), *Cultural politics: Radical movements in modern history* (pp. 235–294). New York: Praeger.

Steffensmeier, D. J. (1978). Crime and the contemporary woman: An analysis of the changing levels of female property crime, 1960–75. *Social Forces, 57,* 556–584.

Steffensmeier, D. J. (1980). Sex differences in patterns of adult crime, 1965–1978. *Social Forces, 58,* 1080–1108.

Steffensmeier, D. J. (1981). Patterns of female property crime, 1960–1978: A postscript. In L. H. Bowker (Ed.), *Women and crime in America* (pp. 59–65). New York: Macmillan.

Steffensmeier, D. J., & Cobb, M. J. (1981). Sex differences in urban arrest patterns, 1934–79. *Social Forces, 61,* 1010–1032.

Stewart, E. A., Schreck, C. J., & Simons, R. L. (2006). "I ain't gonna let no one disrespect me": Does the code of the street reduce or increase violent victimization among African American adolescents? *Journal of Research in Crime and Delinquency, 43,* 427–458.

Stewart, E. A., & Simons, R. L. (2006). Structure and culture in African American adolescent violence: A partial test of the "code of the street" thesis. *Justice Quarterly, 23,* 1–33.

Stoller, N. (2000). *Improving access to health care for California's women prisoners: Executive summary.* Unpublished paper, October. Available at http://www.ucop.edu/cpac/doc uments/stollerpaper.pdf

Strang, H., & Sherman, L. W. (2006). Restorative justice to reduce victimization. In B. C. Welsh & D. P. Farrington (Eds.), *Preventing crime: What works for children, offenders, victims, and places* (pp. 147–160). Dordrecht, The Netherlands: Springer.

Straus, M., Gelles, R. J., & Steinmetz, S. K. (1980). *Behind closed doors: Violence in the American family.* Garden City, NY: Doubleday.

Stretsky, P. B., & Lynch, M. J. (1999). Environmental justice and the prediction of distance to accidental chemical releases in Hillsborough County, Florida. *Social Science Quarterly, 80,* 830–846.

A study of 15 convicted murderers shows that all had once suffered head injuries. (1986, June 3). *New York Times,* p. C8.

Sullivan, D., & Tifft, L. (Eds.). (2006). *Handbook of restorative justice.* New York: Routledge.

Sutherland, E. H. (1937). *The professional thief: By a professional thief.* Chicago: University of Chicago Press.

Sutherland, E. H. (1939). *Principles of criminology* (3rd ed.). Philadelphia: J. B. Lippincott.

Sutherland, E. H. (1940). White-collar criminality. *American Sociological Review, 5,* 1–12.

Sutherland, E. H. (1949). *White collar crime.* New York: Holt, Rinehart & Winston.

Sutherland, E. H. (1973). *On analyzing crime* (K. Schuessler, Ed.). Chicago: University of Chicago Press. (Original work published 1942)

Sutherland, E. H. (1983). *White collar crime: The uncut version.* New Haven, CT: Yale University Press. (Original work published 1949)

Sutherland, E. H., & Cressey, D. R. (1970). *Criminology* (8th ed.). Philadelphia: J. B. Lippincott.

Sutherland, E. H., Cressey, D. R., & Luckenbill, D. F. (1992). *Principles of criminology* (11th ed.). Dix Hills, NY: General Hall.

Swados, H. (Ed.). (1962). *Years of conscience: The muckrakers.* Cleveland, OH: Meridian Books.

Swartz, K. (2010). Anderson, Elijah: Code of the street. In F. T. Cullen & P. Wilcox (Eds.), *Encyclopedia of criminological theory* (pp. 46–51). Thousand Oaks, CA: Sage.

Sweet land of liberty [Special issue]. (1986, Summer). *Newsweek.*

Sykes, G. M. (1974). The rise of critical criminology. *Journal of Criminal Law and Criminology, 65,* 206–213.

Sykes, G. M. (1978). *Criminology.* New York: Harcourt Brace Jovanovich.

Sykes, G. M., & Cullen, F. T. (1992). *Criminology* (2nd ed.). New York: Harcourt Brace Jovanovich.

Sykes, G., & Matza, D. (1957). Techniques of neutralization: A theory of delinquency. *American Sociological Review, 22,* 664–673.

Symons, J. N. (1951). [Discussion]. *American Sociological Review, 16,* 207–208.

Szasz, T. (1987). *Insanity: The idea and its consequences.* New York: John Wiley.

Tannenbaum, F. (1938). *Crime and the community.* New York: Columbia University Press.

Tappan, P. (1947). Who is the criminal? *American Sociological Review, 12,* 96–102.

Taxman, F., Young, D., Byrne, J. M., Holsinger, A., & Anspach, D. (2002). *From prison safety to public safety: Innovations in offender reentry.* Washington, DC: U.S. Department of Justice, National Institute of Justice.

Taylor, C. (2001). The relationship between social and self-control: Tracing Hirschi's criminological career. *Theoretical Criminology, 5,* 369–388.

Taylor, I. (1998). Free markets and the cost of crime: An audit of England and Wales. In P. Walton & J. Young (Eds.), *The new criminology revisited* (pp. 234–258). London: St. Martin's.

Taylor, I., Walton, P., & Young, J. (1973). *The new criminology: For a social theory of deviance.* London: Routledge & Kegan Paul.

Taylor, R. B. (2001). *Breaking away from broken windows: Baltimore neighborhoods and the nationwide fight against crime.* Boulder, CO: Westview.

Thaler, R. H., & Sunstein, C. R. (2008). *Nudge: Improving decisions about health, wealth, and happiness.* New Haven, CT: Yale University Press.

Thapar, A., Langley, K., Fowler, T., Rice, F., Turic, D., Whittinger, N., Aggleton, J., Van den Bree, M.,

Owen, M., & O'Donovan, M. (2005). Catechol O-methyltransferase gene-variant and birth weight predict early-onset antisocial behavior in children with attention-deficit/hyperactivity disorder. *Archives of General Psychiatry, 62,* 1275–1278.

Thomas, C. W., & Bishop, D. M. (1984). The effect of formal and informal sanctions on delinquency: A longitudinal comparison of labeling and deterrence theories. *Journal of Criminal Law and Criminology, 75,* 1222–1245.

Thomas, C. W., & Hepburn, J. R. (1983). *Crime, criminal law, and criminology.* Dubuque, IA: William C. Brown.

Thomas, J., & Boehlefeld, S. (1991). Rethinking abolitionism: "What do we do with Henry" [Review of *The Politics of Redress*]. *Social Justice, 18,* 239–251.

Thomas, W. I. (1907). *Sex and society.* Boston: Little, Brown.

Thomas, W. I. (1923). *The unadjusted girl.* New York: Harper & Row.

Thornberry, T. P. (1987). Toward an interactional theory of delinquency. *Criminology, 25,* 863–891.

Thornberry, T. P. (1996). Empirical support for interactional theory: A review of the literature. In J. D. Hawkins (Ed.), *Delinquency and crime: Current theories* (pp. 198–235). New York: Cambridge University Press.

Thornberry, T. P., & Krohn, M. D. (2005). Applying interactional theory to the explanation of continuity and change in antisocial behavior. In D. P. Farrington (Ed.), *Integrated developmental and life-course theories of offending* (Advances in Criminological Theory, Vol. 14, pp. 183–209). New Brunswick, NJ: Transaction.

Thornberry, T. P., Lizotte, A. J., Krohn, M. D., Smith, C. A., & Porter, P. K. (2003). Causes and consequences of delinquency: Findings from the Rochester Youth Development Study. In T. P. Thornberry & M. D. Krohn (Eds.), *Taking stock of delinquency: An overview of findings from contemporary longitudinal studies* (pp. 11–46). New York: Kluwer Academic/Plenum.

Thrasher, F. M. (1963). *The gang: A study of 1,313 gangs in Chicago.* Chicago: University of Chicago Press. (Original work published 1927)

Tibbetts, S. G. (2003). Selfishness, social control, and emotions: An integrated perspective on criminality. In A. Walsh & L. Ellis (Eds.), *Biosocial criminology: Challenging environmentalism's supremacy* (pp. 83–101). Hauppauge, NY: Nova Science Publishers.

Tierney, J. (1996). *Criminology: Theory and context.* Englewood Cliffs, NJ: Prentice Hall.

Tierney, K. J. (1982). The battered women movement and the creation of the wife beating problem. *Social Problems, 29,* 207–220.

Tittle, C. R. (1975a). Deterrents of labeling? *Social Forces, 53,* 399–410.

Tittle, C. R. (1975b). Labeling and crime: An empirical evaluation. In W. R. Gove (Ed.), *The labeling of deviance: Evaluating a perspective* (pp. 157–179). Beverly Hills, CA: Sage.

Tittle, C. R. (1995). *Control balance: Toward a general theory of deviance.* Boulder, CO: Westview.

Tittle, C. R. (2000). Control balance. In R. Paternoster & R. Bachman (Eds.), *Explaining criminals and crime: Essays in contemporary theory* (pp. 315–334). Los Angeles: Roxbury.

Tittle, C. R. (2004). Refining control balance theory. *Theoretical Criminology, 8,* 395–428.

Tittle, C. R., Bratton, J., & Gertz, M. G. (2003). A test of a micro-level application of shaming theory. *Social Problems, 50,* 592–617.

Tittle, C. R., Ward, D. A., & Grasmick, H. G. (2004). Capacity for self-control and individuals' interest in exercising self-control. *Journal of Quantitative Criminology, 20,* 143–172.

Tonry, M. (1995). *Malign neglect: Race, crime, and punishment in America.* New York: Oxford University Press.

Travis, A. (2006, January 19). Home Office holds up probation privatisation. *The Guardian,* p. 13.

Travis, J. (2002). Invisible punishment: An instrument of social seclusion. In M. Mauer & M. Chesney-Lind (Eds.), *Invisible punishment: The collateral consequences of mass imprisonment* (pp. 15–36). New York: New Press.

Travis, J. (2005). *But they all come back: Facing the challenges of prisoner reentry.* Washington, DC: Urban Institute Press.

Tremblay, R. E., & Craig, W. M. (1995). Developmental crime prevention. In M. Tonry (Ed.), *Crime and justice: A review of research* (Vol. 19, pp. 151–236). Chicago: University of Chicago Press.

A trial that wouldn't end. (1987, June 19). *Newsweek*, pp. 20–21.

Trying to say no. (1986, August 11). *Newsweek*, pp. 14–19.

Tucker, W. H. (1994). *The science and politics of racial research*. Chicago: University of Illinois Press.

Tunnell, K. (1992). *Choosing crime: The criminal calculus of property offenders*. Chicago: Nelson-Hall.

Turk, A. T. (1969a). *Criminality and legal order*. Chicago: Rand McNally.

Turk, A. T. (1969b). Introduction. In W. Bonger (Ed.), *Criminality and economic conditions* (pp. 3–20). Bloomington: Indiana University Press.

Turk, A. T. (1987). Turk and conflict theory: An autobiographical reflection. *The Criminologist, 12*, 3–7.

Turner, J. H. (1978). *The structure of sociological theory*. Homewood, IL: Dorsey.

Turner, M. G., Hartman, J. L., Exum, M. L., & Cullen, F. T. (2007). Examining the cumulative effects of protective factors: Resiliency among a national sample of high-risk youths. *Journal of Offender Rehabilitation, 46*, 81–111.

Turner, M. G., Piquero, A. R., & Pratt, T. C. (2005). The school as a source of self-control. *Journal of Criminal Justice, 33*, 327–339.

Tversky, A., & Kahneman, D. (1974). Judgment under uncertainty: Heuristics and biases. *Science, 185*, 1124–1131.

Tversky, A., & Kahneman, D. (1981). The framing of decisions and the psychology of choice. *Science, 211*, 453–458.

T.V. evangelist resigns, citing sexual blackmail. (1987, March 21). *New York Times*, pp. A1, A33.

20 years after the Kerner Report: Three societies, all separate. (1988, February 29). *New York Times*, p. A13.

Unnever, J. D., Cochran, J. K., Cullen, F. T., & Applegate, B. K. (2010). The pragmatic American: Attributions of crime and the hydraulic relation hypothesis. *Justice Quarterly, 27*, 431–457.

Unnever, J. D., Colvin, M., & Cullen, F. T. (2004). Crime and coercion: A test of core theoretical propositions. *Journal of Research in Crime and Delinquency, 41*, 244–268.

Unnever, J. D., Cullen, F. T., & Agnew, R. (2006). Why is "bad" parenting criminogenic: Implications from rival theories. *Youth Violence and Juvenile Justice, 4*, 1–31.

Unnever, J. D., Cullen, F. T., Mathers, S. A., McClure, T. E., & Allison, M. C. (2009). Racial discrimination and Hirschi's criminological classic: A chapter in the sociology of knowledge. *Justice Quarterly, 26*, 377–406.

Unnever, J. D., Cullen, F. T., & Pratt, T. C. (2003). Parental management, ADHD, and delinquent involvement: Reassessing Gottfredson and Hirschi's general theory. *Justice Quarterly, 20*, 471–500.

Unusual sentence stirs legal dispute. (1987, August 27). *New York Times*, p. A10.

U.S. churches gain members slowly: Rate of increase keeps pace with population growth. (1987, June 15). *New York Times*, p. A13.

U.S. had more than 7,000 hate crimes in 1993, FBI head says. (1994, June 29). *New York Times*, p. A9.

Van Ness, D. W., & Strong, K. H. (1997). *Restoring justice*. Cincinnati, OH: Anderson.

van Swaaningen, R. (1997). *Critical criminology: Visions from Europe*. London: Sage.

van Swaaningen, R. (1999). Reclaiming critical criminology: Social justice and the European tradition. *Theoretical Criminology, 3*, 5–28.

Van Voorhis, P., & Salisbury, E. (2004). Social learning models. In P. Van Voorhis, M. Braswell, & D. Lester (Eds.), *Correctional counseling and rehabilitation* (5th ed., pp. 163–182). Cincinnati, OH: Anderson/LexisNexis.

Vaughan, D. (1983). *Controlling unlawful organizational behavior: Social structure and corporate misconduct*. Chicago: University of Chicago Press.

Vaughan, D. (1996). *The Challenger launch decision: Risky technology, culture, and deviance at NASA*. Chicago: University of Chicago Press.

Vaughan, D. (1997). Anomie theory and organizations: Culture and the normalization of deviance at NASA.

In N. Passas & R. Agnew (Eds.), *The future of anomie theory* (pp. 95–123). Boston: Northeastern University Press.

Vazsonyi, A., Pickering, L. F., Junger, M., & Hessing, D. (2001). An empirical test of a general theory of crime: A four-nation comparative study of self-control and the prediction of deviance. *Journal of Research in Crime and Delinquency, 38,* 91–131.

Vesely, R. (2004). California rebuked on female inmates. December 22. Retrieved April 14, 2006, from http://www.womensenews.org/story/incarceration/041222/california rebuked-female-inmates

Veysey, B. M., & Messner, S. F. (1999). Further testing of social disorganization theory: An elaboration of Sampson and Groves's "community structure and crime." *Journal of Research in Crime and Delinquency, 36,* 156–174.

Victor, B., & Cullen, J. B. (1988). The organizational bases of ethical work climates. *Administrative Science Quarterly, 33,* 101–125.

Vieraitis, L. M., Kovandzic, T. V., & Britto, S. (2008). Women's status and risk of homicide victimization. *Homicide Studies, 12,* 163–176.

Vila, B. J. (1994). A general paradigm for understanding criminal behavior: Extending evolutionary ecological theory. *Criminology, 32,* 311–359.

Vila, B. J. (1997). Human nature and crime control: Improving the feasibility of nurturant strategies. *Politics and the Life Sciences, 16*(1), 16–21.

Vila, B. J., & Cohen, L. E. (1993). Crime as strategy: Testing an evolutionary ecological theory of expropriative behavior. *American Journal of Sociology, 98,* 873–912.

Villettaz, P., Killias, M., & Zoder, I. (2006). *The effects of custodial vs. noncustodial sentences on re-offending: A systematic review of the state of knowledge.* Philadelphia, PA: Campbell Collaboration Crime and Justice Group.

Visher, C. A. (1987). Incapacitation and crime control: Does a "lock 'em up" strategy reduce crime? *Justice Quarterly, 4,* 513–543.

Volavka, J. (1987). Electroencephalogram among criminals. In S. A. Mednick, T. A. Moffitt, & S. A. Stack (Eds.), *The causes of crime: New biological approaches* (pp. 137–145). New York: Cambridge University Press.

Vold, G. B. (1958). *Theoretical criminology.* New York: Oxford University Press.

Vold, G. B., & Bernard, T. J. (1986). *Theoretical criminology* (3rd ed.). New York: Oxford University Press.

Wacquant, L. (2001). Deadly symbiosis: When ghetto and prison meet and mesh. *Punishment and Society, 3,* 95–134.

Wacquant, L. (2009). *Punishing the poor: The neoliberal government of social insecurity.* Durham, NC: Duke University Press.

Walsh, A. (2003a). Intelligence and antisocial behavior. In A. Walsh & L. Ellis (Eds.), *Biosocial criminology: Challenging environmentalism's supremacy* (pp. 105–124). Hauppauge, NY: Nova Science Publishers.

Walsh, A. (2003b). Introduction to the biosocial perspective. In A. Walsh & L. Ellis (Eds.), *Biosocial criminology: Challenging environmentalism's supremacy* (pp. 3–12). Hauppauge, NY: Nova Science Publishers.

Walsh, A. (2009). *Biology and criminology: The biosocial synthesis.* New York: Routledge.

Walsh, A., & Beaver, K. M. (Eds.). (2009). *Biosocial criminology: New directions in theory and research.* New York: Routledge.

Walton, P., & Young, J. (1998). *The new criminology revisited.* New York: St. Martin's.

Wambaugh, J. (1989). *The blooding.* New York: William Morrow.

Ward, D. A., & Tittle, C. R. (1993). Deterrence or labeling: The effects of informal sanctions. *Deviant Behavior, 14,* 43–64.

Ward, J. T., Gibson, C. L., Boman, J., & Leite, W. L. (2010). Assessing the validity of the Retrospective Behavioral Self-Control Scale: Is the general theory of crime stronger than the evidence suggests? *Criminal Justice and Behavior, 37,* 336–357.

Waring, E., Weisburd, D., & Chayet, E. (1995). White-collar crime and anomie. In F. Adler & W. S. Laufer (Eds.), *The legacy of anomie theory: Advances in criminological theory* (Vol. 6, pp. 207–225). New Brunswick, NJ: Transaction.

Warr, M. (1998). Life-course transitions and desistance from crime. *Criminology, 36,* 183–216.

Warr, M. (2002). *Companions in crime: The social aspects of criminal conduct.* Cambridge, UK: Cambridge University Press.

Warr, M., & Stafford, M. (1993). A reconceptualization of general and specific deterrence. *Journal of Research in Crime and Delinquency, 30,* 123–135.

Watergate.Info (2010). *Brief timeline of events.* Retrieved April 3, 2010, from http://www .watergate.info/chronology/brief.shtml

Watt, N. (2010, February 4). Tories criticised over misleading crime figures: Jacqui Smith says Conservatives have been caught bang to rights. Retrieved March 1, 2010, from http://www.guardian.co.uk/politics/2010/ feb/04/tories-criticised-over-misleading crime-figures

Weaver, R. M. (1948). *Ideas have consequences.* Chicago: University of Chicago Press.

Webster, C. M., & Doob, A. N. (2007). Punitive trends and stable imprisonment rates in Canada. In M. Tonry (Ed.), *Crime, punishment, and politics in comparative perspective* (Crime and Justice: A Review of Research (Vol. 36, pp. 297–370). Chicago: University of Chicago Press.

Webster, F. (2004). Cultural studies and sociology at, and after, the closure of the Birmingham School. *Cultural Studies, 18*(6), 847–862.

Weisburd, D., & Waring, E. (with Chayet, E. E.). (2001). *White-collar crime and criminal careers.* Cambridge, UK: Cambridge University Press.

Welsh, B. C. (2003). Economic costs and benefits of primary prevention of delinquency and later offending: A review of the research. In D. P. Farrington & J. W. Coid (Eds.), *Early prevention of adult antisocial behavior* (pp. 318–355). Cambridge, UK: Cambridge University Press.

Welsh, B. C., & Farrington, D. P. (2006). *Preventing crime: What works for children, offenders, victims, and places.* Dordrecht, The Netherlands: Springer.

Welsh, B. C., & Farrington, D. P. (2009). *Making public places safer: Surveillance and crime prevention.* New York: Oxford University Press.

Welsh, B. C., Farrington, D. P., & Sherman, L. W. (Eds.). (2001). *Costs and benefits of preventing crime.* Boulder, CO: Westview.

West, H. C., & Sabol, W. J. (2008). *Prisoners in 2007.* Washington, DC: U.S. Department of Justice, Bureau of Justice Statistics. Available at bjs.ojp.usdoj.gov/content/ pub/pdf/p07.pdf

West, H. C., & Sabol, W. J. (2009). *Prison inmates at midyear 2008.* Washington, DC: U.S. Department of Justice, Bureau of Justice Statistics. Available at bjs.ojp.usdoj.gov/ content/pub/pdf/pim08st.pdf

Western, B. (2006). *Punishment and inequality in America.* New York: Russell Sage.

Whitman, J. Q. (2003). *Harsh justice: Criminal punishment and the widening divide between America and Europe.* New York: Oxford University Press.

Whyte, L. (1957). *The organization man.* New York: McGraw-Hill.

Wife tells jury of love story, then "torture." (1994, January 13). *New York Times,* p. A8.

Wikstrom, P.-O. H. (1995). Preventing city-center street crimes. In M. Tonry & D. P. Farrington (Eds.), Building a safer society: Strategic approaches to crime prevention (Crime and Justice: A Review of Research, Vol. 19, pp. 429–468). Chicago: University of Chicago Press.

Wilcox, P., Land, K. C., & Hunt, S. A. (2003). *Criminal circumstances: A dynamic multicontextual criminal opportunity theory.* New York: Aldine de Gruyter.

Williams, C. J. (2010, February 4). Justice Kennedy laments the state of prisons in California, U.S. *Los Angeles Times.* Retrieved on March 18, 2010, from http:// articles.latimes.com/2010/feb/04/local/ la-me-kennedy4-2010feb04

Williams, K. R., & Hawkins, R. (1986). Perceptual research on general deterrence: A critical review. *Law and Society Review, 20,* 545–572.

Wilson, C. (1984). *A criminal history of mankind.* New York: Carroll and Graf.

Wilson, D. B., Bouffard, L. A., & MacKenzie, D. L. (2005). A quantitative review of structured, group-oriented, cognitive-behavioral programs for offenders. *Criminal Justice and Behavior, 32,* 172–204.

Wilson, E. O. (1975). *Sociobiology.* New York: Knopf.

Wilson, J. Q. (1975). *Thinking about crime.* New York: Vintage.

Wilson, J. Q., & Herrnstein, R. J. (1985). *Crime and human nature.* New York: Simon & Schuster.

Wilson, J. Q., & Kelling, G. L. (1982, March). Broken windows: The police and neighborhood safety. *Atlantic Monthly,* pp. 29–38.

Wilson, W. J. (1987). *The truly disadvantage: The inner city, the underclass, and public policy.* Chicago: University of Chicago Press.

Wilson, W. J. (1996). *When work disappears: The world of the new urban poor.* New York: Knopf.

Wirth, L. (1938). Urbanism as a way of life. *American Journal of Sociology, 44,* 1–24.

Wolff, E. N. (2001, February 12). The rich get richer: And why the poor don't. *American Prospect,* pp. 15–17.

Wolff, K. H. (Ed.). (1964). Introduction. In *The sociology of Georg Simmel.* New York: Free Press.

Wolfgang, M. E. (1973). Cesare Lombroso. In H. Mannheim (Ed.), *Pioneers in criminology* (2nd ed., pp. 232–291). Montclair, NJ: Patterson Smith.

Wolfgang, M. E., & Ferracuti, F. (1982). *The subculture of violence: Toward an integrated theory in criminology.* Beverly Hills, CA: Sage.

Wolfgang, M. E., Figlio, R. M., & Sellin, T. (1972). *Delinquency in a birth cohort.* Chicago: University of Chicago Press.

Wood, P. B., Pfefferbaum, B., & Arneklev, B. J. (1993). Risk-taking and self-control: Social psychological correlates of delinquency. *Journal of Crime and Justice, 16,* 111–130.

World prison brief. (2009). London: King's College London, International Centre for Prison Studies.

Wozniak, J. F. (2003, November). *The relevance of Richard Quinney's writing on peacemaking criminology: Toward personal and social transformation.* Paper presented at the annual meeting of the American Society of Criminology, Denver, CO.

Wozniak, J. F. (2011). Becoming a peacemaking criminologist: The travels of Richard Quinney. In F. T. Cullen, C. L. Jonson, A. J. Myer, & F. Adler (Eds.), *The origins of American criminology* (Advances in Criminological Theory, Vol. 16, pp. 223–244). New Brunswick, NJ: Transaction.

Wozniak, J. F., Braswell, M. C., Vogel, R. E., & Blevins, K. R. (Eds.). (2008). *Transformative justice: Critical and peacemaking theme influenced by Richard Quinney.* Lanham, MD: Lexington Books.

Wright, E. O. (1973). *The politics of punishment: A critical analysis of prisons in America.* New York: Harper & Row.

Wright, J. P., & Beaver, K. M. (2005). Do parents matter in creating self-control in their children? A genetically informed test of Gottfredson and Hirschi's theory of low self-control. *Criminology, 43,* 1169–1202.

Wright, J. P., & Cullen, F. T. (2000). Juvenile involvement in occupational delinquency. *Criminology, 38,* 863–892.

Wright, J. P., & Cullen, F. T. (2004). Employment, peers, and life-course transitions. *Justice Quarterly, 21,* 183–205.

Wright, J. P., Cullen, F. T., & Blankenship, M. B. (1995). The social construction of corporate violence: Media coverage of the Imperial Food Products fire. *Crime & Delinquency, 41,* 20–36.

Wright, J. P., Dietrich, K. N., Ris, M. D., Hornung, R. W., Wessel, S. D., Lanphear B. P., Ho, M., & Rae, M. N. (2008). Association of prenatal and childhood blood lead concentrations with criminal arrests in early adulthood. *PLoS Medicine, 5,* 732–740.

Wright, R. A. (1994). *In defense of prisons.* Westport, CT: Greenwood.

Wright, R. A., & Decker, S. (1994). *Burglars on the job: Streetlife and residential break-ins.* Boston: Northeastern University Press.

Wright, R. A., & Decker, S. (1997). *Armed robbers in action: Stickups and street culture.* Boston: Northeastern University Press.

Wrong, D. (1961). The oversocialized conception of man in modern sociology. *American Sociological Review, 26,* 187–193.

Wuthnow, R. (1991). *Acts of compassion: Caring for others and helping ourselves.* Princeton, NJ: Princeton University Press.

Yang, Y., Glenn, A. L., & Raine, A. (2008). Brain abnormalities in antisocial individuals. *Behavioral Sciences and the Law, 26,* 65–83.

Yar, J., & Penna, S. (2004). Between positivism and post-modernity? *British Journal of Criminology, 44,* 533–549.

Yeager, P. C., & Simpson, S. S. (2009). Environmental crime. In M. Tonry (Ed.), *The Oxford handbook of crime and public policy* (pp. 325–355). New York: Oxford University Press.

Yochelson, S., & Samenow, S. (1976). *The criminal personality.* New York: Jason Aronson.

Yoder, S. A. (1979). Criminal sanctions for corporate illegality. *Journal of Criminal Law and Criminology, 69,* 40–58.

Young, J. (1971). *The drugtakers.* London: Paladin.

Young, J. (1975). Working-class criminology. In I. Taylor, P. Walton, & J. Young (Eds.), *Critical criminology* (pp. 63–95). London: Routledge & Kegan Paul.

Young, J. (1976). Foreword. In F. Pearce (Ed.), *Crime and the powerful* (pp. 11–21). London: Pluto.

Young, J. (1986). The failure of criminology: The need for a radical realism. In J. Young & R. Matthews (Eds.), *Confronting crime* (pp. 4–30). London: Sage.

Young, J. (1988). Radical criminology in Britain. *British Journal of Criminology, 28,* 159–183.

Young, J. (1991). Asking questions of left realism. In B. D. MacLean & D. Milovanovic (Eds.), *New directions in critical criminology* (pp. 15–18). Vancouver, BC: Collective Press.

Young, J. (1992a). [Review of *Feminist perspectives in criminology*]. *Journal of Law and Society, 19,* 289–292.

Young, J. (1992b). Ten points of realism. In J. Young & R. Matthews (Eds.), *Rethinking criminology: The realist debate.* London: Sage.

Young, J. (1999). *The exclusive society.* London: Sage.

Young, J. (2003). Merton with energy, Katz with structure: The sociology of vindictiveness and the criminology of transgression. *Theoretical Criminology, 7*(3), 389–414.

Young, J. (2004). Voodoo criminology and the numbers game. In J. Ferrell, K. Hayward, W. Morrison, & M. Presdee (Eds.), *Cultural criminology unleashed* (pp. 13–28). London: Glasshouse.

Young, J. (2007). *The vertigo of late modernity.* London: Sage.

Young, J., & Brotherton, D. C. (2005, November 15–19). *Social seclusion, deportation, and the transnational order.* Paper presented at the annual meetings of the American Society of Criminology, Toronto.

Young, J., & Matthews, R. (1992). *Rethinking criminology: The realist debate.* London: Sage.

Zhang, S. X., Chin, K. L., & Miller, J. (2007). Women's participation in Chinese transnational human smuggling; A gendered market perspective. *Criminology, 45,* 699–733.

Zimring, F. E. (2006). *The great American crime decline.* New York: Oxford University Press.

Zimring, F. E., & Hawkins, G. (1997). *Crime is not the problem: Lethal violence in America.* New York: Oxford University Press.

Zinn, H. (1964). *SNCC: The new abolitionists.* Boston: Beacon.

Zizek, S. (2002). *Welcome to the desert of the real.* London: Verso.

Zizek, S. (2006a). *How to read Lacan.* London: Granta.

Zizek, S. (2006b). *On belief.* London: Routledge.

Zizek, S. (2006c). *The parallax view.* Boston: MIT Press.

Zizek, S. (2007). *The universal exception.* London: Continuum.

Zizek, S. (2008). *Violence.* London: Profile Books.

Zuckerman, M. (1983). A biological model of sensation seeking. In M. Zuckerman (Ed.), *Biological basis of sensation seeking, impulsivity, and anxiety* (pp. 37–76). Hillsdale, NJ: Lawrence Erlbaum.

Photo Credits

Photo, page 1 (*The Thinker*). © Alan Schein Photography/Corbis.

Photo, page 15 (Lombroso). Public Domain.

Photo, page 39 (Shaw). Chicago Area Project, http://www.chicagoareaproject.org/about.html.

Photo, page 61 (Merton). Courtesy of Columbia University Archives.

Photo, page 88 (Reckless). American Society of Criminology.

Photo, page 109 (Hirschi). Steve Agan Photography.

Photo, page 139 (Braithwaite). John Braithwaite (1951–), Australian National University, reprinted with permission.

Photo, page 166 (Chambliss). Courtesy of William J. Chambliss.

Photo, page 199 (Taylor). Ian Taylor (1945–2001), University of Durham, UK, reprinted with permission.

Photo, page 230 (Adler). Courtesy of Freda Adler.

Photo, page 260 (Sutherland). Edwin H. Sutherland (1883–1950). Public Domain.

Photo, page 295 (Wilson). Courtesy of James Q. Wilson.

Photo, page 328 (Felson). Courtesy of Marcus Felson.

Photo, page 351 (Darwin). Public Domain.

Photo, page 384 (John H. Laub). Courtesy of John H. Laub.

Name Index

Subject Index

About the Authors

J. Robert Lilly is Regents Professor of Sociology/Criminology and Adjunct Professor of Law at Northern Kentucky University. His research interests include the pattern of capital crimes committed by U.S. soldiers during World War II, the "commercial–corrections complex," juvenile delinquency, house arrest and electronic monitoring, criminal justice in the People's Republic of China, the sociology of law, and criminological theory. He has published in *Criminology, Crime & Delinquency, Social Problems, Legal Studies Forum, Northern Kentucky Law Review, Journal of Drug Issues, The New Scholar, Adolescence, Qualitative Sociology, Federal Probation, International Journal of Comparative and Applied Criminal Justice, Justice Quarterly*, and *The Howard Journal*. He has coauthored several articles and book chapters with Richard A. Ball, and he is coauthor of *House Arrest and Correctional Policy: Doing Time at Home* (1988). In 2003 he published *La Face Cachee Des GI's: Les Viols commis par des soldats amercains en France, en Angleterre et en Allemange pendat la Second Guerre mondial*. It was translated into Italian and published (2004) as *Stuppi Di Guerra: Le Violenze Commesse Dai Soldati Americani in Gran Bretagna, Francia e Germania 1942–1945*. It was published in English in 2007 as *Taken by Force: Rape and American GIs in Europe During World War II*. The latter work is part of his extensive research on patterns of crimes and punishments experienced by U.S. soldiers in WWII in the European Theater of War. *The Hidden Face of the Liberators*, a made-for-TV documentary by Program 33 (Paris), was broadcast in Switzerland and France in March 2006 and was a finalist at the International Television Festival of Monte Carlo in 2007. He is the past treasurer of the American Society of Criminology. In 1988, he was a visiting professor in the School of Law at Leicester Polytechnic and was a visiting scholar at All Soul's College in Oxford, England. In 1992, he became a visiting professor at the University of Durham in England. He is currently coeditor of *The Howard Journal of Criminal Justice*.

Francis T. Cullen is Distinguished Research Professor in the School of Criminal Justice at the University of Cincinnati, where he also holds a joint appointment in sociology. He received a Ph.D. (1979) in sociology and education from Columbia University. Professor Cullen has published over 275 works in the areas of criminological theory, corrections, white-collar crime, public opinion, and the measurement of sexual victimization. He is author of *Rethinking Crime and Deviance Theory: The Emergence of a Structuring Tradition* and is coauthor of *Reaffirming Rehabilitation*,

Corporate Crime Under Attack: The Ford Pinto Case and Beyond, Criminology, Combating Corporate Crime: Local Prosecutors at Work, Unsafe in the Ivory Tower: The Sexual Victimization of College Women, and *Correctional Theory: Context and Consequences.* He also is coeditor of *Contemporary Criminological Theory, Offender Rehabilitation: Effective Correctional Intervention, Criminological Theory: Past to Present—Essential Readings, Taking Stock: The Status of Criminological Theory, The Origins of American Criminology,* and the *Encyclopedia of Criminological Theory.* Professor Cullen is a Past President of the American Society of Criminology and of the Academy of Criminal Justice Sciences.

Richard A. Ball is Professor of Administration of Justice at Penn State—Fayette and former Program Head for Administration of Justice for the 12-campus Commonwealth College of Penn State. He is former Chair of the Department of Sociology and Anthropology at West Virginia University, and received his doctorate from Ohio State University in 1965. He has authored several monographs on community power structure and correctional issues and coedited a monograph and a book on white-collar crime. He has authored or coauthored approximately 100 articles and book chapters, including articles in the *American Journal of Corrections, American Sociological Review, The American Sociologist, British Journal of Social Psychiatry, Correctional Psychology, Crime and Delinquency, Criminology, Federal Probation, International Journal of Comparative and Applied Criminal Justice, International Social Science Review, Journal of Communication, Journal of Contemporary Criminal Justice, Journal of Small Business Management, Journal of Psychohistory, Justice Quarterly, Northern Kentucky Law Review, Qualitative Sociology, Rural Sociology, Social Forces, Social Problems, Sociological Focus, Sociological Symposium, Sociology and Social Welfare, Sociology of Work and Occupations, Urban Life, Victimology,* and *World Futures.* He is coauthor of *House Arrest and Correctional Policy: Doing Time at Home* (1988).

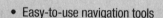